Tianshuang Qiu and Ying Guo
Signal Processing and Data Analysis

Also of interest

ASSD Series
F. Derbel, N. Derbel, O. Kanoun (Eds.), 4 Issues/year
ISSN 2364-7493, e-ISSN 2364-7507

Signals and Systems
G. Li, L. Chang, S.Li, 2015
ISBN 978-3-11-037811-5, e-ISBN 978-3-11-037954-9,
e-ISBN (EPUB) 978-3-11-041684-8

Signals and Systems, vol 1 + 2
W. Zhang, 2017
ISBN 978-3-11-054409-1

Blind Equalization in Neural Networks
L. Zhang, 2017
ISBN 978-3-11-044962-4, e-ISBN 978-3-11-044962-4,
e-ISBN (EPUB) 978-3-11-044967-9

Tianshuang Qiu and Ying Guo

Signal Processing and Data Analysis

—

DE GRUYTER  清華大学出版社
TSINGHUA UNIVERSITY PRESS

Authors

Prof. Tianshuang Qiu
Dalian University of Technology
Faculty of Electronic Information
& Electrical Engineering
2 Linggong Street
Ganjinzi District
116024 Dalian
qiutsh@dlut.edu.cn

Dr. Ying Guo
Shenyang University of Technology
School of Electronic Information
111 Sheliaoxi Road
110870 Shenyang
yingguo@sut.edu.cn

ISBN 978-3-11-046158-9
e-ISBN (PDF) 978-3-11-046508-2
e-ISBN (EPUB) 978-3-11-046513-6

Library of Congress Cataloging-in-Publication Data

Names: Qiu, TianShuang, author. | Guo, Ying, author.
Title: Signal processing and data analysis / Tianshuang Qiu and Ying Guo.
Description: First edition. | Berlin/ Boston : De Gruyter, [2018] | Includes
      bibliographical references and index.
Identifiers: LCCN 2018009548 | ISBN 9783110461589 (softcover : acid-free
      paper) | ISBN 9783110465082 (pdf) | ISBN 9783110465136 (epub)
Subjects: LCSH: Signal processing–Digital techniques.
Classification: LCC TK5102.9 .Q65 2018 | DDC 621.382/20285–dc23 LC record available
at https://lccn.loc.gov/2018009548

Bibliographic information published by the Deutsche Nationalbibliothek
The Deutsche Nationalbibliothek lists this publication in the Deutsche Nationalbibliografie; detailed
bibliographic data are available on the Internet at http://dnb.dnb.de.

© 2018 Walter de Gruyter GmbH, Berlin/Boston
Typesetting: Integra Software Services Pvt. Ltd
Printing and binding: CPI books GmbH, Leck
Cover image: Goja1 / iStock / Getty Images Plus

www.degruyter.com

MIX
Papier aus verantwor-
tungsvollen Quellen
FSC® C083411

Preface

With the fast development of signal processing theory and computer technology, signal processing and data analysis technology is becoming an important research area in information technology. At the same time, its role is increasing and it is receiving increasingly more attention and applications in noninformation engineering areas, such as mechanical engineering, civil engineering, chemical engineering, biomedical engineering, and so on. In turn, the applications in these fields also put forward new projects for the research on signal processing and data analysis.

For graduate students and senior undergraduate students in the areas of nonelectronic information, *Signal Processing and Data Analysis* introduces the fundamental theories and applications of signal processing and data analysis. The purpose of this book is to enable readers to have a basic knowledge of the theories and methods of signal processing and data analysis, and to enable readers to solve the problems of scientific research or engineering in their professional fields related to the signal processing technology.

This book has the following characteristics: first, the content of this book goes step by step from shallower to deeper; second, the content of this book is extensive, and its system is complete; third, it contacts the practice and pays attention to applications. This book is based on a wide range of materials, basically covering the signals and systems, digital signal processing, data processing and error analysis, statistical signal processing, and modern signal processing. Thus, it forms a relatively complete system of its own. On the basis of the basic theory of signal processing and data analysis, this book gives many examples and practical applications. It also combines the algorithms with MATLAB programming to facilitate learning for readers in noninformation areas.

In order to compile this book, the authors conducted a wide range of literature survey and research. On the basis of reference to many outstanding works, literature, and textbooks, this book is completed. It is suitable in colleges and universities as a textbook or as a reference book in the course of signal processing for nonelectronic information fields. It can also be used as a reference by teachers and engineers.

This book contains 14 chapters, including the basic concepts and principles of signals and systems, Fourier theory and frequency analysis of signals and systems, the complex domain analysis of signals and systems with Laplace transform and z-transform, discretization of continuous-time signals and reconstruction of discrete-time signals, the discrete Fourier transform and the fast Fourier transform, digital filter and digital filter design, finite word-length effects in digital signal processing, data error analysis and the preprocessing of signals, the foundation of random signal processing, correlation estimation of random signals and power spectral density estimation, the optimal filtering for random signals, adaptive filtering, signal processing with higher order and fractional order statistics, and selected topics in modern signal processing.

https://doi.org/10.1515/9783110465082-201

The authors of this book are Prof. Tianshuang Qiu (Chapters 1–12) and Associate Prof. Ying Guo (Chapters 13 and 14).

This book is supported by the teaching material construction project from the Graduate School of Dalian University of Technology. The authors would like to express their thanks to the Faculty of Electronic Information and Electrical Engineering and the Graduate School of Dalian University of Technology (DUT) for their powerful support to this book. The authors would also like to thank PhD and master's degree students Shengyang Luan, Peng Wang, Jitong Ma, Tao Liu, Fangxiao Jin, Jiangan Dai, Jingjing Zong, Quan Tian, Ying Zhang, Ruijiao Dai, Yixin Shi, Yongjie Zhu, Zhaofeng Chen, Jiacheng Zhang, Cheng Liu, Yuanting Chou and Hongjie Gao from Dalian University of Technology for their enthusiastic and effective help in computer simulation, material collection, and manuscript proofreading.

Tianshuang Qiu and Ying Guo
Dalian University of Technology, Dalian, China

Contents

1 Basic Concepts and Principles of Signals and Systems

1.1 Introduction

Modern electronic information technology is broad and profound. In addition to its own rigorous and complete theoretical system, its applications are very extensive. From spacecraft satellites and ships to the microcosmic world of particle physics and nanotechnology; from the basic industries of civil, chemical, metallurgy, and machinery engineering to human health-oriented medical diagnosis and treatment technology, everywhere you can see the electronic information technology, not to mention the modern network and communication technology. Although the application field of the electronic information technology is almost ubiquitous, its basic content can be attributed to the concepts of signals and systems. It can be said that the theory of signals and systems is the cornerstone of modern electronic information technology. Any theory and application of modern electronic information technology and its development are related to the two basic concepts.

This chapter introduces the basic concepts of signals and systems. Through a detailed study of the continuous-time and discrete-time signals and systems, we can see that the signals and systems constitute a complete system, which includes a set of language and powerful analysis method for the description and analysis of signals and systems. This description and analysis are closely related to each other.

1.2 Basic Concepts of Signals and Systems

1.2.1 Basic Concept of Signals

In general, the signal is the carrier of information, and the information is the specific content of the signal. In the long-term production and life, humans created a lot of ways for information representation, exchange, and communication. Among them, language communication is the most basic, and the most widely used mode of information exchange and transmission. When a person speaks, the sound waves emitted from the mouth reach the ears of auditors through the air, causing the vibration of the membrane of auditors, so that the auditors can understand the intention of the speaker.

In Chinese history, the light and smoke from beacon towers were used to represent the information of the enemy invasion, passing down a comic story about the cry wolf from King Zhou for a smile of Queen Baosi. In fact, the use of the light and smoke from beacon towers is a kind of optical signal transmission. Moreover, various

https://doi.org/10.1515/9783110465082-001

kinds of electric signals in communications and computer networks can be used to express and transmit information.

In electronic information science and technology, the most commonly used signals are electrical signals, which can be in the form of voltage, current, electric charge, magnetic flux, or electromagnetic wave, changing with time, space, and/or other independent variables. Other forms of signals, such as temperature, pressure, and speed, are usually first converted into electrical signals through sensors.

There are many ways to represent signals. The most common way is to describe signals with mathematical functions and curves. A signal can be described by a mathematical function that varies with time, space, or other variables. For example, a sinusoidal signal can be used to express a signal changing with time in accordance with the law of sine function as follows:

$$x(t) = A \sin(\Omega t + \varphi) \tag{1.1}$$

where t is the time variable, A represents the amplitude of the signal, Ω is the frequency of the signal, and φ expresses the initial phase of the signal. $x(t)$ denotes the signal changing with time t in accordance with the law of sine. This signal can also be represented by a curve. If time t is taken as the independent variable, and $x(t)$ is as the function, we can get the curve of this signal or function point by point, referred to as the signal waveform. Figure 1.1 shows the waveform of $x(t) = A \sin(\Omega t + \varphi)$.

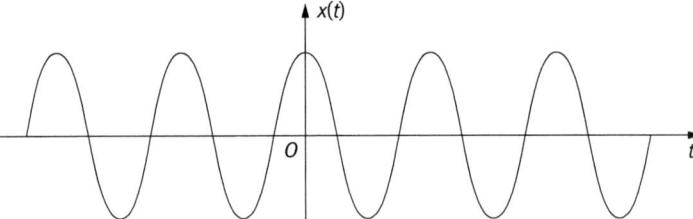

Figure 1.1: The waveform of a sinusoidal signal.

From a mathematical point of view, Figure 1.1 is not a function of the sine curve, but a curve of the cosine function. In fact, in the field of electronic information technology, it is usually not strictly distinguishable between sine and cosine signals. Such signals are often referred to as sinusoidal signals. The curve shown in Figure 1.1 can be considered as a sinusoidal signal with initial phase of $\varphi = \pi/2 = 90°$.

In addition, there are a number of other ways to express a signal, such as the frequency spectrum, the time–frequency distribution, and the spatial representation, and so on. These representations will be introduced in the later part of the book.

Closely related to the concept of the signal are the concepts of signal processing and signal analysis. The so-called signal processing is a kind of transformation and operation to signals, including weakening the useless part in the signal, filtering out the noise and interference, transforming the signal into the form easy to be analyzed and identified, and estimating signal parameters. While the signal analysis is to study the basic performance of signals, including the description, decomposition, transformation, detection, feature extraction of signals, and the signal design.

1.2.2 The Classification of Signals

In order to facilitate the description, analysis, and application of signals, we usually classify signals in different ways. There are many different methods for the classification of signals, including:

1. Deterministic signals and random signals

The deterministic signal is a signal that can be expressed as a function of a certain time variable. If the signal has a certain degree of unpredictable uncertainty, then the signal is called the uncertainty signal or random signal.

Figure 1.2 gives a schematic diagram of a deterministic signal and a random signal.

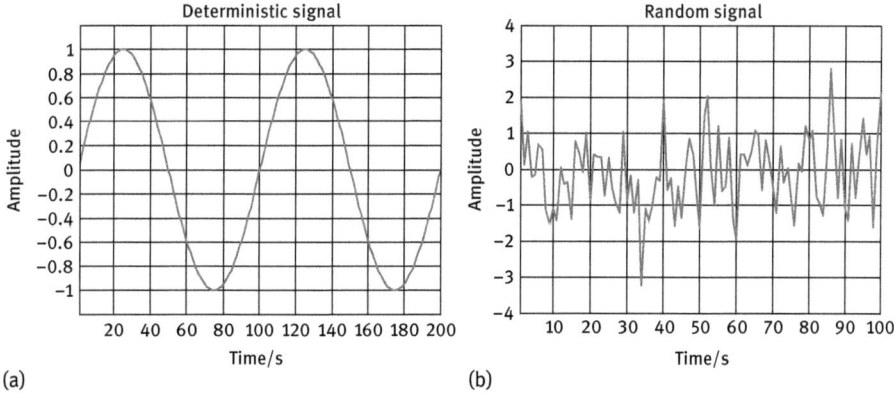

(a)   (b)

Figure 1.2: An example of deterministic and random signals: (a) a deterministic signal and (b) a random signal.

2. Continuous-time signals and discrete-time signals

If a signal is defined for a continuum of values of the independent variables (usually time t), or its independent variable is continuous, then such signal is defined as a continuous-time signal. On the other hand, a discrete-time signal is defined only at discrete times (usually time n), or the independent variable of such signal takes only on a discrete set of values. We usually use $x(t)$ to express a continuous-time signal, while the discrete-time signal is usually denoted by $x(n)$.

Figure 1.3 gives curves of a continuous-time signal and a discrete-time signal.

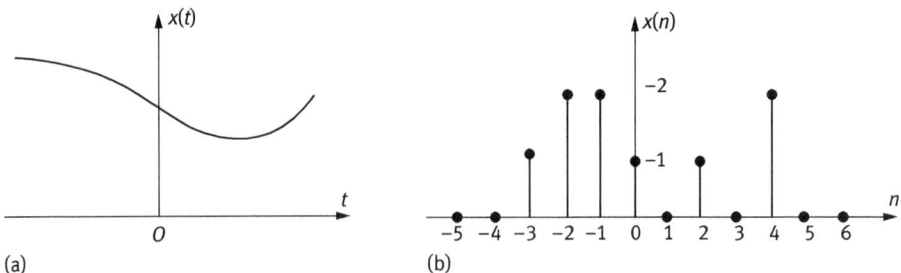

(a)   (b)

Figure 1.3: An example of continuous-time and discrete-time signals: (a) a continuous-time signal; and (b) a discrete-time signal.

3. Periodic signals and aperiodic signals

If signal $x(t)$ meets the following condition:

$$x(t) = x(t + mT), m = \pm 1, \pm 2, \ldots \tag{1.2}$$

then the signal $x(t)$ is called the periodic signal, where T is the fundamental period. $2T$, $3T$, and mT are all periods of the signal.

For discrete-time signals, if $x(n)$ meets the following condition:

$$x(n) = x(n + mN), m = \pm 1, \pm 2, \ldots \tag{1.3}$$

then it is also called the periodic signal, where N is the fundamental period, and $2N$, $3N$, and mN are all periods of the signal too.

Signals that do not satisfy the conditions in eq. (1.2) or eq. (1.3) are all aperiodic signals. Moreover, the periodic signal can be changed into the aperiodic signal if the period $T \to \infty$ (or $N \to \infty$).

Figure 1.4 shows the waveforms of a periodic signal and an aperiodic signal.

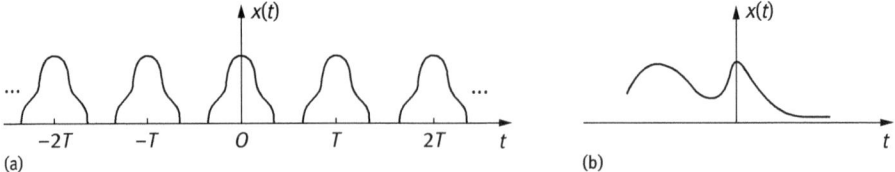

(a)   (b)

Figure 1.4: An example of a periodic signal (a) and an aperiodic signal (b).

4. Energy signals and power signals

A signal with finite energy is called an energy-limited signal, or simply an energy signal. A signal whose power is limited is defined as the power-limited signal or referred to as the power signal. In general, the bounded signal with finite duration is an energy signal, while the continuous-time sinusoidal signal is a typical example of the power signal.

5. One-dimensional signals and multidimensional signals

The signal as a function of one variable is called a one-dimensional signal. $x(t)$ (or $x(n)$) is the commonly used one-dimensional signal, which takes t (or n) as its independent variable. The signal as a function of multiple variables is called a multidimensional signal. Common multidimensional signals include image signal $f(x, y)$, whose independent variables x and y represent the spatial position of the image, and $f(\cdot)$ represents gray or color of the image in the position.

1.2.3 Classical Signals and Their Properties

1. Continuous-time complex exponential signals

The continuous-time complex exponential signal is defined as

$$x(t) = Ce^{st} \tag{1.4}$$

where both parameters C and s are generally complex numbers, defined as $C = |C|e^{j\theta}$ and $s = \sigma + j\Omega$, in which $|C|$ and θ are the magnitude and phase of C, respectively, and σ and Ω represent the real part and imaginary part of s, respectively. According to Euler's relation, the complex exponential signal $x(t)$ can be decomposed into a linear combination of sine and cosine terms as

$$x(t) = Ce^{st} = Ce^{\sigma t}\{\cos \Omega t + j \sin \Omega t\} \tag{1.5}$$

Depending on the values of parameters C and s, the complex exponential signal $x(t)$ can exhibit several different characteristics.

(1) Real exponential signals

If both parameters C and s take real values in eq. (1.4), then $x(t) = Ce^{st}$ represents a real exponential signal. Based on different values of C and s, this signal can be expressed in different forms, including the growing exponential, the decaying exponential, and so on. Figure 1.5 demonstrates four different waveforms of real-valued exponential signals.

If $\sigma = 0$, the real exponential signal $x(t)$ becomes a constant C.

(2) Periodic complex exponential signals

If we let $C = 1$ and $s = j\Omega$ in eq. (1.4), then we have

$$x(t) = e^{j\Omega t} \tag{1.6}$$

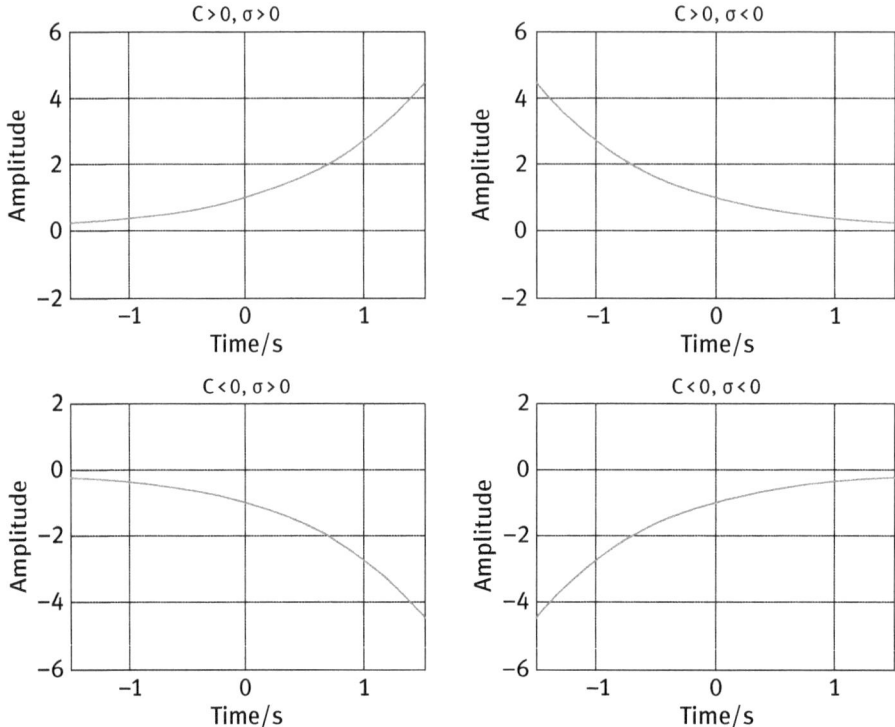

Figure 1.5: Four different forms of real-valued exponential signals.

The signal expressed in eq. (1.6) is a periodic signal. We can assume that the time variable t in $x(t)$ is beginning to change from an arbitrary value. When it changes to a certain value $t + T$, if the following relation is satisfied:

$$x(t + T) = e^{j\Omega(t+T)} = e^{j\Omega t} = x(t) \tag{1.7}$$

then $x(t) = e^{j\Omega t}$ is a periodic signal. As a matter of fact, we can always find such T to have eq. (1.7) satisfied. For example, as long as the following relation exists:

$$T = \frac{2\pi}{\Omega} \tag{1.8}$$

then eq. (1.7) can be rewritten as

$$x(t + T) = e^{j\Omega(t+T)} = e^{j\Omega t}e^{j\Omega T} = e^{j\Omega t}e^{j\Omega\frac{2\pi}{\Omega}} = e^{j\Omega t}$$

Thus $x(t) = e^{j\Omega t}$ is evidently a periodic signal.

(3) General complex exponential signals

Based on Euler's relation, the general form of continuous-time complex exponential signal Ce^{st} shown in eq. (1.4) can be written as follows:

$$Ce^{st} = |C|e^{\sigma t}\cos(\Omega t + \theta) + j|C|e^{\sigma t}\sin(\Omega t + \theta) \tag{1.9}$$

By analyzing eq. (1.9), we can find the following results: if $\sigma > 0$, it corresponds to a sinusoidal signal multiplied by a growing exponential; if $\sigma < 0$, it corresponds to a sinusoidal signal multiplied by a decaying exponential. Both cases are shown in Figure 1.6. If $\sigma = 0$, the real part and imaginary part of the complex exponential signal are all sinusoidal signals.

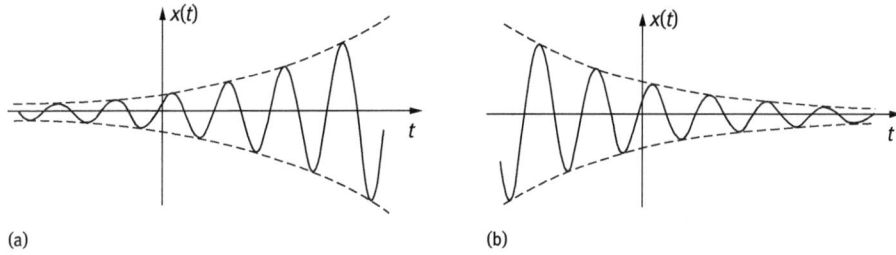

(a)   (b)

Figure 1.6: Growing (a) and decaying (b) sinusoidal signals.

2. Continuous-time sinusoidal signals

The general form of a continuous-time sinusoidal signal can be written as follows:

$$x(t) = A\cos(\Omega t + \varphi) \tag{1.10}$$

where A is the amplitude of the signal, φ is its initial phase whose unit is radians, and Ω is its angular frequency (or radian frequency) whose unit is radians per second (commonly denoted by rad/s). We usually call the angular frequency as the frequency for simplicity.

In eq. (1.5), we see that the complex exponential signal can be expressed as the sinusoidal signal based on Euler's formula. Similarly, we can also change a sinusoidal signal into a complex exponential signal by the Euler's formula as follows:

$$A\cos(\Omega t + \varphi) = \frac{A}{2}e^{j\varphi}e^{j\Omega t} + \frac{A}{2}e^{-j\varphi}e^{-j\Omega t} \tag{1.11}$$

According to Euler's formula, the continuous-time complex exponential signal and the continuous-time sinusoidal signal can be converted to each other. As we know, the continuous-time sinusoidal signal is the most commonly used periodic signal. For a sine signal, its fundamental period, usually denoted as T_0, is defined as its minimal period in its all periods. The frequency corresponding to T_0 is called the fundamental frequency, denoted as Ω_0. In this way, the relation between the frequency and the period represented by eq. (1.8) can be rewritten as

$$T_0 = \frac{2\pi}{\Omega_0} \tag{1.12}$$

3. Discrete-time complex exponential signals

The discrete-time complex exponential signal is a kind of significant discrete-time signal, which plays an important role in signal analysis and processing. The general form of the discrete-time complex exponential signal is defined as

$$x(n) = C\alpha^n \tag{1.13}$$

where both parameters C and α are usually complex. According to the different values of C and α, the complex exponential signal $x(n)$ has the following important forms:

(1) Real exponential form

If both C and α are all real numbers, then $x(n) = C\alpha^n$ is a real exponential signal. According to different real values for C and α, the signal $x(n)$ can be expressed in different forms. In order to simplify the analysis without losing generality, we assume that $C = 1$. Under this condition, if $|\alpha| > 1$, the magnitude of the signal grows exponentially with time n, while if $|\alpha| < 1$, we get a decaying exponential. The form of the signal is also related to the value of α. If α is positive, all the values of $x(n)$ are of the same sign, but if α is negative, the sign of $x(n)$ alternates. The different forms of discrete-time real exponential signals are illustrated in Figure 1.7.

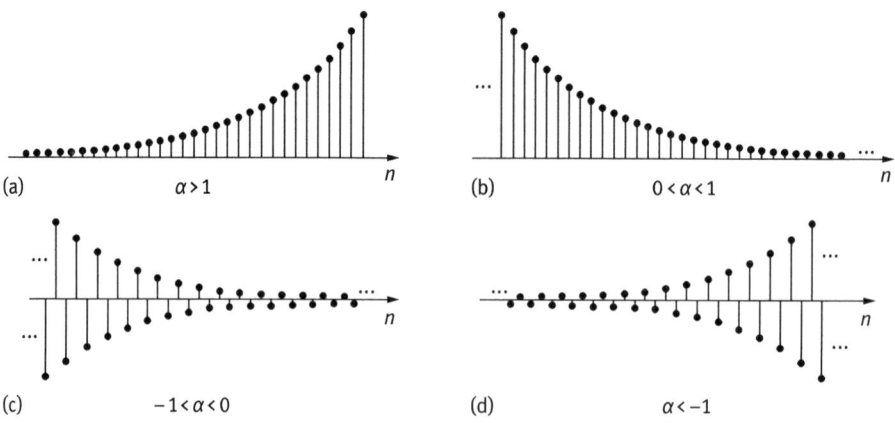

Figure 1.7: Discrete-time real exponential signals. (a) $\alpha > 1$; (b) $0 < \alpha < 1$; (c) $-1 < \alpha < 0$; and (d) $\alpha < -1$.

(2) A special case of complex exponential signals

If we set $C = 1$ and $\alpha = e^{j\omega_0}$ in eq. (1.13), then it yields

$$x(n) = e^{j\omega_0 n} \tag{1.14}$$

where ω_0 is the radian frequency of the signal. The signal shown in the above equation plays a very important role in the signal analysis. There are two main characteristics of the signal, namely, the periodicity with frequency change and the periodicity with time change.

① Periodicity with frequency change

The signal $x(n) = e^{j\omega_0 n}$ is periodic with the change of frequency ω_0 shown as follows:

$$e^{j(\omega_0 + 2k\pi)n} = e^{j\omega_0 n} \cdot e^{j2k\pi n} = e^{j\omega_0 n} \tag{1.15}$$

where k is an arbitrary integer.

② Periodicity with time change

In the following equation, if the ratio between m and N is a rational number,

$$\frac{\omega_0}{2\pi} = \frac{m}{N} \tag{1.16}$$

then $x(n) = e^{j\omega_0 n}$ is a periodic signal. We have

$$x(n + N) = e^{j\omega_0(n + N)} = e^{j\omega_0 n} \cdot e^{j\omega_0 N} = x(n) \tag{1.17}$$

In the above equations, N is the fundamental period of the signal, and $2\pi/N$ is the fundamental frequency.

Example 1.1: A discrete-time signal is given as $x(n) = e^{j7\pi n}$. Try to determine whether the signal is a periodic signal. If it is a period signal, try to find its fundamental period.

Solution: Let N be an integer, then we have

$$x(n + N) = e^{j7\pi(n + N)} = e^{j7\pi n} \cdot e^{j7\pi N}$$

If $N = 2k$, $k = 0$, ± 1, ± 2, $\ldots$, is satisfied, we have $e^{j7\pi N} = 1$. Thus $x(n + N) = e^{j7\pi n} = x(n)$. Therefore, $x(n) = e^{j7\pi n}$ is a periodic signal.

On the other hand, we can determine the periodicity of the signal based on eq. (1.16). From $x(n) = e^{j7\pi n}$, we know that $\omega_0 = 7\pi$, and $\omega_0/2\pi = 7\pi/2\pi = 7/2$ is a rational number. So we can conclude that the signal is a periodic signal.

From eq. (1.16), we get $N = (2\pi/\omega_0)m = (2\pi/7\pi)m$. By taking m as a suitable positive integer to have N as the smallest positive integer, we can get N as the fundamental period. Obviously, when $m = 7$, $N = 2$ is the fundamental period of the signal.

(3) The general complex exponential signals
For the general form of the discrete-time complex exponential signal shown in eq. (1.13), if we write both parameters C and α into the polar coordinate form as

$$C = |C|e^{j\theta}, \quad \alpha = |\alpha|e^{j\omega_0} \tag{1.18}$$

we have

$$x(n) = |C||\alpha|^n \cos(\omega_0 n + \theta) + j|C||\alpha|^n \sin(\omega_0 n + \theta) \tag{1.19}$$

The curves of the real parts (or the imaginary parts) of the signal when $|\alpha| > 1$ and $|\alpha| < 1$ are given, respectively, in Figure 1.8.

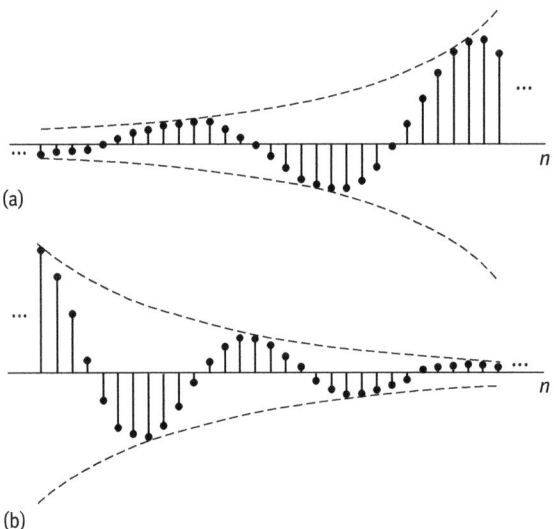

(a)

(b)

Figure 1.8: The curves of the real parts (or the imaginary parts) of the signal: (a) $|\alpha| > 1$ and (b) $|\alpha| < 1$.

4. Discrete-time sinusoidal signals
The general form of discrete-time sinusoidal signal is given as follows:

$$x(n) = A \cos(\omega_0 n + \varphi) \tag{1.20}$$

A discrete-time sinusoidal signal is shown in Figure 1.9.

Different from the continuous-time sinusoidal signal, the discrete-time sinusoidal signal is periodic only when the condition in eq. (1.16) is satisfied.

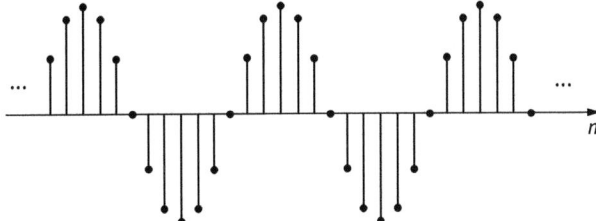

Figure 1.9: A discrete-time sinusoidal signal.

Example 1.2: Given a discrete-time signal $x(n) = \sin(3n)$, try to determine whether the signal is periodic.

Solution: According to eq. (1.16), $x(n) = \sin(3n)$ is an aperiodic signal since $\omega_0/2\pi = 3/2\pi$ is an irrational number.

5. Continuous-time unit impulse and unit step signals

(1) Continuous-time unit step signal

Definition 1.1: A continuous-time unit step signal is defined as

$$u(t) = \begin{cases} 1, & t > 0 \\ 0, & t < 0 \end{cases} \tag{1.21}$$

From Definition 1.1 we see that the unit step signal is not continuous at $t = 0$. Figure 1.10 shows its waveform.

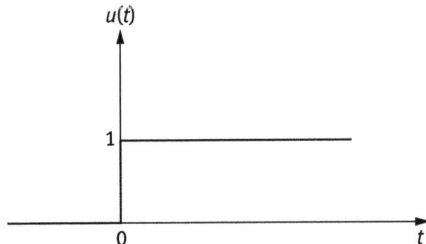

Figure 1.10: The waveform of the continuous-time unit step signal.

The unit step signal is a very important signal in signal analysis and processing, which is usually used to express the nonzero range of a signal, and also to restrict the integrating range.

(2) Continuous-time unit impulse signal
The continuous-time unit impulse signal is usually defined by the following three forms:

Definition 1.2: The continuous-time unit impulse signal is defined as the first-order derivative of the continuous-time unit step signal as follows:

$$\delta(t) = \frac{du(t)}{dt} \tag{1.22}$$

As a matter of fact, $\delta(t)$ is a kind of singularity signal since $u(t)$ is not continuous at $t = 0$, causing its derivative at this time does not exist. For $\delta(t)$, its value is zero when $t \neq 0$, while its value tends to be infinite when $t = 0$.

Definition 1.3: The integration of the continuous-time unit impulse signal in range $(-\infty, +\infty)$ is 1, which is shown as

$$\int_{-\infty}^{\infty} \delta(t)dt = 1, \quad \delta(t) = 0 \ (\text{if } t \neq 0) \tag{1.23}$$

Definition 1.4: The continuous-time unit step signal $u(t)$ can be expressed as a running integral of the unit impulse signal as

$$u(t) = \int_{-\infty}^{t} \delta(\tau)d\tau \tag{1.24}$$

Figure 1.11 shows the waveform of the continuous-time unit impulse signal.

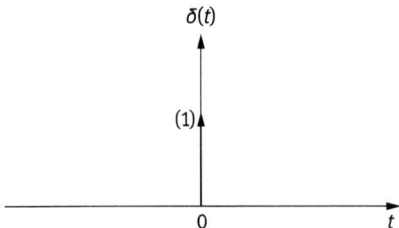

Figure 1.11: The waveform of the continuous-time unit impulse signal.

The continuous-time unit impulse signal plays a very important role in signal analysis and representation, and its main properties and applications include:

① Sampling property

$$x(t)\delta(t) = x(0)\delta(t) \tag{1.25}$$

or

$$x(t)\delta(t - t_0) = x(t_0)\delta(t - t_0) \tag{1.26}$$

The sampling property of the continuous-time unit impulse signal shows that any signal $x(t)$ multiplied by $\delta(t-t_0)$ will result in the value of $x(t)$ at $t=t_0$ multiplied by $\delta(t-t_0)$.

② Sifting property

$$\int_{-\infty}^{\infty} x(t)\delta(t)dt = x(0) \int_{-\infty}^{\infty} \delta(t)dt = x(0) \tag{1.27}$$

or

$$\int_{-\infty}^{\infty} x(t)\delta(t-t_0)dt = x(t_0) \int_{-\infty}^{\infty} \delta(t-t_0)dt = x(t_0) \tag{1.28}$$

where both $x(0)$ and $x(t_0)$ in eqs. (1.27) and (1.28) are all constant signals irrelevant with time t.

Many other properties and applications of the continuous-time impulse signal will be introduced later one by one, including its symmetry and convolution properties.

Example 1.3: Please calculate the following equations:

(1) $\int_{-\infty}^{\infty} e^{-j\omega t}[\delta(t)-\delta(t-t_0)]dt$; (2) $\int_{-\infty}^{\infty} x(t+t_0)\delta(t-t_0)dt$.

Solution:

(1) $\int_{-\infty}^{\infty} e^{-j\omega t}[\delta(t)-\delta(t-t_0)]dt = \int_{-\infty}^{\infty} e^{-j\omega t}\delta(t)dt - \int_{-\infty}^{\infty} e^{-j\omega t}\delta(t-t_0)dt = 1 - e^{-j\omega t_0}$.

(2) $\int_{-\infty}^{\infty} x(t+t_0)\delta(t-t_0)dt = x(2t_0)$.

6. Discrete-time unit impulse and unit step signals

(1) Discrete-time unit step signal

Definition 1.5: The discrete-time unit step signal is defined as

$$u(n) = \begin{cases} 1, & n \geq 0 \\ 0, & n < 0 \end{cases} \tag{1.29}$$

The waveform of the discrete-time unit step signal is shown in Figure 1.12.

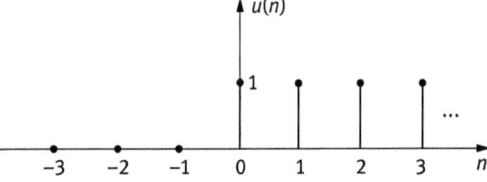

Figure 1.12: The waveform of the discrete-time unit step signal.

(2) Discrete-time unit impulse signal

Definition 1.6: The discrete-time unit impulse signal is defined as

$$\delta(n) = \begin{cases} 1, & n = 0 \\ 0, & n \neq 0 \end{cases} \tag{1.30}$$

The waveform of the discrete-time unit impulse signal is shown in Figure 1.13.

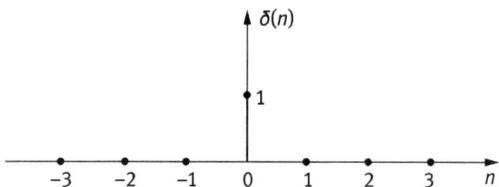

Figure 1.13: The waveform of the discrete-time unit impulse signal.

(3) Properties and applications of the discrete-time unit impulse signal
① Relations with the discrete-time unit step signal
There is a close relationship between the discrete-time unit impulse signal and the discrete-time unit step signal. In fact, the former is the first difference of the latter, shown as

$$\delta(n) = u(n) - u(n-1) \tag{1.31}$$

In turn, the latter is the dynamic sum of the former shown as

$$u(n) = \sum_{m=-\infty}^{n} \delta(m) \tag{1.32}$$

Eq. (1.32) can also be expressed as

$$u(n) = \sum_{k=0}^{\infty} \delta(n-k) \tag{1.33}$$

② Sampling property

Similar to that of the continuous-time unit impulse signal, the sampling property of the discrete-time unit impulse signal can be expressed as

$$x(n)\delta(n-n_0) = x(n_0)\delta(n-n_0) \tag{1.34}$$

③ Sifting property

The sifting property of the discrete-time unit impulse signal is shown as follows:

$$\sum_{n=-\infty}^{\infty} x(n)\delta(n-n_0) = \sum_{n=-\infty}^{\infty} x(n_0)\delta(n-n_0) = x(n_0) \qquad (1.35)$$

It should be noticed that $x(n_0)$ is a constant sequence.

1.2.4 The Operation of Signals

This section introduces basic signal operations, mainly including the independent variable transformations, and the addition and multiplication operations of signals. Because the operational rules and methods for both continuous-time and discrete-time signals are basically the same, we just generally introduce the operations for continuous-time signals. If the rules of operations for both types of signals are not the same, we will introduce them, respectively.

1. Time shift
The time shift of a signal includes two cases, namely, leading and lagging:

$$x(t) \rightarrow x(t-t_0) \qquad (1.36)$$

where t_0 denotes the time shift and the symbol "$\rightarrow$" expresses the signal transformation. In eq. (1.36), $t_0 > 0$ represents a delay of $x(t-t_0)$ with respect to $x(t)$, while $t_0 < 0$ represents a leading ahead of $x(t)$. Figure 1.14 demonstrates the time shift of a signal.

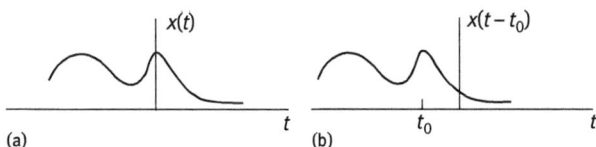

(a)   (b)

Figure 1.14: Time shift of a signal: (a) original signal and (b) time-shifted signal.

2. Time reversal
The time reversal of a signal is defined as follows:

$$x(t) \rightarrow x(-t) \qquad (1.37)$$

An example of the time reversal is the playback of a recorded signal from its rear part shown as (Figure 1.15).

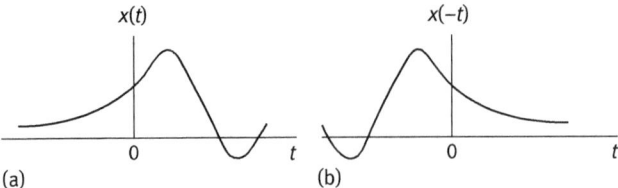

Figure 1.15: Time reversal of a signal: (a) original signal and (b) the signal after time reversal.

3. Time scaling

The time scaling describes the compression and extension of the signal on its time axis:

$$x(t) \rightarrow x(at) \tag{1.38}$$

where $a \neq 0$ is the scaling parameter. The different values of a indicate different scaling transformation forms. $|a| > 1$ indicates a linear compression on its time axis. $|a| < 1$ denotes a linear extension on its time axis. $a = -1$ means a time reversal of the signal. Figure 1.16 demonstrates different time scaling transforms of a signal.

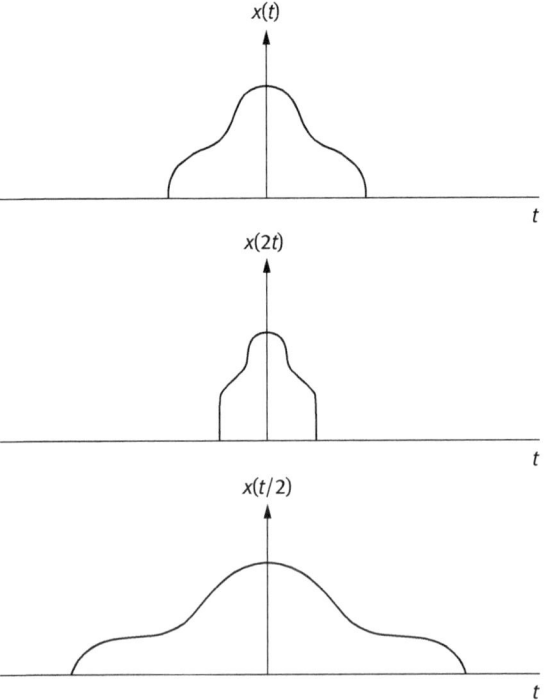

Figure 1.16: Time scaling transformation of a signal. The upper figure shows the original signal, the middle figure shows the time-compressed signal, and the lower figure shows the time-extended signal.

The time shift, time reversal, and time scaling transforms of a signal are often used in combination, describing a comprehensive independent variable transform.

Example 1.4: The waveform of a signal $x(t)$ is given in Figure 1.17(a). Try to depict the waveforms of $x(t+1)$, $x(-t+1)$, $x(\frac{3}{2}t)$, and $x(\frac{3}{2}t+1)$.

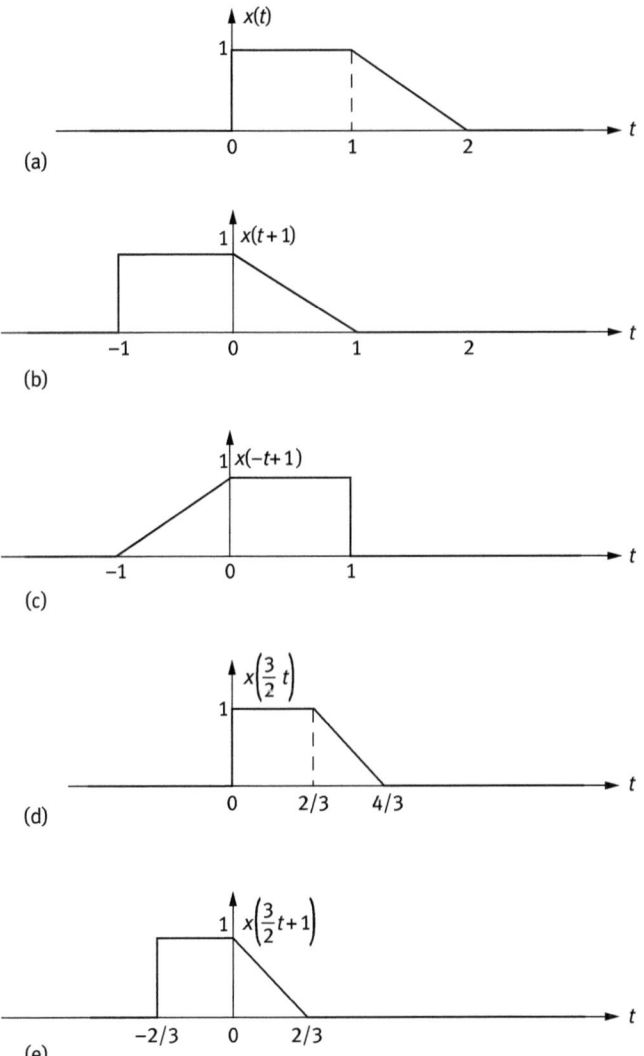

Figure 1.17: An example of comprehensive independent variable transform.

Solution: The waveforms of the four transforms are shown in Figure 1.17(b)–(e) as follows:

4. Signal differential (difference) operation and signal integration (accumulation)
The differential operation and integration of a continuous-time signal are expressed as

$$y(t) = x'(t) = \frac{dx(t)}{dt} \tag{1.39}$$

and

$$y(t) = \int_{-\infty}^{\infty} x(\tau)d\tau \tag{1.40}$$

Similarly, the difference and accumulation of a discrete-time signal are shown as

$$y(n) = x(n) - x(n-1) \tag{1.41}$$

and

$$y(n) = \sum_{m=-\infty}^{n} x(m) \tag{1.42}$$

5. The addition of signals
The addition of signals is the most basic signal operation in signal processing. Suppose that m signals listed as $x_1(t),\ x_2(t),\ \dots, x_m(t)$ attend the addition operation. The result of the addition is shown as

$$y(t) = x_1(t) + x_2(t) + \cdots + x_m(t) \tag{1.43}$$

6. The multiplication of signals
In communication technology and signal processing, the operation of signal multiplication is often encountered. In fact, the multiplication of two signals is usually used to describe the operation of sampling or modulation of signals. Suppose that m signals $x_1(t),\ x_2(t),\ \dots, x_m(t)$ attend the multiplication operation, its result is shown as

$$y(t) = x_1(t)x_2(t) \cdots x_m(t) \tag{1.44}$$

1.2.5 The Basic Concept of Systems

In the field of electronic information technology, the concepts of signals and systems are always inseparable. Signals are always transmitted, stored, processed, or run in systems. It is difficult to complete the information transfer or exchange without a system.

What exactly is the system? Generally speaking, a system is an entirety with a specific function composed of a number of interacting and interdependent things. In electronic and information technology, a system is formally defined as an entity that manipulates one or more signals to accomplish a function, thereby yielding new signals.

On the other hand, the system is hierarchical. A bigger system may contain a number of subsystems. On the contrary, a system may become a subsystem of a larger system.

1.2.6 Classification of Systems and Their Properties

The relationship between signals and systems can be expressed in a variety of ways, such as differential equations, difference equations, system input/output relations, and system diagram. The basic relationship between signals and systems is given in Figure 1.18.

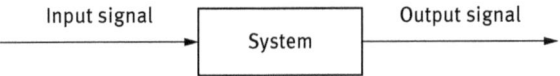

Figure 1.18: The basic relationship between signals and systems.

Various types of systems are briefly introduced as follows.

1. Continuous-time system and discrete time system

If the input and output signals of the system are all continuous-time signals, and the system does not convert the input continuous-time signals into discrete-time signals, such systems are called continuous-time systems. Similarly, if the input and output signals of a system are all discrete-time signals, it is called discrete-time system. Commonly used analog electronic circuits are examples of continuous-time systems, while digital computers are typical discrete-time systems. Actually, the continuous-time system and the discrete-time system are often mixed. For example, in the data acquisition system, an A/D converter is usually used to convert a continuous-time signal into a discrete-time signal or digital signal. Such system is often referred to as a hybrid system.

Continuous-time systems can be described by differential equations, and discrete-time systems are usually described by difference equations.

Example 1.5: Given an RC circuit shown in Figure 1.19. Try to express the relation between $V_C(t)$ and $V_s(t)$.

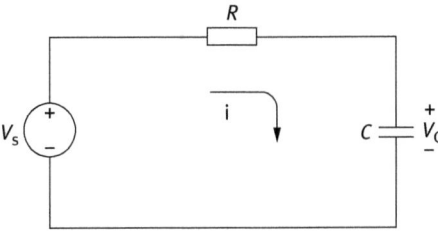

Figure 1.19: An RC circuit.

Solution: According to the Ohm's law, Kirchhoff's law, and the relationship between voltage and current on components, the system differential equation can be listed as

$$\frac{V_s(t)}{R} - \frac{V_C(t)}{R} = C\frac{dV_C(t)}{dt}$$

After arrangement, we get

$$\frac{d}{dt}V_C(t) + \frac{1}{RC}V_C(t) = \frac{1}{RC}V_s(t)$$

In the above equation, the input signal and the output signal are denoted by $V_s(t)$ and $V_C(t)$, respectively.

Example 1.6: Consider the balance of a bank saving account. Suppose $x(n)$ denotes the net deposit of the account, $y(n)$ and $y(n-1)$ denote the balance of this month and last month, and $K = 0.01$ is the interest rate. Then the balance of the account can be expressed as the following difference equation:

$$y(n) = (1+K)y(n-1) + x(n) = 1.01y(n-1) + x(n)$$

2. Memory system and memoryless system
If the output signal of a system depends only on the input signal at the same time, the system is a memoryless system. Conversely, if the output signal of a system depends not only on the input signal at the same time, but also on the input of other moments, the system is referred to as a memory system. Equations (1.45) and (1.46) demonstrate examples for a memoryless system and a memory system:

$$y(t) = 2x(t) + 3 \tag{1.45}$$

$$y(t) = 2x(t+1) + 3x(t-1) \tag{1.46}$$

Examples of common memory systems include the energy storage elements based circuit and the digital computer system. It is precisely because of the memory of the system, the system can process the input signals better.

3. Linear system and nonlinear system
The linear system is a system that possesses the important property of superposition. Otherwise the system is called a nonlinear system. If an input of the linear system consists of the weighted sum of several signals, then its output is the superposition of the responses of the system to each of those signals. If $x_1(t) \rightarrow y_1(t)$ and $x_2(t) \rightarrow y_2(t)$, then we have

$$ax_1(t) + bx_2(t) \rightarrow ay_1(t) + by_2(t) \tag{1.47}$$

where the symbol "$\rightarrow$" denotes the operation of the system.

Example 1.7: The input–output relation of a system is given as $y(t) = t^2 x(t)$. Determine whether the system is a linear system.

Solution: Based on the input–output relation, we have $y_1(t) = t^2 x_1(t)$, $\quad y_2(t) = t^2 x_2(t)$. Let $x_3(t) = a x_1(t) + b x_2(t)$, where a and b are arbitrary constants. Then

$$x_3(t) \rightarrow y_3(t) = t^2 x_3(t) = t^2 [a x_1(t) + b x_2(t)] = a t^2 x_1(t) + b t^2 x_2(t) = a y_1(t) + b y_2(t)$$

Since the output $y_3(t)$ equals to the weighted combination of $y_1(t)$ and $y_2(t)$ (weighted with the same factors with the input signals), the system is a linear system.

4. Time-invariant system and time varying system

If the parameters of a system do not change with time, the system is referred to as the time-invariant system, abbreviated as TI system. On the contrary, if the parameters of a system change with time, then the system is said to be a time varying system, abbreviated as TV system.

Suppose that the input and output signals of a TI system are $x(t)$ and $y(t)$, respectively. If $x(t) \rightarrow y(t)$ is used to express the relation between the input and output signals, then $x(t - t_0) \rightarrow y(t - t_0)$ holds.

Example 1.8: A continuous-time system is given as $y(t) = e^{x(t)}$. Determine whether the system is a TI system.

Solution: ① Let $x_1(t)$ be an arbitrary input signal of the system. The corresponding output signal is $y_1(t) = e^{x_1(t)}$; ② shift $x_1(t)$ to get $x_2(t) = x_1(t - t_0)$; ③ change the input signal into $x_2(t) - x_1(t - t_0)$, and get the output signal $y_2(t) - e^{x_2(t)} - e^{x_1(t - t_0)}$; ④ shift $y_1(t)$ with the same shift as that of $x_1(t)$, and get $y_1(t - t_0) = e^{x_1(t - t_0)}$. Since $y_2(t) = y_1(t - t_0)$, the system is a TI system.

The TI property is one of the most important characteristics of the system. In electronic information area, both linear property and TI property are usually considered together to form a new property, referred to as the linear and TI property. A system satisfying such property is called the linear time-invariant (LTI) system.

5. Invertible system and noninvertible system

A system is said to be invertible if distinct inputs lead to distinct outputs. Conversely, the system is a noninvertible system. If a system is an invertible system, then there must be an inverse system for this system.

6. Causal system and noncausal system

If the output of a system is only related to the input at present and past time, the system is referred to as a causal system. On the contrary, the system is a noncausal system. The bank account balance in Example 1.6 is an example of the causal system.

A causal system is a real-time physically realizable system, which plays an important role in signal analysis and processing, while the noncausal system is not a real-time physically realizable system. As a matter of fact, not all signal processing systems are made up of causal systems. For example, in digital signal processing, noncausal digital filters are often used.

Example 1.9: Try to determine the causality of the following systems:

(1) $y(t) = 3x(t) + 4x(t-1)$; (2) $y(n) = 3x(n) + 4x(n+1)$.

Solution: (1) The system is a causal system because the output of the signal is only related to the input signal at present time and past time.

(2) The system is a noncausal system since the output is related to the input in the future.

7. Stable system and unstable system
If a bounded input signal leads to a bounded output of the system, then the system is stable. Otherwise, the system is unstable.

1.2.7 Basic Analysis Methods for Systems

The system analysis method can be roughly divided into two aspects: the establishment of the system model and the solution of the system model.

1. The establishment of system models
(1) The input–output method for the establishment of system models
The input–output method focuses on the relationship between the excitation and response (i.e., the input and output) of the system, and does not care about the internal variables of the system. This kind of approach is usually used for the single input and single output systems. It is also the focus of this book.

The system model for the input–output method is usually described by the linear constant coefficient differential equation (or difference equation) for the continuous-time LTI system (or the discrete-time LTI system). The system is also described by the unit impulse response for both continuous-time and discrete-time systems.

(2) The state variable method for the establishment of system models
The state variable method describes the states and characteristics of a system with the state equation and output equation. This method can not only describe the relationship between the input and output of the system but also describe the internal states of the system.

2. Solution of the system model
(1) Solving the system model with time domain methods
The time domain method for solving a system model is the most basic method of system analysis. This method directly analyzes and studies the corresponding characteristics of

the system in the time domain. The main advantage of this method is its clear physical concept and meaning. For the system described by an input–output model, the classical method such as the linear constant coefficient differential equation or the difference equation is usually used. The output of the system is also usually obtained by the convolution method. For the system described by the state variable method, the state equation and the output equation can be solved by the linear algebra method.

(2) Solving the system model with transform domain methods
The transform domain method for solving a system model transforms the signal and system in the time domain into a specified transform domain. The main purpose of the transformation is to simplify the calculation. Commonly used transforms include Fourier transform (FT), Discrete-time Fourier transform (DTFT), Laplace transform (LT), and z-transform (ZT). Among them, FT and DTFT are suitable for frequency spectrum analysis of signals and systems, and ZT and LT are usually used to analyze the stability and causality of systems. In addition, other orthogonal transforms, such as discrete Fourier transform (DFT) and fast Fourier transform (FFT) are also widely used in the analysis and processing of signals and systems. Compared with the time domain methods, the transform domain methods are more simple and intuitive, so they have been paid more attention.

1.3 Time Domain Analysis of LTI System and Convolution

1.3.1 The Basic Concept on LTI System

Combining both concepts of linear system and TI system, we get a new concept: LTI system. The so-called LTI system is such a system that satisfies both linear property and TI property. The characteristics of a continuous-time LTI system can be described as follows: if both $x_1(t) \rightarrow y_1(t)$ and $x_2(t) \rightarrow y_2(t)$ are satisfied, then

$$ax_1(t - t_1) + bx_2(t - t_2) \rightarrow ay_1(t - t_1) + by_2(t - t_2) \tag{1.48}$$

Correspondingly, the characteristics of the discrete-time LTI system can also be obtained. The analysis of the LTI system is one of the key contents of this book. This is mainly due to the following two reasons: first, many physical processes have LTI characteristics, and therefore they can be described by LTI systems; second, the analysis of the LTI system can be a theoretical foundation for subsequent signal analysis and processing.

1.3.2 The Time Domain Analysis for Continuous-Time LTI Systems

The analysis of a continuous-time LTI system in the time domain is an issue to find the solution of a linear constant coefficient differential equation. Such kind of differential equation contains excitation and response (input and output) as functions of time, as well as their linear combination of derivative in various orders. Equation (1.49) gives the general form of the N th-order differential equation for an LTI system:

$$\sum_{k=0}^{N} a_k \frac{d^k y(t)}{dt^k} = \sum_{k=0}^{M} b_k \frac{d^k x(t)}{dt^k} \tag{1.49}$$

where $x(t)$ and $y(t)$ denote the input and output signals of the system, a_k and b_k are the weighting factors in the term of the k th-order derivative, M and N denote the order of the input and output, respectively. In general, $N \geq M$ is satisfied.

Generally speaking, in order to solve the differential equation as shown in eq. (1.49), we have to get some additional conditions and transform the system into the form in which the output is expressed by the input. On the other hand, it is usually not easy to solve differential equations. Therefore, this book does not consider the solution of differential equations in the time domain, and in the later chapters, the solution for differential equations will be solved by orthogonal transformations. Readers are suggested to refer to relevant mathematical reference books for the direct solution of differential equations in the time domain.

1.3.3 Convolution of Continuous-Time LTI Systems

The most commonly used method of time domain analysis for a continuous-time LTI system is to use the system unit impulse response to characterize the system. The unit impulse response of an LTI system is defined as the response of the system to the unit impulse signal $\delta(t)$. The unit impulse response is completely determined by the characteristics of the system itself, which is not related to the input of the system.

1. The representation of arbitrary continuous-time signals in terms of impulses
Suppose that the waveform of a continuous-time signal is shown in Figure 1.20. In the figure, the smooth curve represents the waveform of the signal, and the step line is an approximation to the waveform. If we define $\delta_\Delta(t)$ as

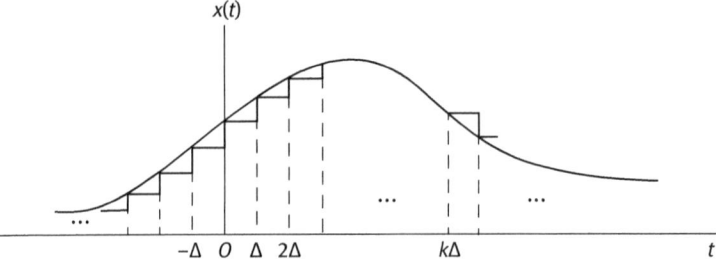

Figure 1.20: The waveform of a continuous-time signal.

$$\delta_\Delta(t) = \begin{cases} \frac{1}{\Delta}, & 0 \leq t \leq \Delta \\ 0, & \text{otherwise} \end{cases} \tag{1.50}$$

where Δ is the step length or interval of the step line. If we let $\Delta \to 0$, we get $\delta_\Delta(t) \to \delta(t)$. Thus the step line tends to be the waveform of $x(t)$, which is shown as

$$x(t) = \lim_{\Delta \to 0} \sum_{k=-\infty}^{+\infty} x(k\Delta)\delta_\Delta(t - k\Delta)\Delta = \int_{-\infty}^{+\infty} x(\tau)\delta(t - \tau)d\tau \tag{1.51}$$

We see from eq. (1.51) that any continuous-time signal can be expressed as a linear combination of unit impulse signal and its time shift.

2. The unit impulse response of continuous-time LTI systems

For a continuous-time LTI system, if the input–output relation is shown as $\delta_\Delta(t) \to h_\Delta(t)$, based on the linear and TI properties, we have

$$\begin{cases} \quad \vdots \\ x(-\Delta)\delta_\Delta(t+\Delta) \to x(-\Delta)h_\Delta(t+\Delta) \\ x(0)\delta_\Delta(t) \to x(0)h_\Delta(t) \\ x(\Delta)\delta_\Delta(t-\Delta) \to x(\Delta)h_\Delta(t-\Delta) \\ \quad \vdots \end{cases} \tag{1.52}$$

The above equation can be interpreted as the response of the LTI system to the signal $\delta_\Delta(t)$ and its weighted time shift. If we take a summation to both sides of eq. (1.52), we get the response of the LTI system to $\delta_\Delta(t)$ and its combination. Since the arbitrary signal $x(t)$ can be expressed as a linear combination of $\delta_\Delta(t)$, the summation of the left side of eq. (1.52) in fact is approximately to show an arbitrary continuous-time signal, while the combination of the right side of eq. (1.52) in fact is the response of the system to arbitrary continuous-time signals. When $\Delta \to 0$, we have $\delta_\Delta(t) \to \delta(t)$ and $h_\Delta(t) \to h(t)$. Thus the summation of eq. (1.52) changes into an integration as

$$y(t) = \int_{-\infty}^{+\infty} x(\tau)h(t - \tau)d\tau \tag{1.53}$$

where $x(t)$ is an arbitrary continuous-time signal as the input of the LTI system, $y(t)$ is the output of the system. $h(t)$ is defined as the unit impulse response of the system which is the response of an LTI system to a unit impulse signal.

3. The convolution of the continuous-time LTI system

In fact, eq. (1.53) is referred to as the convolution integral of an LTI system, denoted as

$$y(t) = x(t) \star h(t) = \int_{-\infty}^{+\infty} x(\tau)h(t-\tau)d\tau \tag{1.54}$$

It is easy to prove that the convolution operation of an LTI system obeys the commutative law, the associative law, and the distributive law, which is shown as

$$x(t) \star h(t) = h(t) \star x(t)$$
$$x(t) \star [h_1(t) \star h_2(t)] = [x(t) \star h_1(t)] \star h_2(t) \tag{1.55}$$
$$[x_1(t) + x_2(t)] \star h(t) = x_1(t) \star h(t) + x_2(t) \star h(t)$$

The convolution is the most commonly used method to solve the output response of an LTI system in time domain. If the unit impulse response $h(t)$ is known, and the input signal $x(t)$ is given, then the output signal $y(t)$ of the system can be calculated according to eq. (1.54).

The basic steps of convolution operation are listed as follows:

Step 1: Change the independent variable t as τ, such as $x(t) \to x(\tau)$ and $h(t) \to h(\tau)$.

Step 2: Select $x(\tau)$ (or $h(\tau)$) and reflect it as $x(-\tau)$ (or $h(-\tau)$). Suppose that we select and reflect $h(\tau)$.

Step 3: Shift $h(-\tau)$ on the time axis by t to get $h(t-\tau)$.

Step 4: Take the multiplication to get $x(\tau)h(t-\tau)$.

Step 5: Take the integration of $\int_{-\infty}^{+\infty} x(\tau)h(t-\tau)d\tau$.

Example 1.10: The unit impulse response of a continuous-time system is given as $h(t) = u(t)$, whose input signal is $x(t) = e^{-2t}u(t)$. Try to find its output signal $y(t)$.

Solution: Change the input signal $x(t) = e^{-2t}u(t)$ as $x(\tau) = e^{-2\tau}u(\tau)$, and rewrite the impulse response as $h(\tau) = u(\tau)$.

Change the unit impulse response as $h(-\tau) = u(-\tau)$.

Get $h(t-\tau) = u(t-\tau)$ from $h(-\tau) = u(-\tau)$. In this example, the integration range can be divided into two parts. If $t < 0$, it means that $u(-\tau)$ shifts to the left; if $t > 0$, $u(-\tau)$ will shift to the right.

Since $x(\tau)$ and $u(t-\tau)$ do not overlap when $t < 0$, the product of the multiplication is zero. When $t > 0$, both signal and system overlap with each other, and the product is $e^{-2\tau}u(\tau)u(t-\tau)$.

Taking the integration to $e^{-2\tau}$ in the range of $(0, t)$, we get $y(t) = \int_0^t e^{-2\tau}d\tau = \frac{1}{2}$ $[1-e^{-2t}]u(t)$.

Example 1.11: Suppose that the input signal and the unit impulse response of an LTI system are given as $h(t) = t[u(t) - u(t-1)]$ and $x(t) = t[u(t) - u(t-1)]$. Try to calculate the output $y(t)$ with a MATLAB program, and plot its waveform.

Solution: The program code is listed as follows:

```
clear all;
for i=1:100
    t(i)=(i-1)/40;  F=@(x)sawtooth(x).*sawtooth(t(i)-x);
      y(i)=quad(F,0,t(i));
end
plot(t,y);  xlabel('t/s');  ylabel('Amplitude');
   title('Convolution result'); function R=sawtooth(t); [M,N]=size(t);
for i=1:N
    if t(i)<0
        R(i)=0;
    elseif t(i)>1
        R(i)=0;
    else
        R(i)=t(i);
    end
end
```

Figure 1.21 shows the result of the convolution operation.

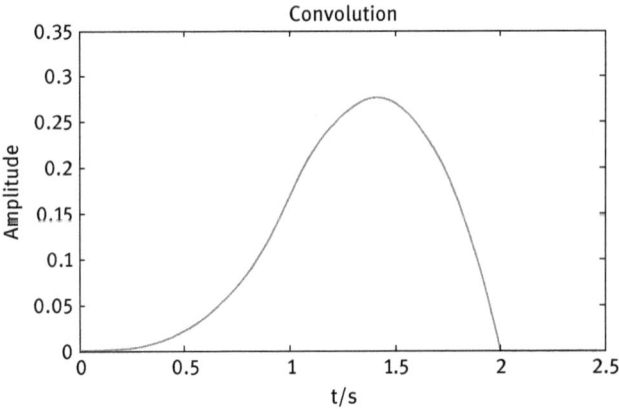

Figure 1.21: The result of the convolution operation.

1.3.4 The Time Domain Analysis for Discrete-Time LTI Systems

Similar to the analysis methods for the continuous-time LTI systems in the time domain, the analysis methods for the discrete-time LTI systems can be summarized as establishing and solving the linear constant coefficient difference equations. Such difference equation consists of a combination of the functions and their various orders of differences. The general form of the difference equation is given as follows:

$$\sum_{k=0}^{N} a_k y(n-k) = \sum_{k=0}^{M} b_k x(n-k) \tag{1.56}$$

where $x(n)$ and $y(n)$ denote the input and output signals of the LTI system, a_k and b_k denote the weight factors in the kth term, M and N are the orders of the input and output, respectively.

The difference equation can be solved by different methods, including the recurrence method and some other methods. However, the above-mentioned methods are not used frequently. Similar to the solution of the continuous-time LTI system, the difference equation for the discrete-time system can be solved by some kind of orthogonal transformation in the transform domain. Of course, the system can also be transformed into the time domain again to analyze and calculate the system through the unit impulse response.

1.3.5 Convolution of Discrete-Time LTI Systems

1. The representation of arbitrary discrete-time signals in terms of impulses
Figure 1.22 shows an arbitrary discrete-time signal $x(n)$.

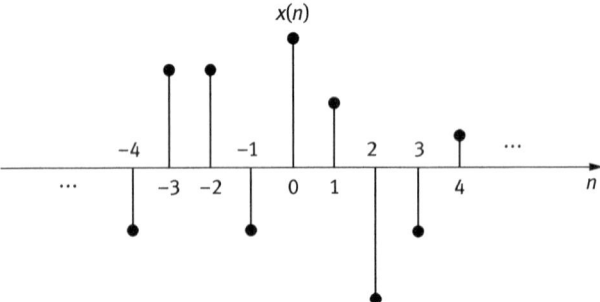

Figure 1.22: An arbitrary discrete-time signal.

It can be seen from the figure that a discrete-time signal $x(n)$ can be expressed as a weighted linear combination of the unit pulse signal $\delta(n)$ and its time shift, which is shown as

$$x(n) = \sum_{k=-\infty}^{+\infty} x(k)\delta(n-k) \tag{1.57}$$

2. The unit impulse response of the discrete-time LTI system
For an arbitrary discrete-time LTI system, if its input and output signals are $\delta(n)$ and $h(n)$, respectively, then based on the linear and TI properties of the LTI system, we have

$$\begin{cases} \quad \vdots \\ x(-1)\delta(n+1) \rightarrow x(-1)h(n+1) \\ \quad x(0)\delta(n) \rightarrow x(0)h(n) \\ x(1)\delta(n-1) \rightarrow x(1)h(n-1) \\ \quad \vdots \end{cases} \tag{1.58}$$

Taking the summation for both sides of the above expression, we get

$$y(n) = \sum_{k=-\infty}^{+\infty} x(k)h(n-k) \tag{1.59}$$

where $x(n)$ and $y(n)$ are the input and output signals of the LTI system, respectively, and $h(n)$ denotes its unit impulse response.

3. The convolution of the discrete-time LTI system

Similar to that of the continuous-time LTI system, the convolution sum of the discrete-time LTI system is defined as

$$y(n) = x(n) \star h(n) = \sum_{k=-\infty}^{+\infty} x(k)h(n-k) \tag{1.60}$$

The convolution sum also obeys the commutative, the associative, and the distributive laws.

The discrete convolution is the most commonly used method for analyzing and solving the discrete-time LTI systems. If the unit impulse response $h(n)$ and the input signal $x(n)$ are given, we can get the output $y(n)$ of the system by eq. (1.60).

Example 1.12: The unit impulse response $h(n)$ and the input signal $x(n)$ of an LTI system are given in Figure 1.23(a) and (b). Try to calculate the output $y(n)$, and plot its waveform.
Solution: According to eq. (1.60), we get

$$y(n) = x(n) \star h(n) = x(0)h(n-0) + x(1)h(n-1)$$
$$= 0.5h(n) + 2h(n-1)$$

From the above equation, we have $y(0) = 0.5$, $y(1) = 2.5$, $y(2) = 2.5$, $y(3) = 2$, and other $y(n)$ are all zeros. So we get $y(n) = [0.5 \quad 2.5 \quad 2.5 \quad 2]$. The waveform of $y(n)$ is plotted in Figure 1.23(c).

Example 1.13: Calculate $y(n)$ in Example 1.12 by MATLAB programming.
Solution: The MATLAB program code is shown as follows:

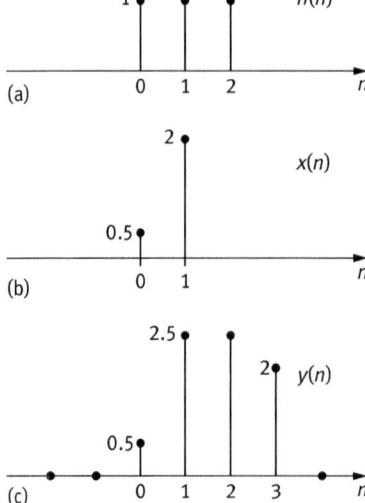

(a)

(b)

(c)

Figure 1.23: Waveforms: (a) the unit impulse response; (b) the input signal; and (c) the output signal.

```
x=[0.5 2]; h=[1 1 1]; y=zeros(1,8); y1=conv(x,h);
for i=1:4
        y(i+2)=y1(i)
end
nn=-2:1:5; stem(nn,y,'filled'); axis([-2 5 0 3]); xlabel('n');
ylabel('Amplitude'); title('Convolution result');
```

The convolution result is demonstrated in Figure 1.24, which is the same as that in Figure 1.23(c).

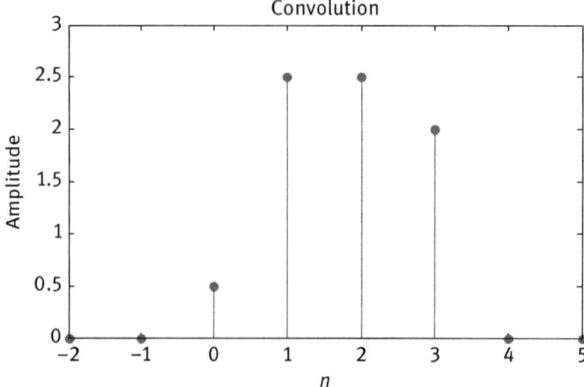

Figure 1.24: The convolution result.

1.4 Basic Properties of LTI Systems

1.4.1 System with and Without Memory

The memory of a system represents the memory capacity to the input signal at previous or later moments. For an LTI system, if it satisfies $h(t) = 0$ when $t \neq 0$ (or $h(n) = 0$ when $n \neq 0$), then it is a memoryless system. If the above condition is not satisfied, then the system is a memory system. A memoryless system can be expressed as

$$h(t) = K\delta(t) \text{ or } h(n) = K\delta(n) \tag{1.61}$$

where K is a nonzero constant. If $K = 1$, the above equation denotes that the response of an LTI system to a unit impulse signal is the unit impulse signal itself. Such system is referred to as the identity system. Thus, the convolution integral and convolution sum can be written as

$$x(t) = x(t) \star \delta(t) \tag{1.62}$$

$$x(n) = x(n) \star \delta(n) \tag{1.63}$$

From the above equations we learnt that the convolution between an arbitrary signal and the unit impulse signal equals the unit impulse signal itself.

1.4.2 Invertibility and Inverse Systems

Figure 1.25 demonstrates the invertibility of an LTI system.

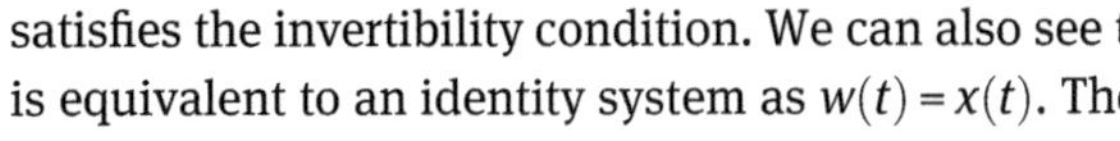

Figure 1.25: The invertibility of an LTI system.

We see from Figure 1.25 that the output of the cascade for two LTI systems equals to the original input, that is, $w(t) = x(t)$. Thus $h_1(t)$ is the inverse system of $h(t)$, and $h(t)$ satisfies the invertibility condition. We can also see that the cascade of $h(t)$ and $h_1(t)$ is equivalent to an identity system as $w(t) = x(t)$. Therefore, we get

$$h(t) \star h_1(t) = \delta(t) \tag{1.64}$$

or equivalently

$$h(n) \star h_1(n) = \delta(n) \tag{1.65}$$

1.4.3 Causality and Causal Systems

The causality is an important characteristic of an LTI system. According to the introduction in Section 1.2.6, the causality indicates that the output at present is only determined by the input at present and past. For an LTI system, the causality of the system can well be represented by the unit impulse response of the system. If the LTI system is a causal system, then it satisfies

$$h(t) = 0, \quad t < 0 \tag{1.66}$$

or

$$h(n) = 0, \quad n < 0 \tag{1.67}$$

This means that, for an LTI system, the system is causal only if all the nonzero values of its unit impulse response are on the right side of its unit impulse response, and vice versa.

1.4.4 Stability and Stable Systems

The system is stable if the bounded input only causes the bounded output. For an LTI system, if its unit impulse response can meet the absolute integrable (or the absolute summable) condition, the system is stable. Both absolute integrable and absolute summable conditions are as follows:

$$\int_{-\infty}^{+\infty} |h(\tau)| d\tau < \infty \tag{1.68}$$

$$\sum_{k=-\infty}^{+\infty} |h(k)| < \infty \tag{1.69}$$

Example 1.14: Determine the stability for the following systems: (1) $h(t) = e^{-2t}u(t)$; (2) $h(n) = u(n)$.
Solution: (1) Substituting $h(t) = e^{-2t}u(t)$ into eq. (1.68) yields

$$\int_{-\infty}^{+\infty} |h(\tau)| d\tau = \int_{0}^{+\infty} |e^{-2\tau}| d\tau = \frac{1}{2} < \infty$$

The absolute integrable condition is satisfied. So the system is a stable system.

(2) Substituting $h(n) = u(n)$ into eq. (1.69) yields

$$\sum_{k=-\infty}^{+\infty} |h(k)| = \sum_{k=0}^{+\infty} |u(k)| = \infty$$

Since the absolute summable condition is not satisfied, the system is an unstable system.

Exercises

1.1 Determine whether or not the following signals are periodic.
(a) $x(t) = 2\sin(t + \pi/4)u(t)$; (b) $x(t) = e^{j2t}$; (c) $x(t) = e^{j2t}u(t)$;
(d) $x(n) = 2\sin(n + \pi/4)$; (e) $x(n) = 2\cos(2\pi n)$; (f) $x(n) = e^{j3\pi n}$

1.2 Find the fundamental period for signal $x(t) = 2\sin(10t + 1) + \sin(4t - 1)$.

1.3 Find the fundamental period for signal $x(n) = 1 + e^{j4\pi n/7} - e^{j2\pi n/5}$.

1.4 The waveform of a continuous-time signal $x(t)$ is shown in Figure E1.1. Draw the waveforms for the following signals: (a) $x(t-1)$; (b) $x(2-t)$; (c) $x(2t+1)$; (d) $x(\frac{2}{3}t - 1)$.

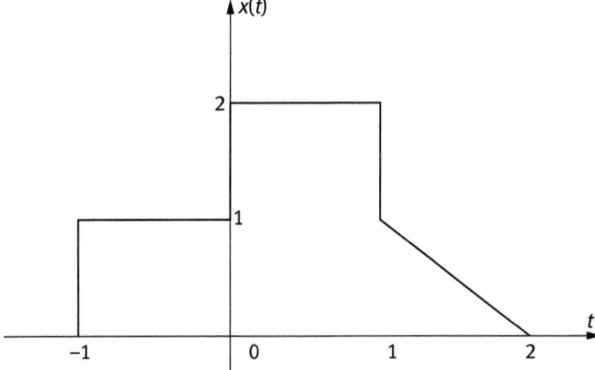

Figure E1.1: Waveform of $x(t)$.

1.5 Determine (1) memory property; (2) TI property; (3) linearity; (4) causality; and (5) stability for the following systems:
(a) $y(t) = x(t-2) + x(1-t)$; (b) $y(t) = x(t/2)$; (c) $y(t) = [\sin(2t)]x(t)$;
(d) $y(n) = x(-n)$; (e) $y(n) = (n+1)x(n)$; (f) $y(n) = x(n-2) - 2x(n-8)$

1.6 Calculate the following convolutions:
(a) $x(t) = e^{-at}u(t)$, $h(t) = e^{-bt}u(t)$, $a \neq b$; (b) $x(t) = e^{-at}u(t)$, $h(t) = e^{-bt}u(t)$,
$a = b$;(c)$x(t) = u(t) - 2u(t-2) + u(t-5)$, $h(t) = e^{2t}u(1-t)$

1.7 Calculate the following convolutions and draw their curves:

(a) $x(n) = \delta(n) + 2\delta(n-1) - \delta(n-3), \quad h(n) = 2\delta(n+1) + 2\delta(n-1);$

(b) $x(n) = \left(\frac{1}{2}\right)^{n-2} u(n-2), h(n) = u(n+2);$

(c) $x(n) = \left(\frac{1}{3}\right)^{-n} u(-n-1), h(n) = u(n-1)$

1.8 Set $x(t) = \begin{cases} 1, & 0 \le t \le 1 \\ 0, & \text{otherwise} \end{cases}$, $h(t) = x(t/a)$. (1) Calculate and draw the convolution of $y(t) = x(t) \star h(t)$; (2) If there is only three discontinuous points in $dy(t)/dt$, find $\alpha = ?$

1.9 The input–output relation of a causal system is given as $y(n) = \frac{1}{4}y(n-1) + x(n)$. Find $y(n)$ if $x(n) = \delta(n-1)$.

1.10 The unit impulse response of an LTI system is shown in Figure E1.2. Over what interval must we know $x(t)$ in order to determine $y(0)$.

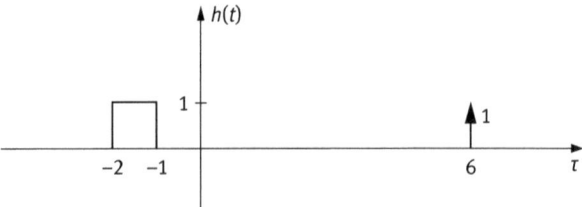

Figure E1.2: The unit impulse response of the given LTI system.

1.11 Try to calculate the convolution and draw its waveform:

$y(t) = [u(t+\pi/2) - u(t-\pi/2)] \star \{\cos[\delta(\sin t)]\}.$

1.12 Given $x(t) = u(t-2), \quad h(t) = e^t u(-t-1)$, calculate $y(t) = x(t) \star h(t)$.

1.13 Try to generate and plot a sinusoidal signal with MATLAB programming.

1.14 Try to generate and plot a triangular wave with MATLAB programming.

1.15 Try to generate and plot a sawtooth wave with MATLAB programming.

1.16 Try to generate and plot a unit impulse signal with MATLAB programming.

1.17 Try to generate and plot a unit step signal with MATLAB programming.

1.18 Try to do the time shift, time reflection, and time scale transforms for the generated sawtooth wave with MATLAB programming, and plot them.

1.19 Try to calculate the convolution $y(n) = x(n) \star h(n)$ with MATLAB programming and plot the waveform of $y(n)$. The input signal and the unit impulse response of the system are given as $x(n) = u(n), \quad h(n) = e^{-3n}u(n)$.

1.20 Try to calculate the convolution $y(t) = x(t) \star h(t)$ with MATLAB programming and plot the waveform of $y(t)$. The input signal and the unit impulse response of the system are given as $x(t) = u(t), \quad h(t) = e^{-3t}u(t)$.

2 Fourier Theory and Frequency Analysis of Signals and Systems

2.1 Introduction

2.1.1 Frequency Analysis of Signals and Systems and a Summary on Fourier Theory

In this chapter, we will introduce systemically theories and methods of signals and systems analysis in the frequency domain. The so-called frequency domain analysis is the analysis and processing of signals and systems in the frequency domain by means of Fourier series (FS) and Fourier transform (FT). It is shown that any periodic function (signal) can be composed of sine and cosine functions (in fact, it can be expressed as the complex exponential signals) with infinite orders, known as Fourier series. FT relaxes the requirement of periodicity, applicable to a broad class of signals.

We see from the previous chapter that an arbitrary continuous-time signal (or an arbitrary discrete-time signal) can be expressed or decomposed as a linear combination of unit impulse signals. Such decomposition is carried out in the time domain, while FS or FT transforms a signal in the time domain into the frequency domain, whose basis functions are sine, cosine, and complex exponential signals. Taking such decomposition is just taking the FS or FT for a given signal.

There are many benefits to the analysis of signals and systems by FS or FT to transform signals or systems from the time domain to the frequency domain.

First, it can be more convenient for us to analyze the frequency characteristics of signals and systems in the frequency domain. We may obtain more information on the spectrum of signal and system; second, the convolution operation in the time domain can be significantly simplified; and third, many problems in signal processing and communication technology can only be interpreted clearly in the frequency domain.

2.1.2 The Development of Fourier Theory

Baron Jean Baptiste Joseph Fourier (1768–1830), the famous French mathematician and physicist, was a member of the French Academy of Sciences. He was born in a tailor's family in Auxerre, a central France city in 1768. He became an orphan when he was 8. After studying in a military school, he became an assistant professor in 1795. He also campaigned in Egypt following Napoleon in 1798. After his return, he was appointed as the stadtholder of province Grenoble.

Fourier's main scientific contribution is that he proposed the theory of FS during his research on heat conduction. FS indicates that any periodic function can be expanded into a sinusoidal series. This theory generalized the existed methods on

https://doi.org/10.1515/9783110465082-002

the trigonometric series proposed by Euler and Bernoulli in some special cases into a general theory, and established a foundation of Fourier analysis. After that, Fourier derived the Fourier integral to deal with the heat conduction problem in an infinite region. All of them greatly promoted the boundary value problem in partial differential equation. Due to his contribution to the theory of heat transfer, he was elected as an academician of Academy of Sciences in Paris in 1817. Later in 1822 he became the permanent secretary of the Academy of Sciences.

The study relating to Fourier theory can be traced back to the ancient Babylonian period when people used this idea of the trigonometric series to predict the movements of the heavenly body. In 1748, Euler used this idea to study the vibration of strings and got a valuable result. In 1753, Bernoulli discovered that the actual motion of a string can be represented by a linear combination of standard oscillation modes. However, the results of the study were strongly criticized by Lagrange, who believed that the trigonometric series cannot be used to indicate the function that there are discontinuous points on it.

In studying the phenomenon of the heat transmission and diffusion, Fourier found that the trigonometric series with harmonic relation is very useful for representing the temperature distribution. He claimed that any periodic signal can be expressed in such a trigonometric series. Based on his study results, Fourier submitted a paper in 1807 for publication. Among the four reviewers, S. F. Lacroix, G. Monge, and P. S. de Laplace were in favor of publication of the paper. However, J. L. Lagrange rejected the paper strongly, and so Fourier's paper did not appear. It was 15 years later, Fourier's theory appeared in the textbook titled *Théorie Analytique de la Chaleur* in 1822.

Although Fourier's theory on trigonometric series for any periodic signals is of very important innovational significance and potential applications, the theory is flawed in its original paper and book. It was not until 1829, several exact constraints on FS were given by P. L. Dirichlet, the theory of FS became more rigorous and complete.

2.1.3 Classification of FS and FT

After nearly 200 years of development, Fourier theory has a great influence on various aspects of modern science and technology. Roughly speaking, Fourier theory can be divided into FS for periodic signals and FT for aperiodic signals, including

1. Fourier series for continuous-time periodic signals, referred to as FS.
2. FS for discrete-time periodic signals, referred to as discrete Fourier series (DFS).
3. Fourier transform mainly for continuous-time aperiodic signals, referred to as FT.
4. FT mainly for discrete-time aperiodic signals, referred to as discrete-time FT (DTFT).

5.   Discrete Fourier transform for discrete-time aperiodic signals, referred to as DFT.
6.   A fast algorithm for DFT, referred to as fast FT.
7.   A kind of time–frequency joint expression method referred to as short-time Fourier transform.
8.   A generalized FT referred to as fractional FT (FRFT).

This book will introduce all kinds of the FS and FT listed above, except for the FRFT.

2.2 FS Representation of Continuous-Time Periodic Signals

2.2.1 Continuous-Time Periodic Signal and Its FS Representation

1. The harmonic relation of periodic signals

The harmonic relation of periodic signals is a set of periodic signals, such as the sinusoidal signal or the complex exponential signal, in which every signal is periodic with a same common period T.

Taking the complex exponential signal $e^{j\Omega t}$ as an example, we see that the necessary condition of the signal as periodic is it satisfies

$$e^{j\Omega T_0} = 1 \tag{2.1}$$

where T_0 is the period of the signal. Therefore, the following equation is also satisfied:

$$\Omega T_0 = 2\pi k, \quad k = 0, \pm 1, \pm 2, \ldots \tag{2.2}$$

The fundamental frequency of a set of harmonically related signals is defined as $\Omega_0 = 2\pi/T_0$. Thus, the set of harmonically related complex exponential signals can be defined as

$$\phi_k(t) = e^{jk\Omega_0 t}, \quad k = 0, \pm 1, \pm 2, \ldots \tag{2.3}$$

in which, if $k = 0$, then $\phi_k(t) = 1$, showing the direct current component; if $k = 1$, then $\phi_k(t) = e^{j\Omega_0 t}$, demonstrating the fundamental component of the signal. Figure 2.1 shows the harmonically related sine signals.

In fact, any continuous-time periodic signal can be written as a linear combination of harmonically related complex exponential signals:

$$x(t) = \sum_{k=-\infty}^{+\infty} a_k e^{jk\Omega_0 t} = \sum_{k=-\infty}^{+\infty} a_k e^{jk(2\pi/T)t} \tag{2.4}$$

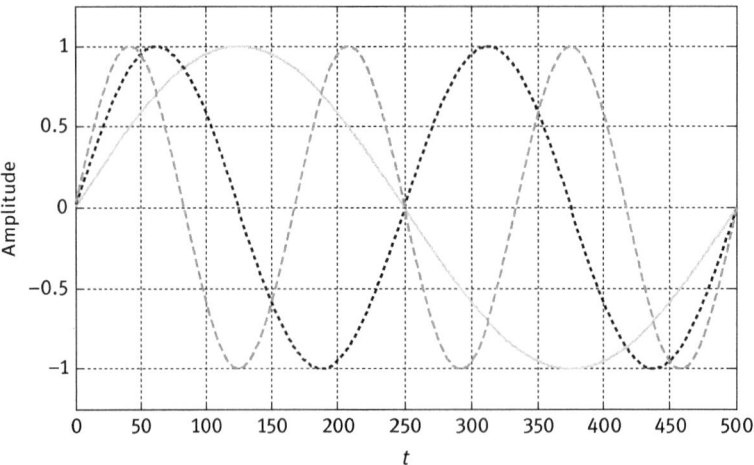

Figure 2.1: The harmonically related signals. Solid line: the fundamental component; dot-dashed line: the second-order harmonics; dashed line: the third-order harmonics.

2. The FS of continuous-time periodic signals

Definition 2.1: The FS of a continuous-time periodic signal is defined as

$$x(t) = \sum_{k=-\infty}^{+\infty} a_k e^{jk\Omega_0 t} = \sum_{k=-\infty}^{+\infty} a_k e^{jk(2\pi/T)t} \tag{2.5}$$

$$a_k = \frac{1}{T}\int_T x(t)e^{-jk\Omega_0 t}\,dt = \frac{1}{T}\int_T x(t)e^{-jk(2\pi/T)t}\,dt \tag{2.6}$$

In the above definitions, eq. (2.5) is the synthesis equation and is also called as the inverse FS; while eq. (2.6) is the analysis equation, referred to as the FS representation. The set of coefficients a_k, $k = 0, \pm 1, \pm 2, \ldots$, are often called the FS coefficients. Obviously, eq. (2.5) is consistent with eq. (2.4), which indicates that the FS of a continuous-time periodic signal is actually a linear combination of a set of complex exponential signals.

3. Dirichlet conditions

Dirichlet conditions point out the conditions that a periodic signal $x(t)$ needs to meet if it takes the transform of FS, including

Condition 1: Over any period, the continuous-time periodic signal $x(t)$ must be absolutely integrable, that is,

$$\int_T |x(t)|\,dt < \infty \tag{2.7}$$

Condition 2: In any finite interval of time, $x(t)$ is of bounded variation. That is to say, there are no more than a finite number of maxima and minima during any single period of the signal.

Condition 3: In any finite interval of time, there are only a finite number of discontinuities. Furthermore, each of these discontinuities is finite.

In fact, Dirichlet conditions complement the imperfection of FS and make the theory of FS representation more rigorous by solving the existence problem. Figure 2.2 shows some examples for which Dirichlet conditions are not satisfied.

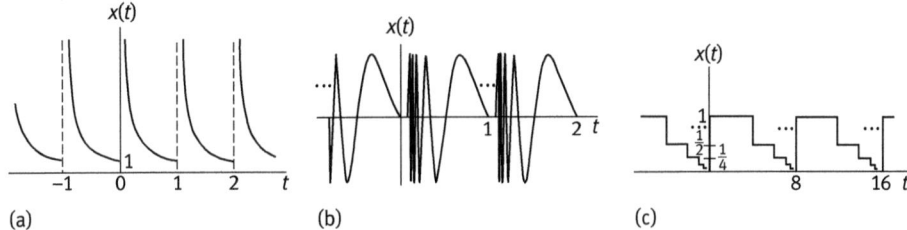

Figure 2.2: Some examples for which Dirichlet conditions are not satisfied: (a) condition 1 is not satisfied; (b) condition 2 is not satisfied; and (c) condition 3 is not satisfied.

In fact, the actual signals encountered in nature and engineering practice all meet Dirichlet conditions. Therefore, their FS representations are existent or convergent. Thus we will not discuss Dirichlet conditions again in detail in the later part of this book, and we will suppose the existence of FS representation and FT.

Example 2.1: Given a periodic signal $x(t) = \begin{cases} 1, & |t| < T_1 \\ 0, & T_1 < |t| < T/2 \end{cases}$. Find its Fourier coefficients a_k.

Solution: According to the expression of $x(t)$, we can draw the waveform of the signal as shown in Figure 2.3.

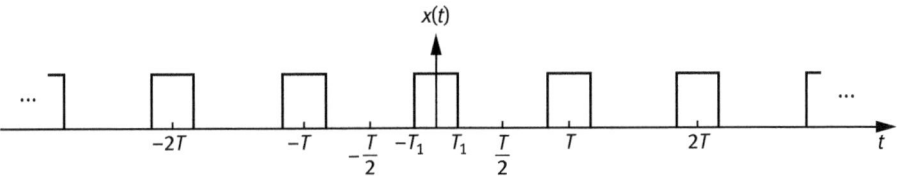

Figure 2.3: The waveform of the given signal.

Substituting the expression of $x(t)$ into eq. (2.6), we get

$$a_k = \frac{1}{T} \int_{-T_1}^{T_1} e^{-jk\Omega_0 t} dt = -\frac{1}{jk\Omega_0 T} e^{-jk\Omega_0 t} \Big|_{-T_1}^{T_1} = \frac{2}{k\Omega_0 T} \left[\frac{e^{jk\Omega_0 T_1} - e^{-jk\Omega_0 T_1}}{2j} \right]$$

$$= \frac{\sin(k\Omega_0 T_1)}{k\pi}, \quad k \neq 0$$

The case of $k = 0$ needs to be recalculated as $a_0 = \left(\frac{1}{T}\right) \int_{-T_1}^{T_1} dt = \left(\frac{2T_1}{T}\right)$.

The coefficients a_k of FS are usually called the spectrum of the continuous-time periodic signals. The so-called spectrum is the distribution of signal components in the frequency domain. Generally speaking, the spectrum can be divided into two parts, the amplitude spectrum and the phase spectrum, expressing the distribution of amplitude and phase of the spectrum in the frequency domain. The amplitude spectrum and phase spectrum represented by FS are shown in eqs. (2.8) and (2.9), respectively:

$$|a_k| = \sqrt{[\mathrm{Re}(a_k)]^2 + [\mathrm{Im}(a_k)]^2} \tag{2.8}$$

$$\sphericalangle a_k = \tan^{-1}\left[\frac{\mathrm{Im}[a_k]}{\mathrm{Re}[a_k]}\right] \tag{2.9}$$

where $\mathrm{Re}[a_k]$ and $\mathrm{Im}[a_k]$ denote the real part and imaginary part of a_k, respectively. $\tan^{-1}(\cdot) = \arctan(\cdot)$ is an antitangent operation. Figure 2.4 shows spectra of the

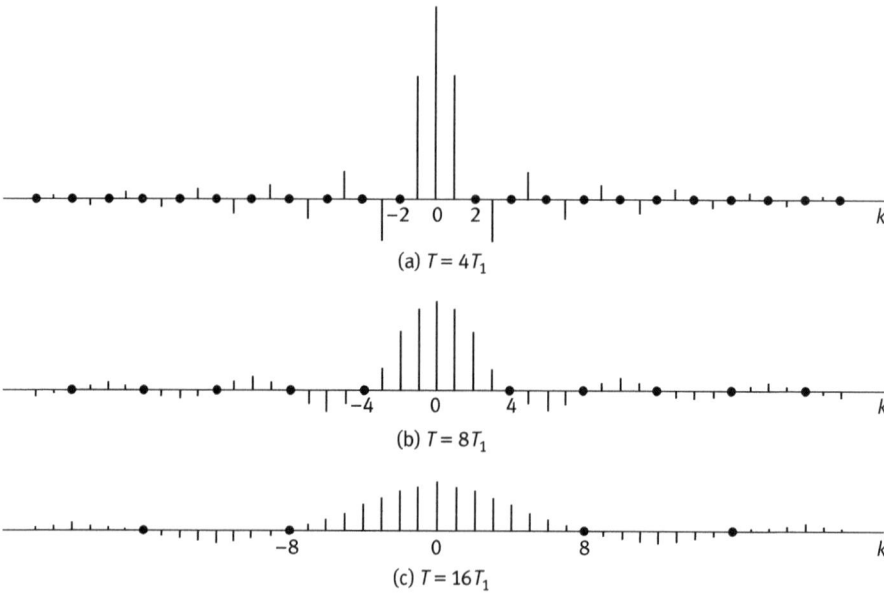

(a) $T = 4T_1$

(b) $T = 8T_1$

(c) $T = 16T_1$

Figure 2.4: The spectra of the signal: (a) $T = 4T_1$; (b) $T = 8T_1$; and (c) $T = 16T_1$.

signal used in Example 2.3 represented by coefficients of FS. In fact, the spectra are the frequency distribution of a_k. According to the relation between T_1 and T, we can draw different spectra.

Obviously, the magnitude of the spectrum becomes smaller and its fluctuation period becomes longer with the decrease of T_1.

Example 2.2: Try to calculate coefficients a_k of FS for signal $x(t) = \cos \Omega_0 t$.
Solution: Based on Euler's equation, $x(t) = \cos \Omega_0 t$ can be rewritten as a complex exponential form as $\cos \Omega_0 t = \frac{1}{2} e^{j\Omega_0 t} + \frac{1}{2} e^{-j\Omega_0 t}$. Comparing it with eq. (2.5), we get

$$a_1 = a_{-1} = \frac{1}{2}, \quad a_k = 0, \text{ if } k \neq \pm 1$$

2.2.2 Properties of FS

FS has a series of significant properties. These properties are important for understanding the concept and theory of FS. They are also useful to simplify the calculation of FS. Table 2.1 presents the main properties of continuous-time FS:

Table 2.1: The main properties of continuous-time Fourier series.

	Property	Continuous-time signals $x(t)$, $y(t)$,Period T, Fundamental frequency $\Omega_0 = 2\pi/T$	Coefficients of Fourier series a_k, b_k
1	Linearity	$Ax(t) + By(t)$	$Aa_k + Bb_k$
2	Time shifting	$x(t - t_0)$	$a_k e^{-jk\Omega_0 t_0} = a_k e^{-jk(2\pi/T)t_0}$
3	Frequency shifting	$e^{jM\Omega_0 t} x(t) = e^{jM(2\pi/T)t} x(t)$	a_{k-M}
4	Conjugation	$x^*(t)$	a_{-k}^*
5	Time reversal	$x(-t)$	a_{-k}
6	Time scale	$x(\alpha t),\ \alpha > 0$ (the period is T/α)	a_k
7	Periodic convolution	$\int_T x(\tau)y(t-\tau)d\tau$	$Ta_k b_k$
8	Multiplication	$x(t)y(t)$	$\sum\limits_{l=-\infty}^{+\infty} a_l b_{k-l}$
9	Differentiation in the time domain	$\frac{dx(t)}{dt}$	$jk\Omega_0 a_k = jk\frac{2\pi}{T} a_k$
10	Conjugate symmetry for real signals	$x(t)$ is a real-valued signal	$\begin{cases} a_k = a_{-k}^* \\ Re[a_k] = Re[a_{-k}] \\ Im[a_k] = -Im[a_{-k}] \\ |a_k| = |a_{-k}| \\ \sphericalangle a_k = -a_{-k} \end{cases}$
11	Parseval's relation		$\frac{1}{T}\int_T |x(t)|^2 dt = \sum\limits_{k=-\infty}^{+\infty} |a_k|^2$

Example 2.3: Suppose that $x(t)$ is a continuous-time periodic signal with period T. Its Fourier coefficients are denoted by a_k. Express the following signals with a_k by means of properties of FS: (1) $x(t - t_0) + x(t + t_0)$; (2) $x(3t - 1)$.

Solution: (1) According to the linear and time shift properties of FS, we have $b_k = 2a_k \cos((2\pi k)/T)t_0$.

(2) Since $x(3t - 1) = x[3(t - 1/3)]$, we know that its period is $T/3$. According to the time scale and time shift properties, we have $c_k = a_k e^{-jk\frac{2\pi}{3T}}$.

Example 2.4: Calculate the Fourier coefficients of the periodic square wave by MATLAB programming. Suppose that the amplitude is $A = 0.5$, and the period and the pulse width are $T = 4$ and $T_1 = 2$, respectively.

Solution: The MATLAB program is given as follows:

```
% Calculate the Fourier coefficients of the periodic square wave, and plot its spectrum.
clc, clear; T=4; width=2; A=0.5;
t1=-T/2:0.001:T/2; ft1=0.5*[abs(t1)<width/2]; t2=[t1-2*T t1-T t1 t1+T t1+2*T];
ft=repmat(ft1,1,5);
figure(1)
subplot(211); plot(t2,ft); axis([-8 8 0 0.8]); xlabel('t'); ylabel('Amplitude');
grid on;
w0=2*pi/T; N=20; K=0:N; ak=zeros(N+1,1);
for k=0:N
    factor1=['exp(-j*t*',num2str(w0),'*',num2str(k),')'];
    f_t=[num2str(A),'*rectpuls(t,2)']; ak(k+1)=quad([f_t,'.*',factor1],-T/2,T/2)/T;
end
kk=0:N; subplot(212); stem(kk',abs(ak),'filled'); xlabel('k'); ylabel('Magnitude');
grid on;
```

Figure 2.5 demonstrates the waveform and spectrum of the periodic square wave.

2.2.3 Other Forms of FS

FS can also be written as other forms besides its complex exponential form.

1. Trigonometric form

$$x(t) = A_0 + \sum_{k=1}^{\infty} A_k \cos k\Omega_0 t + \sum_{k=1}^{\infty} B_k \sin k\Omega_0 t \tag{2.10}$$

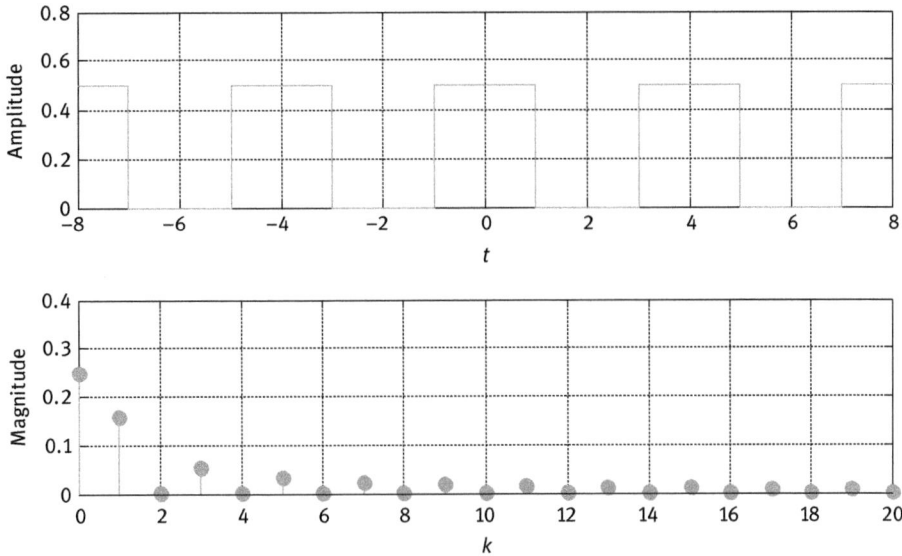

Figure 2.5: The waveform and spectrum of the periodic square wave.

where Fourier coefficients A_0, A_k, and B_k are defined as

$$A_0 = \frac{1}{T} \int_T x(t) dt = a_0$$
$$A_k = \frac{2}{T} \int_T x(t) \cos(k\Omega_0 t) dt$$
$$B_k = \frac{2}{T} \int_T x(t) \sin(k\Omega_0 t) dt$$

2. Magnitude and phase form

$$x(t) = d_0 + \sum_{k=1}^{\infty} d_k \cos(k\Omega_0 t + \phi_k) \qquad (2.11)$$

where Fourier coefficients d_0, d_k, and ϕ_k are defined as

$$d_0 = A_0 = a_0$$
$$d_k = \sqrt{A_k^2 + B_k^2}$$
$$\phi_k = \tan^{-1}(-B_k/A_k)$$

The different forms of FS can be converted to each other. The coefficients of different forms can also be expressed with each other.

2.3 FS Representation of Discrete-Time Periodic Signals

2.3.1 Discrete-Time Periodic Signals and FS

1. The harmonic relations of discrete-time periodic signals

We take the discrete-time complex exponential signal as an example to introduce the harmonic relations of discrete-time periodic signals in this section. Suppose the discrete-time periodic complex exponential signal is shown as follows:

$$x(n) = e^{j\omega_0 n} \tag{2.12}$$

where $\omega_0/2\pi = m/N$ is a rational number and N is the fundamental period of the signal.

A discrete-time harmonic set of complex exponential signal can be constructed as

$$\phi_k(n) = e^{jk\omega_0 n} = e^{jk(2\pi/N)n}, \quad k = 0, \pm 1, \pm 2, \dots \tag{2.13}$$

Every signal in this set is a periodic signal, and the fundamental frequency of every signal is the integral multiple of $2\pi/N$.

It is needed to explain that there are only N signals that are distinct from others in the set of the discrete-time periodic complex exponential signals due to their periodicity in frequency index shown as

$$\phi_k(n) = \phi_{k+rN}(n) \tag{2.14}$$

where r is an arbitrary integer. The above equation indicates when the increment of k is the integral multiple of N, the identical sequence is generated.

In fact, any discrete-time periodic signal can be written as a linear combination of a set of harmonically related signals, shown as follows:

$$x(n) = \sum_{k=\langle N \rangle} a_k \phi_k(n) = \sum_{k=\langle N \rangle} a_k e^{jk\omega_0 n} = \sum_{k=\langle N \rangle} a_k e^{jk(2\pi/N)n} \tag{2.15}$$

where $k = \langle N \rangle$ denotes that the summation takes place inside a period.

2. Discrete FS

Definition 2.2: The discrete-time FS is defined as

$$x(n) = \sum_{k=\langle N \rangle} a_k e^{jk\omega_0 n} = \sum_{k=\langle N \rangle} a_k e^{jk(2\pi/N)n} \tag{2.16}$$

$$a_k = \frac{1}{N} \sum_{n = \langle N \rangle} x(n)e^{-jk\omega_0 n} = \frac{1}{N} \sum_{n = \langle N \rangle} x(n)e^{-jk(2\pi/N)n} \quad (2.17)$$

Similar to the definition of FS for the continuous-time signal in eqs. (2.5) and (2.6), the definition of DFS is given in eqs. (2.16) and (2.17), where eq. (2.16) is referred to as the synthesis equation and eq. (2.17) as the analysis equation. The set of coefficients $\{a_k\}$ are often called FS coefficients. In both equations, $k = \langle N \rangle$ and $n = \langle N \rangle$ denote the summation takes place only inside one period. From the definition of DFS, we see that both $x(n)$ and a_k are all periodic, with the period of N. On the other hand, we also see from eqs. (2.16) and (2.17) that both $x(n)$ and a_k are all linear combinations of the complex exponential harmonic signals.

Example 2.5: The waveform of a given discrete-time periodic square wave signal $x(n)$ is shown in Figure 2.6. Try to determine its DFS coefficients a_k.

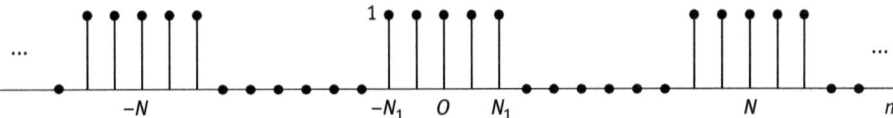

Figure 2.6: The waveform of a given discrete-time periodic square wave signal.

Solution: According to definition (2.17), we get

$$a_k = \frac{1}{N} \sum_{n = -N_1}^{N_1} e^{-jk(2\pi/N)n}$$

Letting $m = n + N_1$, we get $a_k = (1/N) \sum_{m=0}^{2N_1} e^{-jk(2\pi/N)(m - N_1)} = (1/N)e^{jk(2\pi/N)N_1} \sum_{m=0}^{2N_1} e^{-jk(2\pi/N)m}$. Thus

$$a_k = \frac{1}{N} e^{jk(2\pi/N)N_1} \left(\frac{1 - e^{-jk2\pi(2N_1 + 1)/N}}{1 - e^{-jk(2\pi/N)}} \right)$$

$$= \frac{1}{N} \frac{\sin[2\pi k(N_1 + 1/2)/N]}{\sin(\pi k)/N}, \quad k \neq 0, \pm N, \pm 2N, \ldots$$

and

$$a_k = \frac{2N_1 + 1}{N}, \quad k = 0, \pm N, \pm 2N, \ldots$$

Figure 2.7 shows the spectra of the signal when $N = 20$ and 40, respectively, and $2N_1 + 1 = 5$.

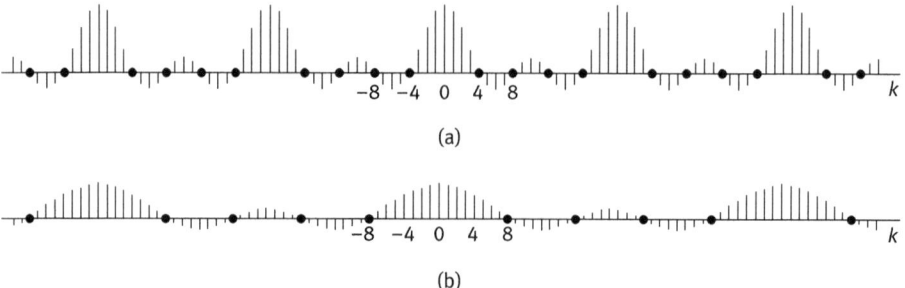

(a)

(b)

Figure 2.7: The spectra of the given signal: (a) $N = 20$ and (b) $N = 40$.

2.3.2 Properties of DFS

The properties of the DFS are similar to those of the FS of continuous-time signals. The main properties of the DFS are given in Table 2.2.

Example 2.6: Suppose we have the following fact about a discrete-time signal $x(n)$: (1) $x(n)$ is periodic with period $N = 6$; (2) $\sum_{n=1}^{6} x(n) = 2$; (3) $\sum_{n=2}^{7} (-1)^n x(n) = 1$; (4) $x(n)$ has the minimum power per period among the set of signals satisfying the above three conditions. Try to determine the sequence.

Solution: Denote the FS of $x(n)$ as a_k. From (2), we conclude that $a_0 = 1/3$. Since $(-1)^n = e^{j\pi n} = e^{j(2\pi/6)3n}$, from (3) we have $a_3 = 1/6$. From Parseval's relation, the average power in $x(n)$ is $P = \sum_{k=0}^{5} |a_k|^2$. Since each nonzero coefficient contributes a positive value to the power P, and since the values of a_0 and a_3 are prespecified, the value of P is minimized by $a_1 = a_2 = a_4 = a_5 = 0$. Thus $x(n) = a_0 + a_3 e^{j\pi n} = \frac{1}{3} + \frac{1}{6}(-1)^n$ is obtained.

2.4 FT of Continuous-Time Signals

2.4.1 From FS to FT

1. The periodicity of signals

In Sections 2.2 and 2.3, we introduced the theories on FS and DFS. We noticed that both signals $x(t)$ and $x(n)$ in the above sections are all periodic signals. The theory of FS shows that any periodic signal can be expressed as a linear combination of harmonically related periodic complex exponential (or sinusoidal) signals.

However, the strictly periodic signal is rarely seen in nature and engineering technology. The vast majority of signals in practice are not periodic. In this way, the practical applications put forward a new requirement to Fourier theory. How to transform an aperiodic signal into the frequency domain is a problem needed to be solved by FT.

Table 2.2: The main properties of the discrete Fourier series.

Properties	Discrete-time periodic signal $x(n)$, $y(n)$, period N, fundamental frequency $\omega_0 = 2\pi/N$	Coefficients of discrete Fourier series a_k, b_k, period N
1	Linearity	$Ax(n) + By(n)$	$Aa_k + Bb_k$
2	Time shifting	$x(n - n_0)$	$a_k e^{-jk\omega_0 n_0} = a_k e^{-jk(2\pi/N)n_0}$
3	Frequency shifting	$e^{jM\Omega_0 n}x(n) = e^{jM(2\pi/N)n}x(n)$	a_{k-M}
4	Conjugation	$x^*(n)$	a^*_{-k}
5	Time reversal	$x(-n)$	a_{-k}
6	Time scaling	$x_{(m)}(n) = \begin{cases} x(n/m), & \text{if } n \text{ is a multiple of } m \\ 0, & \text{if } n \text{ is not a multiple of } m \end{cases}$	$\frac{1}{m} a_k$
7	Periodic convolution	$\sum_{r=(N)} x(r)y(n-r)$	$Na_k b_k$
8	Multiplication	$x(n)y(n)$	$\sum_{r=(N)} a_l b_{k-l}$
9	First difference	$x(n) - x(n-1)$	$\left(1 - e^{-jk(2\pi/N)}\right) a_k$
10	Running sum	$\sum_{k=-\infty}^{n} x(k)$, (finite-valued and periodic only if $a_0 = 0$)	$\frac{1}{1 - e^{-jk(2\pi/N)}} a_k$
11	Conjugate symmetry for real signals	$x(n)$ is a real-valued signal	$\begin{cases} a_k = a^*_k \\ \text{Re}[a_k] = \text{Re}[a_{-k}] \\ \text{Im}[a_k] = -\text{Im}[a_{-k}] \\ \|a_k\| = \|a_{-k}\| \\ \sphericalangle a_k = -\sphericalangle a_{-k} \end{cases}$
12	Parseval's relation	$\frac{1}{N} \sum_{n=(N)} |x(n)|^2 = \sum_{k=(N)} |a_k|^2$	

2. A qualitative analysis of a continuous-time periodic square wave

Suppose a continuous-time periodic square wave is shown as

$$x(t) = \begin{cases} 1, & |t| < T_1 \\ 0, & T_1 < |t| < T/2 \end{cases}$$

where T is the period of the signal. The waveform of $x(t)$ is shown in Figure 2.3 and its Fourier coefficients can be obtained from eq. (2.6) as follows:

$$a_k = \frac{2\sin(k\Omega_0 T_1)}{k\Omega_0 T}$$

We see from the expressions of $x(t)$ and a_k that if we keep T_1 unchanged and extend the period T and finally make $T \to \infty$, then the period of $x(t)$ will increase and tend to be infinity, resulting $x(t)$ as an aperiodic signal. Meanwhile, the fundamental frequency of $x(t)$ becomes smaller and smaller, resulting in the spectrum a_k as a continuous spectrum. Figure 2.8 shows the process.

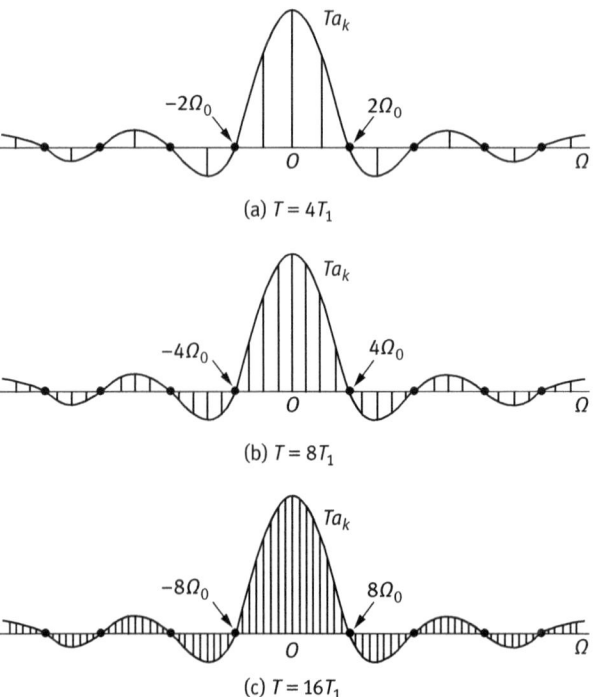

(a) $T = 4T_1$

(b) $T = 8T_1$

(c) $T = 16T_1$

Figure 2.8: The Fourier coefficients and their envelops of a periodic square wave with a fixed T_1: (a) $T = 4T_1$; (b) $T = 8T_1$; and (c) $T = 16T_1$.

3. Further quantitative analysis

In order to distinguish between the periodic and the aperiodic signals, we temporarily use $\tilde{x}(t)$ and $x(t)$ to represent a periodic signal and an aperiodic signal, respectively. According to the definition of continuous FS, we have

$$
\begin{cases}
\tilde{x}(t) = \sum\limits_{k=-\infty}^{+\infty} a_k e^{jk\Omega_0 t} \\
a_k = \frac{1}{T} \int\limits_{-T/2}^{T/2} \tilde{x}(t) e^{-jk\Omega_0 t} dt
\end{cases}
\tag{2.18}
$$

Define an auxiliary function as $x(t) = \begin{cases} \tilde{x}(t), & |t| < T/2 \\ 0, & \text{otherwise} \end{cases}$, and make $T \to \infty$ with a fixed T_1, we get

$$
a_k = \frac{1}{T} \int\limits_{-T/2}^{T/2} x(t) e^{-jk\Omega_0 t} dt = \frac{1}{T} \int\limits_{-\infty}^{+\infty} x(t) e^{-jk\Omega_0 t} dt
$$

Define $T a_k = X(jk\Omega_0)$, then we get $X(jk\Omega_0) = \int_{-\infty}^{\infty} x(t) e^{-jk\Omega_0 t} dt$, or $a_k = (1/T) X(jk\Omega_0)$. Based on the inverse FS, using $\Omega_0 = 2\pi/T$, we get

$$
\tilde{x}(t) = \sum\limits_{k=-\infty}^{+\infty} \frac{1}{T} X(jk\Omega_0) e^{jk\Omega_0 t} = \frac{1}{2\pi} \sum\limits_{k=-\infty}^{+\infty} X(jk\Omega_0) e^{jk\Omega_0 t} \cdot \Omega_0
$$

When $T \to \infty$, we have $\Omega_0 \to d\Omega$ and $k\Omega_0 \to \Omega$. During this progress, the summation will change into the integration. Thus, FS tends to be FT as follows:

$$
\tilde{x}(t) = x(t) = \frac{1}{2\pi} \int\limits_{-\infty}^{+\infty} X(j\Omega) e^{j\Omega t} d\Omega
$$

$$
X(j\Omega) = \int\limits_{-\infty}^{+\infty} x(t) e^{-j\Omega t} dt
$$

2.4.2 FT of Continuous-Time Signals

1. The definition of FT and the concept of spectrum

Definition 2.3: The FT of a continuous-time signal is defined as

$$x(t) = \frac{1}{2\pi} \int_{-\infty}^{+\infty} X(j\Omega)e^{j\Omega t}\,d\Omega \qquad (2.19)$$

$$X(j\Omega) = \int_{-\infty}^{+\infty} x(t)e^{-j\Omega t}\,dt \qquad (2.20)$$

Equations (2.19) and (2.20) are referred to as the FT pair, where the function $X(j\Omega)$ is referred to as the FT of $x(t)$, and eq. (2.19) as the inverse FT. Similar to the FS, eq. (2.19) is also referred to as the synthesis equation, and eq. (2.20) as the analysis equation. Usually, $X(j\Omega)$ is called the spectrum function of a continuous-time signal $x(t)$.

The spectrum $X(j\Omega)$ represents the distribution of frequency components of $x(t)$ in the frequency domain. It is generally a complex function shown as

$$X(j\Omega) = |X(j\Omega)|e^{j\sphericalangle X(j\Omega)} \qquad (2.21)$$

where $|X(j\Omega)|$ is referred to as the magnitude spectrum and $\sphericalangle X(j\Omega)$ is referred to as the phases spectrum.

Example 2.7: Given a signal as $x(t) = e^{-at}u(t)$, $a > 0$. Find its FT $X(j\Omega)$.
Solution: Substituting $x(t) = e^{-at}u(t)$, $a > 0$ into eq. (2.20), we get

$$X(j\Omega) = \int_{0}^{+\infty} e^{-at}e^{-j\Omega t}\,dt = -\frac{1}{a+j\Omega}e^{-(a+j\Omega)t}\Big|_{0}^{+\infty} = \frac{1}{a+j\Omega}, \quad a > 0$$

We can further calculate the magnitude spectrum and phase spectrum from $X(j\Omega)$ as

$$|X(j\Omega)| = \frac{1}{\sqrt{a^2+\Omega^2}} \quad \text{and} \quad \sphericalangle X(j\Omega) = -\tan^{-1}\left(\frac{\Omega}{a}\right)$$

Figure 2.9 demonstrates curves of the magnitude spectrum and the phase spectrum.

Example 2.8: The spectrum of signal $x_1(t)$ is given as $X_1(j\Omega) = \begin{cases} 1, & |\Omega| < W \\ 0, & |\Omega| > W \end{cases}$. Find $x_1(t)$.
Solution: Substituting $X_1(j\Omega)$ into eq. (2.19), we get

$$x_1(t) = \frac{1}{2\pi} \int_{-W}^{W} e^{j\Omega t}\,d\Omega = \frac{\sin Wt}{\pi t} \qquad (2.22)$$

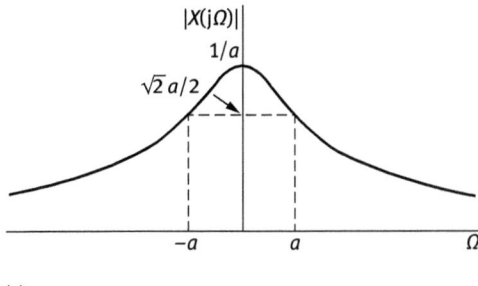

(a)

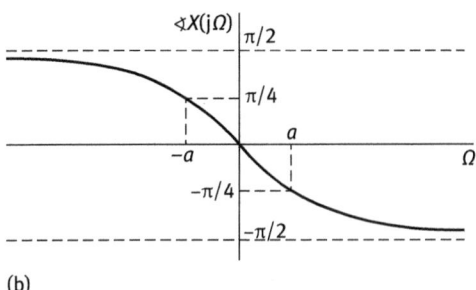

(b)

Figure 2.9: The magnitude spectrum (a) and phase spectrum (b) of the given signal.

Example 2.9: Find FT of $x(t) = e^{-2t}u(t)$ with MATLAB programming and plot its waveform and spectrum.

Solution: The MATLAB program is shown as follows:

```
clc, clear;  syms t w f;  f=exp(-2*t)*sym('heaviside(t)'); F=fourier(f);
figure(1)
subplot(211); ezplot(f,[0:2,0:1.2]); title('Waveform'); ylabel('Amplitude');
subplot(212); ezplot(abs(F),[-10:10]);  title('Magnitude spectrum');
ylabel('Magnitude');
```

The waveform and spectrum are given in Figure 2.10.

Example 2.10: A rectangular pulse signal is given as $x_2(t) = \begin{cases} 1, & |t| < T_1 \\ 0, & |t| > T_1 \end{cases}$. Find its FT.

Solution: Substituting $x_2(t)$ into eq. (2.20), we get

$$X_2(j\Omega) = \int_{-T_1}^{T_1} e^{-j\Omega t}dt = 2\frac{\sin \Omega T_1}{\Omega} \tag{2.23}$$

The function forms in eqs. (2.22) and (2.23) are usually called the sinc function, defined as

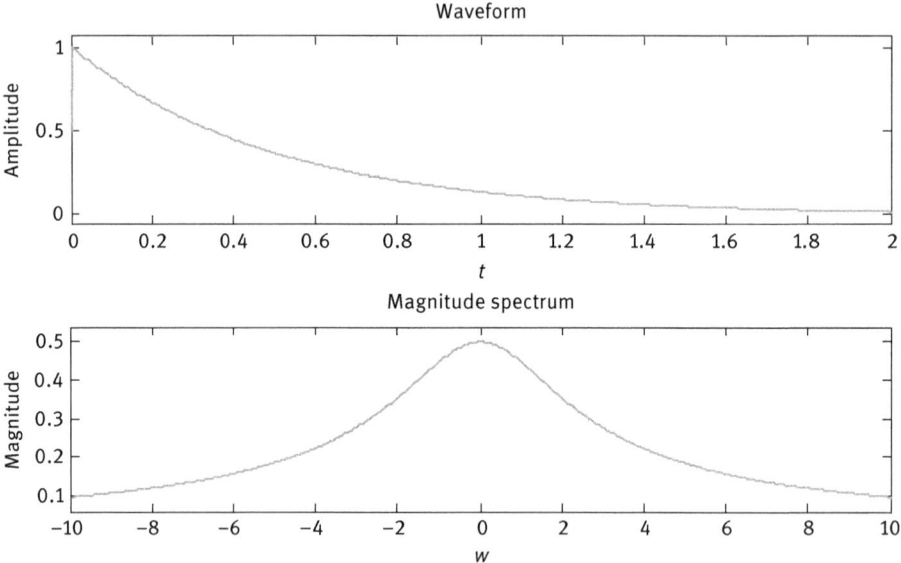

Figure 2.10: The waveform and spectrum of the given signal.

$$\sin c(\theta) = \frac{\sin \pi\theta}{\pi\theta} \tag{2.24}$$

Therefore, both eqs. (2.22) and (2.23) can be rewritten as sinc functions as

$$\frac{\sin Wt}{\pi t} = \frac{W}{\pi} \sin c\left(\frac{Wt}{\pi}\right)$$

$$2\frac{\sin \Omega T_1}{\Omega} = 2T_1 \sin c\left(\frac{\Omega T_1}{\pi}\right)$$

We see from Examples 2.8 and 2.9 that the spectrum of the rectangular pulse signal $x_2(t)$ is a sinc function, while the signal corresponding to the rectangular pulse spectrum $X_1(j\Omega)$ is also a sinc function. Such special relationship is a direct consequence of the duality property for FT shown in Figure 2.11.

We see from Figure 2.11 that as the width W of the spectrum (referred to as bandwidth) increases, the main peak of the signal $x_1(t)$ becomes higher and the width of the first lobe of the signal becomes narrower. In fact, if $W \to \infty$, the spectrum $X_1(j\Omega)$ becomes flat and the corresponding signal converges to an impulse $\delta(t)$. We can also see similar change when T_1 increases and tends to be infinity.

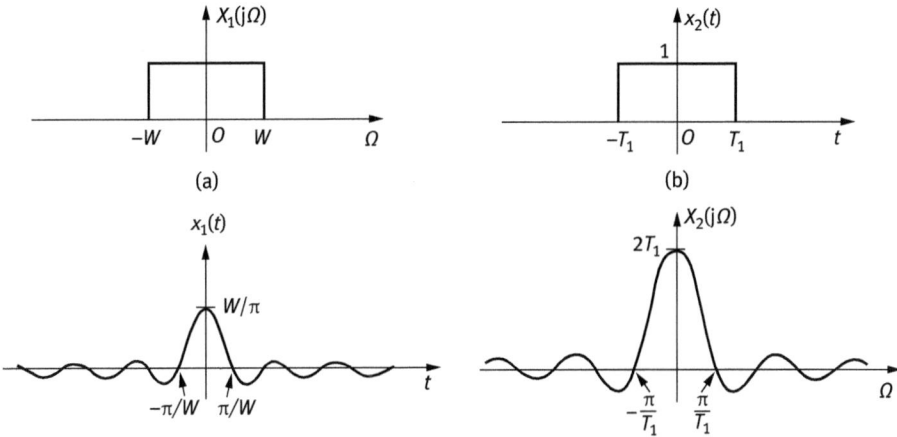

Figure 2.11: An example of duality property for Fourier transform. Sinc function (c) and its Fourier transform (a); rectangular pulse signal (b) and its Fourier transform (d).

2. The FT of periodic signals

Although we get an aperiodic signal by pushing the period of a periodic signal to infinity during the process of our derivation for FT from FS, FT applies not only for nonperiodic signals but also for periodic signals. Equation (2.25) shows the FT for a continuous-time periodic signal:

$$X(j\Omega) = \sum_{k=-\infty}^{+\infty} 2\pi a_k \delta(\Omega - k\Omega_0) \tag{2.25}$$

where a_k are Fourier coefficients. Obviously, $X(j\Omega)$ is a combination of a series of $\delta(\Omega)$ and its frequency shifts.

2.4.3 Properties of FT and Basic FT Pairs

1. Properties of FT

There are some very important properties of the continuous-time FT. The study and mastery of these properties are of great significance for the deep understanding of FT. The main properties of FT are shown in Table 2.3.

Example 2.11: Find FT of the unit step signal $x(t) = u(t)$.
Solution: We cannot get the correct result by using the definition of FT. However, we can find the result by using properties of FT as follows. Set

$$g(t) = \delta(t) \leftrightarrow G(j\Omega) = 1$$

Table 2.3: The main properties of Fourier transform.

	Properties	Continuous-time aperiodic signals $x(t)$, $y(t)$	Fourier transform $X(j\Omega)$, $Y(j\Omega)$
1	Linearity	$ax(t) + by(t)$	$aX(j\Omega) + bY(j\Omega)$
2	Time shifting	$x(t-t_0)$	$e^{-j\Omega t_0} X(j\Omega)$
3	Frequency shifting	$e^{j\Omega_0 t} x(t)$	$X(j(\Omega - \Omega_0))$
4	Conjugation	$x^*(t)$	$X^*(-j\Omega)$
5	Time reversal	$x(-t)$	$X(-j\Omega)$
6	Time scaling	$x(at)$	$\frac{1}{|a|} X(j\frac{\Omega}{a})$
7	Convolution	$x(t)*y(t)$	$X(j\Omega)Y(j\Omega)$
8	Multiplication	$x(t)y(t)$	$\frac{1}{2\pi}X(j\Omega)*Y(j\Omega)$
9	Differentiation	$\frac{dx(t)}{dt}$	$j\Omega X(j\Omega)$
10	Integration	$\int\limits_{-\infty}^{t} x(\tau)d\tau$	$\frac{1}{j\Omega}X(j\Omega) + \pi X(0)\delta(\Omega)$
11	Differentiation in the frequency domain	$tx(t)$	$j\frac{d}{d\Omega}X(j\Omega)$
12	Duality (an example)	$x_1(t) = \begin{cases} 1, & |t| < T_1 \\ 0, & |t| > T_1 \end{cases}$	$X_1(j\Omega) = \frac{2\sin\Omega T_1}{\Omega}$
		$x_2(t) = \frac{\sin Wt}{\pi t}$	$X_2(j\Omega) = \begin{cases} 1, & |\Omega| < W \\ 0, & |\Omega| > W \end{cases}$
13	Conjugate symmetry for real signals	$x(t)$ is a real-valued signal	$\begin{cases} X(j\Omega) = X^*(j\Omega) \\ \mathrm{Re}[X(j\Omega)] = \mathrm{Re}[X(-j\Omega)] \\ \mathrm{Im}[X(j\Omega)] = -\mathrm{Im}[X(-j\Omega)] \\ |X(j\Omega)| = |X(-j\Omega)| \\ \sphericalangle X(j\Omega) = -\sphericalangle X(-j\Omega) \end{cases}$
14	Parseval's relation	$\int_{-\infty}^{+\infty} |x(t)|^2\, dt = \frac{1}{2\pi}\int_{-\infty}^{+\infty} |X(j\Omega)|^2 d\Omega$	

and

$$x(t) = u(t) = \int_{-\infty}^{t} g(\tau)d\tau$$

Taking FT to both sides of the above equation and using the integration property, we get

$$X(j\Omega) = \frac{G(j\Omega)}{j\Omega} + \pi G(0)\delta(\Omega) = \frac{1}{j\Omega} + \pi\delta(\Omega)$$

By using the differentiation property, we can get FT of the unit impulse $\delta(t)$ as

$$\delta(t) = \frac{du(t)}{dt} \leftrightarrow j\Omega\left[\frac{1}{j\Omega} + \pi\delta(\Omega)\right] = 1$$

Example 2.12: Suppose FTs of $x(t)$ and $h(t)$ are $X(j\Omega)$ and $H(j\Omega)$, respectively, satisfying $y(t) = x(t)*h(t)$ and $g(t) = x(3t)*h(3t)$. (1) Try to prove $g(t) = Ay(Bt)$ using properties of FT. (2) Find the values of A and B.

Solution: (1) From $y(t) = x(t)*h(t)$, by using the convolution property of FT, we get $Y(j\Omega) = X(j\Omega)H(j\Omega)$. According to the given condition $g(t) = x(3t)*h(3t)$ and properties of time scaling and convolution of FT, we have

$$G(j\Omega) = \frac{1}{9}X(j\Omega/3)H(j\Omega/3) = \frac{1}{3}\cdot\frac{1}{3}X(j\Omega/3)H(j\Omega/3) = \frac{1}{3}\cdot\frac{1}{3}Y(j\Omega/3)$$

Therefore, we get $g(t) = \frac{1}{3}y(3t)$.

(2) Comparing $g(t) = \frac{1}{3}y(3t)$ and $g(t) = Ay(Bt)$, we get $A = \frac{1}{3}$, $B = 3$.

Example 2.13: Given $x(t) \leftrightarrow X(j\Omega)$, find the FTs for the following signals:

(1) $tx(2t)$; (2) $x(1-t)$; (3) $t\frac{dx(t)}{dt}$; (4) $x(2t-5)$.

Solution: (1) Based on the time scaling property, we have $x(2t) \leftrightarrow \frac{1}{2}X(j\Omega/2)$. By using the frequency differential property, we get

$$tx(2t) \leftrightarrow \frac{j}{2}\frac{dX(j\Omega/2)}{d\Omega}$$

(2) According to the time reversal property, we have $x(-t) \leftrightarrow X(-j\Omega)$. Thus $x(1-t) = x(-(t-1)) \leftrightarrow X(-j\Omega)e^{-j\Omega}$ is obtained.

(3) From the differentiation property $\frac{dx(t)}{dt} \leftrightarrow j\Omega X(j\Omega)$ and the frequency differentiation property, we get

$$t\frac{dx(t)}{dt} \leftrightarrow j\frac{d}{d\Omega}[j\Omega X(j\Omega)] = -X(j\Omega) - \Omega\frac{dX(j\Omega)}{d\Omega}.$$

(4) From the time scaling property, we have $x(2t) \leftrightarrow \frac{1}{2}X(j\Omega/2)$. From the time shifting property we get $x\left(2\left(t - \frac{5}{2}\right)\right) \leftrightarrow \frac{1}{2}X((j\Omega)/2)e^{-j\frac{5}{2}\Omega}$.

Example 2.14: Find $x(t)$ from the given spectra: (1) $\delta(\Omega - 2)$; (2) $\cos(2\Omega)$; (3) $e^{a\Omega}u(-\Omega)$. *Solution:* (1) Since the FT of $1/2\pi$ is $\delta(\Omega)$, from the frequency shifting property, the inverse FT of $\delta(\Omega - 2)$ is $x(t) = (1/2\pi)e^{j2t}$.

(2) Since $\cos\Omega \leftrightarrow \pi[\delta(\Omega + 1) + \delta(\Omega - 1)]$, from the duality property we have

$$2\pi \cos\Omega \leftrightarrow \pi[\delta(t + 1) + \delta(t - 1)]$$

Thus

$$\cos 2\Omega \leftrightarrow \frac{1}{4}\left[\delta\left(\frac{t}{2} + 1\right) + \delta\left(\frac{t}{2} - 1\right)\right]$$

(3) Since $u(t) \leftrightarrow 1/(j\Omega) + \pi\delta(\Omega)$, from the duality property we have $2\pi u(-\Omega) \leftrightarrow (1/jt) + \pi\delta(t)$. Based on the time shifting property, we have

$$e^{a\Omega}u(-\Omega) \leftrightarrow \frac{1}{j2\pi(t + a)} + \frac{1}{2}\delta(t + a)$$

2. Basic FT pairs
Some commonly used FT pairs are given in Table 2.4. If the properties of FT given in Table 2.3 and the transform pairs shown in Table 2.4 are used jointly, the calculation of FT can be more concise.

2.5 Discrete-Time Fourier Transform

2.5.1 Discrete-Time Fourier Transform

1. The concept of DTFT
In Section 2.4, we studied the concept and theory of continuous-time FT. It can be seen that the definition of FT is obtained by pushing the period of a periodic signal to infinity. When the period $T \to \infty$, the discrete frequency $k\Omega_0$ tends to be a continuous frequency Ω. Meanwhile, FT is obtained from FS.

The similar things happen to discrete-time signals. Suppose a discrete-time signal satisfies $x(n) = x(n + N)$. When $N \to \infty$, the periodic signal changes into an aperiodic signal. At the same time, the DFS for discrete-time periodic signals is gradually changed into DTFT for aperiodic signals.

Here we just give the definition of DTFT without a derivation.

Table 2.4: Commonly used Fourier transform pairs

Name of signals	Expressions of $x(t)$	Fourier transforms $X(j\Omega)$
Unilateral exponential signal	$e^{-at}u(t)$, $Re[a]>0$	$\frac{1}{a+j\Omega}$
Bilateral exponential signal	$e^{-a|t|}$, $Re[a]>0$	$\frac{2a}{a^2+\Omega^2}$
Derived signal from the spectrum of unilateral signal	$te^{-at}u(t)$, $Re[a]>0$	$\frac{1}{(a+j\Omega)^2}$
	$\frac{t^{n-1}}{(n-1)!}e^{-at}u(t)$, $Re[a]>0$	$\frac{1}{(a+j\Omega)^n}$
Rectangular pulse signal	$\begin{cases}1,&|t|<T_1\\0,&|t|>T_1\end{cases}$	$2\frac{\sin\Omega T_1}{\Omega}$
Sinc function	$\frac{\sin Wt}{\pi t}$	$\begin{cases}1,&|\Omega|<W\\0,&|\Omega|>W\end{cases}$
Unit impulse signal	$\delta(t)$	1
Unit step signal	$u(t)$	$\frac{1}{j\Omega}+\pi\delta(\Omega)$
Unit impulse with time shift	$\delta(t-t_0)$	$e^{-j\Omega t_0}$
Constant signal	1	$2\pi\delta(\Omega)$
Periodic square wave	$x(t)=\begin{cases}1,&|t|<T_1\\0,&T_1<|t|\le T/2\end{cases}$ $x(t+T)=x(t)$	$\sum_{k=-\infty}^{+\infty}\frac{2\sin k\Omega_0 T_1}{k}\delta(\Omega-k\Omega_0)$
Cosine signal	$\cos(\Omega_0 t)$	$\pi[\delta(\Omega+\Omega_0)+\delta(\Omega-\Omega_0)]$
Sine signal	$\sin(\Omega_0 t)$	$j\pi[\delta(\Omega+\Omega_0)-\delta(\Omega-\Omega_0)]$
Unilateral cosine signal	$\cos(\Omega_0 t)u(t)$	$\frac{\pi}{2}[\delta(\Omega+\Omega_0)+\delta(\Omega-\Omega_0)]+\frac{j\Omega}{\Omega_0^2-\Omega^2}$
Unilateral sine signal	$\sin(\Omega_0 t)u(t)$	$\frac{j\pi}{2}[\delta(\Omega+\Omega_0)-\delta(\Omega-\Omega_0)]+\frac{\Omega_0}{\Omega_0^2-\Omega^2}$
Complex exponential signal	$e^{j\Omega_0 t}$	$2\pi\delta(\Omega-\Omega_0)$
Unit impulse sequence	$\sum_{n=-\infty}^{+\infty}\delta(t-nT)$	$\frac{2\pi}{T}\sum_{k=-\infty}^{+\infty}\delta(\Omega-\frac{2\pi k}{T})$

Definition 2.4: The DTFT is defined as

$$x(n) = \frac{1}{2\pi} \int_{2\pi} X(e^{j\omega}) e^{j\omega n} d\omega \tag{2.26}$$

$$X(e^{j\omega}) = \sum_{n=-\infty}^{+\infty} x(n) e^{-j\omega n} \tag{2.27}$$

where eq. (2.26) is the synthesis equation, referred to as the inverse DTFT, while eq. (2.27) is the analysis equation, referred to as the DTFT. We see from the definition that the spectrum $X(e^{j\omega})$ of the DTFT is continuous and periodic with period 2π.

Example 2.15: Suppose a discrete-time signal is $x(n) = a^n u(n)$, $|a| < 1$. Find its DTFT.

Solution: According to the definition of DTFT, we have

$$X(e^{j\omega}) = \sum_{n=-\infty}^{+\infty} a^n u(n) e^{-j\omega n} = \sum_{n=0}^{+\infty} (ae^{-j\omega})^n = \frac{1}{1-ae^{-j\omega}}$$

Figure 2.12 demonstrates the magnitude spectrum and phase spectrum of $X(e^{j\omega})$.

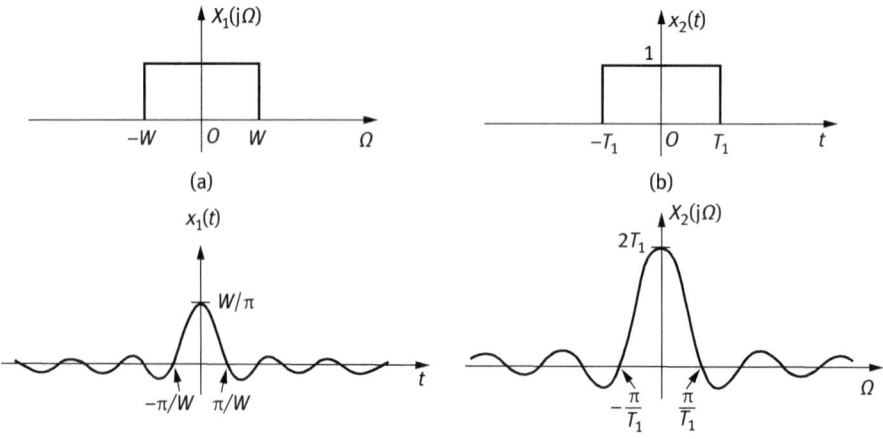

(a)

(b)

Figure 2.12: The magnitude spectrum and phase spectrum in Example 2.15: (a) $a > 0$ and (b) $a < 0$.

Example 2.16: Suppose a discrete-time rectangular pulse is given as

$$x(n) = \begin{cases} 1, & |n| \le N_1 \\ 0, & |n| > N_1 \end{cases}.$$ Find its DTFT.

Solution: According to the definition of DTFT, we get

$$X(e^{j\omega}) = \sum_{n=-\infty}^{+\infty} x(n)e^{-j\omega n} = \sum_{n=-N_1}^{N_1} e^{-j\omega} = \frac{\sin\omega(N_1+1/2)}{\sin(\omega/2)}$$

2. The DTFT for discrete-time periodic signals

As with the FT of continuous-time periodic signals, the DTFT is also suitable for discrete-time periodic signals. The DTFT for discrete-time periodic signals are shown as follows:

$$X(e^{j\omega}) = \sum_{k=-\infty}^{+\infty} 2\pi a_k \delta\left(\omega - \frac{2\pi k}{N}\right) \tag{2.28}$$

where a_k are the FS coefficients of $x(n)$ and N is its fundamental period. It is shown that the DTFT $X(e^{j\omega})$ of a periodic signal $x(n)$ can be obtained by calculating the coefficients a_k of the FS.

2.5.2 Properties of DTFT and Basic DTFT Pairs

1. Properties of DTFT

The same as FT for continuous-time signals, the DTFT also has a series of important properties. The mastery and application of these properties play an important role in study and applications. Table 2.5 presents the main properties of the DTFT.

2. Basic DTFT pairs

Basic DTFT pairs are given in Table 2.6. If the properties of DTFT given in Table 2.5 and the transform pairs shown in Table 2.6 are used jointly, the calculation of DTFT can be more concise.

2.5.3 The Duality in FS and FT

The theory of FS and FT involves three kinds of dualities in time and frequency domains, namely the duality in FT, the duality in DFS, and the duality between DTFT and FS. The three dualities will be introduced in this section.

1. The duality of FT

In Examples 2.8 and 2.9, we introduced the time–frequency duality of FT, in which the spectrum of a rectangular signal is a sinc function, while the rectangular time signal corresponded spectrum is also a sinc function. This is a typical example of the time–frequency duality.

As a matter of fact, this kind of duality can be generalized to the general FT. It means that for any FT pair, there is a certain correspondence between its time and frequency variables exchanged. This is the time–frequency duality of FT.

Table 2.5: Main properties of discrete-time Fourier transform.

Properties	Discrete-time signals $x(n)$, $y(n)$	Discrete-time Fourier transform $X(e^{j\omega})$, $Y(e^{j\omega})$
1	Linearity	$ax(n)+by(n)$	$aX(e^{j\omega})+bY(e^{j\omega})$
2	Time shifting	$x(n-n_0)$	$e^{-j\omega n_0}X(e^{j\omega})$
3	Frequency shifting	$e^{j\omega_0 n}x(n)$	$X(e^{j(\omega-\omega_0)})$
4	Conjugation	$x^*(n)$	$X^*(e^{-j\omega})$
5	Time reversal	$x(-n)$	$X(e^{-j\omega})$
6	Time expansion	$x_{(k)}(n)=\begin{cases}x(n/k), & \text{if } n=\text{multiple of } k\\ 0, & \text{if } n\neq \text{multiple of } k\end{cases}$	$X(e^{jk\omega})$
7	Convolution	$x(n)*y(n)$	$X(e^{j\omega})Y(e^{j\omega})$
8	Multiplication	$x(n)y(n)$	$\frac{1}{2\pi}\int_{2\pi}X(e^{j\theta})Y(e^{j(\omega-\theta)})d\theta$
9	Difference	$x(n)-x(n-1)$	$(1-e^{-j\omega})X(e^{j\omega})$
10	Accumulation	$\sum_{k=-\infty}^{n}x(k)$	$\frac{1}{1-e^{-j\omega}}X(e^{j\omega})+\pi X(e^{j0})\sum_{k=-\infty}^{+\infty}\delta(\omega-2k\pi)$
11	Difference in frequency	$nx(n)$	$j\frac{dX(e^{j\omega})}{d\omega}$
12	Conjugate symmetry for real signals	$x(n)$ is real	$\begin{cases}X(e^{j\omega})=X^*(e^{-j\omega})\\ \mathrm{Re}[X(e^{j\omega})]=\mathrm{Re}[X(e^{-j\omega})]\\ \mathrm{Im}[X(e^{j\omega})]=-\mathrm{Im}[X(e^{-j\omega})]\\ |X(e^{j\omega})|=|X(e^{-j\omega})|\\ \sphericalangle X(e^{j\omega})=-\sphericalangle X(e^{-j\omega})\end{cases}$
13	Parseval's relation	$\sum_{n=-\infty}^{+\infty}|x(n)|^2=\frac{1}{2\pi}\int_{2\pi}|X(e^{j\omega})|^2 d\omega$	

Table 2.6: Basic discrete-time Fourier transform pairs.

	Name of signals	Expressions of $x(n)$	Discrete-time Fourier transforms $X(e^{j\omega})$
1	Complex exponential signal	$e^{j\omega_0 n}$	$\sum_{l=-\infty}^{+\infty} 2\pi\delta(\omega - \omega_0 - 2\pi l)$
2	Cosine signal	$\cos\omega_0 n$	$\sum_{l=-\infty}^{+\infty} \pi[\delta(\omega - \omega_0 - 2\pi l) + \delta(\omega + \omega_0 - 2\pi l)]$
3	Sine signal	$\sin\omega_0 n$	$\frac{1}{j}\sum_{l=-\infty}^{+\infty} \pi[\delta(\omega - \omega_0 - 2\pi l) - \delta(\omega + \omega_0 - 2\pi l)]$
4	Constant sequence	1	$\sum_{l=-\infty}^{+\infty} 2\pi\delta(\omega - 2\pi l)$
5	Unit impulse signal	$\delta(n)$	1
6	Unit step signal	$u(n)$	$\frac{1}{1-e^{-j\omega}} + \sum_{k=-\infty}^{+\infty} \pi\delta(\omega - 2\pi k)$
7	Unit impulse with shift	$\delta(n-n_0)$	$e^{-j\omega n_0}$
8	Rectangular square wave	$x(n) = \begin{cases} 1, & |n| \le N_1 \\ 0, & N_1 < |n| \le N/2 \end{cases}$ $x(n)=x(n+N)$	$\sum_{k=-\infty}^{+\infty} 2\pi a_k\delta\left(\omega - \frac{2\pi k}{N}\right)$
9	Unilateral exponential signal	$a^n u(n), \ |a|<1$	$\frac{1}{1-ae^{-j\omega}}$
10	Rectangular signal	$x(n) = \begin{cases} 1, & |n| \le N_1 \\ 0, & |n| > N_1 \end{cases}$	$\frac{\sin\omega(N_1 + 1/2)}{\sin(\omega/2)}$
11	Sinc function	$\frac{\sin Wn}{\pi n} = \frac{W}{\pi}\sin c\left(\frac{Wn}{\pi}\right)$ $0<W<\pi$	$X(e^{j\omega}) = \begin{cases} 1, & 0 \le |\omega| \le W \\ 0, & W < |\omega| \le \pi \end{cases}$
12	Derived signal from the spectrum of unilateral exponential signal	$(n+1)a^n u(t), \ |a|<1$ $\frac{(n+r-1)!}{n!(r-1)!}a^n u(t), \ |a|<1$	$\frac{1}{(1-ae^{-j\omega})^2}$ $\frac{1}{(1-ae^{-j\omega})^r}$

We can see from the definition of FT shown in eqs. (2.19) and (2.20) that the signal $x(t)$ is continuous and aperiodic, while the spectrum $X(j\Omega)$ is also continuous and aperiodic. This characteristic of FT determines the existence of its duality.

Example 2.17: Given a signal as $g(t) = 2/(1+t^2)$. Find its FT $G(j\Omega)$.
Solution: It is very hard to calculate the FT to the given signal $g(t)$ based on the definition. Under this condition, we usually need to consider whether properties of FT can be used. Unless the duality property, no other properties can be used for this calculation.

By using the dual property to solve the FT, we have to know a spectrum which is in the same form as that of the given signal $g(t)$, and the corresponding signal of the known spectrum. Based on the knowledge on FT pairs, we can get the signal as $x(t) \leftrightarrow X(j\Omega) = 2/(1+\Omega^2)$. Thus, we get

$$x(t) = e^{-|t|} \leftrightarrow X(j\Omega) = \frac{2}{1+\Omega^2}$$

The inverse FT of $X(j\Omega)$ can be expressed as

$$x(t) = e^{-|t|} = \frac{1}{2\pi} \int_{-\infty}^{+\infty} \left(\frac{2}{1+\Omega^2}\right) e^{-j\Omega t} d\Omega$$

Multiplying 2π in both sides of the above equation, and replacing t by $-t$, we obtain

$$2\pi e^{-|t|} = \int_{-\infty}^{+\infty} \left(\frac{2}{1+\Omega^2}\right) e^{-j\Omega t} d\Omega$$

Then interchanging the names of variables t and Ω, we find that

$$2\pi e^{-|\Omega|} = \int_{-\infty}^{-\infty} \left(\frac{2}{1+t^2}\right) e^{-j\Omega t} dt$$

Thus the FT of $g(t)$ is shown as

$$g(t) = \int_{-\infty}^{+\infty} \left(\frac{2}{1+t^2}\right) e^{-j\Omega t} dt = 2\pi e^{-|\Omega|}$$

2. The duality of DFS

We can see from the definition of DFS shown in eqs. (2.16) and (2.17) that both time domain signal $x(n)$ and FS coefficients a_k are all discrete and periodic, resulting in a duality property between $x(n)$ and a_k.

In fact, both time shifting and the frequency shifting properties in the DFS satisfy the duality relation, shown as

$$x(n-n_0) \longleftrightarrow a_k e^{-jk(2\pi/N)n_0}$$
$$e^{jM(2\pi/N)n} \longleftrightarrow a_{k-M}$$

3. The duality between DTFT and FS

We can see from the definitions of the DTFT and FS in eqs. (2.26), (2.27), (2.5), and (2.6) that both $x(n)$ in DTFT and a_k in FS are all discrete and aperiodic, which construct a set of duality relations, while both $X(e^{j\omega})$ in DTFT and $x(t)$ in FS are all continuous and periodic, which form another set of duality relations.

Example 2.18: Try to find the spectrum of $x(n) = \frac{\sin(\pi n/2)}{\pi n}$ by using the duality property between DTFT and FS.

Solution: We must first identify a continuous-time signal $g(t)$ with period $T = 2\pi$ and Fourier coefficients $a_k = x(k) = \frac{\sin(\pi k/2)}{\pi k}$. According to the basic FT pairs, we find

$$g(t) = \begin{cases} 1, & |t| \le T_1 \\ 0, & T_1 < |t| \le \pi \end{cases} \quad \longleftrightarrow \quad \frac{\sin(\pi k/2)}{\pi k} = a_k$$

satisfying the requirement. Consequently, if we take $T_1 = \pi/2$, then we obtain $a_k = x(k)$. Thus we have

$$a_k = F\{g(t)\} = \frac{\sin(\pi k/2)}{\pi k} = \frac{1}{2\pi} \int_{-\pi}^{\pi} g(t) e^{-jk\Omega_0 t} dt = \frac{1}{2\pi} \int_{-\pi/2}^{\pi/2} e^{-jkt} dt$$

where $\Omega_0 = (2\pi/T) = 1$ and $g(t) \equiv 1$ in the range of $[-(\pi/2), \pi/2]$. If we rename k as n, and t as ω, we have

$$x(n) = \frac{\sin(\pi n/2)}{\pi n} = \frac{1}{2\pi} \int_{-\pi/2}^{\pi/2} e^{-j\omega n} d\omega$$

Replacing n by $-n$ on both sides of the above equation, and noticing that $\frac{\sin(\pi n/2)}{\pi n}$ is even, we obtain

$$x(n) = \frac{\sin(\pi n/2)}{\pi n} = \frac{1}{2\pi} \int_{-\pi/2}^{\pi/2} e^{j\omega n} d\omega$$

where

$$X(e^{j\omega}) = \begin{cases} 1, & |\omega| \leq \pi/2 \\ 0, & \pi/2 < |\omega| \leq \pi \end{cases}$$

2.6 Frequency Domain Analysis of Signals and Systems

In earlier sections of this chapter we introduced the fundamental theories and methods in detail of four kinds of FT and FS, including FS for continuous-time periodic signals, DFS for discrete-time periodic signals, FT for continuous-time aperiodic signals, and DTFT for discrete-time aperiodic signals.

In fact, in electronic information technology, these four kinds of Fourier analysis methods are mainly used to represent the characteristics of signals and systems in the frequency domain. In this section, we will introduce further the representation, the characteristics, and the operation of signals and systems in the frequency domain.

2.6.1 The Spectral Representation of Signals

1. The concept and representation of spectrum

For a given a continuous-time signal $x(t)$, periodic or aperiodic, we can get FS coefficients a_k or FT $X(j\Omega)$, respectively. Similarly, for a given discrete-time signal $x(n)$, periodic or aperiodic, we can get DFS coefficients a_k or DTFT $X(e^{j\omega})$, respectively. As a matter of fact, the above four types of FTs a_k, $X(j\Omega)$, a_k, and $X(e^{j\omega})$ are all spectra of given signals. The so-called spectra are in fact the frequency components distributed in the frequency domain.

Due to different types of signals and different transform methods used, these four types of spectra have different characteristics, in which a_k obtained from FS are discrete and aperiodic. The frequency variable k is simplified from $k\Omega_0$, where Ω_0 is referred to as the fundamental frequency. a_k obtained from DFS are discrete and periodic. Its frequency variable k is simplified from $k\omega_0$, where ω_0 is also referred to as the fundamental frequency. $X(j\Omega)$ obtained from FT is continuous and aperiodic. Its frequency variable Ω is called analog radian frequency (usually called frequency), with the unit of rad/s. In fact, the concepts of frequency and radian frequency can be connected by $\Omega = 2\pi F$, where F is really the frequency with the unit of Hz. $X(e^{j\omega})$ obtained from DTFT is continuous and periodic. Its frequency variable is ω, referred to as the digital frequency with the unit of rad. Generally, the analog frequency and digital frequency are connected by $\omega = \Omega T$, where T is the sampling interval. Based on the above facts, we can get the following general rules shown in Figure 2.13:

Figure 2.14 demonstrates the waveforms and spectra of various types of signals.

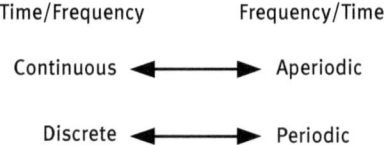

Figure 2.13: The time/frequency domain relations among various Fourier transforms.

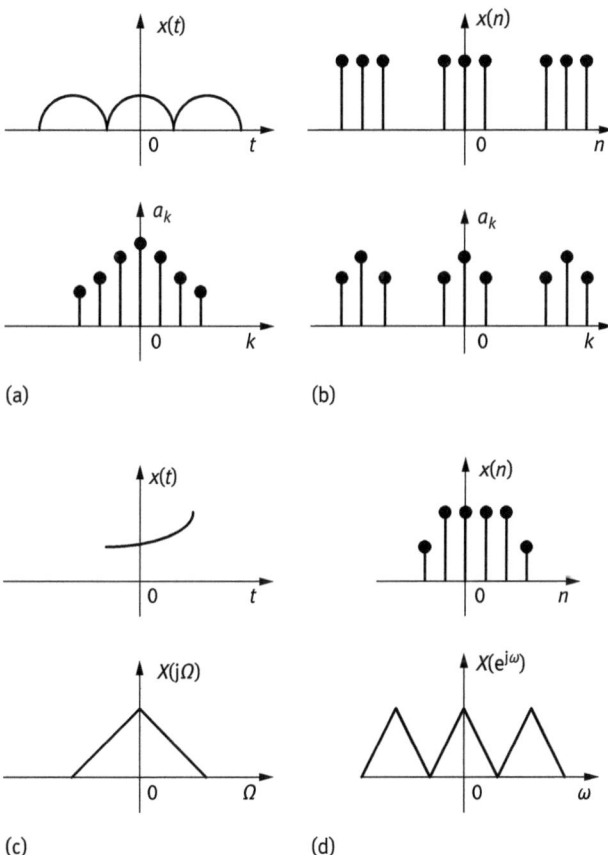

Figure 2.14: The waveforms and spectra of various types of signals: (a) continuous-time periodic signal; (b) discrete-time periodic signal; (c) continuous-time aperiodic signal; and (d) discrete-time aperiodic signal.

2. The magnitude and phase of the spectrum

The spectrum of a signal is generally a complex function. For example, the spectrum of signal $x(t)$ obtained from FT is expressed as

$$X(j\Omega) = |X(j\Omega)|e^{j\sphericalangle X(j\Omega)}$$

(2.29)

where $|X(j\Omega)|$ is the magnitude spectrum representing the magnitude of various components, while $\sphericalangle X(j\Omega)$ is the phase spectrum, representing the phase information of the spectrum.

In engineering practice, we seem to concern more about the magnitude spectrum of a signal. For example, in radio and television systems, we need to specify a channel (or frequency range) for a specific signal. And the specific frequency range is determined by both modulation of the signal and the bandwidth, as well as other frequency characteristics of the signal. In communication systems, similar to those in radio and television systems, we also need to distribute frequency for different users, with a more complicated situation.

On the other hand, the characteristics of the phase spectrum cannot be ignored in many cases. In fact, the phase spectrum contains a large amount of information about a signal. In some cases, even if the magnitude spectrum of the signal remains the same and only its phase spectrum changes, the waveform of the signal will also be an unexpected change. Please consider the superposition of three sinusoidal signals as an example:

$$x(t) = 1 + \frac{1}{2}\cos(2\pi t + \phi_1) + \cos(4\pi t + \phi_2) + \frac{2}{3}\cos(6\pi t + \phi_3)$$

The waveform changes of the superposition of three sinusoidal signals are given in Figure 2.15.

We see from Figure 2.15 that when the phases of the three sine signals take a linear relation, it results in a time shift of the original signal only; and when the phases of the three signals are nonlinear, the impact of the nonlinear phase is significant, and it even makes the signal unrecognizable.

2.6.2 The Frequency Analysis of Linear Time-Invariant systems

1. The frequency response of linear time-invariant (LTI) systems
In the first chapter we learnt that the output response $y(t)$ of an LTI system is obtained by the convolution of the input signal $x(t)$ and the unit impulse response $h(t)$ of the system, shown as

$$y(t) = h(t)*x(t)$$

(2.30)

Based on the convolution property of FT, the above equation can be rewritten as

$$Y(j\Omega) = H(j\Omega)X(j\Omega)$$

(2.31)

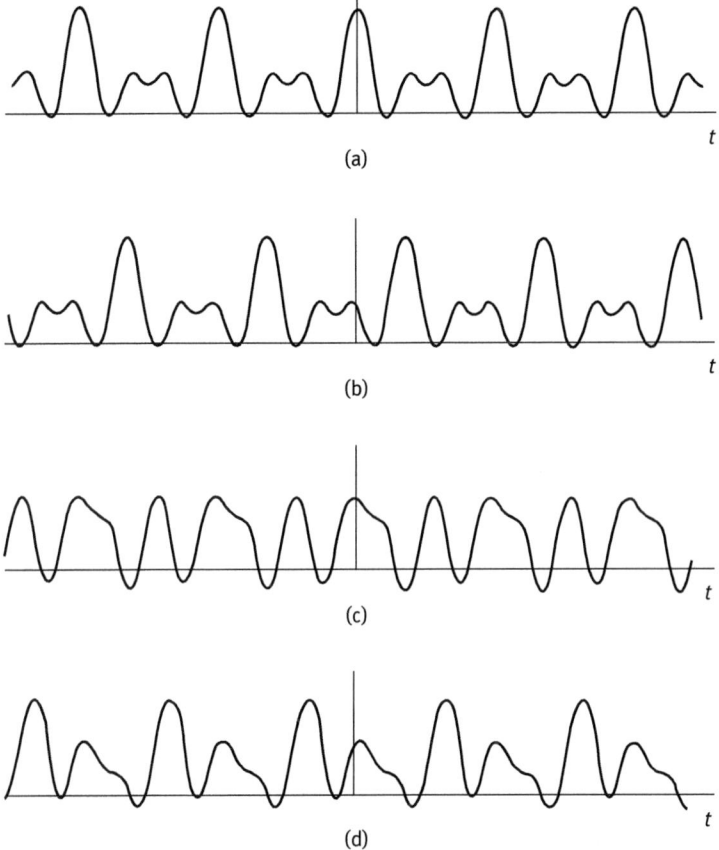

Figure 2.15: The effects of the phase to signals: (a) zero phase; (b) linear phase, $\phi_1 = 4$, $\phi_2 = 8$, $\phi_3 = 12$; (c) nonlinear phase, $\phi_1 = 6$, $\phi_2 = -2.7$, $\phi_3 = 0.93$; and (d) nonlinear phase, $\phi_1 = 1.2$, $\phi_2 = 4.1$, $\phi_3 = -7.02$.

where $X(j\Omega)$, $Y(j\Omega)$ are the spectra of the input and output signals, and $H(j\Omega)$ is the frequency response of the system, also referred to as the transfer function of the system. Obviously, the calculation complexity of the operation is significantly reduced by FT.

The frequency response $H(j\Omega)$ is a complex function. It can be written in the form of magnitude and phase as

$$H(j\Omega) = |H(j\Omega)|e^{j \sphericalangle H(j\Omega)} \qquad (2.32)$$

where $|H(j\Omega)|$ is the magnitude spectrum or gain of the system, $\sphericalangle H(j\Omega)$ represents the phase characteristics of the system. Based on eqs. (2.31) and (2.32), we have

$$|Y(j\Omega)| = |H(j\Omega)||X(j\Omega)| \tag{2.33}$$

$$\sphericalangle Y(j\Omega) = \sphericalangle H(j\Omega) + \sphericalangle X(j\Omega) \tag{2.34}$$

It is obvious that the effect of an LTI system to the input signal is multiplying $|X(j\Omega)|$ by $|H(j\Omega)|$. At the same time, the LTI system also adds $\sphericalangle X(j\Omega)$ by $\sphericalangle H(j\Omega)$, changing the relative relations among various frequency components. If the change to the input signal is expected, it indicates that the system can usually be used as a filter for filtering or processing input signals. Conversely, if the change is not desired, then it may be caused by the distortion of the system.

2. Linear phase and nonlinear phase
If the phase spectrum of a system satisfies

$$\sphericalangle H(j\Omega) = -\Omega t_0 \tag{2.35}$$

then the system is a linear phase system. A typical linear phase system is shown as

$$H(j\Omega) = e^{-j\Omega t_0} \tag{2.36}$$

The unit impulse response of the system is a unit impulse $\delta(t - t_0)$. The effect of the system to an input signal is delaying the signal with t_0, resulting in $y(t) = x(t - t_0)$. Such system is a rather ideal system for communications if the delay t_0 is acceptable. It can be seen the whole characteristics of the input signal are maintained except for the delay.

The system will be a nonlinear system if the linear phase property shown in eq. (2.35) is not satisfied. The nonlinear phase system may affect the input signal significantly.

3. The concept of filters
The filter is a widely used concept in the field of electronic information technology, especially in signal processing. A so-called filter is a circuit or algorithm that has a processing effect on the signal. In fact, the concept of the system introduced in this book can be seen as a generalized filter. The purpose of these systems is to do some processing for an input signal, so that the output signal is more suitable for the need of users.

The filtering is a concept closely related to the filter. The so-called filtering, in a narrow sense, is to filter the signal in the specific frequency band with a filter. Filtering is an important means of suppression and elimination of noise and interference. However, with the progress of the modern signal processing technology, the concepts on filter and filtering are not strictly in accordance with the frequency band. The introduction on modern filters will be given in later chapters of this book. The generalized concept of filtering should be the process of suppressing or eliminating all useless signals.

4. The ideal low-pass filter

The classical filter is mainly based on the frequency characteristics of signals. Generally, the classical filter can be divided into different types, including the low-pass filter, the high-pass filter, the band-pass filter, the band-stop filter, and the all-pass filter. Figure 2.16 shows the frequency characteristics of different ideal classical filters.

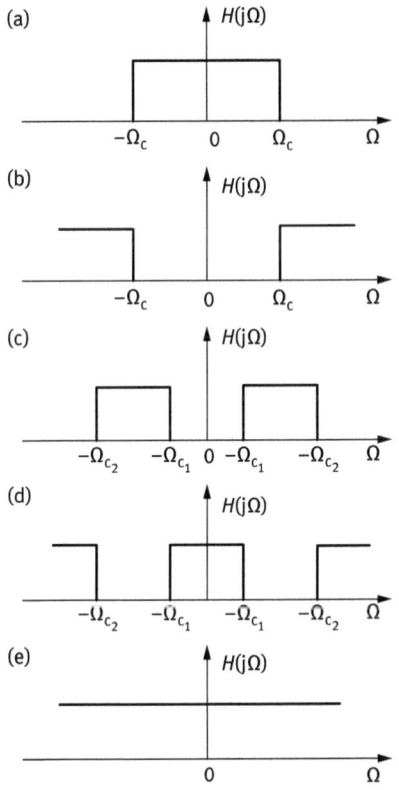

Figure 2.16: The frequency responses of the classical ideal filters: (a) low-pass filter; (b) high-pass filter; (c) band-pass filter; (d) band-stop filter; and (e) all-pass filter.

We see from Figure 2.16 that the ideal low-pass filter allows all components below the cutoff frequency Ω_c in the input signal unhindered passing it through, and removes all the components higher than Ω_c. Similarly, other ideal filters (high-pass, band-pass, and band-stop filters) allow the part of components in the input signal passing them through with a boundary of the cutoff frequency, and prohibit the rest components. The ideal all-pass filter allows all components in the input signal passing it through without any change.

The classical ideal filter is often used to describe the idealized system characteristics. However, in practice, the ideal filter cannot be physically implemented in real time. We usually use a nonideal filter to approximate the ideal filter. Next, we take the ideal low-pass filter as an example. The frequency response of the ideal low-pass filter is shown in eq. (2.37).

$$H(j\Omega) = \begin{cases} 1, & |\Omega| \le \Omega_c \\ 0, & |\Omega| > \Omega_c \end{cases} \tag{2.37}$$

The phase of the ideal low-pass filter is a constant zero since $H(j\Omega)$ is a real function of Ω. The corresponding unit impulse response is shown as

$$h(t) = \frac{\sin \Omega_c t}{\pi t} \tag{2.38}$$

Another common ideal low-pass filter is the linear phase low-pass filter, shown as

$$|H(j\Omega)| = \begin{cases} 1, & |\Omega| \le \Omega_c \\ 0, & |\Omega| > \Omega_c \end{cases} \tag{2.39}$$

$$\sphericalangle H(j\Omega) = -\alpha\Omega \tag{2.40}$$

The magnitude spectrum and phase spectrum of both ideal low-pass filters are given in Figure 2.17.

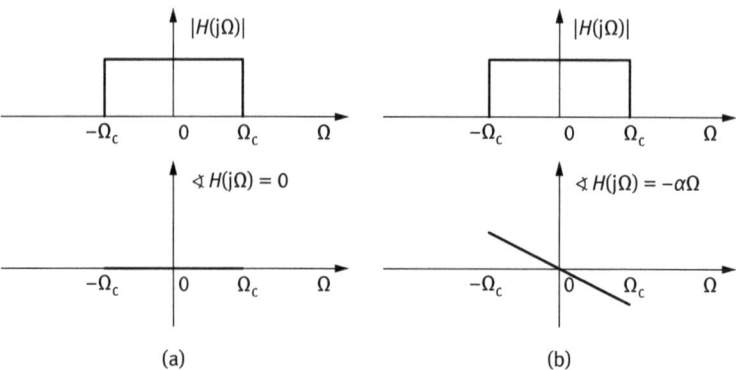

Figure 2.17: The frequency responses of ideal low-pass filters: (a) ideal low-pass filter with zero phase;(b) ideal low-pass filter with a linear phase.

The unit impulse responses of both ideal low-pass filters are shown in Figure 2.18.
From Figure 2.18, we can see that the unit impulse response of either ideal low-pass filter is a noncausal system, so it cannot be implemented physically in real time.

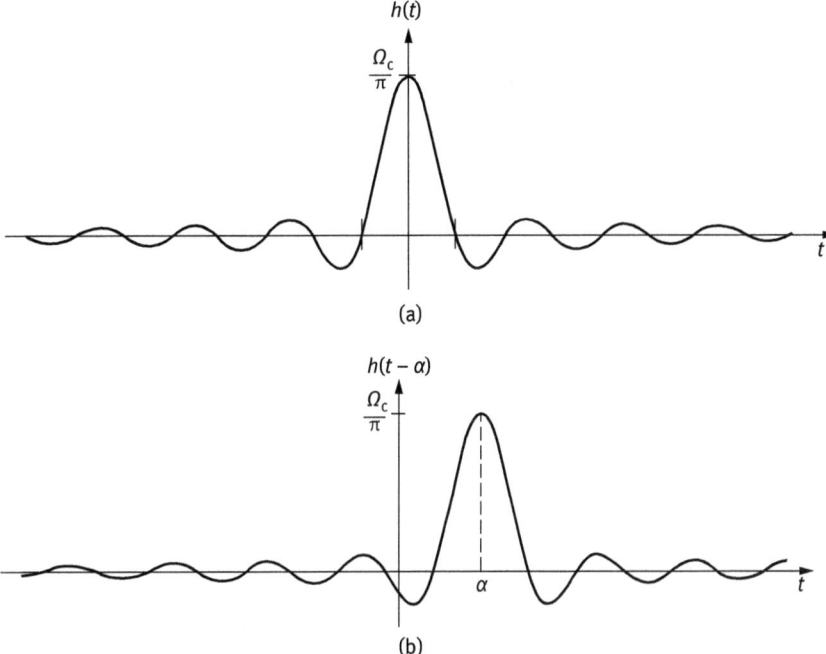

Figure 2.18: The unit impulse responses of ideal low-pass filters: (a) ideal low-pass filter with zero phase and (b) ideal low-pass filter with a linear phase.

This is the first drawback of the ideal low-pass filter. The second drawback is the fluctuation of its unit impulse response, which is the basic feature of the sinc function. The sinc function will sustain a damping oscillation with time t. Such oscillation will cause a sustained vibration of the unit step response of the system, which is not wanted in signal processing.

5. Nonideal filters

The role of a filter is to separate the useful signal from useless signals, such as noises and interferences. In the actual signal processing, filters used are usually non-ideal filters, since the shortcomings of ideal filters. Figure 2.19 shows the frequency response of a conventional nonideal low-pass filter.

In Figure 2.19, the frequency and magnitude are shown in the horizontal axis and vertical axis, respectively. The dashed line in the figure shows the frequency response of the low-pass filter. Ω_p and Ω_s are the cut-off frequencies of the passband and the stopband, respectively. The part between Ω_p and Ω_s is called the transition band. δ_1 and δ_2 represent the tolerances of passband and stopband, respectively.

We can see from Figure 2.19 that unlike the ideal filter, the passband of the non-ideal filter is not strictly flat, existing certain permissible ripple. The stopband of the filter is not strictly forbidden, existing a permissible stopband ripple, too.

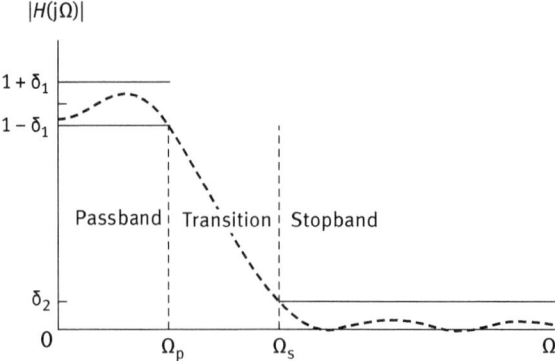

Figure 2.19: The frequency response of a conventional nonideal low-pass filter.

Detailed discussion on the filter design and the performance will be given in later chapters of this book.

2.6.3 Bode Plots

Bode plot is a semi-log coordinate graph showing the frequency response of a system, whose frequency axis is in logarithmic scale, and its longitudinal axis expresses the magnitude of the system in a linear dimension. Bode plot is usually divided into two types: the magnitude–frequency plot and the phase–frequency plot. Bode plot is commonly used in the electronic information technology.

1. Bode plot of the first-order system
Suppose that the unit impulse response of the first-order LTI system is

$$h(t) = \frac{1}{\tau}e^{-t/\tau}u(t) \tag{2.41}$$

where the parameter τ is referred to as the system time constant for controlling the response speed of the system. Taking FT to $h(t)$, we can get the frequency response (or the transfer function) $H(j\Omega)$ of the system as

$$H(j\Omega) = \frac{1}{j\Omega\tau + 1} \tag{2.42}$$

Taking the logarithm to $|H(j\Omega)|$ of base 10 and then times 20, we get

$$20\log_{10}|H(j\Omega)| = -10\log_{10}[(\Omega\tau)^2 + 1] \ (\text{dB}) \tag{2.43}$$

Analyzing the above equation further, we get

$$20\log_{10}|H(j\Omega)| \approx \begin{cases} 0, & \Omega\tau \ll 1 \\ -20\log_{10}(\Omega) - 20\log_{10}(\tau), & \Omega\tau \gg 1 \end{cases} \tag{2.44}$$

Equations (2.43) and (2.44) show that the log magnitude of the first-order system is approximately 0 dB in the low-frequency band, and a linear function of $\log_{10}\Omega$ in the high-frequency band, which is sometimes referred to as a "20 dB per decade" asymptote.

The similar log asymptotes for the phase spectrum of $H(j\Omega)$ can also be obtained as

$$\sphericalangle H(j\Omega) = -\tan^{-1}(\Omega\tau) = \begin{cases} 0, & \Omega \le 0.1/\tau \\ -(\pi/4)[\log_{10}(\Omega\tau) + 1], & 0.1/\tau \le \Omega \le 10/\tau \\ -\pi/2, & \Omega \ge 0.1/\tau \end{cases} \tag{2.45}$$

The Bode plot for a given first-order system is shown in Figure 2.20.

In Figure 2.20(a), there is a point of intersection for both asymptotes at $\Omega = 1/\tau$, referred to as the break frequency. According to eq. (2.43), the attenuation in magnitude at this point is about -3 dB. Because of the fact, this point is sometimes called the 3 dB point, and the bandwidth at this point is defined as the 3 dB bandwidth.

2. The Bode plot of the second-order system
The second-order LTI system is represented by the second-order constant coefficient differential equation as

$$\frac{d^2y(t)}{dt^2} + 2\zeta\Omega_n \frac{dy(t)}{dt} + \Omega_n^2 y(t) = \Omega_n^2 x(t) \tag{2.46}$$

where parameters ζ and Ω_n are the damping ratio and the undamped natural frequency, respectively. Taking FT to eq. (2.46), we get the transfer function of the system as

$$H(j\Omega) = \frac{\Omega_n^2}{(j\Omega)^2 + 2\zeta\Omega_n(j\Omega) + \Omega_n^2} \tag{2.47}$$

Figure 2.21 shows the Bode plot of the second-order system, showing the situation in different damping ratios.

2.6.4 Conditions for Distortionless Transmission and Physical Realization of Systems

1. Distortionless transmission conditions
The concept of distortionless transmission condition for a system indicates that the output signal is the same as the input signal, except the changes in amplitude and

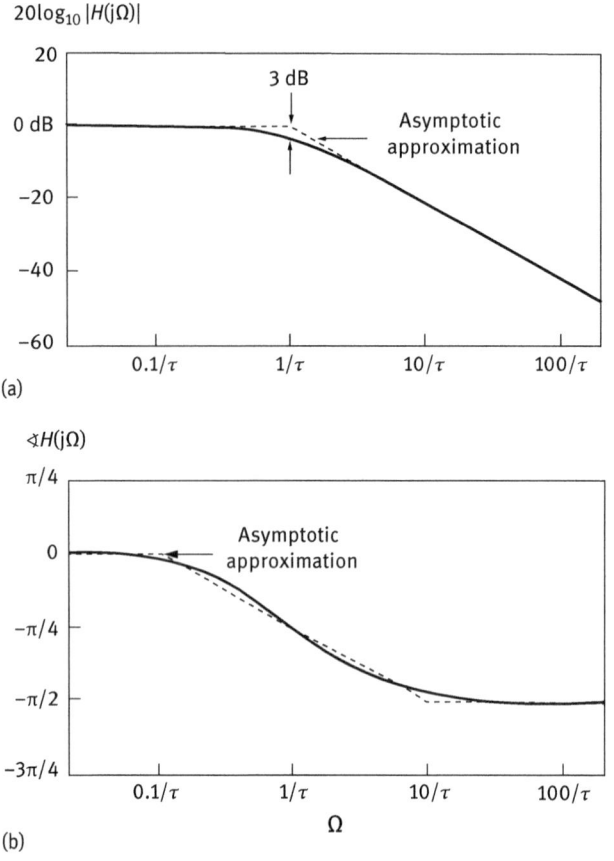

Figure 2.20: The Bode plot of a first-order system: (a) Bode plot of magnitude and (b) Bode plot of phase.

time shift. If the input and output signals are $x(t)$ and $y(t)$, respectively, then the distortionless transmission condition for the system is shown as

$$y(t) = Kx(t - t_0) \tag{2.48}$$

where K is a constant and t_0 represents the time shift. Taking FT to eq. (2.48), we can get the transfer function of the system as

$$H(j\Omega) = Ke^{-j\Omega t_0} \tag{2.49}$$

The above equation shows the requirement for the frequency response of the system under the distortionless transmission condition. Taking the inverse FT, we can get the unit impulse response of the system as

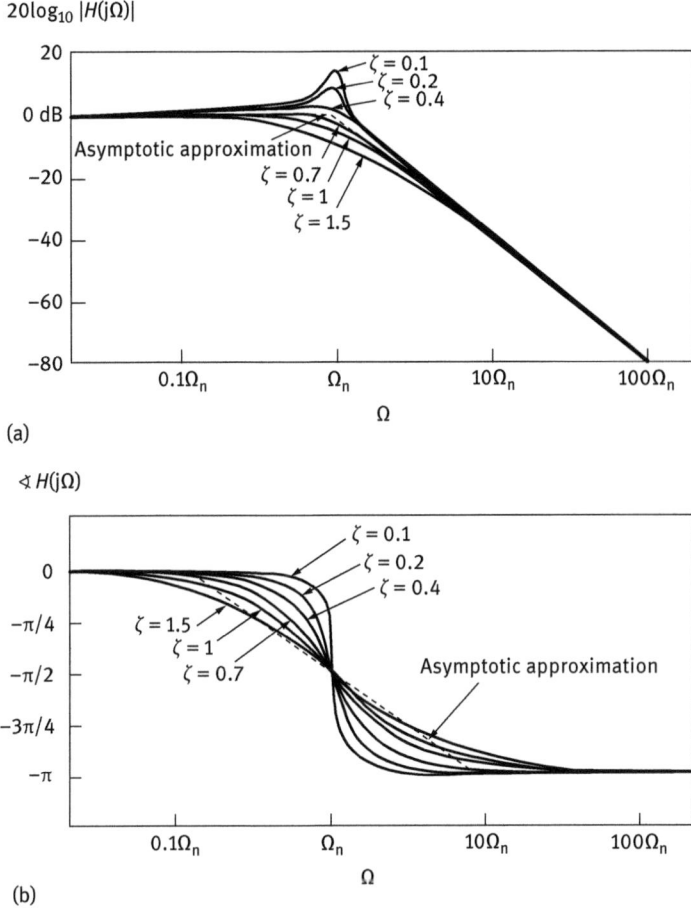

Figure 2.21: The Bode plot of a second-order system with different damping ratios: (a) Bode plot of magnitude and (b) Bode plot of phase.

$$h(t) = K\delta(t - t_0) \tag{2.50}$$

It is obvious that $h(t)$ is a unit impulse signal with a delay t_0 and a gain K.

2. Physical realization conditions

A physically realizable system has to meet the following two conditions. First, the system needs to satisfy the causality condition as

$$h(t) = 0, \quad \text{if } t < 0 \tag{2.51}$$

Second, the system needs to satisfy Paley–Wiener criterion as

$$\int\limits_{-\infty}^{+\infty} \frac{|\ln\{|H(j\Omega)|\}|}{1+\Omega^2} d\Omega < \infty \tag{2.52}$$

Paley–Wiener criterion is a necessary condition for a system to be a physical realization system.

✐ Exercises

2.1 Try to explain the characteristics of FS for the continuous-time periodic signal.

2.2 Try to explain the characteristics of DFS for the discrete-time periodic signal.

2.3 Try to explain the characteristics of FT.

2.4 Try to explain the characteristics of DTFT.

2.5 Given a continuous-time periodic signal $x(t) = 2 + \cos\left(\frac{2\pi}{3}t\right) + \sin\left(\frac{5\pi}{3}t\right)$. Find its fundamental frequency Ω_0 and the FS coefficients a_k.

2.6 Suppose the fundamental frequency is $\Omega_0 = \pi$. Find FS coefficients a_k of the continuous-time signal $x(t) = \begin{cases} 1.5, & 0 \le t < 1 \\ -1.5 & 1 \le t < 2 \end{cases}$.

2.7 Suppose that the period of discrete-time periodic signals $x_1(n)$ and $x_2(n)$ is N. Their corresponding Fourier coefficients are a_k and b_k, respectively. We also know that $a_0 = a_3 = \frac{1}{2}a_1 = \frac{1}{2}a_2 = 1$, and $b_0 = b_1 = b_2 = b_3 = 1$. Try to determine the FS coefficients c_k of $g(n) = x_1(n)x_2(n)$, based on the multiplication property of FS.

2.8 Find the FS coefficients of the following signals:
(1) $x(t) = e^{-t}$, $-1 < t < 1$. The period is 2; (2) $x(t) = \begin{cases} \sin\pi t, & 0 \le t \le 2 \\ 0, & 2 < t \le 4 \end{cases}$. The period is 4.

2.9 Given the FS coefficients a_k, the period is 4. Find the continuous-time periodic signal $x(t)$.
(1) $a_k = \begin{cases} 0, & k = 0 \\ (j)^k \frac{\sin k\pi/4}{k\pi}, & \text{otherwise} \end{cases}$; (2) $a_k = (-1)^k \frac{\sin k\pi/8}{2k\pi}$.

2.10 Suppose that the FS coefficients a_k are given for the following problems. The period of a_k is 8. Find the discrete-time periodic signal $x(n)$.
(1) $a_k = \cos\left(\frac{k\pi}{4}\right) + \sin\left(\frac{3k\pi}{4}\right)$; $a_k = \begin{cases} \sin\left(\frac{k\pi}{3}\right), & 0 \le k \le 6 \\ 0, & k = 7 \end{cases}$; (3) a_k is given in Figure E2.1.

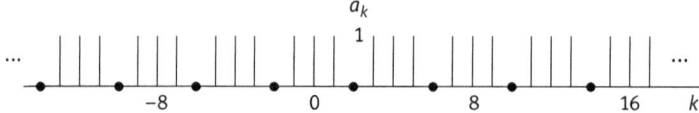

Figure E2.1: The given a_k.

2.11 Calculate the FT for the following signals, and plot their magnitude spectrum.
(1) $x(t) = e^{-2(t-1)}u(u-1)$; (2) $x(t) = e^{-2|t-1|}$.

2.12 Calculate the inverse FT for the following spectra:

(1) $X(j\Omega) = 2\pi\delta(\Omega) + \pi\delta(\Omega - 4\pi) + \pi\delta(\Omega + 4\pi)$; (2) $X(j\Omega) = \begin{cases} 2, & 0 \le \Omega \le 2 \\ -2, & -2 \le \Omega < 0 \\ 0, & |\Omega| > 2 \end{cases}$

2.13 Given $x(t) \leftrightarrow X(j\Omega)$. Try to express the FT of the following signals using $X(j\Omega)$:
(1) $x_1(t) = x(1-t) + x(-1-t)$; (2) $x_2(t) = x(3t-6)$; (3) $x_3(t) = \frac{d^2}{dt^2}[x(t-2)]$

2.14 Given $e^{-|t|} \leftrightarrow \frac{2}{1+\Omega}$, (1) find FT of $te^{-|t|}$; (2) based on the result of (1), find FT of $\frac{4t}{(1+t^2)^2}$.

2.15 The frequency response of a causal LTI system is given as $H(j\Omega) = \frac{1}{j\Omega+3}$. For a specific input signal $x(t)$, the output of the system is $y(t) = e^{-3t}u(t) - e^{-4t}u(t)$. Find $x(t)$.

2.16 Given signal $x_0(t) = \begin{cases} e^{-t}, & 0 \le t \le 1 \\ 0, & \text{otherwise} \end{cases}$, try to calculate FT $X_0(j\Omega)$ of $x_0(t)$ and to express the FTs for the following signals shown in Figure E2.2.

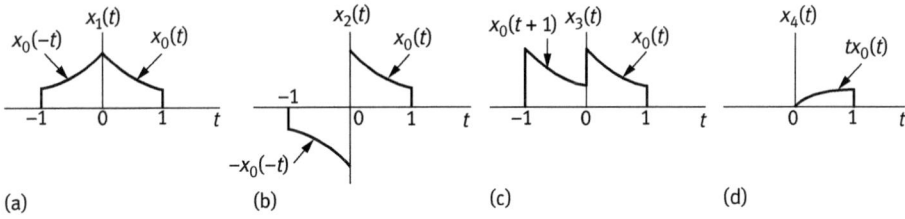

Figure E2.2: The waveforms of the given signals.

2.17 Calculate the following convolutions based on the convolution property and FT:
(1) $x(t) = te^{-2t}u(t)$, $h(t) = e^{-4t}u(t)$; (2) $x(t) = te^{-2t}u(t)$, $h(t) = te^{-4t}u(t)$;
(3) $x(t) = e^{-t}u(t)$, $h(t) = e^{t}u(-t)$.

2.18 The input and output signals of an LTI system are given as $x(t) = [e^{-t} + e^{-3t}]u(t)$, and $y(t) = [2e^{-t} - 2e^{-4t}]u(t)$.
(1) Find the frequency response of the system.
(2) Determine the unit impulse response of the system.

2.19 The phase function of an all-pass system is given as $\triangleleft H(j\Omega) = -\frac{\pi}{2}\text{sgn}(\Omega) = \begin{cases} \pi/2, & \Omega < 0 \\ -\pi/2, & \Omega > 0 \end{cases}$. Find the unit impulse response of the system.

2.20 The input signal of an LTI system is $\delta(t)$, and its output is a triangular pulse with bottom width of τ, that is,

$$y(t) = \begin{cases} 1 - \frac{2|t|}{\tau}, & |t| < \tau/2 \\ 0, & |t| \ge \tau/2 \end{cases}$$

Find the frequency response of the system.

2.21 A low-pass filter $H(j\Omega)$ has an ideal magnitude response and a nonlinear phase characteristic shown as $H(j\Omega) = H_1(j\Omega)e^{-j\Delta\varphi(\Omega)}$, in which $H_1(j\Omega)$ is an ideal low-pass filter, and it meets $\Delta\varphi(\Omega) \ll 1$, shown as

$$\Delta\varphi(\Omega) = a_1 \sin\left(\frac{\Omega}{\Omega_c}\right) + a_2 \sin\left(\frac{2\Omega}{\Omega_c}\right) + \cdots + a_m \sin\left(\frac{m\Omega}{\Omega_c}\right)$$

Find the unit impulse response of the system and compare it with the ideal low-pass filter.

2.22 Calculate the following DTFTs:
(1) $x(n) = \left(\frac{1}{2}\right)^{n-1} u(n-1)$; (2) $x(n) = \left(\frac{1}{2}\right)^{|n-1|}$; (3) $x(n) = \delta(n-1) + \delta(n+1)$

2.23 For $-\pi \le w < \pi$, calculate the DTFT of the following signals:
(1) $x(n) = \sin\left(\frac{\pi}{3}n + \frac{\pi}{4}\right)$; (2) $x(n) = 2 + \cos\left(\frac{\pi}{6}n + \frac{\pi}{8}\right)$

2.24 Given $X(e^{jw}) = \frac{1}{1-e^{-jw}}\left(\frac{\sin 3w/2}{\sin w/2}\right) + 5\pi\delta(w)$, $-\pi < w \le \pi$. Find $x(n)$.

2.25 A discrete-time signal satisfies the following conditions: (a) $x(n) = 0$, $n > 0$; (b) $x(0) > 0$; (c) $Im[X(e^{jw})] = \sin w - \sin 2w$; (d) $\frac{1}{2\pi}\int_{-\pi}^{\pi}|X(e^{jw})|^2 dw = 3$. Find $x(n)$.

2.26 A system with the unit impulse response $h_1(n) = \left(\frac{1}{3}\right)^n u(n)$ connects with a causal LTI system $h_2(n)$ in parallel. The frequency response of the paralleled system is
$X(e^{jw}) = \frac{-12 + 5e^{-jw}}{12 - 7e^{-jw} + e^{-j2w}}$. Find $h_2(n)$.

2.27 A discrete-time signal $x(n)$ is shown in Figure E2.3, whose DTFT is represented as $X(e^{jw})$. Find the following results without the calculation of $X(e^{jw})$:

(1) $X(e^{j0})$; (2) $\sphericalangle X(e^{jw})$; (3) $\int_{-\pi}^{\pi} X(e^{jw})dw$; (4) $X(e^{j\pi})$; (5) $\int_{-\pi}^{\pi} |X(e^{jw})|^2 dw$.

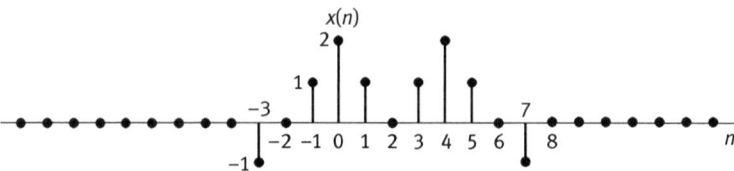

Figure E2.3: The waveform of the given signal $x(n)$.

2.28 Suppose the DTFT of the signal $x(n)$ is $X(e^{jw})$. Try to derive the following DTFT with $X(e^{jw})$.
(1) $Re[x(n)]$; (2) $x^*(-n)$.

2.29 Try to prove the convolution property of the DTFT: $x(n)*h(n) \leftrightarrow X(e^{jw})H(e^{jw})$.

2.30 The frequency response of a discrete-time LTI system is $H(e^{jw}) = |H(e^{jw})|e^{j\sphericalangle H(e^{jw})}$. Its real-valued unit impulse response is $h(n)$. If we send an input signal $x(n) = \sin(w_0 n + \phi_0)$ to the system, its output is expressed as $y(n) = |H(e^{jw_0})|x(n-n_0)$. If both $\sphericalangle H(e^{jw_0})$ and w_0 are correlated with a special relation, try to find the special relation.

2.31 The frequency response of a causal and stable LTI system is $H(j\Omega) = \frac{1-j\Omega}{1+j\Omega}$. Try to prove $|H(j\Omega)| = A$, and find the value of A.

2.32 Try to depict the Bode plots of the following systems:

(1) $H(j\Omega) = 40 \cdot \frac{j\Omega + 0.1}{j\Omega + 40}$; (2) $H(j\Omega) = 0.04 \cdot \frac{j\Omega + 50}{j\Omega + 0.2}$.

2.33 Try to depict the Bode plots of the following second-order systems:

(1) $H(j\Omega) = \frac{250}{(j\Omega)^2 + 50.5j\Omega + 25}$; (2) $H(j\Omega) = 0.02\left(\frac{j\Omega + 50}{(j\Omega)^2 + 0.2j\Omega + 1}\right)$.

2.34 Suppose that the frequency response of the low-pass differentiator is shown in Figure E2.4. When the input signals are $x_1(t) = \cos(2\pi t + \theta)$ and $x_2(t) = \cos(4\pi t + \theta)\lhd$, respectively, try to find the output signals $y_1(t)$ and $y_2(t)$.

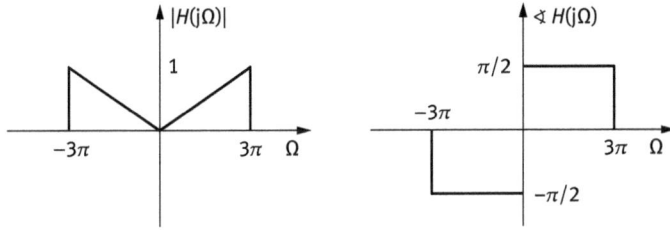

Figure E2.4: The frequency response of the low-pass differentiator.

2.35 A periodic square wave signal is given, whose period is 4, its pulse width is 2, and its amplitude is 1. Try to find its FS coefficients with MATLAB programming, and plot its magnitude spectrum and phase spectrum.

2.36 Try to find the FT of signal $x(t) = e^{-5t}u(t)$ with MATLAB programming, and plot its magnitude spectrum and phase spectrum.

2.37 Try to find the FT of signal $x(t) = \begin{cases} 1, & |t| < 1 \\ 0, & |t| > 1 \end{cases}$ with MATLAB programming, and plot its magnitude spectrum and phase spectrum.

3 The Complex Frequency Domain Analysis of Signals and Systems with Laplace Transform and z-Transform

3.1 Introduction

In Chapter 2, we systematically introduced basic concepts and theories of Fourier analysis, including Fourier series, discrete Fourier series, Fourier transform, and discrete-time Fourier transform (DTFT). In fact, the analysis of signals and systems by using Fourier theories is often referred to as the frequency domain analysis. The most important feature of the frequency domain analysis is that the frequency characteristics of signals and systems can be clearly obtained, and the operation can be significantly simplified.

This chapter will introduce the analysis theories and methods of signals and systems in complex frequency domain systematically. The so-called complex frequency domain analysis (or the complex frequency analysis) is to analyze signals and systems in the complex frequency domain via Laplace transform and/or z-transform. Similar to the frequency analysis method, the complex frequency analysis of signals and systems can greatly simplify the calculation of the analysis, and such kind of analysis, to some extent, contains the contents of frequency analysis of signals and systems. It is convenient for us not only to understand the frequency characteristics of signals and systems, but also to grasp the causality and the stability of systems by using the complex frequency analysis. Therefore, the complex frequency domain analysis for signals and systems is of great importance.

The complex frequency domain, referred to as the s domain, is obtained via the Laplace transform for continuous-time signals and systems. While for discrete-time signals and systems, the complex frequency domain is obtained via the z-transform, referred to as the z-domain. Complex frequency variables s and z are defined as eqs. (3.1) and (3.2), respectively:

$$s = \sigma + j\Omega \tag{3.1}$$

$$z = re^{j\omega} \tag{3.2}$$

where σ and Ω are the real and imaginary parts of s, respectively. r and ω are the magnitude and the angle of z, respectively. It is obvious that Fourier transform is the Laplace transform with a zero real part, and DTFT is the z-transform with magnitude 1.

There are a lot of useful properties for Laplace transform and z-transform. These properties provide some methods for signal and system analysis, which cannot be done by Fourier transform. For example, if the signal is unbounded, its Fourier

https://doi.org/10.1515/9783110465082-003

transform does not exist. However, for Laplace transform, we can select an appropriate convergence region to have the Laplace transform converged. So in a sense, the complex frequency domain analysis of signals and systems is a broader tool than the Fourier analysis.

3.2 The Laplace Transform

The Laplace transform is a kind of useful integral transform in engineering mathematics. The role of the Laplace transform is to transform a continuous-time signal or system into the complex frequency domain (or s domain), which may simplify the operation of the analysis for signals and systems.

3.2.1 The Definition of the Laplace Transform

1. From Fourier transform to Laplace transform
The definition of Fourier transform given in Chapter 2 is as follows:

$$X(j\Omega) = \int_{-\infty}^{+\infty} x(t)e^{-j\Omega t}dt \tag{3.3}$$

For some signals whose amplitude may increase with time, their Fourier transform may not be converged. For example, the Fourier transform of $x(t) = e^{at}$, $a > 0$ does not exist.

In order to have more signals converged, the Laplace transform introduces an attenuation factor $e^{-\sigma t}$, in which σ is an arbitrary real number. Multiplying $x(t)$ by $e^{-\sigma t}$, we then take Fourier transform to the product $x_1(t) = e^{-\sigma t}x(t)$. Thus the convergence is easily satisfied, which is shown as

$$X_1(j\Omega) = \int_{-\infty}^{+\infty} e^{-\sigma t}x(t)e^{-j\Omega t}dt = \int_{-\infty}^{+\infty} x(t)e^{-(\sigma + j\Omega)t}dt \tag{3.4}$$

Taking $\sigma + j\Omega$ in eq. (3.4) as a new variable s, we get the definition of the bilateral Laplace transform.
Definition 3.1: The bilateral Laplace transform of a continuous-time signal $x(t)$ is given as

$$X(s) = \int_{-\infty}^{+\infty} x(t)e^{-st}dt \tag{3.5}$$

$$x(t) = \frac{1}{2\pi j} \int\limits_{\sigma - j\infty}^{\sigma + j\infty} X(s)e^{st}ds \tag{3.6}$$

Eqs. (3.5) and (3.6) are the bilateral Laplace transform and its inverse transform, respectively. They are usually denoted as

$$X(s) = \mathscr{L}[x(t)], \qquad x(t) = \mathscr{L}^{-1}[X(s)] \tag{3.7}$$

The bilateral Laplace transform is also represented as $x(t) \leftrightarrow X(s)$, where "$\leftrightarrow$" denotes the Laplace transform and its inverse transform. In order to distinguish between Fourier transform and Laplace transform, we can use "$\overset{\mathscr{L}}{\leftrightarrow}$" to show the Laplace transform. In Laplace transform, $s = \sigma + j\Omega$ is a complex variable.

2. Calculation based on the definition

Example 3.1: Try to calculate both Fourier transform and Laplace transform for $x(t) = e^{-at}u(t)$.

Solution: The Fourier transform is

$$X(j\Omega) = \int\limits_{-\infty}^{+\infty} x(t)e^{-j\Omega t}dt = \int\limits_{0}^{+\infty} e^{-at}e^{-j\Omega t}dt = \frac{1}{j\Omega + a}, \quad a > 0$$

The Laplace transform is

$$X(s) = \int\limits_{-\infty}^{+\infty} x(t)e^{-st}dt = \int\limits_{0}^{+\infty} e^{-(s+a)t}dt = \int\limits_{0}^{+\infty} e^{-(\sigma+a)t}e^{-j\Omega t}dt = \frac{1}{s+a}, \quad \mathrm{Re}[s] > -a$$

In the calculation of Laplace transform for the above example, we in fact take $\sigma + a > 0$ as a condition for its convergence. In Laplace transform, $\sigma + a > 0$ is usually represented as $\mathrm{Re}[s] > -a$, where $\mathrm{Re}[s]$ denotes the real part of variable s. The zeros are defined as the s values causing $X(s) = 0$, while poles are defined as the s values causing $X(s) \to \infty$. The figure representing the characteristics of Laplace transform in s-plane with zeros and poles is referred to as the pole–zero plot.

Example 3.2: Find the Laplace transform of $x(t) = -e^{-at}u(-t)$.

Solution: $X(s) = -\int_{-\infty}^{+\infty} e^{-at}e^{-st}u(-t)dt = \int_{-\infty}^{0} e^{-(s+a)t}dt = \frac{1}{s+a}$, $\mathrm{Re}[s] < -a$.

We see from the comparison between Examples 3.1 and 3.2 that different signals in either example have the same Laplace transform expression. However, the value range of s is different in both examples. In Example 3.1, the value range of s making Laplace transform converged is $\mathrm{Re}[s] > -a$, while such value range in Example 3.2 is $\mathrm{Re}[s] < -a$. Generally, the range of values of s for which the integral in eq. (3.5) converges is referred

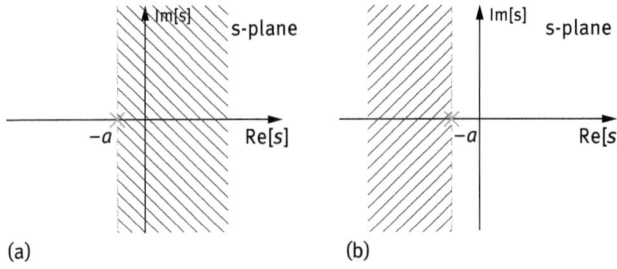

(a) (b)

Figure 3.1: The schematic diagrams of the Laplace transform: (a) ROC of Example 3.1 and (b) ROC of Example 3.2.

to as the region of convergence (ROC) of the Laplace transform. Figure 3.1 shows the schematic diagrams corresponding to Examples 3.1 and 3.2 in the s-plane, where the shadow parts are the ROCs of Laplace transforms. The symbol "×" denotes the poles of the transform. The Re[s] (or σ) is the real part of s, the Im[s] (or jΩ) represents the imaginary part of s.

3.2.2 Properties of the ROC for the Laplace Transform

The properties of the ROC of the Laplace transform $X(s)$ are listed as follows.
Property 3.1: The ROC of $X(s)$ consists of strips paralleled to the jΩ-axis in the s-plane.
Property 3.2: For rational Laplace transforms, the ROC does not contain any poles.
Property 3.3: If $x(t)$ is of finite duration and is absolutely integrable, then the ROC is the entire s-plane.
Property 3.4: If the Laplace transform $X(s)$ of $x(t)$ is rational, then its ROC is bounded by poles or extends to infinity. In addition, no poles of $X(s)$ are contained in the ROC.
Property 3.5: If the Laplace transform $X(s)$ of $x(t)$ is rational, then if $x(t)$ is right sided, the ROC is the region in the s-plane to the right of the rightmost pole. If $x(t)$ is left sided, the ROC is the region in the s-plane to the left of the leftmost pole.

The so-called rational Laplace transform above indicates that the Laplace transform $X(s)$ is a rational fractional function.

Example 3.3: Given $x(t) = \begin{cases} e^{-at}, & 0 < t < T \\ 0, & \text{othewise} \end{cases}$. Find the Laplace transform $X(s)$.

Solution: From the definition of Laplace transform, $X(s) = \int_0^T e^{-at} e^{-st} dt = \frac{1}{s+a}[1 - e^{-(s+a)T}]$.

Based on Property 3.3, the ROC of the Laplace transform of $x(t)$ should be an entire s-plane, since $x(t)$ is a finite duration signal. By analyzing $X(s)$ we see that there is a pole at $s = -a$, which is not in accordance with that property. In fact, there is a zero at the same place with the pole. We can use L'Hôpitals rule at $s = -a$ for the $\frac{0}{0}$ form and get

$$\lim_{s \to -a} X(s) = \lim_{s \to -a} \left\{ \frac{1}{s+a} [1 - e^{-(s+a)T}] \right\} = \lim_{s \to -a} T e^{-aT} e^{-sT} = T$$

As a result, Property 3.3 is still satisfied.

3.2.3 The Inverse Laplace Transform

The definition of the inverse Laplace transform is given again as follows:

$$x(t) = \frac{1}{2\pi j} \int_{\sigma - j\infty}^{\sigma + j\infty} X(s) e^{st} ds \tag{3.8}$$

It indicates that a signal $x(t)$ can be represented using a weighted integral of a complex exponential signal. Generally speaking, the calculation of eq. (3.6) needs a contour integral in the complex plane. We usually do not do that. The inverse Laplace transform is usually calculated by the partial fraction and residue methods.

1. The partial fraction for solving the inverse Laplace transform
The partial fraction method for solving the inverse Laplace transformation is suitable for such condition, where $X(s)$ is a rational function. The basic idea of this method is to expand the rational function into a combination of lower order terms, and then to find the inverse Laplace transform for each of the lower order terms.

Suppose that there is not any higher-order pole in $X(s)$, and the order of the numerator is not higher than that of the denominator, then $X(s)$ can be expanded as

$$X(s) = \sum_{i=1}^{m} X_i(s) = \sum_{i=1}^{m} \frac{A_i}{s + a_i} \tag{3.9}$$

Every ROC in $X_i(s)$ can be determined if the ROC of $X(s)$ is known. Every $x_i(t)$ can be obtained by taking the inverse Laplace transform of $X_i(s)$. In fact, there are two possibilities of the inverse Laplace transform for each $X_i(s) = A_i/(s + a_i)$: If the ROC is to the right of the pole at $s = -a_i$, then the inverse Laplace transform is $X_i(s) = A_i/(s + a_i) \leftrightarrow A_i e^{-a_i t} u(t) = x_i(t)$, a right-sided signal. If the ROC is to the left of the pole at $s = -a_i$, then the inverse Laplace transform is $X_i(s) = A_i/(s + a_i) \leftrightarrow -A_i e^{-a_i t} u(-t) = x_i(t)$, a left-sided signal. Summing all of $x_i(t)$ up, we get the inverse Laplace transform of $x(t)$ as

$$x(t) = \sum_{i=1}^{m} x_i(t) \tag{3.10}$$

Example 3.4: The Laplace transform is given as $X(s) = 1/((s+1)(s+2))$, $Re[s] > -1$.
Find $x(t)$.

Solution: Calculate the partial fraction of $X(s)$ as

$$X(s) = \frac{1}{(s+1)(s+2)} = \frac{A}{s+1} + \frac{B}{s+2}$$

The method of undetermined coefficients is used to determine the values of A and B.
We get

$$X(s) = \frac{1}{s+1} - \frac{1}{s+2}$$

For the first term $X_1(s) = 1/(s+1)$ of $X(s)$, the corresponding signal $x_1(t)$ is a
right-sided signal, since the ROC is to the right side of $s = -1$. Therefore, we
get $x_1(t) = \mathcal{L}^{-1}[X_1(s)] = e^{-t}u(t)$. For the second term $X_2(s) = -(1/(s+2))$ of $X(s)$,
the corresponding signal $x_2(t)$ is also a right-sided signal, since the ROC is to
the right side of $s = -2$, which is shown as $x_2(t) = \mathcal{L}^{-1}[X_2(s)] = -e^{-2t}u(t)$. Thus,
the inverse Laplace transform of $X(s)$ is as follows:

$$x(t) = x_1(t) + x_2(t) = [e^{-t} - e^{-2t}]u(t), \quad Re[s] > -1$$

If the Laplace transform $X(s)$ in Example 3.4 changes as $X(s) = 1/((s+1)$
$(s+2))$, $Re[s] < -2$, then its ROC is to the left side of both poles. The corresponding
inverse Laplace transform is as

$$x(t) = [-e^{-t} + e^{-2t}]u(-t), \quad Re[s] < -2$$

If $X(s)$ in Example 3.4 changes to $X(s) = (1/((s+1)(s+2)))$, $-2 < Re[s] < -1$,
then its ROC is to the left side of the pole at $s = -1$, and to the right side
of the pole at $s = -2$. The corresponding inverse Laplace transform is as
follows:

$$x(t) = -e^{-t}u(-t) - e^{-2t}u(t), \quad -2 < Re[s] < -1$$

Example 3.5: Find the inverse Laplace transform of $X(s) = \frac{s^2 + 4s + 7}{(s+1)(s+2)}$, $Re[s] > -1$.
Solution: $X(s)$ can be rewritten as $X(s) = 1 + \frac{4}{s+1} - \frac{3}{s+2}$. Taking the inverse Laplace
transform, we get $x(t) = \delta(t) + 4e^{-t}u(t) - 3e^{-2t}u(t)$.

2. The residue method for solving the inverse Laplace transform

For a causal signal $x(t)$ we have the following result according to the residue theorem
in the theory of complex function:

$$\frac{1}{2\pi j}\oint_C X(s)e^{st}ds = \sum_i \text{Res}_i \tag{3.11}$$

where the curve integral on the left side of the above equation is in the s-plane along the closed curve C bypassing all of the poles of $X(s)$. The right-hand side of the above equation represents the summation of all residues of $X(s)$ inside the curve C. Comparing eq. (3.11) and the definition of the inverse Laplace transform shown in eq. (3.6), we see that to take advantage of residue theorem to calculate the inverse Laplace transformation, we need to add an integral path to form a closed curve in the definition of eq. (3.6). A circular with an infinite radius is usually considered as shown in Figure 3.2.

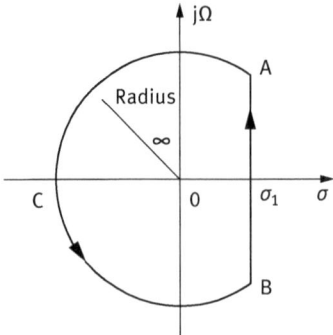

Figure 3.2: The integral path of $X(s)$.

The following equation should also be satisfied:

$$\int_{ACB} X(s)e^{st}ds = 0 \tag{3.12}$$

Therefore, the inverse Laplace transform of $X(s)$ is transformed into the residue operation, simplifying the calculation of the inverse Laplace.

If $X(s)$ is a rational function, its residue of the first-order pole $s = p_i$ is as follows:

$$\text{Res}_i = [(s - p_i)X(s)e^{st}]_{s = p_i} \tag{3.13}$$

The kth-order residue of $X(s)$ is shown as

$$\text{Res}_i = \frac{1}{(k-1)!}\left[\frac{d^{k-1}}{ds^{k-1}}(s-p_i)^k X(s)e^{st}\right]_{s = p_i} \tag{3.14}$$

We see from eqs. (3.13) and (3.14) that the residue method is the same as the partial fraction method. Moreover, the residue method can be used to calculate the inverse Laplace transform for not only the rational function but also the irrational function. It is important to note that we need to decompose $X(s)$ into a combination of a polynomial and a proper fraction (see Example 3.5).

Example 3.6: Find the inverse Laplace transform of $X(s) = \left((s+2)/\left(s(s+3)(s+1)^2\right)\right)$ with the residue method.

Solution: There are two first-order poles $p_1 = 0$, $p_2 = -3$, and one second-order pole $p_3 = -1$ in $X(s)$. The residues on every pole are calculated as

$$\mathrm{Res}_1 = [(s-p_1)X(s)e^{st}]_{s=p_1=0} = \left[\frac{s+2}{(s+3)(s+1)^2}e^{st}\right]_{s=p_1=0} = \frac{2}{3}$$

$$\mathrm{Res}_2 = \left[(s+3)\frac{s+2}{s(s+3)(s+1)^2}e^{st}\right]_{s=p_2=-3} = \frac{1}{12}e^{-3t}$$

$$\mathrm{Res}_3 = \frac{1}{(k-1)!}\left[\frac{d^{k-1}}{ds^{k-1}}(s-p_3)^k X(s)e^{st}\right]_{s=p_3=-1}$$

$$= \frac{1}{(2-1)!}\left[\frac{d}{ds}(s+1)^2\frac{s+2}{s(s+3)(s+1)^2}e^{st}\right]_{s=p_3=-1} = -\frac{1}{2}\left(t+\frac{3}{2}\right)e^{-t}$$

Thus the inverse Laplace transform is

$$x(t) = \sum_{i=1}^{3}\mathrm{Res}_i = \left[\frac{2}{3}+\frac{1}{12}e^{-3t}-\frac{1}{2}\left(t+\frac{3}{2}\right)e^{-t}\right]u(t)$$

3.2.4 Properties of the Laplace Transform and Basic Laplace Transform Pairs

1. Properties of the Laplace transform

Similar to the Fourier transform, the Laplace transform has a series of important properties. It is very important for us to study and master these properties, so as to understand the theory and solve the problems of the Laplace transformation. The main properties of the Laplace transform are presented in Table 3.1.

Example 3.7: Find the Laplace transform of $x(t) = te^{-at}u(t)$.

Solution: Since $e^{-at}u(t) \leftrightarrow (1/(s+a))$, $\mathrm{Re}[s] > -a$, based on the differentiation property of the Laplace transform, we get

Table 3.1: The main properties of the Laplace transform.

	Properties	Continuous-time signals $x(t)$, $x_1(t)$ $x_2(t)$	Laplace transforms $X(s)$, $X_1(s)$, $X_2(s)$	ROC R, R_1, R_2
1	Linearity	$ax_1(t) + bx_2(t)$	$aX_1(s) + bX_2(s)$	At least $R_1 \cap R_2$
2	Time scaling	$x(t - t_0)$	$e^{-st_0}X(s)$	R
3	Shift in s-domain	$e^{s_0 t}x(t)$	$X(s - s_0)$	Shifted version of R
4	Time scaling	$x(at)$	$\frac{1}{|a|}X(\frac{s}{a})$	$\frac{R}{a}$
5	Conjugation	$x^*(t)$	$X^*(s^*)$	R
6	Convolution	$x_1(t)*x_2(t)$	$X_1(s)X_2(s)$	At least $R_1 \cap R_2$
7	Differentiation	$\frac{dx(t)}{dt}$	$sX(s)$	At least R
8	Integration	$\int_{-\infty}^{t} x(\tau)d\tau$	$\frac{1}{s}X(s)$	At least $R_1 \cap \{\mathrm{Re}[s] > 0\}$
9	Differentiation in s-domain	$-tx(t)$	$\frac{d}{ds}X(s)$	R
10	Initial and final theorems	If $x(t) = 0$ for $t < 0$, and $x(t)$ contains no impulse or higher-order singularities at $t = 0$, then $x(0^+) = \lim\limits_{s\to\infty} sX(s)$, $\lim\limits_{t\to\infty} x(t) = \lim\limits_{s\to 0} sX(s)$		

$$te^{-at}u(t) \longleftrightarrow -\frac{d}{ds}\left[\frac{1}{s+a}\right] = \frac{1}{(s+a)^2}, \quad \mathrm{Re}[s] > -a$$

2. Basic Laplace transform pairs

The basic Laplace transform pairs are listed in Table 3.2.

3.3 The Analysis of Continuous-Time Signals and Systems in Complex Frequency Domain

Section 3.2 introduces the basic concept principle of the Laplace transform. As you can see, there are many similarities between Laplace transform and Fourier transform. In fact, the Laplace transform is in a sense a generalized Fourier transform, or the Fourier transform is a special case of the Laplace transform when $\sigma = 0$. Fourier transform is suitable for the analysis of the frequency characteristics of signals and systems, while the Laplace transform is usually used to solve the linear differential equation and to analyze the causality and stability of a system. This section will introduce with emphasis on the analysis of LTI systems using Laplace transform, and introduces the unilateral Laplace transform and its applications.

Table 3.2: Basic Laplace transform pairs.

	Signals	Expressions of $x(t)$	Laplace transform $X(s)$	ROC
1	Unit impulse signal	$\delta(t)$	1	Entire s-plane
2	Unit step signal	$u(t)$	$\frac{1}{s}$	Re$[s] > 0$
3	Reversed unit step signal	$-u(-t)$	$\frac{1}{s}$	Re$[s] < 0$
4	Single side power function	$\frac{t^{n-1}}{(n-1)!}u(t)$	$\frac{1}{s^n}$	Re$[s] > 0$
5	Reversed single side power function	$-\frac{t^{n-1}}{(n-1)!}u(-t)$	$\frac{1}{s^n}$	Re$[s] < 0$
6	Single side exponential signal	$e^{-at}u(t)$	$\frac{1}{s+a}$	Re$[s] > -a$
7	Reversed single side exponential signal	$-e^{-at}u(-t)$	$\frac{1}{s+a}$	Re$[s] < -a$
8	Derived from the spectrum of single side exponential signal	$\frac{t^{n-1}}{(n-1)!}e^{-at}u(t)$	$\frac{1}{(s+a)^n}$	Re$[s] > -a$
		$-\frac{t^{n-1}}{(n-1)!}e^{-at}u(-t)$	$\frac{1}{(s+a)^n}$	Re$[s] < -a$
9	Unit impulse with time shift	$\delta(t-T)$	e^{-sT}	Entire s plane
10	Single side cosine signal	$[\cos \Omega_0 t]u(t)$	$\frac{s}{s^2 + \Omega_0^2}$	Re$[s] > 0$
11	Single side sine signal	$[\sin \Omega_0 t]u(t)$	$\frac{\Omega_0}{s^2 + \Omega_0^2}$	Re$[s] > 0$
12	Single side growing/ decaying cosine signal	$[e^{-at}\cos \Omega_0 t]u(t)$	$\frac{s+a}{(s+a)^2 + \Omega_0^2}$	Re$[s] > -a$
13	Single side growing/ decaying sine signal	$[e^{-at}\sin \Omega_0 t]u(t)$	$\frac{\Omega_0}{(s+a)^2 + \Omega_0^2}$	Re$[s] > -a$

3.3.1 The Laplace Transform of Differential Equations and System Functions

In general, a linear time-invariant (LTI) system can be represented with the Nth-order linear constant coefficients differential equation shown as follows:

$$\sum_{k=0}^{N} a_k \frac{d^k y(t)}{dt^k} = \sum_{k=0}^{M} b_k \frac{d^k x(t)}{dt^k} \tag{3.15}$$

where M and N are the orders of the input and output terms of the equation. Generally, $N \geq M$ is satisfied. b_k and a_k represent the weight factors of the input and output terms, respectively. Taking Laplace transform to the above equation, we get

$$\left(\sum_{k=0}^{N} a_k s^k\right) Y(s) = \left(\sum_{k=0}^{M} b_k s^k\right) X(s) \tag{3.16}$$

The system function $H(s)$ is defined as

$$H(s) = \frac{Y(s)}{X(s)} = \frac{\sum_{k=0}^{M} b_k s^k}{\sum_{k=0}^{N} a_k s^k} \tag{3.17}$$

The system function described by a linear constant coefficients differential equation is always rational. From eq. (3.17) we can obtain the following information. First, the zeros and poles can be obtained by letting the numerator and denominator as zero, respectively; Second, this equation indicates the input and output relation of the LTI system. From eq. (3.17) we have

$$Y(s) = H(s)X(s) \tag{3.18}$$

where $X(s)$ and $Y(s)$ represent the Laplace transforms of the input and output signals of the system, respectively. In fact, eq. (3.18) is the convolution property of the Laplace transform. In time domain, it is written as

$$y(t) = h(t) \star x(t) \tag{3.19}$$

In the above equations, both $h(t)$ and $H(s)$ are a pair of Laplace transform. Third, if we set $\sigma = 0$ (or $s = j\Omega$) in $H(s)$, we can get the frequency response (or the transfer function) $H(j\Omega)$ of the system.

Example 3.8: A causal LTI system is described by the following differential equation:

$$\frac{dy(t)}{dt} + 5y(t) = 2x(t)$$

(1) Find its system function $H(s)$; (2) find its unit impulse response $h(t)$; and (3) determine the stability of the system.

Solution: (1) Taking the Laplace transform to both sides of the given equation, we have

$$sY(s) + 5Y(s) = 2X(s)$$

And then we get

$$H(s) = \frac{Y(s)}{X(s)} = \frac{2}{s+5}, \quad \mathrm{Re}[s] > -5$$

(2) Taking the inverse Laplace transform to $H(s)$, we get

$$h(t) = \mathscr{L}^{-1}[H(s)] = \mathscr{L}^{-1}\left[\frac{2}{s+5}\right] = 2e^{-5t}u(t)$$

(3) Since the unit impulse response $h(t)$ is causal and bounded, the system is stable.

3.3.2 Causality and Stability Analysis of LTI Systems

In Chapter 1, we introduced how to determine the stability of causality for a system in time domain. Here we will see that the causality and stability of LTI systems can be judged more effectively based on the Laplace transform.

1. The causality of the LTI system
A causal LTI system should satisfy the following relation:

$$h(t) = 0, \quad t < 0 \tag{3.20}$$

It means that the unit impulse response of a causal system should satisfy strictly the right-hand side condition. According to the discussion in Section 3.2.2, we see that the ROC of the system function $H(s)$ of $h(t)$ should be a right half of the s-plane.

Property 3.6 (necessary condition): The ROC associated with the system function for a causal system is a right-half s-plane.
Property 3.7 (necessary and sufficient condition): For a system with a rational function, the causality of the system is equivalent to the ROC being the right-half s-plane to the right of the rightmost pole.

Example 3.9: Try to determine the causality of following systems:
(1) $H_1(s) = \frac{1}{s+1}$, $\text{Re}[s] > -1$; (2) $H_2(s) = \frac{-2}{s^2+1}$, $-1 < \text{Re}[s] < 1$;
(3) $H_3(s) = \frac{e^s}{s+1}$, $\text{Re}[s] > -1$.
Solution:
   (1) Since the given system $H_1(s)$ is a rational system and its ROC is a right-half s-plane, we can conclude based on Property 3.7 that the system is a causal system.
   (2) The given system $H_2(s)$ is a rational system. However, it is not a causal system since its ROC is not a right-half s-plane. So the system is a noncausal system. We can get the same conclusion based on the unit impulse response.
   (3) Although the ROC of $H_3(s)$ is a right-half s-plane, it cannot be determined as a causal system since its system function is not a rational function. If we take the inverse Laplace transform to $H_3(s)$, we get $h_3(t) = e^{-(t+1)}u(t+1)$. Obviously, the system is a noncausal system.

2. The stability of the LTI system

The stability of a system can be determined in time domain. For an LTI system, the stability corresponds to the absolute integrability of its unit impulse response:

$$\int_{-\infty}^{+\infty} |h(\tau)| d\tau < \infty \tag{3.21}$$

We can also determine the stability of a system in s-domain.

Property 3.8: An LTI system is stable if and only if the ROC of its system function $H(s)$ includes the jΩ-axis.

Property 3.9: A causal system with rational system function $H(s)$ is stable if and only if all of the poles of $H(s)$ lie in the left half of the s-plane.

Example 3.10: Try to determine the stability of following systems:

1. $H_1(s) = \frac{s+2}{(s+1)(s-2)}$, $\text{Re}[s] > 2$; (2) $H_2(s) = \frac{1}{s+2}$, $\text{Re}[s] > -2$.

   Solution: (1) Since $H_1(s)$ is a causal system and its ROC is to the right of the rightmost pole $p = 2$ (see Figure 3.3(a)), the system is an unstable system.

2. Since $H_2(s)$ is a causal system and its ROC includes the jΩ-axis (see Figure 3.3(b)), the system is a stable system.

   From $h_1(t)$ and $h_2(t)$, the inverse Laplace transforms of $H_1(s)$ and $H_2(s)$, we can get the same conclusions.

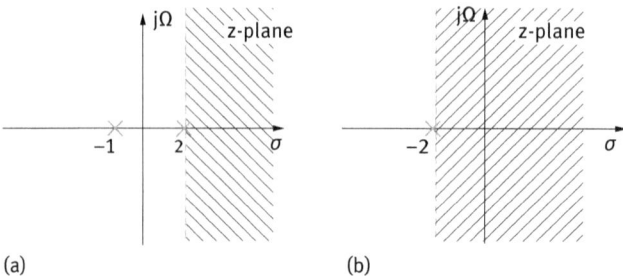

(a)   (b)

Figure 3.3: The ROC and zero–pole plot of (a) system $H_1(s)$ and (b) system $H_2(s)$.

3.3.3 Unilateral Laplace Transform and Its Applications

The Laplace transform defined in eq. (3.5) is, in fact, the bilateral Laplace transform. Its integral interval in the time domain is $(-\infty, +\infty)$, corresponding to the continuous-time signals defined in $(-\infty, +\infty)$. However, in nature and engineering practice many signals encountered are causal. For causal signals, we just need to calculate the integration between 0 and infinity in Laplace transform. That is the unilateral Laplace transform, which is useful to analyze the differential equations with nonzero initial conditions.

Definition 3.2: The unilateral Laplace transform for the continuous-time signal is defined as

$$X_u(s) = \int_0^{+\infty} x(t)e^{-st}dt \tag{3.22}$$

where 0^- denotes any impulse signals or the higher-order singularity signals are included inside the integral interval. Since the integration of the unilateral Laplace transform is always taken in the interval of $t > 0$, its ROC always corresponds to a right half of the s-plane.

Comparing the bilateral Laplace transform $X(s)$ with the unilateral Laplace transform $X_u(s)$, we see that both Laplace transforms are the same for causal signals. However, the bilateral and unilateral Laplace transforms are not the same for non-causal signals.

Example 3.11: A continuous-time signal is given as $x(t) = e^{-a(t+1)}u(t+1)$. Find its bilateral and unilateral Laplace transforms, respectively.

Solution: For bilateral Laplace transform, since $e^{-at}u(t) \leftrightarrow (1/(s+a))$, $\mathrm{Re}[s] > -a$, we get $e^{-a(t+1)}u(t+1) \leftrightarrow e^s/(s+a)$, $\mathrm{Re}[s] > -a$ by using the time shifting property. Thus

$$X(s) = \frac{e^s}{s+a}, \quad \mathrm{Re}[s] > -a$$

For unilateral Laplace transform, we have

$$X_u(s) = \int_{0^-}^{+\infty} e^{-a(t+1)}u(t+1)e^{-st}dt = \int_{0^-}^{+\infty} e^{-a}e^{-(s+a)t}dt = \frac{e^{-a}}{s+a}, \quad \mathrm{Re}[s] > -a$$

It is obvious that $X(s)$ and $X_u(s)$ are different.

The major properties of the unilateral Laplace transform are the same as those of the bilateral Laplace transform, such as the linearity, time scaling, shift in s-domain, conjugation, differentiation in s-domain, and the initial and final theorems. However, there are some different properties between the unilateral Laplace transform and the bilateral Laplace transform, such as differentiation and integration in the time domain. Table 3.3 shows the main properties of the unilateral Laplace transform, which are different from the bilateral Laplace transform.

The ROC of the unilateral Laplace transform is always a right half of the s-plane.

Table 3.3: Basic properties of the unilateral Laplace transform.

	Properties	Continuous-time signal $x(t)$	Unilateral Laplace transform $X_u(s)$
1	Differentiation in time domain	$\frac{dx(t)}{dt}$	$sX_u(s) - x(0^-)$
2	Higher-order differentiation in time domain	$\frac{d^k x(t)}{dt^k}$	$s^k X_u(s) - s^{k-1}x(0^-) - s^{k-2}x'(0^-) - \cdots - x^{(k-1)}(0^-)$
3	Integration in time domain	$\int\limits_{0^-}^{t} x(\tau)d\tau$	$\frac{1}{s}X_u(s)$

Example 3.12: Calculate the following unilateral Laplace transform:

(1) $x(t) = \delta(t) + e^t u(t)$;   (2) $x(t) = [e^{-t} - e^{-2t}]u(t)$.

Solution: (1) The unilateral Laplace transform of $x(t)$ is the same as its bilateral Laplace transform since it is a causal signal. $X_u(s)$ is shown as

$$X_u(s) = 1 + \frac{1}{s-1} = \frac{s}{s-1}, \quad \mathrm{Re}[s] > 1$$

(2) $x(t)$ here is also a causal signal. Its unilateral Laplace transform is shown as

$$X_u(s) = \frac{1}{(s+1)(s+2)}, \quad \mathrm{Re}[s] > -1$$

Example 3.13: Try to solve the linear constant coefficients differential equation with the non-zero initial condition by using the unilateral Laplace transform. The differential equation and initial conditions are given as

$$\frac{d^2 y(t)}{dt^2} + 3\frac{dy(t)}{dt} + 2y(t) = x(t), \quad y(0^-) = b, \quad y'(0^-) = c, \quad x(t) = au(t)$$

Solution: Taking the unilateral Laplace transform to both sides of the differential equation, we have

$$s^2 Y_u(s) - bs - c + 3sY_u(s) - 3b + 2Y_u(s) = \frac{a}{s}$$

After arrangement, we have

$$Y_u(s) = \frac{b(s+3)}{(s+1)(s+2)} + \frac{c}{(s+2)(s+2)} + \frac{a}{s(s+2)(s+2)}$$

The last term in the above equation, in fact, is the system response when its initial condition is zero, which is referred to as the zero-state response; while the first two terms on the right side of the above equation are actually the system response when its input is zero, which is known as the zero-input response. As a result, the total system response is the summation of the zero-state response and the zero-input response. We have

$$Y_u(s) = \frac{1}{s} - \frac{1}{s+1} + \frac{3}{s+2}, \quad \mathrm{Re}[s] > 0$$

Calculating the inverse Laplace transform, we get

$$y(t) = [1 - e^{-t} + 3e^{-2t}]u(t)$$

3.4 The z-Transform

The z-transform is a mathematical transformation of discrete-time signals. The role of the z-transform is the same as that of the Laplace transform in continuous-time signals and systems. The z-transform is an important tool for the analysis of LTI systems, and it is widely used in digital signal processing and other areas.

3.4.1 The Definition of the z-Transform

1. The definition of the z-transform
Definition 3.3: The z-transform of the discrete-time signal $x(n)$ is defined as

$$X(z) = \sum_{n=-\infty}^{+\infty} x(n)z^{-n} \tag{3.23}$$

Its inverse transform is defined as

$$x(n) = \frac{1}{2\pi j} \oint X(z)z^{n-1}dz \tag{3.24}$$

where z is a complex variable, and it is usually written as

$$z = re^{j\omega} \tag{3.25}$$

in which $r = |z|$ is the magnitude of z, and ω is its angle or phase. $\oint$ denotes the curve integral counterclockwise around a closed circle centered on the origin radius in r.

The bilateral z-transform can be denoted as

$$X(z) = \mathscr{Z}[x(n)] \tag{3.26}$$

or

$$x(n) = \mathscr{Z}^{-1}[X(z)] \tag{3.27}$$

The bilateral z-transform is also represented as $x(n) \leftrightarrow X(z)$ or $x(n) \overset{\mathscr{Z}}{\leftrightarrow} X(z)$.

2. The relations between z-transform and DTFT
Substituting eq. (3.25) into eq. (3.23), we get

$$X(re^{j\omega}) = \sum_{n=-\infty}^{+\infty} x(n)(re^{j\omega})^{-n} = \sum_{n=-\infty}^{+\infty} \{x(n)r^{-n}\}e^{-j\omega n} \tag{3.28}$$

We see from the above equation that $X(z) = X(re^{j\omega})$ is actually the DTFT of $x(n)$ multiplied by r^{-n}, which is shown as

$$X(re^{j\omega}) = \mathscr{F}\{x(n)r^{-n}\} \tag{3.29}$$

The real-valued exponential r^{-n} may be increased or decreased, depending on the value of r. If $r = 1$, then the z-transform reduces to the DTFT. Figure 3.4 shows the complex z-plane, where the abscissa axis represents the real part of z, expressed as $\text{Re}[\cdot]$, and the ordinate axis represents the imaginary part of z, expressed as $\text{Im}[\cdot]$. If $r = 1$ or $z = e^{j\omega}$, it corresponds to a circle in the z-plane, which is referred to as the unit circle, shown in Figure 3.4. Therefore, the DTFT in fact is a z-transform on the unit circle. We can also say that the z-transform is an expansion of the DTFT on the entire z-plane.

Similar to the Laplace transform, the z-transform has also its problem on the ROC.

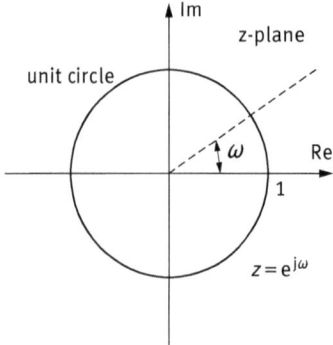

Figure 3.4: Complex z-plane and the unit circle.

3. Examples for calculating the z-transform

Example 3.14: Try to calculate the bilateral z-transform of signal $x(n) = a^n u(n)$.

Solution: Substituting $x(n) = a^n u(n)$ into the definition (3.23), we get

$$X(z) = \sum_{n=-\infty}^{+\infty} a^n u(n) z^{-n} = \sum_{n=0}^{+\infty} (az^{-1})^n$$

For convergence of $X(z)$, it is required that $\sum_{n=0}^{+\infty} |az^{-1}|^n < \infty$. Consequently, the ROC is the range of values of z for which $|az^{-1}| < 1$, or equivalently $|z| > |a|$. Then

$$X(z) = \sum_{n=0}^{+\infty} (az^{-1})^n = \frac{1}{1 - az^{-1}} = \frac{z}{z - a}, \quad |z| > |a|$$

where $|z| > |a|$ is the ROC of the z-transform of this signal.

In the given signal, if $a = 1$, then $x(n)$ becomes a unit step signal. Its z-transform is

$$X(z) = \frac{1}{1 - z^{-1}}, \quad |z| > 1$$

The same as that of the Laplace transform of continuous-time signals, the concepts of zeros and poles of the z-transform are also very important. A zero is a point in the z-plane that makes $X(z) = 0$, while a pole is a point in the z-plane that makes $X(z) \rightarrow \infty$. In the above example, $z = 0$ is a zero, and $z = a$ is a pole. Figure 3.5 shows the ROC and zero and pole on the z-plane.

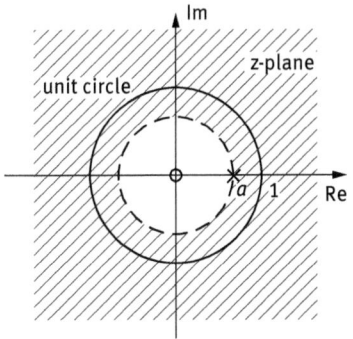

Figure 3.5: The ROC and zero and pole on the z-plane.

Example 3.15: Calculate the z-transform of $x(n) = -a^n u(-n-1)$.

Solution: Substituting $x(n)$ into the definition of z-transform, we have

$$X(z) = -\sum_{n=-\infty}^{+\infty} a^n u(-n-1)z^{-n} = -\sum_{n=-\infty}^{-1} a^n z^{-n} = -\sum_{n=1}^{+\infty} a^{-n} z^n = 1 - \sum_{n=0}^{+\infty} (a^{-1}z)^n$$

If $|a^{-1}z| < 1$ (or $|z| < |a|$) is satisfied, then $X(z)$ converges as

$$X(z) = 1 - \frac{1}{1-a^{-1}z} = \frac{1}{1-az^{-1}}, \quad |z| < |a|$$

Comparing the results between Examples 3.14 and 3.15, we see that both expressions of z-transform are the same, while their ROCs are different. So we need to pay a special attention to the ROC for the z-transform. When we try to express a result of the z-transform, we have to mark the ROC as well as its expression. A z-transform result is incomplete without its ROC. The pole–zero plot with the ROC is shown in Figure 3.6.

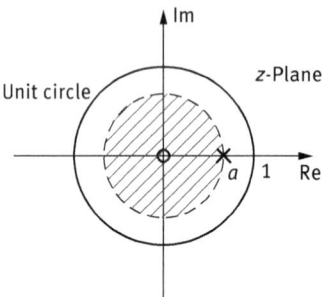

Figure 3.6: The pole–zero plot and the ROC on the z-plane.

3.4.2 The Properties of ROC for the z-Transform

The main properties of the ROC of $X(z)$ are presented in this section without proof.
Property 3.10: The ROC of $X(z)$ consists of a ring in the z-plane centered about the origin.
Property 3.11: The ROC does not contain any poles.
Property 3.12: If $x(n)$ is of finite duration, then the ROC is the entire z-plane, except possibly $z = 0$ and/or $z = \infty$.
Property 3.13: If the z-transform $X(z)$ is rational, then its ROC is bounded by poles or extends to infinity.
Property 3.14: If the z-transform $X(z)$ is rational and if $x(n)$ is right sided, then the ROC is the region in the z-plane outside the outermost pole. Furthermore, if $x(n)$ is causal, then the ROC also includes $z = \infty$.

Property 3.15: If the z-transform $X(z)$ is rational and if $x(n)$ is left sided, then the ROC is the region in the z-plane inside the innermost nonzero pole. Furthermore, if $x(n)$ is anticausal, then the ROC also includes $z = 0$.

Example 3.16: Suppose the z-transform of $x(n)$ is $X(z) = 1/\left((1 - \frac{1}{2}z^{-1})(1 - 2z^{-1})\right)$. (1) Try to analyze the zero–pole of $X(z)$; (2) plot the possible ROCs of $X(z)$.
Solution: (1) $X(z)$ has a second-order zero at $z = 0$, and it has two poles at $z = 1/2$ and $z = 2$, respectively.
(2) There are three possibilities for the ROCs of $X(z)$, since it has two different poles, including $|z| > 2$; $|z| < 1/2$, and $1/2 < |z| < 2$. The ROCs are plotted in Figure 3.7.

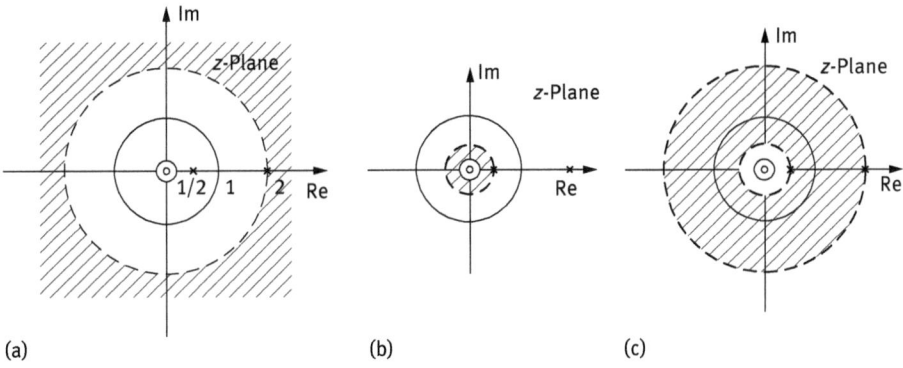

(a)   (b)   (c)

Figure 3.7: Different ROCs. (a) $|z| > 2$; (b) $|z| < 1/2$; and (c) $1/2 < |z| < 2$.

3.4.3 The Inverse z-Transform

The definition of the inverse z-transform is given in eq. (3.24). We usually do not calculate the inverse z-transform with that definition. A few methods for calculating the inverse z-transform are introduced in this section.

1. The partial fraction for solving the inverse z-transform
Suppose that the z-transform $X(z)$ of $x(n)$ can be expressed as a linear combination of the first-order terms as

$$X(z) = \sum_{i=1}^{m} \frac{A_i}{1 - a_i z^{-1}} \tag{3.30}$$

where the inverse z-transform of every term can be obtained directly. In fact, the inverse z-transform of every term in eq. (3.30) may have two different forms as follows: if the ROC is outside the pole $z = a_i$, then the corresponding inverse z-transform is

$x_i(n) = \mathscr{Z}^{-1}\{A_i/(1 - a_i z^{-1})\} = A_i a_i^n u(n)$; if the ROC is inside the pole $z = a_i$, then the corresponding inverse z-transform is $x_i(n) = \mathscr{Z}^{-1}\{A_i/(1 - a_i z^{-1})\} = -A_i a_i^n u(-n-1)$.

In fact, the calculation with the partial fraction method for the inverse z-transform is not limited in the combination of the first-order terms. It can also be used for other forms of z-transforms.

Example 3.17: Find the inverse z-transform of $X(z) = \dfrac{3 - \frac{5}{6}z^{-1}}{(1 - \frac{1}{4}z^{-1})(1 - \frac{1}{3}z^{-1})}$, $\quad |z| > \frac{1}{3}$.

Solution: Two poles $z = 1/3$ and $z = 1/4$ are associated with $X(z)$. Since $X(z)$ is a rational function and its ROC is outside the outermost pole, the corresponding $x(n)$ is a right-sided signal. Expanding $X(z)$ as $X(z) = \dfrac{1}{1 - \frac{1}{4}z^{-1}} + \dfrac{2}{1 - \frac{1}{3}z^{-1}}$, and taking the

inverse z-transform to every term of $X(z)$, we have $x(n) = \left(\frac{1}{4}\right)^n u(n) + 2\left(\frac{1}{3}\right)^n u(n)$.

Example 3.18: Find the inverse z-transform of $X(z)$. The expression of $X(z)$ is the same as that in Example 3.17, but its ROC is (1) $|z| < \frac{1}{4}$, or (2) $\frac{1}{4} < |z| < \frac{1}{3}$.

Solution: (1) The discrete-time signal $x(n)$ is a left-sided signal since the ROC of $X(z)$ is inside the innermost pole. Then

$$x(n) = -\left(\frac{1}{4}\right)^n u(-n-1) - 2\left(\frac{1}{3}\right)^n u(-n-1)$$

(2) Since the ROC of $X(z)$ is outside the pole $z = 1/4$, and inside the pole $z = 1/3$, the corresponding signal should be a two-sided signal, where the signal corresponding to $z = 1/4$ is a right-sided signal, while the signal corresponding to $z = 1/3$ is a left-sided signal. That is

$$x(n) = \left(\frac{1}{4}\right)^n u(n) - 2\left(\frac{1}{3}\right)^n u(-n-1)$$

2. The residue method for solving the inverse z-transform

The inverse z-transform can be expressed as a summation of residues on every pole based on the residue theory:

$$x(n) = \frac{1}{2\pi j} \oint X(z) z^{n-1} dz = \sum_m \text{Res}[X(z)z^{n-1}]_{z = z_m} \qquad (3.31)$$

where Res denotes the residue of the pole, and z_m is the pole of $X(z)z^{n-1}$.

If $X(z)z^{n-1}$ has a first-order pole at $z = z_m$, then its residue is

$$\text{Res}[X(z)z^{n-1}]_{z = z_m} = [(z - z_m)X(z)z^{n-1}]_{z = z_m} \qquad (3.32)$$

If $X(z)z^{n-1}$ has a kth-order pole at $z = z_m$, then its residue is

$$\text{Res}[X(z)z^{n-1}]_{z=z_m} = \frac{1}{(k-1)!}\left[\frac{d^{k-1}}{dz^{k-1}}(z-z_m)^k X(z)z^{n-1}\right]_{z=z_m} \qquad (3.33)$$

Example 3.9: Find the inverse z-transform of $X(z) = \frac{1}{(1-z^{-1})(1-\frac{1}{2}z^{-1})}$, $\quad |z| > 1$.

Solution: Since $X(z)$ has two first-order poles at $z_1 = 1$ and $z_2 = 1/2$, we have

$$\text{Res}[X(z)z^{n-1}]_{z=z_1} = \left[(z-0.5)\frac{1}{(1-z^{-1})(1-\frac{1}{2}z^{-1})}z^{n-1}\right]_{z=1} = 2$$

$$\text{Res}[X(z)z^{n-1}]_{z=z_2} = \left[(z-0.5)\frac{1}{(1-z^{-1})(1-\frac{1}{2}z^{-1})}z^{n-1}\right]_{z=0.5} = -\left(\frac{1}{2}\right)^n$$

As a result,

$$x(n) = \left[2-\left(\frac{1}{2}\right)^n\right]u(n)$$

3. The power series method for solving the inverse z-transform

Since the z-transform of $x(n)$ is in fact the power series of z^{-1}, we can get the inverse z-transform from the coefficients of the power series expansion of $X(z)$ inside its ROC.

Suppose that $X(z)$ is a rational function. $X(z)$ can be written as a ratio of the numerator $N(z)$ and the denominator $D(z)$ as

$$X(z) = \frac{N(z)}{D(z)} \qquad (3.34)$$

If the ROC of $X(z)$ satisfies $|z| > R_1$, then $x(n)$ is a causal signal, and both $N(z)$ and $D(z)$ can be sorted in the decreasing order of z (or perhaps in the increasing order of z^{-1}). If the ROC of $X(z)$ satisfies $|z| < R_2$, then $x(n)$ is a left-sided signal, and both $N(z)$ and $D(z)$ can be sorted in the increasing order of z (or in the decreasing order of z^{-1}), in which R_1 and R_2 represent the circles corresponding to the borders of the ROCs. By using the long division, we can get $x(n)$.

Example 3.20: Try to find the inverse z-transform by using the long division of $X(z) = 1/(1 - az^{-1})$, $\quad |z| > |a|$.

Solution: The long division of $X(z)$ is given as

$$1 - az^{-1} \overline{\smash{\big)}\, 1} \quad \dfrac{1 + az^{-1} + a^2z^{-2} + \cdots}{}$$

$$\underline{1 - az^{-1}}$$
$$az^{-1}$$
$$\underline{az^{-1} - a^2z^{-2}}$$
$$a^2z^{-2}$$
$$\vdots$$

Then $X(z)$ can be rewritten as

$$X(z) = \frac{1}{1 - az^{-1}} = 1 + az^{-1} + a^2z^{-2} + \cdots$$

The series above is converged since $|az^{-1}| < 1$. Comparing the above series with the definition of the z-transform eq. (3.23), we get $x(n) = 0$, if $n < 0$, and $x(0) = 1$, $x(1) = a$, $x(2) = a^2$, ... Thus

$$x(n) = a^n u(n)$$

3.4.4 Properties of the z-Transform and Basic z-Transform Pairs

1. Properties of the z-transform

The main properties of the z-transform are listed in Table 3.4.

Example 3.21: Given the z-transform $\frac{1}{1-z^{-1}}$ of the unit step signal $u(n)$, find the z-transform of $x(n) = nu(n)$.

Solution: From the property of differentiation in the z-domain, we have

$$\mathcal{Z}[nu(n)] = -z\frac{d}{dz}\mathcal{Z}[u(n)] = -z\frac{d}{dz}\left(\frac{1}{1-z^{-1}}\right) = \frac{z^{-1}}{(1-z^{-1})^2}$$

Example 3.22: Find the inverse z-transform of $X(z) = \ln(1 + az^{-1})$, $|z| > |a|$.

Solution: From the property of differentiation in the z-domain, we have

$$\mathcal{Z}[nx(n)] = -z\frac{d}{dz}X(z) = \frac{az^{-1}}{1+az^{-1}}, \quad |z| > |a|$$

Since $a(-a)^n u(n) \leftrightarrow \frac{a}{1+az^{-1}}$, $|z| > |a|$, we have the following result by using the time shift property:

$$a(-a)^{n-1}u(n-1) \leftrightarrow \frac{az^{-1}}{1+az^{-1}}, \quad |z| > |a|$$

Table 3.4: The main properties of the z-transform.

	Properties	Discrete-time signals $x(n)$, $x_1(n)$ $x_2(n)$	z-Transform $X(z)$, $X_1(z)$, $X_2(z)$	ROC R, R_1, R_2
1	Linearity	$ax_1(n) + bx_2(n)$	$aX_1(z) + bX_2(z)$	At least $R_1 \cap R_2$
2	Time shifting	$x(n - n_0)$	$z^{-n_0}X(z)$	R, except for the possible addition or deletion of the origin
3	Scaling in the z-domain	$e^{j\omega_0 n}x(n)$	$X(e^{-j\omega_0}z)$	R
		$z_0^n x(n)$	$X\left(\frac{z}{z_0}\right)$	$\lvert z_0 \rvert R$
		$a^n x(n)$	$X(a^{-1}z)$	Scaled version of R
4	Time reversal	$x(-n)$	$X(z^{-1})$	R^{-1}
5	Time expansion	$x_{(k)}(n) = \begin{cases} x(r), & n = rk \\ 0, & n \neq rk \end{cases}$	$X(z^k)$	$R^{1/k}$
6	Conjugation	$x^*(n)$	$X^*(z^*)$	R
7	Convolution	$x_1(n)*x_2(n)$	$X_1(z)X_2(z)$	At least $R_1 \cap R_2$
8	First difference	$x(n) - x(n-1)$	$(1-z^{-1})X(z)$	At least $R_1 \cap \{\lvert z \rvert > 0\}$
9	Accumulation	$\sum\limits_{k=-\infty}^{n} x(k)$	$\frac{1}{1-z^{-1}}X(z)$	At least $R_1 \cap \{\lvert z \rvert > 1\}$
10	Differentiation in the z-domain	$nx(n)$	$-z\frac{dX(z)}{dz}$	R
11	Initial theorem	$x(n) = 0$, if $n < 0$	$x(0) = \lim\limits_{z\to\infty} X(z)$	
12	Final theorem		$\lim\limits_{n\to\infty} x(n) = \lim\limits_{z\to 1}(z-1)X(z)$	

Thus

$$x(n) = \frac{-(-a)^n}{n} u(n-1)$$

2. Basic z-transform pairs
The basic z-transform pairs are listed in Table 3.5.

3.5 The Analysis of Discrete-Time Signals and Systems in Complex Frequency Domain

Section 3.4 introduced the basic concept and principle of the z-transform. As we can see, there are some similarities between the z-transform and the DTFT. In fact, the z-transform is an extension of the DTFT, or the DTFT is a special case of the z-transform when $\lvert z \rvert = 1$. For discrete-time system analysis, the

Table 3.5: Basic z-transform pairs.

	Signals	$x(n)$	$X(z)$	ROC
1	Unit impulse signal	$\delta(n)$	1	Entire z-plane
2	Unit step signal	$u(n)$	$\frac{1}{1-z^{-1}}$	$\lvert z \rvert > 1$
3	Reversed unit step signal	$-u(-n-1)$	$\frac{1}{1-z^{-1}}$	$\lvert z \rvert < 1$
4	Unit impulse with time shift	$\delta(n-m)$	z^{-m}	Entire z-plane, except 0 or ∞
5	Single side exponential signal	$a^n u(n)$	$\frac{1}{1-az^{-1}}$	$\lvert z \rvert > \lvert a \rvert$
6	Reversed single side exponential signal	$-a^n u(-n-1)$	$\frac{1}{1-az^{-1}}$	$\lvert z \rvert < \lvert a \rvert$
7	Derived from the spectrum of single side exponential signal	$na^n u(n)$	$\frac{az^{-1}}{(1-az^{-1})^2}$	$\lvert z \rvert > \lvert a \rvert$
		$-na^n u(-n-1)$	$\frac{az^{-1}}{(1-az^{-1})^2}$	$\lvert z \rvert < \lvert a \rvert$
8	Single side cosine signal	$[\cos \omega_0 n]u(n)$	$\frac{1-[\cos \omega_0]z^{-1}}{1-[2\cos \omega_0]z^{-1}+z^{-2}}$	$\lvert z \rvert > 1$
9	Single side sine signal	$[\sin \omega_0 n]u(n)$	$\frac{[\sin \omega_0]z^{-1}}{1-[2\cos \omega_0]z^{-1}+z^{-2}}$	$\lvert z \rvert > 1$
10	Single side growing/ decaying cosine signal	$[r^n \cos \omega_0 n]u(n)$	$\frac{1-[r\cos \omega_0]z^{-1}}{1-[2r\cos \omega_0]z^{-1}+r^2z^{-2}}$	$\lvert z \rvert > r$
11	Single side growing/ decaying sine signal	$[r^n \sin \omega_0 n]u(n)$	$\frac{[r\sin \omega_0]z^{-1}}{1-[2r\cos \omega_0]z^{-1}+r^2z^{-2}}$	$\lvert z \rvert > r$

DTFT is suitable for the frequency analysis of signals and systems, while the z-transform can be used to analyze the causality and stability of systems as well as the frequency characteristics. In this section, we will introduce the system analysis with z-transform, and introduce the unilateral z-transform briefly.

3.5.1 The z-Transform of Difference Equations and System Functions

The general form difference equation of the discrete-time LTI system is shown as

$$\sum_{k=0}^{N} a_k y(n-k) = \sum_{k=0}^{M} b_k x(n-k) \tag{3.35}$$

where M and N are the order of the input and output terms, respectively, b_k and a_k denote the weight coefficients of the input and output terms, respectively. Taking the z-transform to both sides of the equation, we get

$$\sum_{k=0}^{N} a_k z^{-k} Y(z) = \sum_{k=0}^{M} b_k z^{-k} X(z) \tag{3.36}$$

The system function $H(z)$ is defined as

$$H(z) = \frac{Y(z)}{X(z)} = \frac{\sum_{k=0}^{M} b_k z^{-k}}{\sum_{k=0}^{N} a_k z^{-k}} \tag{3.37}$$

The system function of an LTI system represented by a linear constant coefficient difference equation is always rational. Based on the system function, we can get further information:

First, the zeros and poles of the system can be obtained by the numerator and denominator of eq. (3.37). Moreover, we can determine the causality and stability of the system based on the poles and ROCs.

Second, we can get the following relation:

$$Y(z) = H(z)X(z) \tag{3.38}$$

It is in fact the convolution property of the z-transform. In the time domain we have the convolution relation as

$$y(n) = h(n) * x(n) \tag{3.39}$$

The convolution property changes the convolution between the input signal and the system in the time domain into the multiplication in the z-domain, reducing the calculation complexity significantly.

Third, we can get the frequency response (or the transfer function) of the discrete-time system by letting $|z| = 1$ (or $z = e^{j\omega}$) in $H(z)$ for further frequency analysis.

Example 3.23: The difference equation of a discrete-time causal system is given as $y(n) - by(n-1) = x(n)$. Suppose that the input signal is $x(n) = a^n u(n)$ with a rest initial condition. (1) Find the system function $H(z)$. (2) Find the unit impulse response $h(n)$. (3) Find the output signal $y(n)$. (4) Analyze the stability of the system.

Solution: (1) Taking the z-transform to both sides of the given equation, we get

$$Y(z) - bz^{-1}Y(z) = X(z)$$

Furthermore, we get the system function as

$$H(z) = \frac{Y(z)}{X(z)} = \frac{1}{1 - bz^{-1}}, \quad |z| > |b|$$

(2) Taking the inverse z-transform to $H(z)$, and concerning its causality, we have

$$h(n) = \mathscr{Z}^{-1}\left\{\frac{1}{1-bz^{-1}}\right\} = b^n u(n)$$

(3) From $X(z) = \mathscr{Z}\{a^n u(n)\} = \frac{1}{1-az^{-1}}$, $|z| > |a|$, we have

$$Y(z) = H(z)X(z) = \frac{1}{(1-bz^{-1})(1-az^{-1})} = \frac{1}{a-b}\left[\frac{a}{1-az^{-1}} - \frac{b}{1-bz^{-1}}\right], |z| > \max(|a|, |b|)$$

Taking the inverse z-transform to $Y(z)$, if $a \neq b$ is satisfied, we have

$$y(t) = \mathscr{Z}^{-1}\left\{\frac{1}{a-b}\left[\frac{a}{1-az^{-1}} - \frac{b}{1-bz^{-1}}\right]\right\} = \frac{1}{a-b}[a^{n+1} - b^{n+1}]u(n), \quad a \neq b$$

If $a = b$, then $Y(z) = H(z)X(z) = \frac{1}{(1-az^{-1})^2}$, $|z| > |a|$ is obtained. We need to take the z-transform for such expression again. That is suggested to be done by the readers.

(4) Under the condition of $a \neq b$, if $|b| < 1$ is satisfied, the system is a stable system based on its unit impulse response $h(n) = b^n u(n)$; if $|b| > 1$ is satisfied, then the system is an unstable system; if $|b| = 1$, the system is referred to as a critical stable system.

3.5.2 Causality and Stability Analysis of LTI Systems

The concepts of the causality and stability of the system have been introduced in Chapter 1. In this section, we will see that the causality and stability can be determined more efficiently with the system function $H(z)$ in the z-domain.

1. The causality of the discrete-time LTI system
A discrete-time causal LTI system should satisfy

$$h(n) = 0, \quad n < 0 \tag{3.40}$$

According to the discussion in Section 3.4.2 on the convergence of the z-transform, we learnt that the ROC of a causal system $H(z)$ (corresponding to $h(n)$) is the exterior of a circle in the z-plane. Then we have following properties:

Property 3.16: A discrete-time LTI system is causal if and only if the ROC of its system function is the exterior of a circle, including infinity.

Property 3.17: A discrete-time LTI system with rational system function $H(z)$ is causal if and only if (a) the ROC is the exterior of a circle outside the outermost pole and (b) with

$H(z)$ expressed as a ratio of polynomials in z, the order of the numerator cannot be greater than the order of the denominator.

Example 3.24: Try to determine the causality of the given system $H(z) = \frac{z^3 + 3z^2 + 2z + 1}{z^2 + 5z + 2}$.
Solution: From the expression of $H(z)$ we see that the order of the numerator is greater than that of the denominator. According to Property 3.17, we conclude that the system is not causal.

Example 3.25: Determine the causality of the system $H(z) = \frac{2z^2 - \frac{5}{2}z}{z^2 - \frac{5}{2}z + 1}$, $\quad |z| > 2$.

Solution: There are two poles for the system, that is, $z_1 = \frac{1}{2}$, $z_2 = 2$. Since the ROC of the system is outside the outermost pole, and the order of the numerator of $H(z)$ is not greater than that of the denominator, the system is a causal system.

2. The stability of the LTI system

To determine the stability of an LTI system in the time domain, the following absolute summable condition needs to be satisfied:

$$\sum_{k=-\infty}^{+\infty} |h(k)| < \infty \tag{3.41}$$

In the z-transform domain, the stability can be determined by the system function $H(z)$.
Property 3.18: An LTI system is stable if and only if the ROC of its system function $H(z)$ includes the unit circle $|z| = 1$.
Property 3.19: A causal LTI system with rational system function $H(z)$ is stable if and only if the poles of $H(z)$ lie inside the unit circle.

Example 3.26: Determine the stability of the system $H(z) = \frac{2z^2 - \frac{5}{2}z}{z^2 - \frac{5}{2}z + 1}$, $\quad |z| > 2$.

Solution: Since the ROC of the system does not contain the unit circle, the system is not stable.

Example 3.27: Try to determine the stability condition for the causal system $H(z) = \frac{1}{1 - az^{-1}}$, $\quad |z| > |a|$.
Solution: We see from the system function that there is one pole at $z = a$. In order to have the system stable, the pole has to be inside the unit circle.

3.5.3 The Block Diagram Representation for Discrete-Time Systems

The input and output signals of an LTI system can be connected by its unit impulse response. The general form of the system block diagram is shown in Figure 3.8.

Figure 3.8 demonstrates the block diagrams for both continuous-time and discrete-time LTI systems in time and transform domains. This section mainly introduces the block diagram in transform domain for discrete-time LTI systems.

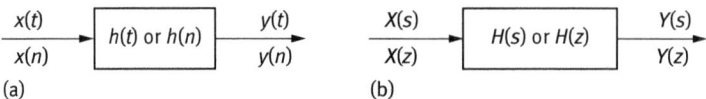

Figure 3.8: The block diagram of an LTI system: (a) in time domain and (b) in transform domain.

1. Basic operations of systems

The system block diagram is based on the basic operations of signals and systems. Generally speaking, there are three kinds of signal operations, including the addition, the multiplication by the coefficient, and the unit delay. The block diagrams of the three operations are given in Figure 3.9.

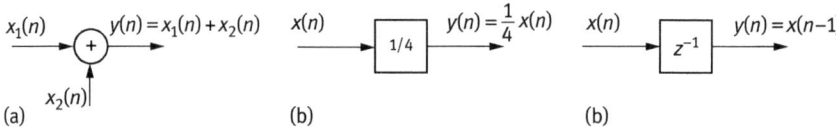

Figure 3.9: Three kinds of signal operations: (a) addition operation; (b) multiplication by coefficient; and (c) unit delay.

2. Interconnection of systems

There are three kinds of interconnection forms for the LTI system, including the cascade, the parallel, and the feedback. Figure 3.10 shows the block diagrams for the three kinds of interconnection.

Suppose the entire system function after the interconnection in Figure 3.10 is $H(z) = Y(z)/X(z)$. For the cascade system in Figure 3.10(a), we have

$$H(z) = H_1(z) \cdot H_2(z) \tag{3.42}$$

For the parallel system shown in Figure 3.10(b), its system function is

$$H(z) = H_1(z) + H_2(z) \tag{3.43}$$

For the feedback system shown in Figure 3.10(c), its system function is

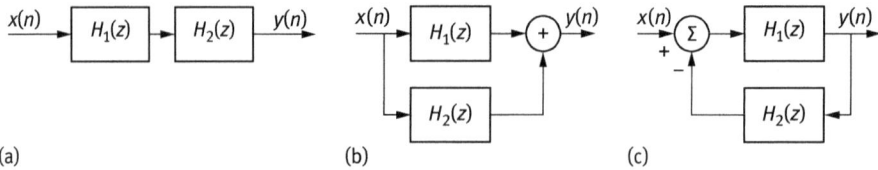

Figure 3.10: Three kinds of interconnection: (a) cascade; (b) parallel; and (c) feedback.

$$H(z) = \frac{H_1(z)}{1 + H_1(z)H_2(z)} \qquad (3.44)$$

where $H_1(z)$ is the forward path and $H_2(z)$ is the feedback path.

3. Block diagrams of the first-order and second-order systems

Suppose that a causal LTI system is given as $H(z) = 1/(1 + az^{-1})$, its system block diagram is shown in Figure 3.11, where $H_1(z) = 1$ and $H_2(z) = az^{-1}$.

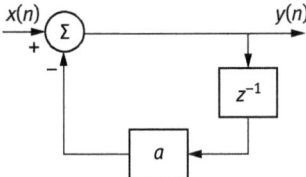

Figure 3.11: The block diagram of a first-order feedback system.

The system function of a causal LTI system is given as $H(z) = (1 + bz^{-1})/(1 + az^{-1})$. This system can be represented as the cascade of a feedback system $1/(1 + az^{-1})$ and a forward system $1 + bz^{-1}$, which is shown in Figure 3.12.

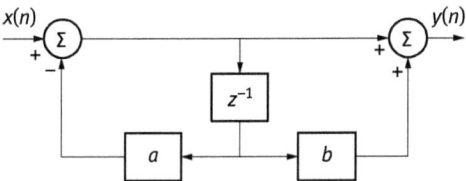

Figure 3.12: A first-order feedback and forward system.

If a second-order system is given as $H(z) = 1/(1 + a_1 z^{-1} + a_2 z^{-2})$, then its system block diagram is shown as Figure 3.13.

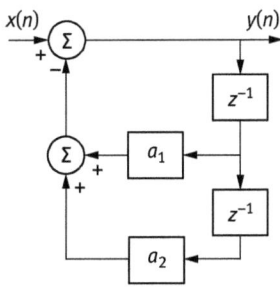

Figure 3.13: A second-order feedback system.

Suppose a second-order causal LTI system is shown as $H(z) = (1 + b_1 z^{-1} + b_2 z^{-2})/(1 + a_1 z^{-1} + a_2 z^{-2})$. This system can be represented as a cascade of a second-order feedback system $1/(1 + a_1 z^{-1} + a_2 z^{-2})$ and a second-order forward system $1 + b_1 z^{-1} + b_2 z^{-2}$ is shown in Figure 3.14.

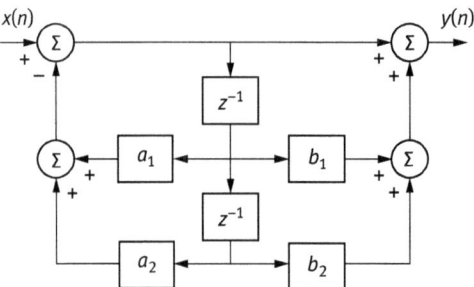

Figure 3.14: A second-order feedback and forward system.

It is important to note that in the above diagrams, both input and feedback signals are operated in a subtraction. If we change the subtraction into the addition operation, then the polarity of the coefficient in each feedback branch should also be changed.

The system block diagrams given above are all in the direct form. As a matter of fact, the system block diagram can also be realized in different ways, including the cascade form and the parallel form. Readers are suggested to refer to relevant references.

3.5.4 Unilateral z-Transform and Its Applications

The definition of the z-transform given in eq. (3.23), in fact, is the bilateral z-transform, whose summation interval is $(-\infty, +\infty)$. Here we give the definition of the unilateral z-transform.

Definition 3.4: The unilateral z-transform of the discrete-time signal $x(n)$ is defined as

$$X_u(z) = \sum_{n=0}^{+\infty} x(n) z^{-n} \tag{3.45}$$

It is denoted as

$$X_u(z) = \mathscr{U}\mathscr{L}[x(n)] \tag{3.46}$$

The summation interval of the unilateral z-transform is $[0, +\infty)$, no matter whether $x(n)$ is zero or not when $n < 0$. Due to the unilateral z-transform is always calculating the z-transform for the causal part of a signal, its ROC is always an exterior of a circle on the z-plane. In addition, the calculation of the unilateral inverse z-transform is basically the same as that of the bilateral z-transform.

Example 3.28: Suppose that the discrete-time signal is $x(n) = \left(\frac{1}{2}\right)^{n+1} u(n+1)$. (1) Find its bilateral z-transform; (2) find its unilateral z-transform.
Solution: (1) According to the definition and properties of the bilateral z-transform, we have

$$X(z) = \frac{z}{1 - \frac{1}{2}z^{-1}}, \quad |z| > \frac{1}{2}$$

(2) According to the definition of the unilateral z-transform, we have

$$X_u(z) = \sum_{n=0}^{+\infty} x(n)z^{-n} = \sum_{n=0}^{+\infty} \left(\frac{1}{2}\right)^{n+1} z^{-n} = \frac{1/2}{1 - \frac{1}{2}z^{-1}}, \quad |z| > \frac{1}{2}$$

It is obvious that the results are not the same.

Many properties of the unilateral z-transform are the same as those of the bilateral z-transform, such as the linearity, the conjugation, the convolution, the scaling in the z-domain, and the differentiation in the z-domain, and so on. However, there are some distinct properties in the unilateral z-transform, as shown in Table 3.6.

Table 3.6: Distinct properties of the unilateral z-transform.

	Properties	Discrete-time signal $x(n)$	Unilateral z-transform $X_u(s)$
1	Time delay	$x(n-1)$	$z^{-1}X_u(z) + x(-1)$
2	Time advance	$x(n+1)$	$zX_u(z) - zx(0)$
3	First-order difference	$x(n) - x(n-1)$	$(1-z^{-1})X_u(z) - x(-1)$

The reason that the ROC of the unilateral z-transform is not shown in Table 3.6 is that the ROC of the unilateral z-transform is always the exterior of a circle on the z-plane. The unilateral z-transform is a powerful tool to solve the difference equation with nonzero initial conditions.

Example 3.29: The difference equation of a causal system is shown as $y(n) + 3y(n-1) = x(n)$ with the initial condition $y(-1) = 1$. The input signal of the system is $x(n) = 8u(n)$. Try to find the output signal of the system.
Solution: Taking the unilateral z-transform to both sides of the difference equation, and using the linearity and time scaling properties, we have

$$Y_u(z) + 3 \times 1 + 3z^{-1}Y_u(z) = \frac{8}{1-z^{-1}}$$

Solving the equation for $Y_u(z)$, we get $Y_u(z) = -\frac{3 \times 1}{1+3z^{-1}} + \frac{8}{(1+3z^{-1})(1-z^{-1})}$
$= \frac{3}{1+3z^{-1}} + \frac{2}{1-z^{-1}}$. Taking the unilateral inverse z-transform to $Y_u(z)$, we get

$$y(n) = [3(-3)^n + 2]u(n)$$

Exercises

3.1 Try to explain the characteristics of the Laplace transform.

3.2 Try to explain the characteristics of the unilateral Laplace transform.

3.3 Compare the bilateral Laplace transform and the unilateral Laplace transform.

3.4 Explain the characteristics of the z-transform.

3.5 Explain the characteristics of the unilateral z-transform.

3.6 Compare the bilateral z-transform and the unilateral z-transform.

3.7 Calculate the following Laplace transforms, determine their ROC, and plot the zero–pole plot.
(1) $x(t) = e^{-2t}u(t) + e^{-3t}u(t)$; (2) $x(t) = e^{-4t}u(t) + e^{-5t}(\sin 5t)u(t)$;
(3) $x(t) = e^{2t}u(-t) + e^{3t}u(-t)$;(4) $x(t) = te^{2|t|}$; (5) $x(t) = |t|e^{-2|t|}$;
(6) $x(t) = |t|e^{2t}u(-t)$.

3.8 Find the following inverse Laplace transforms:
(1) $X(s) = \frac{1}{s^2+9}$, $\text{Re}[s] > 0$; (2) $X(s) = \frac{s}{s^2+9}$, $\text{Re}[s] < 0$;
(3) $X(s) = \frac{s+1}{(s+1)^2+9}$, $\text{Re}[s] < -1$;(4) $X(s) = \frac{s+2}{s^2+7s+12}$, $-4 < \text{Re}[s] < -3$;
(5) $X(s) = \frac{s+1}{s^2+5s+6}$, $-3 < \text{Re}[s] < -2$.

3.9 The differential equations are given as $\frac{dx(t)}{dt} = -2y(t) + \delta(t)$ and $\frac{dy(t)}{dt} = 2x(t)$, where both $x(t)$ and $y(t)$ are right-sided signals. Try to find $X(s)$ and $Y(s)$, and their ROCs.

3.10 The unit impulse response of an LTI system is $h(t) = e^{-2t}u(t)$. Its input signal is $x(t) = e^{-t}u(t)$.
(1) Find the Laplace transforms of $x(t)$ and $h(t)$; (2) find $Y(s)$ based on the convolution property; (3) find $y(t)$ from $Y(s)$; (4) verify the result of (3) by $x(t) \star h(t)$.

3.11 Try to prove if $x(t) = x(-t)$, its Laplace transform satisfies $X(s) = X(-s)$.

3.12 Try to prove if $x(t) = -x(-t)$, its Laplace transform satisfies $X(s) = -X(-s)$.

3.13 Prove whether the following statements are correct:
1. The Laplace transform of $t^2u(t)$ does not converge at any region on the s-plane.
2. The Laplace transform of $e^{t^2}u(t)$ does not converge at any region on the s-plane.

3. The Laplace transform of $e^{j\Omega_0 t}u(t)$ does not converge at any region on the s-plane.

4. The Laplace transform of $e^{j\Omega_0 t}$ does not converge at any region on the s-plane.

5. The Laplace transform of $|t|$ does not converge at any region on the s-plane.

3.14 Find the following z-transforms:

(1) $x(n) = \delta(n+2)$; (2) $x(n) = \delta(n-2)$; (3) $x(n) = (-1)^n u(n)$;

(4) $x(n) = (1/2)^{n+1}u(n+3)$.

3.15 Find the inverse z-transform for the following $X(z)$.

(1) $X(z) = \frac{1-z^{-1}}{1-\frac{1}{4}z^{-2}}$, $|z| > \frac{1}{2}$; (2) $X(z) = \frac{1-z^{-1}}{1-\frac{1}{4}z^{-2}}$, $|z| < \frac{1}{2}$;

(3) $X(z) = \frac{z^{-1}-1/2}{1-\frac{1}{2}z^{-1}}$, $|z| > \frac{1}{2}$;(4) $X(z) = \frac{z^{-1}-1/2}{1-\frac{1}{2}z^{-1}}$, $|z| < \frac{1}{2}$.

3.16 Prove whether the following systems are causal systems:

(1)$H(z) = \frac{1-\frac{3}{4}z^{-1}+\frac{1}{2}z^{-2}}{z^{-1}(1-\frac{1}{2}z^{-1})(1-\frac{1}{3}z^{-1})}$; (2) $H(z) = \frac{z-\frac{1}{2}}{z^2+\frac{1}{2}z-\frac{3}{16}}$.

3.17 Given $a^n u(n) \leftrightarrow \frac{1}{1-az^{-1}}$, $|z| > |a|$. Find the inverse z-transform of

$X(z) = \frac{1-\frac{1}{3}z^{-1}}{(1-z^{-1})(1+2z^{-1})}$, $|z| > 2$.

3.18 The discrete-time signal is given as $y(n) = x_1(n+3) * x_2(-n+1)$, where $x_1(n) = (\frac{1}{2})^n u(n)$ and $x_2(n) = (\frac{1}{3})^n u(n)$. Find $Y(z)$ based on the property of the z-transform.

3.19 The difference equation is given as $y(n) = y(n-1) + y(n-2) + x(n-1)$.

1. Find the system function $H(z)$, plot its zero–pole plot, and point out its ROC;

2. Find its unit impulse response;

3. If the system is unstable, find the unit impulse response that makes the difference equation stable;

3.20 Solve the difference equation $y(n) - \frac{1}{2}y(n-1) + \frac{1}{4}y(n-2) = x(n)$ with the z-transform. The input signal is $x(n) = (\frac{1}{2})^n u(n)$.

3.21 A causal LTI system is shown in Figure E3.1. (1) Find the difference equation of the system; (2) determine the stability of the system.

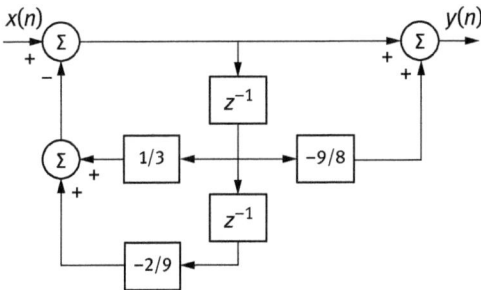

Figure E3.1: The given causal LTI system.

3.22 Find the unilateral z-transform for the following signals:

(1) $x(n) = \delta(n+5)$; (2) $x(n) = 2^n u(-n) + \left(\frac{1}{4}\right)^n u(n-1)$; (3) $\left(-\frac{1}{3}\right)^{n-2} u(n-2)$.

3.23 Find the output signals for the following systems using the unilateral z-transform:

1. $y(n) + 3y(n-1) = x(n)$, $x(n) = \left(\frac{1}{2}\right)^n u(n)$, $y(-1) = 1$;
2. $y(n) - \frac{1}{2}y(n-1) = x(n) - \frac{1}{2}x(n-1)$, $x(n) = u(n)$, $y(-1) = 1$;
3. $y(n) - \frac{1}{2}y(n-1) = x(n) - \frac{1}{2}x(n-1)$, $x(n) = u(n)$, $y(-1) = 0$.

3.24 Suppose $x(n) = x(-n)$, and the rational z-transform of $x(n)$ is $X(z)$.

1. Prove $X(z) = X(1/z)$ based on the definition of the z-transform;
2. If a pole of $X(z)$ is at $z = z_0$, prove that it will have another pole at $z = 1/z_0$;
3. Verify the result of (2) based on $\delta(n+1) + \delta(n-1)$.

3.25 A causal LTI system is given as $H(z) = \frac{1}{1-az^{-1}}$.

1. Write down the difference equation of the system.
2. Draw the system block diagram.
3. Find the frequency response of the system, and plot the magnitude–frequency curves when $a = 0$, 0.5, 1, respectively.

4 Discretization of Continuous-Time Signals and Reconstruction of Discrete-Time Signals

4.1 Introduction

1. Start from the communications

With the rapid development of computer technology and digital signal processing, modern communications are becoming increasingly more the indispensable means in daily life. Figure 4.1 shows the block diagram of the basic communication system.

We take the telephone system as an example. As shown in Figure 4.1, the speech voice via a microphone is transformed into an electric signal, and then it is amplified, sampled, coded, and modulated in the transmitter before it is transmitted. Through the channel, the signal is received by the receiver. After that, the signal is demodulated, decoded, and reconstructed in the destination for use.

In the modern mobile communication technology, the original speech signal in the transmission has experienced transformation many times. From the perspective of continuous-time signal and discrete-time signal analyses, at the sending end, the system transforms the continuous-time voice signal into a discrete-time signal (or a digital signal); at the receiving end, the digital signal is then transformed into a continuous-time signal (i.e., the voice signal again). In the communication process, the discretization of the continuous-time signal and the reconstruction of the discrete-time signal are two important operations.

The modern computer technology has become the main technical support for signal processing, while most signals encountered in nature and engineering are continuous-time signals, such as temperature, pressure, sound, vibration, natural images, and electrocardiogram (ECG) and electroencephalography (EEG) generated from a human body. So, there is a gap between processed objects and processing means for digital signal processing. Therefore, converting a continuous-time signal into a discrete-time signal (or a digital signal) becomes an inevitable choice.

2. Whether information lost after signal discretization

The concepts of the continuous-time and discrete-time signals have been introduced in Chapter 1. It is generally believed that the discrete-time signal is the discretization of the continuous-time signal. Such process is usually called sampling. Since the discrete-time signal is a sampling of the continuous-time signal, it may lose (or discard) a lot of data or information. We need to ensure whether the discrete-time signal keeps all the information in the original continuous-time signal. The sampling theorem introduced in this chapter will answer this question. In fact, as long as the continuous-time signal is of limited bandwidth and the sampling interval is dense enough, the information in the original continuous-time signal will not be lost. As a

https://doi.org/10.1515/9783110465082-004

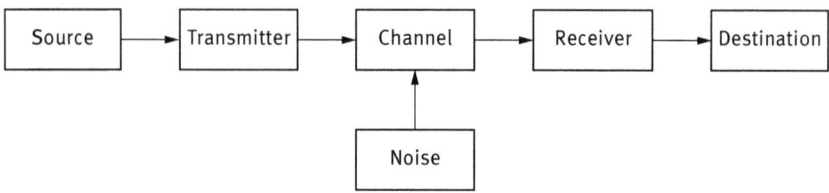

Figure 4.1: The basic communication system.

result, the original continuous-time signal based on the discrete-time signal perfectly can be restored.

3. How to do discretization and reconstruction

The basic method of the discretization of a continuous-time signal is the sampling to the continuous-time signal in accordance with the sampling theorem. The reconstruction of a discrete-time signal can be thought as the inverse transformation of signal sampling, including the low-pass filtering and other signal recovery methods. In engineering practice, the interpolation and fitting of the discrete-time signal are commonly used methods.

4.2 Sampling and the Sampling Theorem for Continuous-Time Signals

4.2.1 Ideal Sampling Based on the Unit Impulse Train and the Sampling Theorem

1. Unit impulse train

The unit impulse train is the periodic prolongation of the unit impulse signal $\delta(t)$, defined as

$$p(t) = \sum_{n=-\infty}^{+\infty} \delta(t - nT) \tag{4.1}$$

where T is the period of the signal. Taking Fourier transform to $p(t)$, we have

$$P(j\Omega) = \frac{2\pi}{T} \sum_{k=-\infty}^{+\infty} \delta\left(\Omega - \frac{2\pi k}{T}\right) = \frac{2\pi}{T} \sum_{k=-\infty}^{+\infty} \delta(\Omega - k\Omega_s) \tag{4.2}$$

We see from the above equation that the Fourier transform of the unit impulse train is also an impulse train with period $\Omega_s = 2\pi/T$. Figure 4.2 shows the unit impulse train and its Fourier transform.

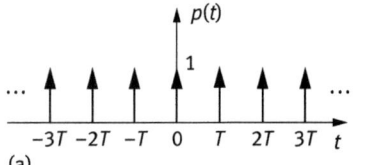 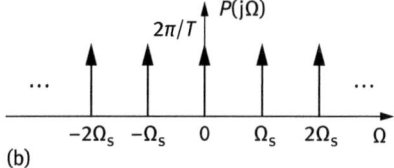

Figure 4.2: The unit impulse train and its Fourier transform: (a) unit impulse train and (b) Fourier transform of the unit impulse train.

2. Impulse train sampling

The impulse train sampling for a continuous-time signal is given in Figure 4.3, in which $x(t)$ is an arbitrary continuous-time signal, $p(t)$ is the unit impulse train used as a sampling sequence, and $x_p(t)$ is the sampled signal. Based on the figure, we have

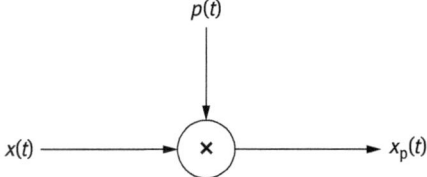

Figure 4.3: The impulse train sampling for a continuous-time signal.

$$x_p(t) = x(t)p(t) \tag{4.3}$$

In fact, $x_p(t)$ is still an impulse sequence, and its amplitude (intensity) is equal to the sample value of $x(t)$ at an interval spaced by T, that is

$$x_p(t) = \sum_{n=-\infty}^{+\infty} x(nT)\delta(t-nT) \tag{4.4}$$

Figure 4.4 demonstrates the waveform change during the sampling process.

3. Frequency analysis of the sampling process

Next we will analyze the sampling process in the frequency domain. By taking Fourier transform on both sides of eq. (4.3), we obtain

$$X_p(j\Omega) = \frac{1}{2\pi}[X(j\Omega) \star P(j\Omega)] \tag{4.5}$$

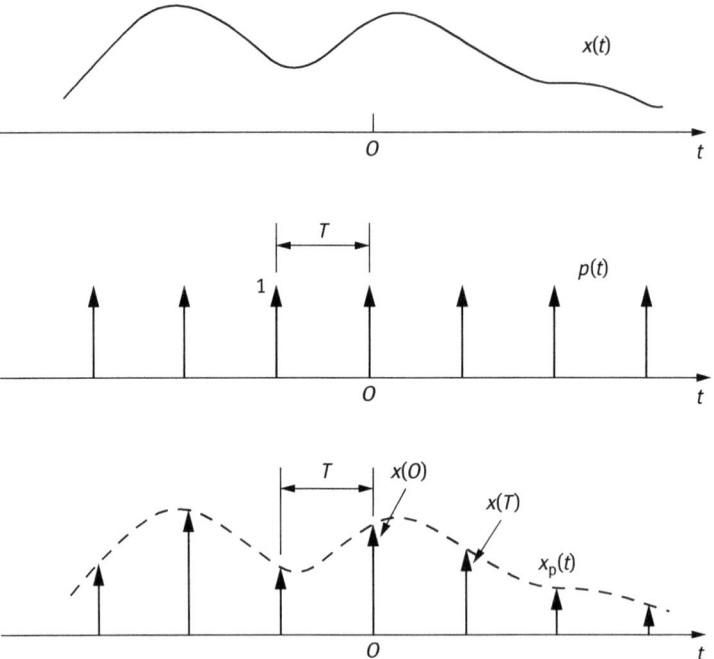

Figure 4.4: The waveform change during the sampling process: (a) the waveform of the original continuous-time signal, (b) the unit impulse train, and (c) the sampled signal.

Substituting eq. (4.2) into eq. (4.5), we have

$$X_p(j\Omega) = \frac{1}{T} \sum_{k=-\infty}^{+\infty} X[j(\Omega - k\Omega_s)] \tag{4.6}$$

The above equation indicates that the spectrum $X_p(j\Omega)$ of $x_p(t)$ is a periodic function of Ω consisting of a superposition of shifted replicas of $X(j\Omega)$, scaled by $1/T$. Specifically, if we set $k=0$ and $T=1$, then $X_p(j\Omega)$ changes to $X(j\Omega)$ in eq. (4.6). This demonstrates that $X_p(j\Omega)$ contains all the information of $X(j\Omega)$. Figure 4.5 shows the frequency analysis of the sampling process with the impulse train, in which Ω_M is the maximal spectrum component of the original signal and Ω_s is the sampling frequency. We see from Figure 4.5 that $X_p(j\Omega)$ is the periodic prolongation of $X(j\Omega)$. We can reconstruct $X(j\Omega)$ from $X_p(j\Omega)$ by using an ideal low-pass filter, shown as the dashed line in Figure 4.5(c). In fact, an implicit condition is included in Figure 4.5, i.e., the sampling frequency Ω_s should be greater than two times the maximal frequency component Ω_M. If the condition is not satisfied, the spectrum of the sampling process will be shown as in Figure 4.6.

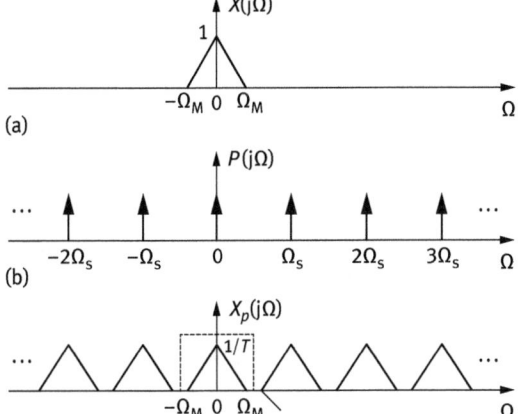

Figure 4.5: Frequency representation of the impulse train sampling: (a) spectrum of the original signal, (b) spectrum of the unit impulse train, and (c) spectrum of the sampled signal.

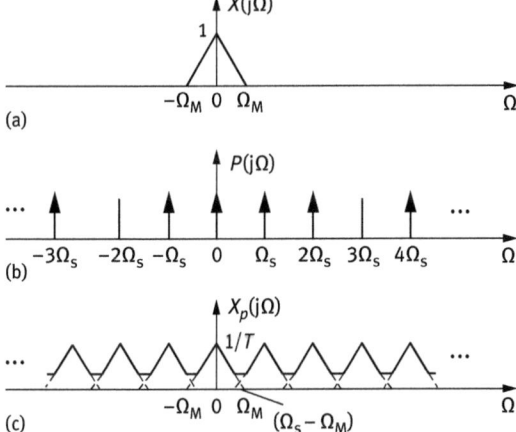

Figure 4.6: The frequency alias in sampling: (a) spectrum of the original signal, (b) spectrum of the impulse train, and (c) frequency alias.

It is obvious that if $\Omega_s > 2\Omega_M$ is not satisfied, the spectrum in adjacent periods will be overlapped, referred to as frequency alias. Thus, we are not able to reconstruct $X(j\Omega)$ from $X_p(j\Omega)$ with an ideal low-pass filter.

4. The sampling theorem

Theorem 4.1 (sampling theorem): Let $x(t)$ be a bandlimited signal with $X(j\Omega) = 0$ for $|\Omega| > \Omega_M$. Then, $x(t)$ is uniquely determined by its samples $x(nT)$, $n = 0$, ± 1, ± 2, $\dots$, if

$$\Omega_s > 2\Omega_M \tag{4.7}$$

where $\Omega_s = 2\pi/T$.

In Theorem 4.1, $\Omega_s = 2\Omega_M$ is usually called the Nyquist frequency. The sampling theorem tells us that the sampled signal $x_p(t)$ or $x(nT)$, $n = 0$, ± 1, ± 2, ..., can be used to resume the original continuous-time signal $x(t)$ as long as both conditions are satisfied: first, the original continuous-time signal is bandlimited; second, the sampling frequency should be greater than two times the maximal frequency component of the original signal.

The sampling theorem is a powerful tool for converting the continuous-time signal to the discrete-time signal. It ensures that the sampled signal contains entire information of the original signal without distortion. It provides a reliable guarantee for the subsequent digital signal processing and computer processing. In practical engineering applications, the sampling frequency is usually taken as

$$\Omega_s = (5 \sim 10)\Omega_M \tag{4.8}$$

5. Reconstruction of the ideal sampled signal

It can be seen from Figure 4.5 that for the spectrum $X_p(j\Omega)$ sampled by the ideal impulse train, the original signal $x(t)$ can be resumed from $X_p(j\Omega)$ by using an ideal low-pass filter. Figure 4.7 demonstrates a block diagram which consists of an ideal sampling system cascaded with an ideal low-pass filter $H(j\Omega)$.

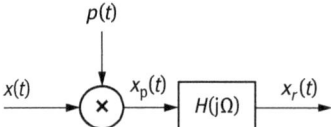

Figure 4.7: An ideal sampling and reconstruction system.

In Figure 4.7, the ideal low-pass filter $H(j\Omega)$ is used to resume the ideal sampled signal $x_p(t)$, and $x_r(t)$ denotes the reconstructed signal. Figure 4.8 gives the frequency analysis of the process of the signal sampling and reconstruction.

In practice, due to the noncausality and the volatility characteristics of the ideal filter in the time domain, we usually replace the ideal filter with another nonideal low-pass filter.

6. Alias and anti-aliasing prefiltering

Figure 4.6 shows the aliasing when the sampling theorem is not satisfied. In practical engineering applications, we have to pay attention to the aliasing when we try to convert the continuous-time signal into a discrete-time one. Once the frequency aliasing appears, the signal distortion is inevitable and could not be resumed to the original continuous-time signal without distortion no matter what method is applied.

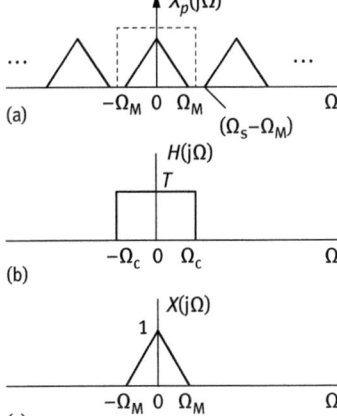

Figure 4.8: The spectrum analysis of the sampling and reconstructing process: (a) spectrum of the sampled signal, (b) ideal low-pass filter, and (c) spectrum of the reconstructed signal.

In order to avoid the frequency alias, the signal sampling process should meet the requirements of the sampling theorem. The sampling theorem actually contains two requirements: first, the signal to be sampled must be bandlimited; second, the sampling frequency must be high enough to meet $\Omega_s > 2\Omega_M$. In practice, we usually preprocess the signal before it is sampled by using an anti-alias filter, by setting the cutoff frequency of the filter which is equal to that of the highest component to be reserved in the signal and get rid of all frequency components higher than the cutoff frequency. It seems that some information is lost, but in fact the sampling result is better than the one with an alias.

4.2.2 Sampling with a Zero-Order Hold for Continuous-Time Signals

1. Concept of zero-order hold

We introduced the sampling with a unit impulse train, analyzed the spectra in the frequency domain, and introduced the sampling theorem in Section 4.2.1. As a matter of fact, the sampling and reconstruction introduced before are all under ideal conditions, including the ideal impulse train and the ideal low-pass filter. In practical engineering applications, such kind of ideal state is not easy to be obtained. For example, the ideal unit impulse train cannot be produced, and the ideal low-pass filter cannot be implemented in real time. For this reason, the zero-order hold technology is used for the discretization of continuous-time signals.

The so-called zero-order hold technology converts the continuous-time signal into a step-like signal with a zero-order hold filter, as demonstrated in Figure 4.9.

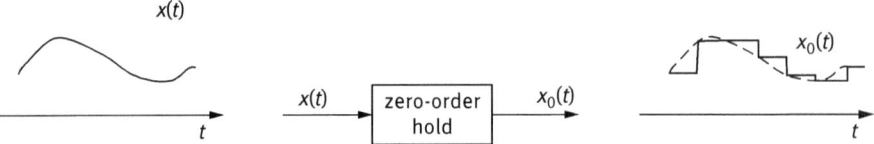

Figure 4.9: Sampling utilizing a zero-order hold.

As shown in Figure 4.9, the system samples $x(t)$ at a given instant and holds that value until next instant at which a sample is taken. The unit impulse response of the zero-order hold is shown in Figure 4.10.

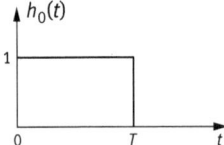

Figure 4.10: The unit impulse response of the zero-order hold.

2. Sampling with a zero-order hold
The block diagram of a sampling system with a zero-order hold is given in Figure 4.11.

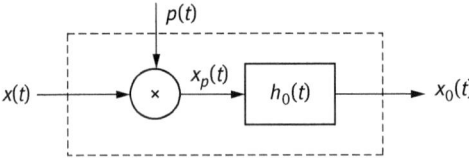

Figure 4.11: The sampling system block diagram with the zero-order hold.

As can see from Figure 4.11, the zero-order hold sampling system is, in fact, a unit impulse train sampling system cascaded with a zero-order hold filter. Although the ideal impulse train cannot be obtained in practice, the entire structure of such cascade form can be implemented. The waveforms of $x_0(t)$ and $x(t)$ have already been given in Figure 4.9.

3. The reconstruction of the zero-order hold sampled signals
The following questions need to be answered: whether the zero-order hold sampling will lose information? Whether we can reconstruct $x(t)$ with $x_0(t)$ without distortion? Now we give some analysis as follows. Suppose we connect a reconstruction filter $h_r(t)$ cascaded with the zero-order hold sampling system, whose transfer function is $H_r(j\Omega)$. The entire system is shown in Figure 4.12.

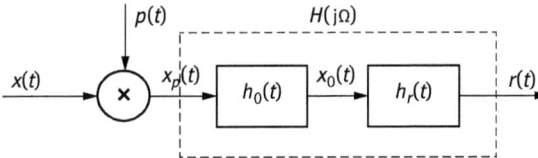

Figure 4.12: The zero-order hold sampling system and signal reconstruction.

The signal reconstruction of the above system is conducted by both the cascade of the zero-order hold sampling system and the reconstruction filter $h_r(t)$, as shown in Figure 4.12. The output signal $r(t)$ is the reconstructed signal from the zero-order hold signal $x_0(t)$. Our purpose is to let $r(t)$ equal to or approximately equal to $x(t)$ by designing $h_r(t)$ or $H_r(j\Omega)$. As labeled with the dashed line in Figure 4.12, both $h_0(t)$ and $h_r(t)$ can be named as $h(t)$ or $H(j\Omega)$ as a whole system. If $H(j\Omega)$ is an ideal low-pass filter, the output signal of the whole system should be the same as the original signal $x(t)$.

How to design the reconstruction filter $h_r(t)$ or $H_r(j\Omega)$? In fact, as long as we let

$$H(j\Omega) = H_0(j\Omega)H_r(j\Omega) \tag{4.9}$$

as an ideal low-pass filter, $r(t)$ will be equal to the original input signal $x(t)$. As the unit impulse response of the zero-order hold filter $h_0(t)$ is known, its transfer function is obtained as

$$H_0(j\Omega) = e^{-j\Omega T/2}\left[\frac{2\sin(\Omega T)/2}{\Omega}\right] \tag{4.10}$$

Therefore, the transfer function of the reconstruction filter is shown as

$$H_r(j\Omega) = \frac{H(j\Omega)}{H_0(j\Omega)} = \frac{e^{j\Omega T/2}H(j\Omega)}{\frac{2\sin(\Omega T)/2}{\Omega}} \tag{4.11}$$

Provided the cutoff frequency of the reconstruction filter designed reasonably, the original signal $x(t)$ can be reconstructed without distortion.

4.3 The Interpolation and Fitting for Discrete-Time Signals

4.3.1 The Interpolation for Discrete-Time Signals

1. The interpolation and its classification
The so-called signal interpolation is to supplement a continuous curve based on given samples and estimate other approximate values on the curve, passing through

all given samples. Interpolation is an important method of approximation of a function, and it is also an important tool of signal reconstruction.

There are many interpolation methods, including the following:

(1) Polynomial interpolation

This is the most commonly used interpolation method. If $n + 1$ discrete samples on a plane are given, we need to find an N th-order polynomial curve passing through all of the samples. The frequently used polynomial interpolations are Lagrange interpolation and Newton interpolation.

(2) Hermite interpolation

It is also called the interpolation with the derivative. This method requires that the interpolation curve should not only pass through all given samples but also have the same slope with the original curve.

(3) Piecewise/spline interpolation

This interpolation is used to avoid fluctuations in higher-order interpolation. The cubic spline interpolation is a typical piecewise interpolation.

(4) Trigonometric function interpolation

If the signal to be interpolated is a periodic signal with 2π as its period, the N th-order trigonometric polynomial is generally used as an interpolation function. The sinc function is a typically used interpolation function.

2. Bandlimited interpolation of an ideal low-pass filter and signal reconstruction

In Section 4.2.1, we introduced the unit-impulse train-based sampling technology and the recovery problem from the sampled signal. If the sampling theorem is satisfied, the original continuous-time signal $x(t)$ can be reconstructed with an ideal low-pass filter without distortion. We also analyzed the frequency characteristics for the process. Here, we will give a further analysis in time domain and lead to the concept of the bandlimited interpolation with the low-pass filter and the interpolation formula.

From Figure 4.7, the output signal $x_r(t)$ of an ideal sampling and reconstruction system is shown as

$$x_r(t) = x_p(t) \star h(t) \tag{4.12}$$

where $h(t)$ denotes the impulse response of an ideal low-pass filter. Substituting (4.4) into (4.12), we obtain

$$x_r(t) = \sum_{n = -\infty}^{+\infty} x(nT)h(t - nT) \tag{4.13}$$

eq. (4.13) is usually called the interpolation formula, since it describes how to interpolate $x(nT)$ as a continuous-time curve $x_r(t)$ by using an ideal low-pass filter. The unit impulse response of the ideal low-pass filter is shown as

$$h(t) = \frac{\Omega_c T \sin(\Omega_c t)}{\pi \Omega_c t} \tag{4.14}$$

The interpolation formula can be written as

$$x_r(t) = \sum_{n=-\infty}^{+\infty} x(nT) \frac{\Omega_c T}{\pi} \frac{\sin[\Omega_c(t-nT)]}{\Omega_c(t-nT)} \tag{4.15}$$

In the above equations, Ω_c represents the cutoff frequency of the ideal low-pass filter and T is the sampling period. Figure 4.13 demonstrates the process of signal reconstruction by an ideal low-pass filter.

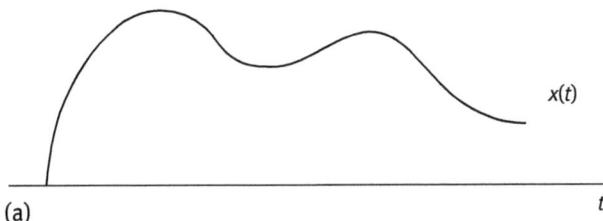

(a)

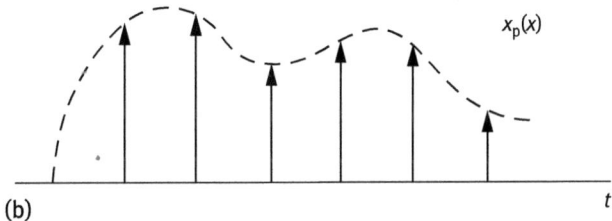

(b)

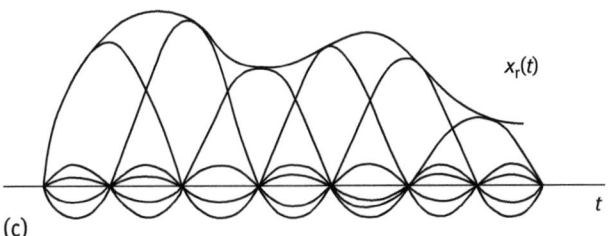

(c)

Figure 4.13: The bandlimited interpolation by an ideal low-pass filter: (a) original bandlimited signal, (b) sampled signal, and (c) reconstruction by an ideal low-pass filter.

In fact, the zero-order hold and the linear interpolation all belong to the low-pass filter interpolation method. The role of the zero-order interpolation is actually to insert some samples with the same value as the previous sample, while the linear interpolation is, in fact, to connect adjacent sample with a straight line.

3. Lagrange interpolation

Lagrange interpolation is the main method in polynomial interpolation. The basic idea of such interpolation is to approximate the discrete data samples with a polynomial function.

Set x_0, x_1, ..., x_N be $N+1$ dissimilar real-valued numbers, referred to as interpolation points or base points. Suppose $a = x_0 < x_1 < \cdots < x_N = b$. The core of the polynomial interpolation or Lagrange interpolation is to find an algebraic polynomial $p(x)$, satisfying $p(x_i) = y_i$, $i = 0, 1, \ldots, N$. The Nth-order Lagrange polynomial is defined as

$$p_N(x) = y_0 l_0(x) + y_1 l_1(x) + \cdots + y_N l_N(x) \tag{4.16}$$

where $l_i(x)$, $i = 0, 1, \cdots, N$, is called the basic Lagrange polynomial. The Lagrange polynomial is usually expressed as

$$p_N(x) = \sum_{i=0}^{N} y_i \prod_{\substack{j=0 \\ j \neq i}}^{N} \frac{(x - x_j)}{(x_i - x_j)}, \quad i = 0, 1, \ldots, N \tag{4.17}$$

In fact, the discrete-time signal can be considered as a set of signal amplitudes of the discrete data points on a plane, with the time t as its transverse axis. In order to adapt to the contents in this section better, we consider to transform the interpolation on the XOY plane into the interpolation to the discrete-time signal. In this way, eq. (4.17) can be rewritten as

$$p_N(t) = \sum_{i=0}^{N} x_i \prod_{\substack{j=0 \\ j \neq i}}^{N} \frac{(t - t_j)}{(t_i - t_j)}, \quad i = 0, 1, \ldots, N \tag{4.18}$$

(1) The first-order Lagrange interpolation

Two base points (t_0, x_0) and (t_1, x_1) in the unknown function $x = x(t)$ are given. We try to find the Lagrange interpolation polynomial passing through those given points. For this purpose, we take $N = 1$ in eq. (4.18) and obtain the first-order Lagrange interpolation polynomial as follows:

$$p_1(t) = x_0 \frac{t - t_1}{t_0 - t_1} + x_1 \frac{t - t_0}{t_1 - t_0} = x_0 + \frac{x_1 - x_0}{t_1 - t_0}(t - t_0) \tag{4.19}$$

It is obvious that eq. (4.19) is a linear function passing through two given base points. The MATLAB program and the result of interpolation are given in Example 4.1 and Figure 4.14(a), respectively.

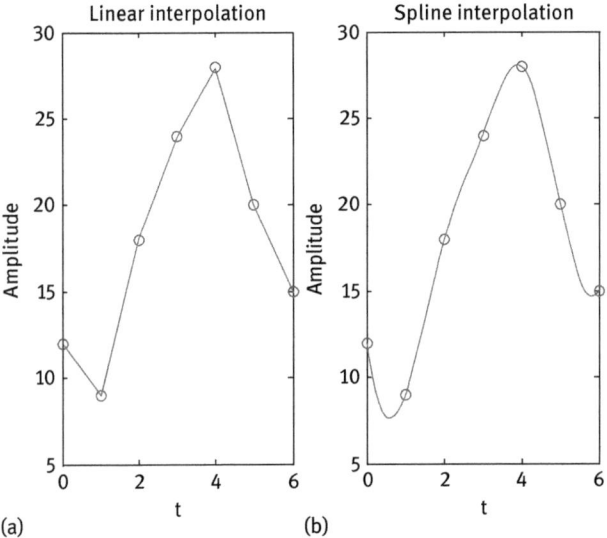

(a)

(b)

Figure 4.14: The results of interpolation for a discrete-time signal: (a) linear interpolation and (b) spline interpolation.

(2) The second-order Lagrange interpolation

Three base points (t_0, x_0), (t_1, x_1), and (t_2, x_2) in the unknown function $x = x(t)$ are given. The interpolation polynomial passing through the three base points is

$$p_2(t) = x_0 \frac{(t - t_1)(t - t_2)}{(t_0 - t_1)(t_0 - t_2)} + x_1 \frac{(t - t_0)(t - t_2)}{(t_1 - t_0)(t_1 - t_2)} + x_2 \frac{(t - t_0)(t - t_1)}{(t_2 - t_0)(t_2 - t_1)}$$

If the three base points are not in the same line, then $p_2(t)$ is a parabolic function passing through the three given base points.

4. Spline interpolation

When the number of base points is large, it is not good to use a single polynomial interpolation. The interpolation precision can be effectively improved by a piecewise

interpolation. A new problem caused by the piecewise interpolation is the discontinuity of the derivative at the junction of segments. In order to solve this problem, the
theory and method of spline interpolation generated.

Definition 4.1 (spline function): Suppose $N+1$ data points (t_0, x_0), (t_1, x_1), $\ldots$,
(t_N, x_N) satisfy $a = x_0 < x_1 < \cdots < x_N = b$. If function $S(t)$ meets the conditions defined
next, then $S(t)$ is defined as the third-order spline function of the above orderly data
points.

(1) $S(t)$ (is referred to as $S_i(t)$) is the third-order polynomial of variable t on every
     small interval $[t_i, t_{i+1}]$.
(2) $S_i(t_i) = x_i$, $i = 0, 1, 2, \ldots, N$.
(3) Both $S'(t)$ and $S''(t)$ are continuous on interval $[a, b]$.

For the interpolation of $N+1$ given data points, $S(t)$ is composed of N third-order
polynomial functions, resulting in $4N$ undetermined coefficients. The total $4N-2$
equations may be established for the above three constraint conditions. The other
two equations can be complemented by the boundary conditions of $S(t)$ at end points
a and b. Thus, the spline interpolation function can be determined, and the signal
interpolation can be completed.

Example 4.1: A discrete-time signal is given as $x(n) = [12, 9, 18, 24, 28, 20, 15]$. Try to
interpolate it using the linear interpolation and spline interpolation. Show their
resulting curves.

Solution: MATLAB programs for the two methods are listed as follows:

```
clear;
x=0:1:6; y=[12 9 18 24 28 20 15]; a=length(y); % The linear interpolation
y1=interp1(x,y,a,'linear'); xi=0:1/7200:7; y1i=interp1(x,y,xi,'linear');
subplot(121); plot(x,y,'o' ,xi,y1i);
xlabel('t'); ylabel('Amplitude'); axis([0 6 5 30]); title('Linear interpolation'); % The spline Segment
y2=interp1(x,y,a,'spline'); y2i=interp1(x,y,xi,'spline');
subplot(122); plot(x,y,'o',xi,y2i);
xlabel('t'); ylabel('Amplitude'); title('Spline interpolation'); axis([0 6 5 30]);
```

In Figure 4.14, the open circle "o" denotes the given (or base) data points, and the
solid line represents the results of the interpolation. It is clear that the spline interpolation is smoother than the linear interpolation.

4.3.2 The Fitting for Discrete-Time Signals

1. The concept of data fitting

The interpolation introduced in the previous subsection is an efficient method to
realize the reconstruction of discrete-time signals. On the other hand, the precision of

interpolation depends on the accuracy of base points, since the interpolation function must pass strictly through all of the base points. If the noise in the signal is significant, or the original data obtained are contaminated by a large error, the interpolation curve will introduce significant error. As a matter of fact, a more reasonable approach is to require the interpolation curve closing to all of the base points, but not to require it passing strictly through the base points. Such curve is called the fitting curve, and the process is referred to as the data fitting.

There are many types of data fitting methods: the most commonly used are the linear fitting and polynomial fitting based on the least squares criterion as well as the spline function fitting.

2. Least squares-based linear fitting
Suppose a set of data is given as (t_i, x_i), $i = 1, 2, \ldots, N$. The least squares linear fitting is to find the least squares line $f(t) = At + B$, so that the coefficients A and B guarantee that the squared residual error function $e(A, B)$ shown in eq. (4.21) reaches the minimum value. That is,

$$e(A, B) = \sum_{i=1}^{N} [f(t_i) - x_i]^2 = \sum_{i=1}^{N} [At_i + B - x_i]^2 \rightarrow \min \tag{4.21}$$

By letting the partial derivative of $e(A, B)$ with respect to A and B be zero

$$\frac{\partial e(A, B)}{\partial A} = 0, \quad \frac{\partial e(A, B)}{\partial B} = 0 \tag{4.22}$$

we have

$$\begin{cases} A\sum_{i=1}^{N} t_i^2 + B\sum_{i=1}^{N} t_i = \sum_{i=1}^{N} t_i x_i \\ A\sum_{i=1}^{N} x_i + NB = \sum_{i=1}^{N} x_i \end{cases} \tag{4.23}$$

The above equation is called the linear normal equation. We can get the coefficients A and B from it, and then we can get the fitting line.

3. Least squares-based polynomial fitting
Suppose a set of discrete data is given as (t_i, x_i), $i = 1, 2, \ldots, N$. The basic idea of the least squares-based polynomial fitting is to find the fitting function $f(t) = c_1 t^K + c_2 t^{K-1} + \cdots + c_{K+1}$, to have its squared residue error reaching the minimum as

$$e(c) = \sum_{i=1}^{N} [f(t_i) - x_i]^2 \rightarrow \min \tag{4.24}$$

where K is the order of the polynomial function and $\mathbf{c} = [c_1, c_2, \ldots, c_{K+1}]$. Letting $\frac{\partial e(\mathbf{c})}{\partial c_k} = 0$, $k = 1, 2, \ldots, K+1$, we get $K+1$ undetermined coefficients c_k, $k = 1, 2, \ldots$, $K+1$, and get the least squares fitting function $f(t)$.

Example 4.2: The discrete data are given as $t = [0.1, 0.4, 0.5, 0.6, 0.9]$ and $x(t) = [0.63, 0.94, 1.05, 1.43, 2.05]$. Please complete the least squares-based linear, second-order, and third-order fittings with MATLAB programming.
Solution: The MATLAB program is listed as follows (Figure 4.15):

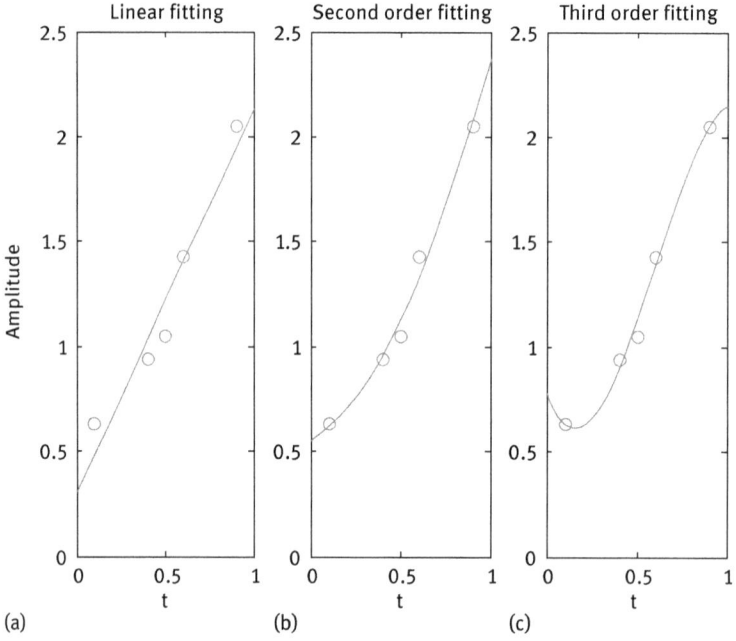

Figure 4.15: The least squares fittings: (a) linear fitting, (b) second-order fitting, and (c) third orderfitting.

```
clear;
x0=[0.1 0.4 0.5 0.6 0.9]; y0=[0.63 0.94 1.05 1.43 2.05];
p1=polyfit(x0,y0,1); p2=polyfit(x0,y0,2); p3=polyfit(x0,y0,3);
x=0:0.01:1.0; y1=polyval(p1,x); y2=polyval(p2,x); y3=polyval(p3,x);
figure(1); subplot(1,3,1); plot(x,y1,x0,y0,'o'); axis([0 1 0 2.5]);
xlabel('t'); ylabel('Amplitude'); title('Linear fitting'); text(0.75,0.25,'(a)');
subplot(1,3,2); plot(x,y2,x0,y0,'o'); axis([0 1 0 2.5]);
xlabel('t'); text(0.75,0.25,'(b)'); title('Second order fitting')
subplot(1,3,3); plot(x,y3,x0,y0,'o'); axis([0 1 0 2.5]);
xlabel('t'); text(0.75,0.25,'(c)'); title('Third order fitting');
```

4.3.3 Error Analysis of Interpolation and Fitting

1. Error analysis of interpolation
From the analysis in Section 4.3.1, we see that the error exists outside the base points. The error of the polynomial interpolation $e_N(t)$ is defined as

$$e_N(t) = p_N(t) - x(t) \tag{4.25}$$

where $p_N(t)$ is the interpolation polynomial. $e_N(t)$ can be used to analyze the deviation degree between the interpolation curve and the real signal.

Example 4.3: Suppose the continuous-time signal is $x(t) = \frac{1}{1+t^2}$. Try to interpolate in the interval of $[-5, 5]$ with 7 and 11 base points, respectively, and analyze the errors.
Solution: According to eq. (4.18), we can calculate the interpolation polynomials $p_6(t)$ and $p_{10}(t)$ and obtain the errors $e_6(t)$ and $e_{10}(t)$, respectively. The interpolation results are shown in Figure 4.16.

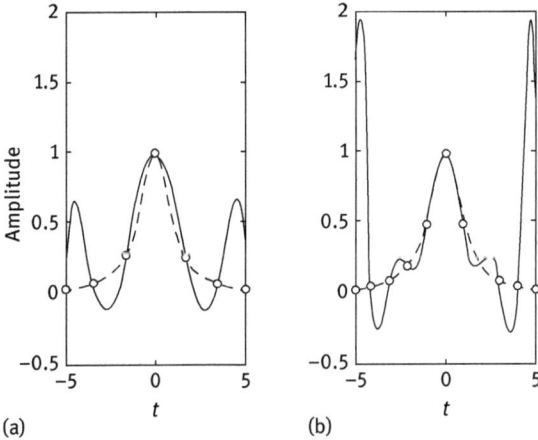

(a)   (b)

Figure 4.16: Error analysis of the polynomial interpolation: (a) 7-point interpolation and (b) 11-point interpolation.

In Figure 4.16, the dashed lines are the waveforms of the original signals, and the solid lines are the interpolated curves. We see that in the middle of the signal, both solid and dashed lines are almost overlapped; while at the ends, both curves deviate significantly. Furthermore, the errors tend to be serious as the interpolation order increases. This is called the Runge phenomenon.

In practice, we generally use the following methods to overcome the error caused by the Runge phenomenon. The first method is to use lower order (such

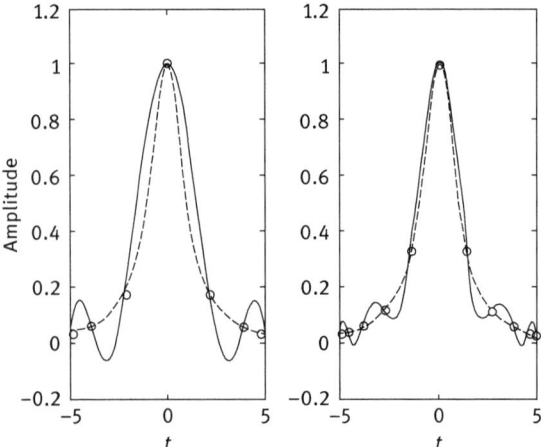

Figure 4.17: The interpolation with Chebyshev method: (a) 7-point interpolation and (b) 11-point interpolation.

as the third or lower) to do the interpolation in piecewise form, such as the spline interpolation. The second method is to use fewer points in the middle part of the signal and to use more points at its ends. Chebyshev is a typically used method shown in Figure 4.17.

2. Error analysis of data fitting
Suppose a set of discrete data is given as (t_0, x_0), (t_1, x_1), ..., (t_N, x_N). The fitting residue between the fitting function and the data points e_i is defined as

$$e_i = x_i - f(t_i), \quad i = 1, 2, \ldots, N \qquad (4.26)$$

where $f(t)$ denotes the fitting function. If N residues are all small, then the fitting is good. We generally use the following criteria to evaluate the performance of the fitting curve.

(1) The criterion of the maximum error

$$E_\infty(f) = \max\{|f(t_i) - x_i|\}, \quad i = 1, 2, \ldots, N \qquad (4.27)$$

(2) The criterion of the average error

$$E_1(f) = \frac{1}{N} \sum_{i=1}^{n} |f(t_i) - x_i| \qquad (4.28)$$

(3)   The criterion of the mean square error

$$E_2(f) = \left\{ \frac{1}{N} \sum_{i=1}^{n} [f(t_i) - x_i]^2 \right\}^{1/2} \qquad (4.29)$$

In the above-mentioned three criteria, the criterion of the mean square error is more valuable, and it is generally used as the optimal criterion.

Exercises

4.1   Summarize the basic methods of the discretization or the sampling of continuous-time signals.

4.2   Summarize the basic methods of the reconstruction or signal interpolation and fitting of discrete-time signals.

4.3   Describe the basic requirements of the sampling theorem.

4.4   Describe the characteristics of the sampling with a unit impulse train.

4.5   Describe the features of the zero-order hold sampling.

4.6   Explain the concepts of interpolation and fitting.

4.7   Explain the basic principle of Lagrange polynomial interpolation.

4.8   Explain the principle of the least squares fitting.

4.9   A real-valued signal is given as $x(t)$. When the sampling frequency is $\Omega_s = 10000\pi$, what should be the bandwidth of the signal if the sampling theorem is satisfied?

4.10 Suppose an ideal unit impulse train is used to sample the continuous-time signal $x(t)$, and an ideal low-pass filter is used to reconstruct the original signal. Which sampling period can be used to guarantee the reconstruction without distortion?
(1) $T = 0.5 \times 10^{-3}$; (2) $T = 2 \times 10^{-3}$; (3) $T = 10^{-4}$.

4.11 Try to determine the sampling frequency of the following signals:
(1) $x(t) = 1 + \cos(2,000\pi t) + \sin(4,000\pi t)$; (2) $x(t) = \frac{\sin(4,000\pi t)}{\pi t}$;
(3) $x(t) = \left(\frac{\sin(4,000\pi t)}{\pi t} \right)^2$.

4.12 Suppose the Nyquist frequency of $x(t)$ is Ω_0. Determine the Nyquist frequency of the following signals:
(1) $x(t) + x(t-1)$; (2) $\frac{dx(t)}{dt}$; (3) $x^2(t)$; (4) $x(t) \cos \Omega_0 t$.

4.13 A signal is given as $x(t) = \left(\frac{\sin 50\pi t}{\pi t} \right)^2$. The signal is sampled with a sampling frequency $\Omega_s = 150\pi$, and the Fourier transform of the sampled signal $g(t)$ is $G(j\Omega)$. In order to guarantee $G(j\Omega) = 75X(j\Omega)$, $|\Omega| \le \Omega_0$, where $X(j\Omega)$ is the Fourier transform of $x(t)$, what is the maximum value of Ω_0?

4.14 Suppose the Fourier transform of $x(t)$ is $X(j\Omega)$. A unit impulse train is used to sample the signal, with the sampling period $T = 10^{-4}$. Check whether the following $x(t)$ or $X(j\Omega)$ can be reconstructed without distortion based on the sampling theorem.

(1) $X(j\Omega) = 0$, $|\Omega| > 5,000\pi$; (2) $X(j\Omega) = 0$, $|\Omega| > 15,000\pi$;

(3) $\text{Re}[X(j\Omega)] = 0$, $|\Omega| > 5,000\pi$;

(4) $x(t)$ is a real-valued signal, and $X(j\Omega) = 0$, $\Omega > 5,000\pi$;

(5) $x(t)$ is a real-valued signal, and $X(j\Omega) = 0$, $\Omega < -15,000\pi$;

(6) $X(j\Omega) \star X(j\Omega) = 0$, $|\Omega| > 15,000\pi$; (7) $|X(j\Omega)| = 0$, $\Omega > 5000\pi$.

4.15 Suppose both $x_1(t)$ and $x_2(t)$ are bandlimited signals, whose spectra satisfy $X_1(j\Omega) = 0$, $|\Omega| > 1000\pi$ and $X_2(j\Omega) = 0$, $|\Omega| > 2000\pi$. We take the sampling with a unit impulse train to $y(t) = x_1(t) \star x_2(t)$. Try to give the range of the sampling period T, which can guarantee the original $y(t)$ to be resumed from the sampled signal.

4.16 A set of discrete data is given as $x(n) = [1, 9, 8, 17, 20, 17, 15]$. Please complete the linear, second-order polynomial, and spline interpolations with MATLAB programming, and plot the results.

4.17 The data set is the same as that in Exercise 4.16. Please complete the least squares linear, second-order, and third-order fittings with MATLAB programming, and plot the results.

5 Discrete Fourier Transform and Fast Fourier Transform

5.1 Introduction

In Chapter 2, we introduced the theory and methods of Fourier analysis, including Fourier series (FS), discrete Fourier series (DFS), Fourier transform (FT), and discrete-time Fourier transform (DTFT). Why do we need to study discrete Fourier transform (DFT) in this chapter again? Is there any difference between DFT and Fourier theory introduced in Chapter 2?

Since the 1960s, with the development of computer and information technologies, digital signal processing, as a new theory and technology, is widely used in speech, image, radar, sonar, geological exploration, communication system, automatic control, remote sensing, telemetry, aerospace, biomedical information detection and processing, and many other areas.

In the digital signal processing, we always hope that the signal processed is discrete or digital in both time and frequency domains. On the other hand, we also hope that the signal processed is aperiodic with finite duration in both time and frequency domains. However, if we check the four types of Fourier series and Fourier transforms introduced in Chapter 2, we see that none of them satisfies the above requirements. It was under such background that the discrete Fourier transform generated. The DFT provides a powerful tool for digital signal processing, and further enriches and develops the theory of Fourier transform.

The discrete Fourier transform converts a finite-duration signal into a bandlimited spectrum. The signal and spectrum are all discrete and aperiodic in both time and frequency domains.

This chapter introduces in detail the fundamental theory of the DFT, including the definition, properties, the relations with other Fourier analysis methods, and other problems. This chapter will also introduce the fast algorithm of the DFT, referred to as the fast Fourier transform (FFT). The study of this chapter will make readers master the theories of Fourier transform and get some skills for applications.

5.2 Discrete Fourier Transform (DFT)

5.2.1 A Brief Review of Fourier Analysis

This section briefly reviews the four different Fourier series and Fourier transforms introduced in Chapter 2 for the further introduction of the discrete Fourier transform.

https://doi.org/10.1515/9783110465082-005

1. Fourier series for a continuous-time periodic signal

The FS for a continuous-time periodic signal $x(t)$, which satisfies Dirichlet conditions, is shown as

$$
a_k = \frac{1}{T} \int_T x(t) e^{-jk\Omega_0 t} dt = \int_T x(t) e^{-jk(2\pi/T)t} dt
$$
$$
x(t) = \sum_{k=-\infty}^{+\infty} a_k e^{jk\Omega_0 t} = \sum_{k=-\infty}^{+\infty} a_k e^{jk(2\pi/T)t}
$$

(5.1)

2. Fourier series for a discrete-time periodic signal

The DFS for a discrete-time periodic signal $x(n)$ is shown as

$$
a_k = \frac{1}{N} \sum_{n=\langle N \rangle} x(n) e^{-jk\omega_0 n} = \frac{1}{N} \sum_{n=\langle N \rangle} x(n) e^{-jk(2\pi/N)n}
$$
$$
x(n) = \sum_{k=\langle N \rangle} a_k e^{jk\omega_0 n} = \sum_{k=\langle N \rangle} a_k e^{jk(2\pi/N)n}
$$

(5.2)

3. Fourier transform for a continuous-time signal

The FT for a continuous-time signal $x(t)$, which satisfies Dirichlet conditions, is shown as

$$
X(j\Omega) = \int_{-\infty}^{+\infty} x(t) e^{-j\Omega t} dt
$$
$$
x(t) = \frac{1}{2\pi} \int_{-\infty}^{+\infty} X(j\Omega) e^{j\Omega t} d\Omega
$$

(5.3)

4. Discrete-time Fourier transform for a discrete-time signal

The DTFT of a discrete-time signal $x(n)$ is shown as

$$
X(e^{j\omega}) = \sum_{n=-\infty}^{+\infty} x(n) e^{-j\omega n}
$$
$$
x(n) = \frac{1}{2\pi} \int_{2\pi} X(e^{j\omega}) e^{j\omega n} d\omega
$$

(5.4)

For the time and frequency relations among the four kinds of Fourier series and transforms, as well as their corresponding curves, readers can refer to Figures 2.13 and 2.14 in Chapter 2.

It is necessary to note that in the above four kinds of Fourier series and Fourier transforms, FS corresponds to a continuous-time periodic signal, with a discrete aperiodic spectrum; DFS corresponds to a discrete-time periodic signal, with a discrete and periodic spectrum; FT corresponds to a continuous-time nonperiodic (may include periodic) signal, with a continuous and nonperiodic spectrum; DTFT corresponds to a discrete-time nonperiodic (may include periodic) signal, with a continuous and periodic spectrum.

As mentioned in Section 5.1, none of the four Fourier series and Fourier transforms reviewed above satisfies the requirements of digital signal processing with computers. In fact, the digital signal processing technology generally requires that signals and their spectra are both digitalized and limited in length. Therefore, we need to have a new kind of Fourier transform. In such Fourier transform, both signal and its spectrum are digitalized with a finite length. It is just the discrete Fourier transform, referred to as DFT.

5.2.2 From Discrete Fourier Series to Discrete Fourier Transform

1. The derivation of DFT and its definition

From the analysis of the four Fourier series and Fourier transforms, we see that only DFS is discrete in both time and frequency domains, denoted by $x(n)$ and a_k, respectively. Furthermore, we know that both $x(n)$ and a_k in DFS are periodic of period N. For convenient illustration, we denote the discrete-time periodic signal and its discrete periodic spectrum with $\tilde{x}(n)$ and $\tilde{a}_k$, while $x(n)$ and a_k are used to express the finite-length discrete signal and its finite-length discrete spectrum, respectively, in this subsection.

Define symbols $R_N(n)$ and $R_N(k)$ to demonstrate the rectangular sequence as

$$R_N(n) = \begin{cases} 1, & 0 \le n \le N-1 \\ 0, & \text{other } n \end{cases} \quad \text{or} \quad R_N(k) = \begin{cases} 1, & 0 \le k \le N-1 \\ 0, & \text{other } k \end{cases} \tag{5.5}$$

It is obvious that $R_N(n)$ (or $R_N(k)$) is a combination of two unit step signals. Since the finite-length sequence $x(n)$ (or a_k) can be considered as a particular period of a periodic sequence $\tilde{x}(n)$ (or $\tilde{a}_k$), the following notations are defined for convenience:

$$\tilde{x}(n) = x(n \text{ modulo } N) = x((n))_N \tag{5.6}$$

$$\tilde{a}_k = a_{(k \text{ modulo } N)} = a_{(k)_N} \tag{5.7}$$

where $((n))_N$ (or $(k)_N$) denotes the operation for taking the remainder of n for modulo N (or taking the remainder of k for modulo N). Thus, by using $R_N(n)$ and $((n))_N$ (or the corresponding $R_N(k)$ and $(k)_N$), we can describe the relations between the finite-length sequences $x(n)$ (or a_k) and the periodic sequences $\tilde{x}(n)$ (or $\tilde{a}_k$) as

$$x(n) = \tilde{x}(n)R_N(n) = x((n))_N R_N(n) \tag{5.8}$$

$$a_k = \tilde{a}_k R_N(k) = a_{(k)_N} R_N(k) \tag{5.9}$$

Definition 5.1 (discrete Fourier transform): The definition of the discrete Fourier transform for a finite-length discrete-time signal (referred to as a finite-length sequence) is defined as

$$X(k) = \text{DFT}[x(n)] = \sum_{n=0}^{N-1} x(n)e^{-j\frac{2\pi}{N}nk} = \sum_{n=0}^{N-1} x(n)W_N^{nk}, \quad k = 0, 1, \ldots, N-1 \quad (5.10)$$

$$x(n) = \text{IDFT}[X(k)] = \frac{1}{N}\sum_{k=0}^{N-1} X(k)e^{j\frac{2\pi}{N}nk} = \frac{1}{N}\sum_{k=0}^{N-1} X(k)W_N^{-nk}, \quad n = 0, 1, \ldots, N-1 \quad (5.11)$$

Equations (5.10) and (5.11) are the discrete Fourier transform and its inverse transform, in which $x(n)$ is a discrete sequence of length N and $X(k)$ is its spectrum, which is also a discrete sequence of length N. In both equations, W_N is defined as

$$W_N = e^{-j\frac{2\pi}{N}} \quad (5.12)$$

In practice, we can get the spectrum $X(k)$ by taking the DFT operation for sequence $x(n)$ of length N. As a matter of fact, $x(n)$ and $X(k)$ can be considered as one period of $\tilde{x}(n)$ and $\tilde{a}_k$, respectively. Thus, for any sequence $x(n)$ with a finite duration, we can calculate its spectrum with a computer conveniently.

2. The graphical explanation of DFT

Table 5.1 demonstrates graphically the derivation of the discrete Fourier transform, which is important for us to further understand the relations among the DFT and other Fourier series and Fourier transforms.

$x(n)$ and $X(k)$ can be considered as the sampled sequence of $x(t)$ and $X(j\Omega)$, respectively. As we know that both $x(t)$ and $X(j\Omega)$ are connected by FT, both $x(n)$ and $X(k)$ are connected by DFT.

3. Relations among DFT and other Fourier transforms and Fourier series

(1) The relation between DFT and FT
     As mentioned above, the discrete-time signal $x(n)$ can be considered as the sampled $x(t)$, while the discrete spectrum $X(k)$ can be considered as the sampled $X(j\Omega)$, which is the spectrum of $x(t)$.
(2) The relation between DFT and DFS
     To some degree, the DFT $X(k)$ is obtained from DFS $\tilde{a}_k$ by taking its fundamental period.
(3) The relation between DFT and DTFT as well as the z-transform

Equation (5.13) shows expressions of the z-transform, DTFT, and DFT of a discrete-time signal $x(n)$:

Table 5.1: The graphical explanation of DFT.

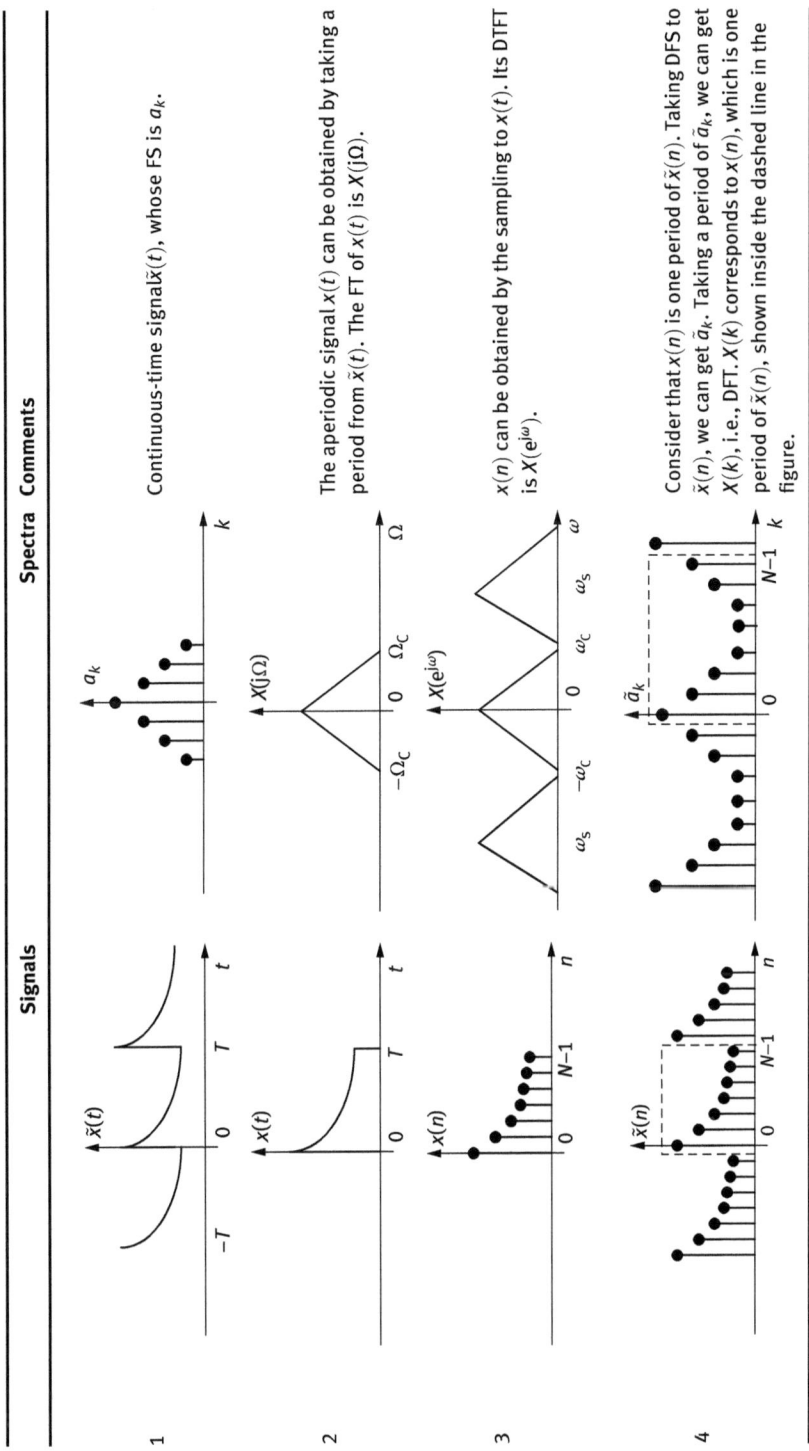

	Signals	Spectra	Comments
1	$\tilde{x}(t)$	a_k	Continuous-time signal $\tilde{x}(t)$, whose FS is a_k.
2	$x(t)$	$X(j\Omega)$	The aperiodic signal $x(t)$ can be obtained by taking a period from $\tilde{x}(t)$. The FT of $x(t)$ is $X(j\Omega)$.
3	$x(n)$	$X(e^{j\omega})$	$x(n)$ can be obtained by the sampling to $x(t)$. Its DTFT is $X(e^{j\omega})$.
4	$\tilde{x}(n)$	$\tilde{a}_k$	Consider that $x(n)$ is one period of $\tilde{x}(n)$. Taking DFS to $\tilde{x}(n)$, we can get $\tilde{a}_k$. Taking a period of $\tilde{a}_k$, we can get $X(k)$, i.e., DFT. $X(k)$ corresponds to $x(n)$, which is one period of $\tilde{x}(n)$, shown inside the dashed line in the figure.

$$X(z) = \sum_{n=0}^{N-1} x(n)z^{-n} = \sum_{n=0}^{N-1} x(n)(re^{j\omega})^{-n}$$

$$X(e^{j\omega}) = \sum_{n=0}^{N-1} x(n)e^{-j\omega n} = X(z)|_{z=e^{j\omega}} \tag{5.13}$$

$$X(k) = \sum_{n=0}^{N-1} x(n)e^{-j\frac{2\pi}{N}nk} = X(e^{j\omega})|_{\omega=\frac{2\pi}{N}k}$$

Figure 5.1 demonstrates the relations among the DFT, DTFT, and z-transform on the z-plane.

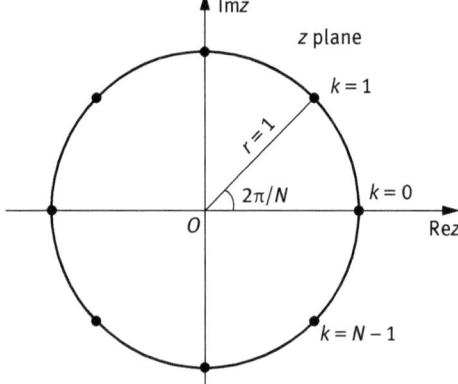

Figure 5.1: The relations among the DFT, DTFT, and z-transform.

As shown in Figure 5.1, $X(z)$ is defined on the entire z-plane and $X(e^{j\omega})$ is defined on the unit circle of the z-plane, while $X(k)$ is only defined on some discrete points with an equal interval on the unit circle. In other words, $X(e^{j\omega})$ is the z-transform defined on the unit circle, while $X(k)$ is the sampled $X(e^{j\omega})$ with an equal sampling interval.

On the other hand, both $X(z)$ and $X(e^{j\omega})$ can be expressed by $X(k)$ as

$$X(z) = \sum_{n=0}^{N-1}\left[\frac{1}{N}\sum_{k=0}^{N-1}X(k)e^{j\frac{2\pi}{N}nk}\right]z^{-n} = \frac{1-z^{-N}}{N}\sum_{k=0}^{N-1}\frac{X(k)}{1-e^{j\frac{2\pi}{N}k}z^{-1}} \tag{5.14}$$

$$X(e^{j\omega}) = \sum_{k=0}^{N-1}X(k)\frac{1-e^{-j\omega N}}{N[1-e^{j\frac{2\pi}{N}k}e^{-j\omega}]} = \frac{1-e^{-j\omega N}}{N}\sum_{k=0}^{N-1}\frac{X(k)}{1-e^{j\frac{2\pi}{N}k}e^{-j\omega}} \tag{5.15}$$

We can see from the above two equations that the continuous spectrum $X(e^{j\omega})$ can be obtained from $X(k)$ by interpolation. $X(z)$ can also be expressed by the discrete spectrum $X(k)$.

Example 5.1: Try to generate three discrete-time sinusoidal signals $x_1(n)$, $x_2(n)$, and $x_3(n)$ with MATLAB programming, whose frequencies are $f_1 = 25$ Hz, $f_2 = 50$

Hz, and $f_3 = 120$ Hz, respectively. The amplitudes of the three signals are $A_1 = 1$, $A_2 = 0.7$, and $A_3 = 0.4$, respectively. Please calculate the spectrum of $x(n) = x_1(n) + x_2(n) + x_3(n)$ and draw its waveform and spectrum.

Solution: The MATLAB program is shown as follows:

```
clear;
n=400; nn=1:n; nn2=0:n/2-1; f1=25; f2=50; f3=120; x1=zeros(n,1);
  x2=zeros(n,1);
x1=sin(2*pi*f1/n.*nn'); x2=0.7*sin(2*pi*f2/n.*nn'); x3=0.4*sin(2*pi*f3/
  n.*nn');
xx=zeros(n,1); xf=zeros(n,1); xx=x1+x2+x3;
% Calculate DFT based on its definition
for ii=1:n
     for jj=1:n
     w(jj)=exp((-j*2*pi*(jj-1)*(ii-1))/n);
     xf(ii)=xf(ii)+xx(jj)*w(jj);
     end
end
% Draw its waveform and spectrum
figure(1); subplot(211); plot(xx); axis([0 n/2 -2 2]);
xlabel('n'); ylabel('Amplitude'); title('Combined signal');
subplot(212); stem(nn2,abs(xf(1:n/2)),'filled');
xlabel('f/Hz'); ylabel('Magnitude');
title('Spectrum of the combined signal');
```

Figure 5.2 shows the combined signal (upper figure) and its spectrum (lower figure) calculated with DFT.

5.2.3 Properties of the Discrete Fourier Transform

Property 5.1: Linearity
If both $x_1(n)$ and $x_2(n)$ are discrete sequences of length N and their DFTs are $X_1(k)$ and $X_2(k)$, respectively, then

$$\mathrm{DFT}[ax_1(n) + bx_2(n)] = aX_1(k) + bX_2(k) \qquad (5.16)$$

Property 5.2: Orthogonality
Suppose matrix $\boldsymbol{W}_N$ and vectors $\boldsymbol{X}_N$ and $\boldsymbol{x}_N$ are defined as follows:

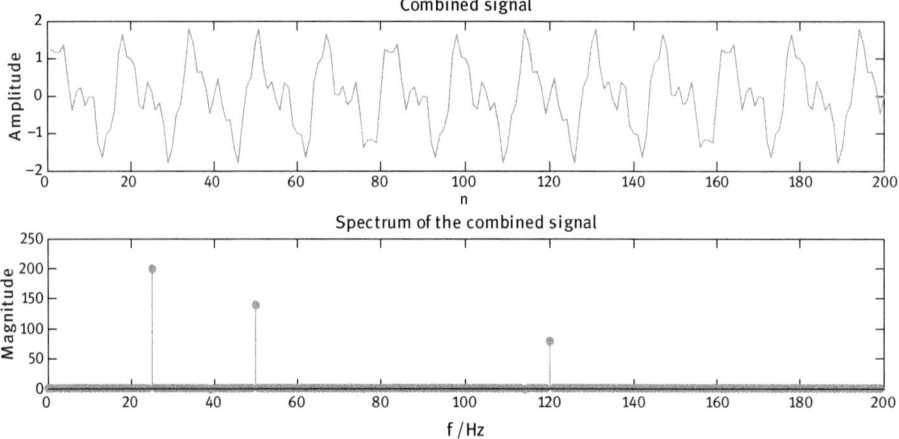

Figure 5.2: The combined signal (upper figure) and its spectrum (lower figure) calculated with DFT.

$$\boldsymbol{W}_N = [W_N^{nk}] = \begin{bmatrix} W_N^0 & W_N^0 & W_N^0 & \cdots & W_N^0 \\ W_N^0 & W_N^1 & W_N^2 & \cdots & W_N^{N-1} \\ W_N^0 & W_N^2 & W_N^4 & \cdots & W_N^{2(N-1)} \\ \vdots & \vdots & \vdots & \ddots & \vdots \\ W_N^0 & W_N^{N-1} & W_N^{2(n-1)} & \cdots & W_N^{(N-1)(N-1)} \end{bmatrix}$$ (5.17)

$$\boldsymbol{X}_N = [\,X(0) \quad X(1) \quad \cdots \quad X(N-1)\,]^{\mathrm{T}}$$ (5.18)

$$\boldsymbol{x}_N = [\,x(0) \quad x(1) \quad \cdots \quad x(N-1)\,]^{\mathrm{T}}$$ (5.19)

Then the DFT can be written as a matrix form

$$\boldsymbol{X}_N = \boldsymbol{W}_N \boldsymbol{x}_N$$ (5.20)

Since

$$\boldsymbol{W}_N^* \boldsymbol{W}_N = \sum_{k=0}^{N-1} W^{mk} W^{-nk} = N\boldsymbol{I} \Rightarrow \begin{cases} N, & m=n \\ 0, & m \neq n \end{cases}$$

$\boldsymbol{W}_N^*$ and $\boldsymbol{W}_N$ are orthogonal (where the symbol "*" denotes the complex conjugation). That means $\boldsymbol{W}_N$ is an orthogonal matrix, and the DFT is an orthogonal transform. Thus, the inverse DFT can be written as

$$\boldsymbol{x}_N = \boldsymbol{W}_N^{-1} \boldsymbol{X}_N = \frac{1}{N} \boldsymbol{W}_N^* \boldsymbol{X}_N$$ (5.21)

Property 5.3: Circular shift of a sequence

Because the finite-duration sequence $x(n)$ of length N is one period of the periodic sequence $\tilde{x}(n)$, the time shift for $x(n)$ should be similar to the time shift of the whole periodic sequence $\tilde{x}(n)$, i.e., the sequence $x(n)$ leaves the interval 0 to $N-1$ at one end, and it enters at the other end. The circular shift is demonstrated in Figure 5.3.

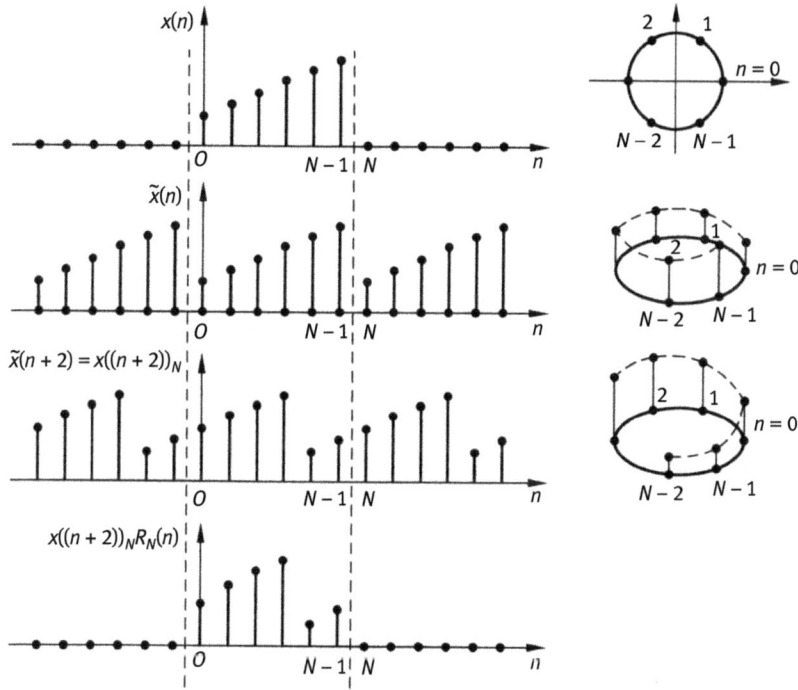

Figure 5.3: The circular shift of the sequence $x(n)$.

If the sequence $x(n)$ of length N is shifted with m sampling intervals shown as $x_m(n) = x((n+m))_N R_N(n)$, then we obtain

$$\text{DFT}[x_m(n)] = W_N^{-mk} X(k) \tag{5.22}$$

Proof: Suppose the DFT of $x_m(n)$ is $X_m(k)$. According to the definition of DFT, we have

$$X_m(k) = \sum_{n=0}^{N-1} x_m(n) W_N^{nk} = \sum_{n=0}^{N-1} x((n+m))_N W_N^{nk}$$

Set the variable substitution as $n+m = r$, and suppose $x((n+m))_N R_N(n)$ is denoted as $x(n+m)$. Thus, we have

$$X_m(k) = \sum_{r=m}^{N-1+m} x(r) W_N^{(r-m)k} = W_N^{-mk} \left[\sum_{r=m}^{N-1} x(r) W_N^{kr} + \sum_{r=N}^{N-1+m} x(r) W_N^{kr} \right]$$

Considering the shift of $x(n)$ in the circular shift, the above equation can be written as

$$X_m(k) = W_N^{-mk} \left[\sum_{r=m}^{N-1} x(r) W_N^{kr} + \sum_{r=0}^{m-1} x(r) W_N^{kr} \right] = W_N^{-mk} X(k)$$

Property 5.4: Symmetry properties
    1. If $x(n)$ is a complex sequence and its DFT is $X(k)$, then

$$\text{DFT}[x^*(n)] = X^*((-k))_N R_N(n) \tag{5.23a}$$

The above equation can also be written as

$$\text{DFT}[x^*(n)] = X^*(-k) \tag{5.23b}$$

    2. If $x(n)$ is a real-valued sequence, then its DFT $X(k)$ has just the circular conjugate symmetric component as

$$\begin{aligned}
&X(k) = X^*((-k))_N R_N(k) = X^*((N-k))_N R_N(k) \\
&X_R(k) = X_R((-k))_N R_N(k) = X_R((N-k))_N R_N(k) \\
&X_I(k) = -X_I((-k))_N R_N(k) = -X_I((N-k))_N R_N(k) \\
&|X(k)| = |X((N-k))_N| R_N(k) \\
&\arg[X(k)] = -\arg[X((-k))_N] R_N(k)
\end{aligned} \tag{5.24a}$$

The above equations can also be written as

$$\begin{aligned}
&X^*(k) = X(-k) = X(N-k) \\
&X_R(k) = X_R(-k) = X_R(N-k) \\
&X_I(k) = -X_I(-k) = -X_I(N-k) \\
&|X(k)| = |X(N-k)| \\
&\arg[X(k)] = -\arg[X(-k)]
\end{aligned} \tag{5.24b}$$

where $X_R(k)$ and $X_I(k)$ denote the real and imaginary parts of $X(k)$, respectively, and $\arg[X(k)]$ denotes the phase of $X(k)$.
    3. If $x(n)$ is a real and even sequence, then $X(k)$ is a real and even sequence.
    4. If $x(n)$ is a real and odd sequence, then $X(k)$ is an imaginary and odd sequence.
    Table 5.2 shows the symmetry relations of DFT. These properties can be used to reduce the calculating complex of DFT.
Property 5.5: Parseval's theorem

Table 5.2: The symmetry properties of DFT.

$x(n)$ [or $X(k)$]	$X(k)$ [or $x(n)$]
Even symmetry	Even symmetry
Odd symmetry	Odd symmetry
Real valued	Even in the real part
	Odd in the imaginary part
Imaginary valued	Odd in the real part
	Even in the imaginary part
Real valued and even symmetry	Real valued and even symmetry
Real valued and odd symmetry	Imaginary valued and odd symmetry
Imaginary valued and even symmetry	Imaginary valued and even symmetry
Imaginary valued and odd symmetry	Real valued and odd symmetry

$$\sum_{n=0}^{N-1} |x(n)|^2 = \frac{1}{N} \sum_{k=0}^{N-1} |X(k)|^2 \tag{5.25}$$

In essence, Parseval's theorem indicates the relationship of conservation of signal energy between the time domain and the transform domain.

Property 5.6: Circular convolution

Set $x_1(n)$ and $x_2(n)$ as finite-duration sequences of length N, and

$$DFT[x_1(n)] = X_1(k), \qquad DFT[x_2(n)] = X_2(k)$$

If $Y(k) = X_1(k)X_2(k)$, then the circular convolution between $x_1(n)$ and $x_2(n)$ is as follows:

$$
\begin{aligned}
y(n) = IDFT[Y(k)] &= \left[\sum_{m=0}^{N-1} x_1(m)x_2((n-m))_N \right] R_N(n) \\
&= \left[\sum_{m=0}^{N-1} x_2(m)x_1((n-m))_N \right] R_N(n)
\end{aligned}
\tag{5.26}
$$

Table 5.3 shows the commonly used properties of the DFT.

Example 5.2: Try to do the DFT for two real-valued sequences by using the conjugate property of DFT.

Solution: Let $x_1(n)$ and $x_2(n)$ be real-valued sequences of length N, satisfying $DFT[x_1(n)] = X_1(k)$ and $DFT[x_2(n)] = X_2(k)$, respectively. Construct $x_1(n)$ and $x_2(n)$ as a complex sequence as $y(n) = x_1(n) + jx_2(n)$; then we obtain

$$
\begin{aligned}
DFT[y(n)] = Y(k) &= DFT[x_1(n) + jx_2(n)] = DFT[x_1(n)] + jDFT[x_2(n)] \\
&= X_1(k) + jX_2(k)
\end{aligned}
$$

Table 5.3: The commonly used properties of the DFT.

Properties	Sequence $x(n)$, $x_1(n)$, $x_2(n)$	DFT $X(k)$, $X_1(k)$, $X_2(k)$
1 Linearity	$ax_1(n) + bx_2(n)$	$aX_1(k) + bX_2(k)$
2 Circular shift in time domain	$x((n+m))_N R_N(n)$	$W_N^{-mk} X(k)$
3 Duality	$X(n)$	$Nx((-k))_N R_N(n)$
4 Circular shift in frequency domain	$W_N^{nl} x(n)$	$X((k+l))_N R_N(k)$
5 Circular convolution	$\displaystyle\sum_{m=0}^{N-1} x_1(m)x_2((n-m))_N R_N(n)$	$X_1(k)X_2(k)$
6 Circular correlation	$\displaystyle r_{x_1x_2}(m) = \sum_{n=0}^{N-1} x_2^*(n)x_1((n+m))_N R_N(m)$	$X_1(k)X_2^*(k)$
7 Multiplication	$x_1(n)x_2(n)$	$\displaystyle\frac{1}{N}\sum_{l=0}^{N-1} X_1(l)X_2((k-l))_N R_N(k)$
8 Conjugation	$x^*(n)$	$X^*((N-k))_N R_N(k)$
9 Time reverse	$x((-n))_N R_N(n)$	$X((N-k))_N R_N(k)$
10 Time reverse and conjugate	$x^*((-n))_N R_N(n)$	$X^*(k)$
11 Real part	$Re[x(n)]$	$\frac{1}{2}\left[X(k)+X^*((N-k))_N\right]R_N(k)$
12 Imaginary part	$jIm[x(n)]$	$\frac{1}{2}\left[X(k)-X^*((N-k))_N\right]R_N(k)$
13 Even part	$\frac{1}{2}\left[x(n)+x^*((N-n))_N\right]R_N(n)$	$Re[X(k)]$
14 Odd part	$\frac{1}{2}\left[x(n)-x^*((N-n))_N\right]R_N(n)$	$jIm[X(k)]$
15 Symmetry	$x(n)$ is an arbitrary real-valued sequence	$X(k)=X^*((N-k))_N R_N(k)$ $X_R(k)=Re[X((N-k))_N]R_N(k)$ $X_I(k)=-Im[X((N-k))_N]R_N(k)$ $|X(k)|=|X((N-k))_N|R_N(k)$ $arg[X(k)]=-arg[X((-k))_N]R_N(k)$
16 Parseval's theorem		$\displaystyle\sum_{n=0}^{N-1}|x(n)|^2 = \frac{1}{N}\sum_{k=0}^{N-1}|X(k)|^2$

Since $x_1(n) = \text{Re}[y(n)]$ and $x_2(n) = \text{Im}[y(n)]$, we obtain the following from Properties 11 and 12 in Table 5.3:

$$X_1(k) = \text{DFT}\{\text{Re}[y(n)]\} = \frac{1}{2}\left[Y(k) + Y^*\left((N-k)\right)_N\right]R_N(k)$$

$$X_2(k) = \text{DFT}\{\text{Im}[y(n)]\} = \frac{1}{2j}\left[Y(k) - Y^*\left((N-k)\right)_N\right]R_N(k)$$

After obtaining $Y(k)$ via DFT, we can further obtain $X_1(k)$ and $X_2(k)$ from the above two equations. The calculation complexity of such method is much less than that by calculating $X_1(k)$ and $X_2(k)$ individually.

5.3 Some Topics in the Theory of DFT

In this section, some problems related to the theories and applications of DFT are introduced, including the resolution of DFT and its parameter selection, the zero-padding for DFT and the relative resolution, the time duration and frequency band of signals, the spectral leakage caused by signal truncation, the fence effect, and the frequency aliasing. At the end of this section, we will briefly introduce the concept of the two-dimensional Fourier transform.

5.3.1 Frequency Aliasing

In Section 4.2 of the previous chapter, we introduced the concept and cause of frequency aliasing. In this section, we will further study this problem associated with the DFT. Based on the sampling theorem, the sampling of a continuous-time signal should satisfy $\Omega_s > 2\Omega_M$, or the sampling period T should satisfy

$$T = \frac{2\pi}{\Omega_s} < \frac{\pi}{\Omega_M} = \frac{1}{2F_M} \tag{5.27}$$

where $F_M = \frac{\Omega_M}{2\pi}$ denotes the maximum frequency component in the signal. If we suppose the sampling interval in the frequency sequence as F_0, then the corresponding period for the periodic time-domain signal or the length of the signal T_0 should satisfy

$$T_0 = \frac{1}{F_0} \tag{5.28}$$

From eqs. (5.27) and (5.28), we see a fact as follows. If we want to increase the maximum frequency component F_M, then the sampling interval T in the time domain will decrease, and also the sampling frequency $\Omega_s = 2\pi F_s$ will increase. Since the sampling number (or the length of the sequence) satisfies

$$\frac{F_s}{F_0} = \frac{T_0}{T} = N \tag{5.29}$$

if we increase F_s and fix N unchanged, then F_0 will also increase, resulting in the decrease of the resolution of DFT. Otherwise, if we want to increase the resolution of DFT, we need to increase T_0, resulting in the increase of T when N is fixed. If we want to avoid the aliasing, we have to decrease the maximal frequency component F_M. The only way to get both rather high F_M and rather high resolution is to increase the length N of the data as follows:

$$N = \frac{F_s}{F_0} > \frac{2F_M}{F_0} \tag{5.30}$$

Example 5.3: The frequency resolution of a given signal is $F_0 \leq 10$ Hz, and the maximal frequency component is $F_M = 4$ kHz. Try to determine (1) the minimum length T_0 of the signal, (2) the maximum sampling interval T, and (3) the minimum sampling number N.

Solution: (1) The minimum length T_0 can be determined by the frequency resolution F_0. Since $1/F_0 = 1/10 = 0.1(\text{s})$, T_0 can be determined as

$$T_0 \geq 1/F_0 = 0.1 \text{ (s)}$$

(2) Based on the sampling theorem $\Omega_s > 2\Omega_M$ or $F_s > 2F_M$, we have

$$T < \frac{1}{2F_M} = \frac{1}{2 \times 4 \times 10^3} = 0.125 \times 10^{-3} \text{ (s)}$$

(3) The minimum sampling number N satisfies

$$N > \frac{2F_M}{F_0} = \frac{2 \times 4 \times 10^3}{10} = 800$$

5.3.2 Spectrum Leakage

In practice, the signal to be processed is sometimes very lengthy. However, the signal to be processed in digital signal processing can only be of a finite length, or the sequence can only be of finite duration. Therefore, we have to conduct the truncation to a long sequence. Generally, the truncation is operated by multiplying a window function by the long sequence. The concept and applications of window functions will be introduced in the later chapters.

Given a discrete-time signal $x_1(n)$, whose spectrum is $X_1(k)$ or $X_1(e^{j\omega})$, we can obtain a finite-length signal $x_2(n)$ by taking the truncation with a rectangular window $w(n)$, yielding the spectrum as $X_2(k)$ or $X_2(e^{j\omega})$. The signal and window and the corresponding spectra are shown in Figure 5.4.

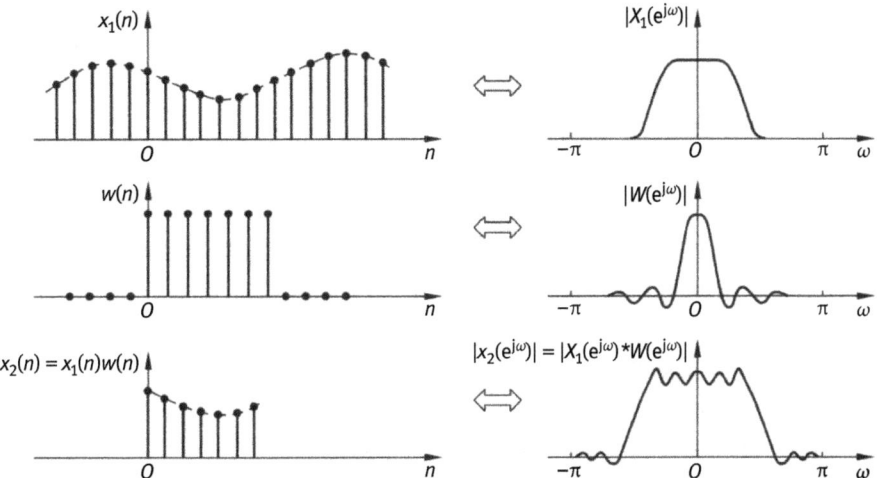

Figure 5.4: The spectrum leakage caused by signal truncation. Top figure: the original sequence and its spectrum. Middle figure: the rectangular window and its spectrum. Bottom figure: the truncated sequence and its spectrum.

As shown in Figure 5.4, the multiplication between $x_1(n)$ and $w(n)$ corresponds to the truncation to $x_1(n)$. The length of the truncated signal $x_2(n)$ is the same as that of $w(n)$. Based on the multiplication property of Fourier transform, the multiplication in the time domain corresponds to the convolution of the spectra in the frequency domain. Thus, the spectrum $X_2(e^{j\omega})$ is broaden relative to $X_1(e^{j\omega})$. Such broaden is referred to as spectrum leakage.

The spectrum leakage is harmful to signal analysis and processing. First, the leaked spectrum cannot reflect the reality of the original spectrum characteristics. Second, due to the periodicity of the spectrum for the discrete-time signal, the leakage may cause frequency aliasing. Therefore, we should find ways to reduce the spectrum leakage in digital signal processing. There are some methods commonly used to reduce the spectrum leakage, including the use of long signal sequence and the use of nonrectangular windows, such as triangle window and raised-cosine window, to make the side lobe smaller in the spectrum.

5.3.3 Fence Effect

Since the spectrum of DFT is discrete, it seems that we watch a scene through a "fence" when we conduct the signal analysis using such spectrum. Such effect is referred to as the "fence effect." The disadvantage of the fence effect is that it may cause some useful components in the spectrum being omitted, so as to influence the outcome of signal analysis and processing.

The basic method to reduce the fence effect is to make the sampling from the continuous spectrum to the discrete spectrum more intensive, that is, the number of points N in the spectrum sequence should be increased. If the date length in the time domain does not increase, the increase of N is equal to padding some zeros at the end of the time-domain data. Since the frequency interval in the DFT spectrum is $\omega_0 = 2\pi/N$, ω_0 will decrease if N is increased, resulting in a more dense spectrum.

5.3.4 Frequency Resolution and Parameter Selection in DFT

1. The concept of resolution

The resolution is a commonly used concept in signal and image processing areas. There are two basic meanings: one is the ability to distinguish two adjacent signal peaks or spectral peaks, the other is the size of the data sample interval in discrete signal or discrete spectrum. In image processing, the resolution is generally used to demonstrate the total number of pixels in an image.

In the area of digital signal processing, the resolution can be divided into two aspects: the time resolution and the frequency resolution, indicating the time interval or frequency band watched by a window. The narrower the window, the higher its resolution.

2. The frequency resolution in FT and DTFT

(1)　The frequency resolution in FT

Given a continuous-time signal $x_T(t)$ whose length is T_L seconds, its spectrum is $X_T(j\Omega)$. The frequency resolution of $X_T(j\Omega)$ is

$$\Delta F = \frac{\Delta\Omega}{2\pi} = \frac{1}{T_L} \, (\text{Hz}) \tag{5.31}$$

Therefore, the minimal frequency interval distinguishable of $X_T(j\Omega)$ will be $1/T_L$ (Hz).

(2)   The frequency resolution in DTFT

If we sample the continuous-time signal $x_T(t)$ with the sampling frequency $F_s = 1/T$, the discrete-time signal $x_M(n)$ with finite duration can be obtained. We can also consider that $x_M(n)$ is obtained from an infinite discrete-time signal $x(n)$ by taking a window $w(n)$ whose time duration is $M = T_L/T$. There are $M = N = T_L/T$ date samples in $x_M(n)$ during T_L seconds. Thus, the frequency resolution in DTFT is

$$\Delta f = \frac{\Delta \omega}{2\pi} = \frac{F_s}{M} \tag{5.32}$$

Since

$$\Delta f = \frac{F_s}{M} = \frac{1}{MT} = \frac{1}{T_L} \tag{5.33}$$

we see that the frequency resolution is inversely proportional to the signal length T_L; the same as that in eq. (5.31). It is obvious that the resolution can be improved if we increase the data length M of the signal.

3. The resolution in DFT

Based on the definition of the DFT, the frequency interval in the spectrum $X_N(k)$ of $x_N(n)$ by using DFT is

$$\Delta f = \frac{F_s}{N} \tag{5.34}$$

This is also a form of frequency resolution. If we want to improve the frequency resolution further, we can increase the data length N to reduce Δf. Furthermore, if the sequence $x_N(n)$ is obtained from $x_M(n)$ and M is fixed, there are still two methods to increase the data length N. One is to increase the density of the frequency bins, and the other is to pad zeros at the end of $x_M(n)$ to make the sequence reaching the desired length N.

However, the truth frequency resolution is not improved by using both above methods, since the effective data length T or M does not increase. This means that it is impossible to separate two nearly fused peaks in the spectrum only by using the zero-padding if the data length is short. Therefore, Δf in the DFT is usually called the "calculated frequency"; it may not reflect the true resolution of the spectrum.

We see from the above discussion that the frequency resolution is inversely proportional to the signal length. If the data length remains unchanged, the selection of different window functions will cause different frequency resolutions and different frequency leakages.

On the other hand, the development of new signal processing algorithms may further improve the frequency resolution of signals. For example, modern spectral estimation methods introduced in later chapters may get a higher frequency resolution due to the implicit extrapolation of finite data.

Example 5.4: A discrete-time signal $x(n)$ is composed of three sinusoidal signals shown as follows:

$$x(n) = \sin(2\pi f_1 n) + \sin(2\pi f_2 n) + \sin(2\pi f_3 n)$$

where $f_1 = 2\,\text{Hz}$, $f_2 = 2.02\,\text{Hz}$, and $f_3 = 2.07\,\text{Hz}$. The sampling frequency is set as $F_s = 10\,\text{Hz}$, and the data length is set as $N_1 = 128$, $N_2 = 256$, $N_3 = 512$. Try to draw the DFT spectra of the signal in the three different data length conditions using MATLAB programming.

Solution: If the data length is $N_1 = 128$, we have $\Delta f = F_s/N_1 = 10/128 = 0.078\text{Hz}$. Since $f_2 - f_1 = 2.02 - 2 = 0.02\,\text{Hz} < \Delta f$ and $f_3 - f_2 = 2.07 - 2.02 = 0.05\,\text{Hz} < \Delta f$, the three peaks in the spectrum cannot be separated. The results are shown in Figure 5.5(a).

If the data length is $N_2 = 256$, then $\Delta f = F_s/N_2 = 10/256 = 0.039\,\text{Hz}$. It is obvious that the resolution of DFT is better than $f_3 - f_2 = 0.05\,\text{Hz}$, but it still cannot distinguish $f_2 - f_1 = 0.02\,\text{Hz} < \Delta f$. The results for $N_2 = 256$ are shown in Figure 5.5(b). We can tell two different peaks in the spectrum.

If the data length is $N_3 = 512$, then $\Delta f = F_s/N_2 = 10/512 = 0.019\,\text{Hz}$. At this time, the three peaks in the spectrum can all be distinguished, as shown in Figure 5.5(c).

The MATLAB program is shown as follows:

```
% Calculate the spectrum by DFT and draw its waveforms and spectra.
clear
k=2; n=256*k; nn=1:n;
f1=2; f2=2.02; f3=2.07; fs=10;
x1=zeros(n,1); x2=zeros(n,1); x3=zeros(n,1);
x1=sin(2*pi*f1/fs.*nn'); x2=sin(2*pi*f2/fs.*nn');
x3=sin(2*pi*f3/fs.*nn');
xx=x1+x2+x3; xxf=abs(fft(xx));
ff=1:n; ff1=0:(fs/2)/(n/2):(fs/2)-(fs/2)/(n/2); ff1=0:fs/n:fs/2-fs/n;
figure(1)
subplot(211); plot(xx); title('Waveform of the signal'); xlabel('n');
   ylabel('Amplitude'); axis([0 n -3 3])
subplot(212); plot(ff1',xxf(1:n/2)); title('Spectrum of the signal');
   xlabel('f/Hz'); ylabel('Magnitude'); axis([0 fs/2 0 600])
% The calculation of DFT is conducted by the fast Fourier transform.
```

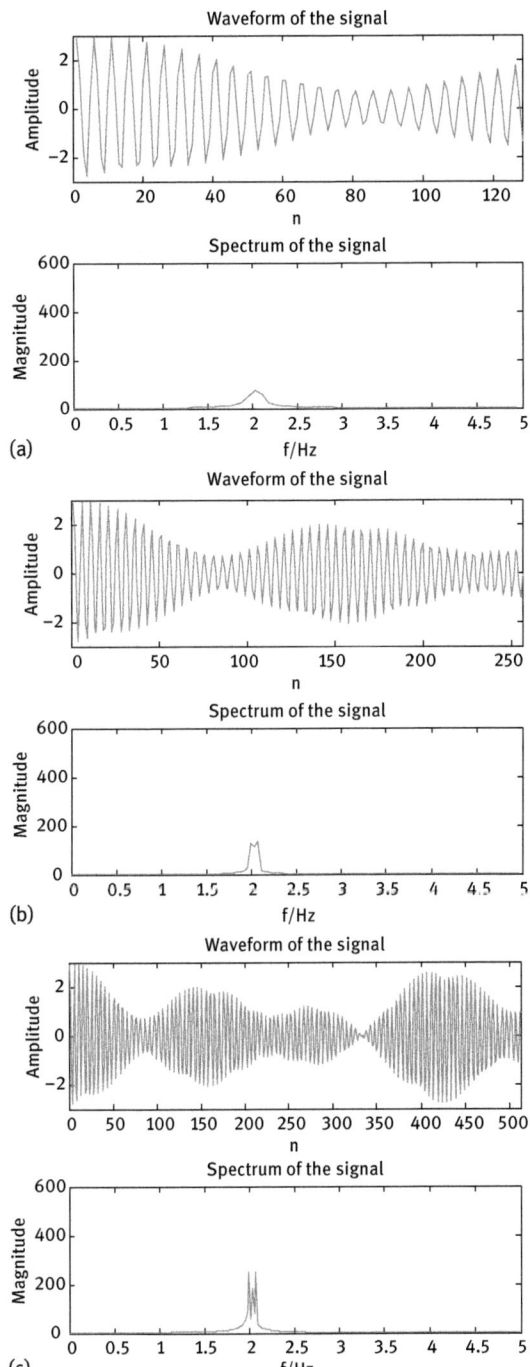

Figure 5.5: Frequency resolution of DFT: (a) $N_1 = 128$, (b) $N_2 = 256$, and (c) $N_3 = 512$.

4. The parameter selection in DFT

The resolution is an important concept in signal analysis and processing. Actually, in addition to the frequency resolution, there are some other problems related to the concept of resolution, including the time resolution, the restriction between frequency resolution and time resolution, and the resolution selection problem.

Corresponding to the frequency resolution, the time resolution refers to the ability of the adjacent signal peaks in the time domain to be distinguished.

For a fast-changing signal, we usually want to get a good time resolution and ignore its frequency resolution; while for a slow-changing signal, we usually want to get a good frequency resolution and ignore its time resolution. How to choose the appropriate time resolution and frequency resolution according to the characteristics of the signal will be given in a later chapter.

5.3.5 Zero-padding

The significances for zero-padding operation in signal processing are mainly as follows. It can improve the calculated resolution of the DFT spectrum, and it can result in a high calculating efficiency. When we calculate the linear convolution with DFT or FFT, we usually need to conduct the zero-padding for signal sequences.

Although zero-padding is important for digital signal processing, it cannot improve the "physical resolution" of the spectrum since it does not increase the information of the signals.

Example 5.5: A discrete-time signal $x(n)$ is composed of three sinusoidal signals; the same as that in Example 5.4. The three frequencies are $f_1 = 2\,\text{Hz}$, $f_2 = 2.02\,\text{Hz}$, and $f_3 = 2.07\,\text{Hz}$, and the sampling frequency is $F_s = 10\,\text{Hz}$. The data length is $N = 256$. Try to draw the spectrum of the signal with MATLAB programming. Then conduct the zero-padding for $x(n)$ to have $N = 1024$. Draw the spectrum of $x(n)$ again.
Solution: The MATLAB program is shown as follows:

```
clear;
k=1; n=256; nn=1:n;
f1=2; f2=2.02; f3=2.07; fs=10;
x1=zeros(n,1); x2=zeros(n,1); x3=zeros(n,1);
x1=sin(2*pi*f1/fs.*nn'); x2=sin(2*pi*f2/fs.*nn');
x3=sin(2*pi*f3/fs.*nn');
xx=x1+x2+x3; xxf=abs(fft(xx)); ff=1:n; ff1=0:fs/n:fs/2-fs/n;
figure(1);
subplot(211); plot(xx); title('Waveform of the signal'); xlabel('n');
   ylabel('Amplitude'); axis([0 n -3 3]);
subplot(212); plot(ff1',xxf(1:n/2));
```

```
title('Spectrum of the signal'); xlabel('f/Hz'); ylabel('Magnitude');
   axis([0 fs/2 0 600]);
% Zeropadding for the signal sequence
m=1; n1=1024*m; bb=zeros(n1,1); xxb=zeros(n1,1); xx1=zeros(n1,1);
   xx1(1:n)=xx;
xxb=xx1+bb; xxfb=abs(fft(xxb)); ffb=0:fs/n1:fs/2-fs/n1;
figure(2);
subplot(211); plot(xx); title('Waveform of the signal'); xlabel('n');
   ylabel('Amplitude'); axis([0 n -3 3]);
subplot(212); plot(ff1',xxf(1:n/2));
title('Spectrum of the signal'); xlabel('f/Hz'); ylabel('Magnitude');
   axis([0 fs/2 0 600]);
```

Figure 5.6 shows the waveforms of the signal without and with zero-padding and their spectra, respectively. Obviously, the spectrum of the zero-padded signal seems to be of better resolution, since there is an increase of the frequency bins in its spectrum. However, the spectrum without zero-padded signal can still distinguish two peaks. It means that the physical resolution does not improve by zero-padding.

5.3.6 Time Width and Bandwidth

The so-called time width is the time duration of a signal. The frequency width is usually called the bandwidth. Based on the time scaling property of Fourier transform, if

$$x(t) \leftrightarrow X(j\Omega) \tag{5.35}$$

then

$$x(at) \leftrightarrow \frac{1}{|a|} X(j\frac{\Omega}{a}) \tag{5.36}$$

The above equation means that both time width and bandwidth cannot be decreased simultaneously. For example, the time width of a rectangular signal is in $(-T_1, T_1)$, whose spectrum is a sinc function. If T_1 is a finite number, the spectrum of the sinc function will spread to $(-\infty, +\infty)$. If $T_1 \to \infty$, the sinc function will tend to be a $\delta(\cdot)$ function, and if $T_1 \to 0$, the sinc function will be a flat horizontal line in the frequency domain.

The relation between the time width and the bandwidth can also be expressed with the product of time width and bandwidth.

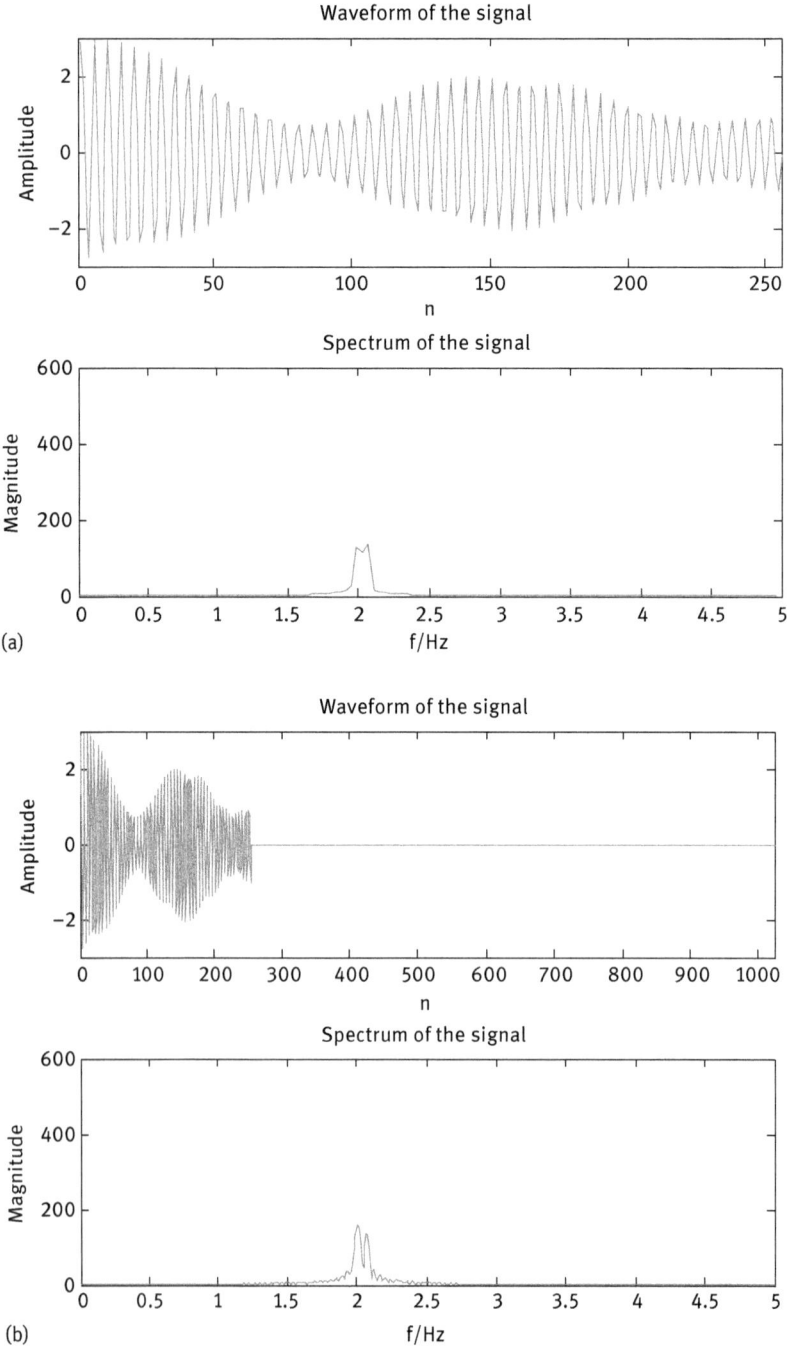

Figure 5.6: Waveforms and spectra of the signal without and with zero-padding: (a) without zero-padding and (b) with zero-padding.

Definition 5.2 (The product of time width and bandwidth): The equivalent time width (TW) and the equivalent bandwidth (FW) of the signal $x(n)$ in mean square sense are as follows:

$$(TW)^2 = \frac{\sum\limits_{n=-\infty}^{\infty} n^2 |x(n)|^2}{\sum\limits_{n=-\infty}^{\infty} |x(n)|^2} \tag{5.37}$$

$$(FW)^2 = \frac{\int_{-F_S/2}^{F_S/2} f^2 |X(f)|^2 df}{\int_{-F_S/2}^{F_S/2} |X(f)|^2 df} \tag{5.38}$$

Then the product of time width and bandwidth is restricted by the following relation:

$$(TW) \cdot (FW) \geq \frac{1}{4\pi} \tag{5.39}$$

The above relation is also called the uncertainty principle of the product of time width and bandwidth.

5.4 A Brief Introduction to the Two-Dimensional Fourier Transform

With the development of the image processing, the two-dimensional (2D) Fourier transform becomes increasingly more important. This section will introduce briefly the definition, properties, and basic applications of the 2D Fourier transform.

Suppose a 2D discrete-time signal (or a 2D sequence) $x(n_1, n_2)$ is the function of discrete spatial variables n_1 and n_2. Generally, almost all problems in one-dimensional signal processing can be extended parallelly to 2D signal processing, such as the signal expression, sampling and transform, system analysis and fast algorithms, and so on. However, since a 2D function is a function with double variables, we will encounter some special problems in extending one-dimensional signal processing to the 2D signal processing.

5.4.1 Commonly Used Two-Dimensional Discrete Sequences

1. The unit impulse sequence

The 2D unit impulse sequence is defined as

$$\delta(n_1, n_2) = \begin{cases} 1, & n_1 = n_2 = 0 \\ 0, & \text{others} \end{cases} \tag{5.40}$$

If $\delta(n)$ is a one-dimensional unit impulse sequence, then

$$\delta(n_1, n_2) = \delta(n_1)\delta(n_2) \tag{5.41}$$

If $x(n_1, n_2) = x(n_1)x(n_2)$ is satisfied, then $x(n_1, n_2)$ is referred to as a separable 2D sequence.

2. The unit step sequence

The 2D unit step sequence is defined as

$$u(n_1, n_2) = \begin{cases} 1, & n_1 \geq 0, n_2 \geq 0 \\ 0, & \text{others} \end{cases} \tag{5.42}$$

Since $u(n_1, n_2) = u(n_1)u(n_2)$, the 2D unit step sequence is a separable sequence.

3. The exponential sequence

The 2D exponential sequence is shown as

$$x(n_1, n_2) = \alpha^{n_1}\beta^{n_2}, \quad n_1 = -\infty \sim +\infty, \; n_2 = -\infty \sim +\infty \tag{5.43}$$

It is obvious that the 2D exponential sequence is a separable sequence. If $\alpha = e^{j\omega_1}$ and $\beta = e^{j\omega_2}$, then

$$x(n_1, n_2) = e^{j\omega_1 n_1}e^{j\omega_2 n_2} = e^{j(\omega_1 n_1 + \omega_2 n_2)} = \cos(\omega_1 n_1 + \omega_2 n_2) + j\sin(\omega_1 n_1 + \omega_2 n_2) \tag{5.44}$$

is a 2D complex exponential or complex sinusoidal sequence, connected by Euler's relation.

5.4.2 The Definitions of Two-Dimensional Z-Transform and Discrete-Time Fourier Transform

1. The definition of the 2D z-transform

Definition 5.3 (The 2D z-transform): The 2D z-transform of the 2D sequence $x(n_1, n_2)$ is defined as

$$X(z_1, z_2) = \sum_{n_1 = -\infty}^{\infty} \sum_{n_2 = -\infty}^{\infty} x(n_1, n_2) z_1^{-n_1} z_2^{-n_2} \tag{5.45}$$

Its inverse transform is defined as

$$x(n_1, n_2) = \frac{1}{(2\pi j)^2} \int_{C_1} \int_{C_2} X(z_1, z_2) z_1^{n_1 - 1} z_2^{n_2 - 1} dz_1 dz_2 \tag{5.46}$$

The region of convergence (ROC) of $X(z_1, z_2)$ is defined as the region in which $X(z_1, z_2)$ is converged.

2. The 2D discrete-time Fourier transform

Definition 5.4 (The 2D discrete-time Fourier transform): The 2D discrete-time Fourier transform (2D-DTFT) of the 2D sequence $x(n_1, n_2)$ is defined as

$$X(e^{j\omega_1}, e^{j\omega_2}) = \sum_{n_1 = -\infty}^{\infty} \sum_{n_2 = -\infty}^{\infty} x(n_1, n_2) e^{-j\omega_1 n_1} e^{-j\omega_2 n_2} \tag{5.47}$$

Its inverse transform is defined as

$$x(n_1, n_2) = \frac{1}{4\pi^2} \int_{-\pi}^{\pi} \int_{-\pi}^{\pi} X(e^{j\omega_1}, e^{j\omega_2}) e^{j\omega_1 n_1} e^{j\omega_2 n_2} d\omega_1 d\omega_2 \tag{5.48}$$

5.4.3 Properties of Two-Dimensional Discrete-Time Fourier Transform

Suppose the 2D-DTFTs of 2D sequences $x(n_1, n_2)$ and $y(n_1, n_2)$ are shown as $x(n_1, n_2) \leftrightarrow X(e^{j\omega_1}, e^{j\omega_2})$ and $y(n_1, n_2) \leftrightarrow Y(e^{j\omega_1}, e^{j\omega_2})$, respectively. The properties of 2D-DTFT are as follows:

Property 5.7: Linearity

$$[ax(n_1, n_2) + by(n_1, n_2)] \leftrightarrow aX(e^{j\omega_1}, e^{j\omega_2}) + bY(e^{j\omega_1}, e^{j\omega_2}) \tag{5.49}$$

Property 5.8: Convolution in the time domain

$$x(n_1, n_2) \star y(n_1, n_2) \leftrightarrow X(e^{j\omega_1}, e^{j\omega_2}) Y(e^{j\omega_1}, e^{j\omega_2}) \tag{5.50}$$

Property 5.9: Multiplication in the time domain

$$x(n_1, n_2) y(n_1, n_2) \leftrightarrow \frac{1}{4\pi^2} \int_{-\pi}^{\pi} \int_{-\pi}^{\pi} X(e^{j\theta_1}, e^{j\theta_2}) Y(e^{j(\omega_1 - \theta_1)}, e^{j(\omega_2 - \theta_2)}) d\theta_1 d\theta_2 \tag{5.51}$$

Property 5.10: Time shifting

$$x(n_1 - m_1, n_2 - m_2) \leftrightarrow X(e^{j\omega_1}, e^{j\omega_2})e^{-j\omega_1 m_1}e^{-j\omega_2 m_2} \tag{5.52}$$

Property 5.11: Frequency shifting

$$e^{j\theta_1 n_1}e^{j\theta_2 n_2}x(n_1, n_2) \leftrightarrow X(e^{j(\omega_1 - \theta_1)}, e^{j(\omega_2 - \theta_2)}) \tag{5.53}$$

Property 5.12: Differentiation in the frequency domain

$$-jn_1 x(n_1, n_2) \leftrightarrow \frac{\partial X(e^{j\omega_1}, e^{j\omega_2})}{\partial \omega_1}, \quad -jn_2 x(n_1, n_2) \leftrightarrow \frac{\partial X(e^{j\omega_1}, e^{j\omega_2})}{\partial \omega_2} \tag{5.54}$$

Property 5.13: Parseval's relation

$$\sum_{n_1 = -\infty}^{\infty} \sum_{n_2 = -\infty}^{\infty} |x(n_1, n_2)|^2 \leftrightarrow \frac{1}{4\pi^2} \int_{-\pi}^{\pi} \int_{-\pi}^{\pi} |X(e^{j\omega_1}, e^{j\omega_2})|^2 d\omega_1 d\omega_2 \tag{5.55}$$

Property 5.14: Separability

$$x_1(n_1)x_2(n_2) \leftrightarrow X_1(e^{j\omega_1})X_2(e^{j\omega_2}) \tag{5.56}$$

5.4.4 Two-Dimensional Discrete Fourier Transform

1. The definition of the 2D discrete Fourier transform

Definition 5.5 (The 2D discrete Fourier transform): Suppose the 2D discrete-time signal $x(n_1, n_2)$ $(n_1 = 0, 1, \ldots, N_1 - 1; n_2 = 0, 1, \ldots, N_2 - 1)$ is a finite-length sequence; the 2D discrete Fourier transform (2D-DFT) is defined as

$$X(k_1, k_2) = \text{DFT}[x(n_1, n_2)] = \sum_{n_1 = 0}^{N_1 - 1} \sum_{n_2 = 0}^{N_2 - 1} x(n_1, n_2)e^{-j\frac{2\pi}{N_1}n_1 k_1}e^{-j\frac{2\pi}{N_2}n_2 k_2}$$
$$k_1 = 0, 1, \ldots, N_1 - 1, \quad k_2 = 0, 1, \ldots, N_2 - 1 \tag{5.57}$$

Its inverse transform is defined as

$$x(n_1, n_2) = \frac{1}{N_1 N_2} \sum_{k_1 = 0}^{N_1 - 1} \sum_{k_2 = 0}^{N_2 - 1} X(k_1, k_2)e^{j\frac{2\pi}{N_1}n_1 k_1}e^{j\frac{2\pi}{N_2}n_2 k_2}$$
$$n_1 = 0, 1, \ldots, N_1 - 1, \quad n_2 = 0, 1, \ldots, N_2 - 1 \tag{5.58}$$

Similar to the one-dimensional DFT, both $x(n_1, n_2)$ and $X(k_1, k_2)$ can be considered as periodic. The period in the direction of n_1 and k_1 is N_1, while the period in the direction of n_2 and k_2 is N_2.

2. The relation between 2D-DFT and 2D-DTFT

The relation between 2D-DFT and 2D-DTFT can be connected by the following equation:

$$X(k_1, k_2) = X(e^{j\omega_1}, e^{j\omega_2})\Big|_{\omega_1 = \frac{2\pi}{N_1}k_1, \ \omega_2 = \frac{2\pi}{N_2}k_2} \tag{5.59}$$

3. The concept of the 2D frequency

The 2D sequence $x(n_1, n_2)$ expresses the distribution of the signal in the 2D space. Therefore, the independent variables n_1 and n_2 are the sampling in spatial distances. The 2D frequencies ω_1 and ω_2 (or 2D discrete frequencies k_1 and k_2) demonstrate the changing rate of the signal with respect to the spatial distances. For 2D-DTFT, the period of the spatial frequency is still 2π. For 2D-DFT, the period of the spatial frequencies is N_1 and N_2, respectively.

5.4.5 Applications of Two-Dimensional Discrete Fourier Transform

Example 5.6: A 2D sequence is given as $x(n_1, n_2) = \begin{cases} 0.8, & n_1 = n_2 = 0 \\ 0.4, & n_1 = \pm 1, \ n_2 = 0 \\ 0.4, & n_1 = 0, \ n_2 = \pm 1 \\ 0, & \text{others} \end{cases}$, whose

waveform is shown in Figure 5.7(a). Try to find its 2D-DTFT.
Solution: From the definition of 2D-DTFT, we obtain

$$X(e^{j\omega_1}, e^{j\omega_2}) = 0.8 + 0.4(e^{j\omega_1} + e^{-j\omega_1}) + 0.4(e^{j\omega_2} + e^{-j\omega_2})$$
$$= 0.8 + 0.8 \cos \omega_1 + 0.8 \cos \omega_2$$

The spectrum of $X(e^{j\omega_1}, e^{j\omega_2})$ is shown in Figure 5.7(b).

Example 5.7: A 2D sequence is given as $x(n_1, n_2) = \begin{cases} 1, & 0 \le n_1 \le N_1 - 1, \ 0 \le n_2 \le N_2 - 1 \\ 0, & \text{others} \end{cases}$.

Find $X(e^{j\omega_1}, e^{j\omega_2})$.

Solution: Based on the definition of 2D-DTFT, we obtain

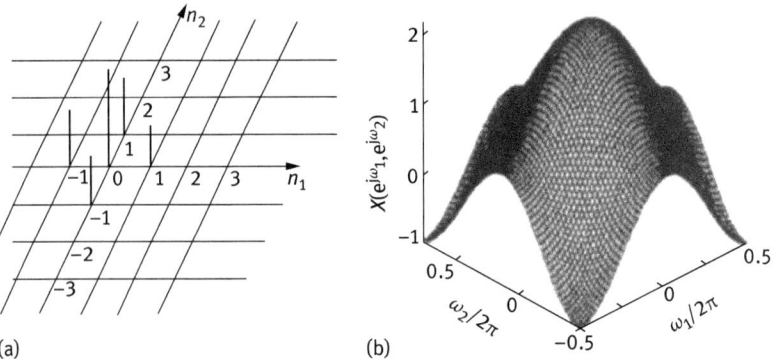

Figure 5.7: The 2D sequence and its 2D-DTFT: (a) the 2D sequence and (b) the spectrum of the 2D sequence.

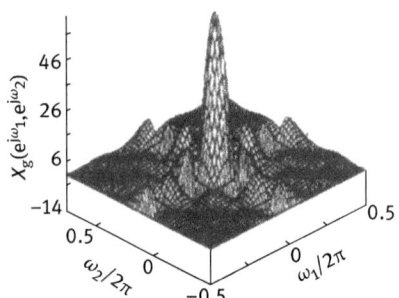

Figure 5.8: The spectrum of Example 5.7.

$$X(e^{j\omega_1}, e^{j\omega_2}) = \sum_{n_1=0}^{N_1-1} \sum_{n_2=0}^{N_2-1} e^{-j\omega_1 n_1} e^{-j\omega_2 n_2} = \frac{1-e^{-j\omega_1 N_1}}{1-e^{-j\omega_1}} \cdot \frac{1-e^{-j\omega_2 N_2}}{1-e^{-j\omega_2}}$$

$$= e^{-j[\omega_1(N_1-1)+\omega_2(N_2-1)]/2} \frac{\sin\left(\frac{\omega_1 N_1}{2}\right) \sin\left(\frac{\omega_2 N_2}{2}\right)}{\sin\left(\frac{\omega_1}{2}\right) \sin\left(\frac{\omega_2}{2}\right)}$$

The 2D spectrum is shown in Figure 5.8.

Example 5.8: Try to construct a 2D window function based on the one-dimensional Hamming window $w(n) = 0.54 - 0.46\cos\left(\frac{2\pi}{N-1}n\right)$, $n = 1, 2, \ldots, N-1$. The rule of the construction is $w(n_1, n_2) = w(n_1)w(n_2)$.

Solution: From the given construction rule, we have

$$w(n_1, n_2) = w(n_1)w(n_2) = \left[0.54 - 0.46\cos\left(\frac{2\pi}{N-1}n_1\right)\right]\left[0.54 - 0.46\cos\left(\frac{2\pi}{N-1}n_2\right)\right]$$

The waveform $w(n_1, n_2)$ and the spectrum $W(e^{j\omega_1}, e^{j\omega_2})$ are shown in Figure 5.9.

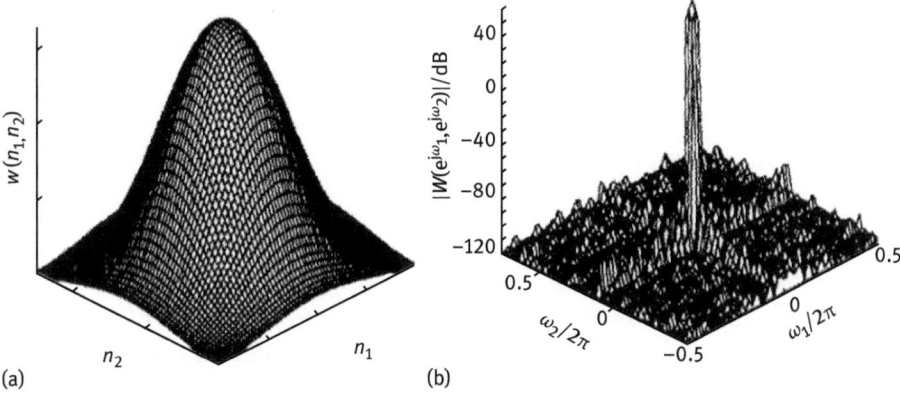

Figure 5.9: The waveform and spectrum of a 2D Hamming window function.

5.5 The Fast Fourier Transform (FFT)

5.5.1 The appearance of FFT

The discrete Fourier transform (DFT) is developed to adapt to the digital signal processing with a digital computer. The DFT can easily transform a discrete-time signal into a discrete spectrum, and the convolution calculation can be done faster by the DFT.

The main shortcoming of the DFT is its large calculation quantity. For example, the complex multiplication of DFT for a sequence with data length N is N^2. If $N = 1024$, the number of complex multiplication is $N^2 = 1,048,576$. To compute such large calculation is very difficult for early computers.

In 1965, J. W. Cooley and J. W. Tukey published a pioneering paper titled "An algorithm for the machine computation of complex Fourier series," in the *Mathematics of Computation* journal, and proposed an effective fast algorithm for the calculation of DFT. After continuous improvement by researchers, it becomes the FFT algorithm, greatly reducing the computation of DFT. For the sequence of data length N, its complex multiplication is reduced to $\frac{N}{2}\log_2 N$. If $N = 1024$, its calculation quantity is just 5,120, much less than that of DFT.

5.5.2 Problems for Calculating DFT Directly and Possible Improvement

1. The calculation analysis of DFT

The DFT and its inverse transform (IDFT) of a sequence with data length N are given again as follows:

$$X(k) = \text{DFT}[x(n)] = \sum_{n=0}^{N-1} x(n) W_N^{nk}, \quad k = 0, 1, \ldots, N-1 \tag{5.60}$$

$$x(n) = \text{IDFT}[X(k)] = \frac{1}{N} \sum_{k=0}^{N-1} X(k) W_N^{-nk}, \quad n = 0, 1, \ldots, N-1 \tag{5.61}$$

where W_N is defined as $W_N = e^{-j\frac{2\pi}{N}}$. In the following discussion, we consider DFT only in eq. (5.60) since the difference between DFT and IDFT is just a constant factor $\frac{1}{N}$. Suppose that $x(n)$, $X(k)$, and W_N are all complex; the calculation of DFT for one value of $X(k)$ needs N complex multiplication and $N-1$ complex addition operations. Thus, the calculation for entire $X(k)$ needs N^2 complex multiplication and $N(N-1)$ complex addition operations.

We see from the above analysis that the complex computation of the DFT is proportional to N^2. Such huge computation is usually not acceptable for digital signal processing.

2. The periodicity and symmetry in DFT

We can see from the calculation of DFT that W_N^{nk} has some properties, such as conjugation and periodicity. Such properties can be used to reduce the computational complexity.

(1)   The conjugation of W_N^{nk}

$$\left(W_N^{nk} \right)^* = \left(e^{-j\frac{2\pi}{N}nk} \right)^* = e^{j\frac{2\pi}{N}nk} = W_N^{-nk} \tag{5.62}$$

(2)   The periodicity of W_N^{nk}

$$W_N^{nk} = W_N^{(n+N)k} = W_N^{n(k+N)} \tag{5.63}$$

(3)   The reducible property of W_N^{nk}

$$W_N^{nk} = W_{mN}^{mnk}, \qquad W_N^{nk} = W_{N/m}^{nk/m} \tag{5.64}$$

And from the above relations, we can also obtain

$$W_N^{n(N-k)} = W_N^{(N-n)k} = W_N^{-nk}, \qquad W_N^{N/2} = -1, \qquad W_N^{(k+N/2)} = -W_N^k \tag{5.65}$$

3. The improvement in the calculation of DFT

By using the symmetry, the periodicity, and the reducible property, we can combine some terms in the calculation of DFT to reduce the computation. On the other hand, we can also separate a long sequence into some short sequences to reduce the computation further.

The so-called fast Fourier transform is just proposed and developed based on the above properties. The FFT algorithm can basically be divided into two types: the decimation-in-time algorithm and the decimation-in-frequency algorithm.

5.5.3 Decimation-in-Time Radix-2 FFT Algorithm

1. The principle of the algorithm

A sequence $x(n)$ is given whose data length N satisfies

$$N = 2^M \tag{5.66}$$

where M is a positive integer. If a sequence does not satisfy the relation, we usually conduct zero-padding to have such relation satisfied. We consider computing $X(k)$ by separating $x(n)$ into two $N/2$ point sequences. One consists of the even-numbered points expressed as $x(2r)$, $r = 0, 1, \ldots, N/2-1$, and the other consists of the odd-numbered points expressed as $x(2r+1)$. Thus, we have

$$
\begin{aligned}
X(k) = \text{DFT}[x(n)] &= \sum_{n=0}^{N-1} x(n) W_N^{nk} = \sum_{r=0}^{N/2-1} x(2r) W_N^{2rk} + \sum_{r=0}^{N/2-1} x(2r+1) W_N^{(2r+1)k} \\
&= \sum_{r=0}^{N/2-1} x(2r) W_{N/2}^{rk} + W_N^k \sum_{r=0}^{N/2-1} x(2r+1) W_{N/2}^{rk}
\end{aligned}
\tag{5.67}
$$

where $W_{N/2} = e^{-j\frac{2\pi}{N/2}} = e^{-j\frac{4\pi}{N}} = W_N^2$. If we let

$$X_1(k) = \sum_{r=0}^{N/2-1} x(2r) W_{N/2}^{rk}, \quad k = 0, 1, \ldots, N/2-1 \tag{5.68}$$

$$X_2(k) = \sum_{r=0}^{N/2-1} x(2r+1) W_{N/2}^{rk}, \quad k = 0, 1, \ldots, N/2-1 \tag{5.69}$$

then

$$X(k) = X_1(k) + W_N^k X_2(k), \quad k = 0, 1, \ldots, N/2-1 \tag{5.70}$$

where both $X_1(k)$ and $X_2(k)$ are $N/2$-point DFTs. On the other hand, eq. (5.70) has just expressed the first-half information of $X(k)$. With periodicity $W_{N/2}^{rk} = W_{N/2}^{r(k+N/2)}$, we can obtain

$$X_1(k+N/2) = \sum_{r=0}^{N/2-1} x(2r) W_{N/2}^{r(k+N/2)} = \sum_{r=0}^{N/2-1} x(2r) W_{N/2}^{rk} = X_1(k), \quad k = 0, 1, \ldots, N/2-1$$

$$(5.71)$$

$$X_2(k+N/2) = \sum_{r=0}^{N/2-1} x(2r+1) W_{N/2}^{r(k+N/2)} = \sum_{r=0}^{N/2-1} x(2r+1) W_{N/2}^{rk} = X_2(k), \quad k = 0, 1, \ldots, N/2-1$$

$$(5.72)$$

The above relations demonstrate that the second half of $X_1(k)$ and $X_2(k)$ are the same as their first half ones, respectively. From

$$W_N^{(k+N/2)} = W_N^k W_N^{N/2} = -W_N^k \tag{5.73}$$

we have

$$X(k+N/2) = X_1(k) - W_N^k X_2(k), \quad k = 0, 1, \ldots, N/2-1 \tag{5.74}$$

Thus, the combination of eqs. (5.70) and (5.74) can be used to express the DFT perfectly. As a matter of fact, when we calculate the N-point DFT, we only need to calculate $X_1(k)$ and $X_2(k)$ in the range between 0 and $N/2-1$.

The calculation of eqs. (5.70) and (5.74) can be expressed with the butterfly signal flow graph. If no coefficient is shown on the branch, it means that the coefficient of the branch is 1 (Figure 5.10).

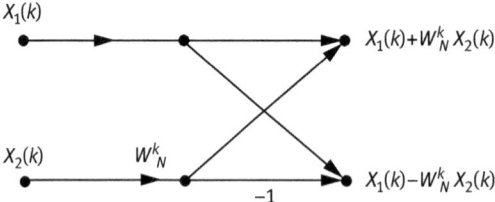

Figure 5.10: The butterfly structure of the decimation-in-time radix-2 FFT algorithm.

The butterfly structure of a signal flow graph is a common expression in the FFT algorithm. It is very clear how the signal $x(n)$ is evolving into $X(k)$ step by step. It can also be clear to reflect the computation of the FFT algorithm.

Based on the butterfly structure, $x(n)$ of data length N can be divided into two sub-signals of data length $N/2$. And then the DFT can be completed for both sub-signals, as shown in Figure 5.11.

The most important benefit of dividing an N-point DFT into two $N/2$-point DFTs is the significant reduction in computation. Furthermore, since $N = 2^M$ is satisfied, $N/2$

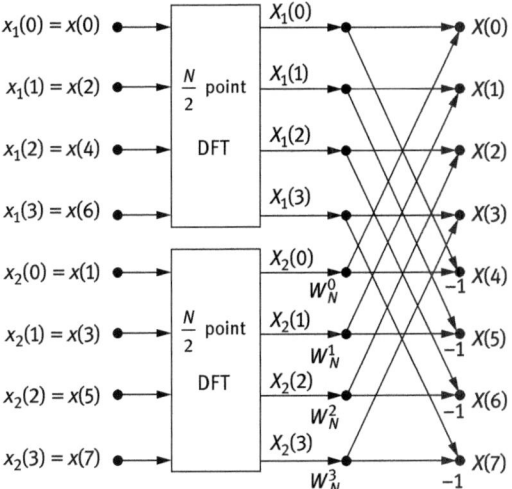

Figure 5.11: An N-point DFT is divided into two $N/2$-point DFTs in the decimation-in-time radix-2 FFT algorithm.

should still be an even number. Then, it can be further divided into two subsequences. Thus,

$$
\begin{aligned}
X_1(k) &= \sum_{l=0}^{N/4-1} x(4l) W_{N/2}^{2lk} + \sum_{l=0}^{N/4-1} x(4l+2) W_{N/2}^{(2l+1)k} \\
&= \sum_{l=0}^{N/4-1} x(4l) W_{N/4}^{lk} + W_{N/2}^{k} \sum_{l=0}^{N/4-1} x(4l+2) W_{N/4}^{lk}, \quad k = 0, 1, \ldots, N/4-1
\end{aligned}
\tag{5.75}
$$

Let

$$
X_3(k) = \sum_{l=0}^{N/4-1} x(4l) W_{N/4}^{lk}, \quad k = 0, 1, \ldots, N/4-1
\tag{5.76}
$$

$$
X_4(k) = \sum_{l=0}^{N/4-1} x(4l+2) W_{N/4}^{lk}, \quad k = 0, 1, \ldots, N/4-1
\tag{5.77}
$$

We have the following results:

$$
X_1(k) = X_3(k) + W_{N/2}^{k} X_4(k), \quad k = 0, 1, \ldots, N/4-1
\tag{5.78}
$$

$$
X_1(k+N/4) = X_3(k) - W_{N/2}^{k} X_4(k), \quad k = 0, 1, \ldots, N/4-1
\tag{5.79}
$$

Similarly,

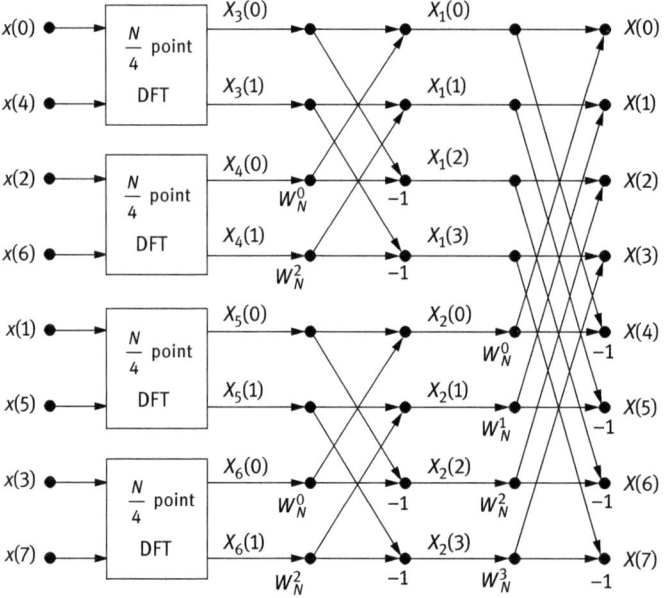

Figure 5.12: An $N = 8$-point DFT is divided into four $N/4$-point DFTs in the decimation-in-time radix-2 FFT algorithm.

$$X_2(k) = X_5(k) + W_{N/2}^k X_6(k), \quad k = 0, 1, \ldots, N/4 - 1 \tag{5.80}$$

$$X_2(k + N/4) = X_5(k) - W_{N/2}^k X_6(k), \quad k = 0, 1, \ldots, N/4 - 1 \tag{5.81}$$

in which

$$X_5(k) = \sum_{l=0}^{N/4-1} x(4l + 1) W_{N/4}^{lk}, \quad k = 0, 1, \ldots, N/4 - 1 \tag{5.82}$$

$$X_6(k) = \sum_{l=0}^{N/4-1} x(4l + 3) W_{N/4}^{lk}, \quad k = 0, 1, \ldots, N/4 - 1 \tag{5.83}$$

If $N = 8$, then $X_3(k)$, $X_4(k)$, $X_5(k)$, and $X_6(k)$ do not need to be divided further, since they are all two-point DFTs. If the data length of $x(n)$ is greater than $N = 8$, for example $N = 16$, 32, 64 or N is a more higher power of 2, then $x(n)$ needs to be divided further based on the above rules, until reaching the two-point DFT. Figure 5.12 demonstrates a sketch map for an eight-point DFT.

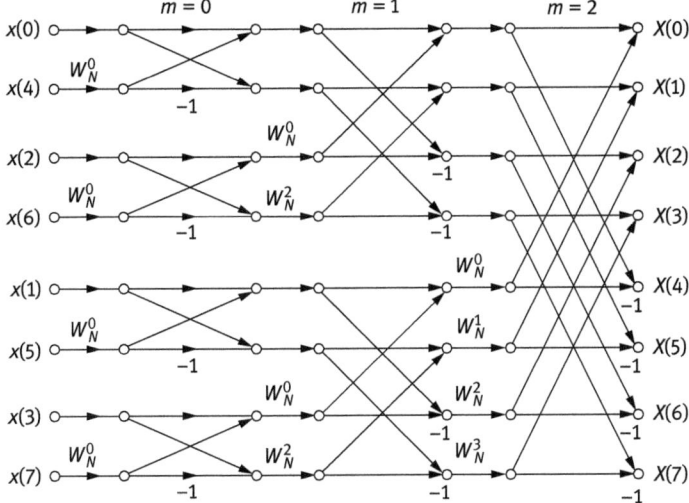

Figure 5.13: The signal flowchart of an eight-point FFT (decimation in time).

Figure 5.13 shows the signal flowchart of an eight-point FFT, which demonstrates a procedure of an eight-point discrete-time signal $x(n)$ evolving into an eight-point spectrum $X(k)$.

2. The features of the algorithm

(1) The computation

We see from Figure 5.13 that when the data length satisfies $N = 2^M$, there are M stages of butterfly structures with $N/2$ operations in each stage. We know that there are one complex multiplication and two additions in each butterfly structure; thus, the total complex multiplication m_F and addition a_F are shown as follows:

$$m_F = \frac{N}{2}M = \frac{N}{2}\log_2 N \tag{5.84}$$

$$a_F = NM = N\log_2 N \tag{5.85}$$

The computation ratio between DFT and FFT is shown as follows:

$$\frac{N^2}{\frac{N}{2}M} = \frac{N^2}{\frac{N}{2}\log_2 N} = \frac{2N}{\log_2 N} \tag{5.86}$$

Table 5.4 shows the computation comparison between DFT and FFT algorithms.

Table 5.4: A comparison of the calculation quantity between DFT and FFT algorithms with same data length.

N	DFT N^2	FFT $\frac{N}{2}\log_2 N$	Ratio of computation $N^2/\left(\frac{N}{2}\log_2 N\right)$
2	4	1	4.0
4	16	4	4.0
8	64	12	5.4
16	256	32	8.0
32	1,024	80	12.8
64	4,096	192	21.4
128	16,384	448	36.6
256	65,536	1,024	64.0
512	262,144	2,304	113.8
1,024	1,048,576	5,120	204.8
2,048	4,194,304	11,264	372.4

(2) The concepts on "stage" and "line"

In the eight-point FFT algorithm shown in Figure 5.13, $m = 0$, $m = 1$, and $m = 2$ on the top of the figure express the stages in the signal flowchart. We can see that there are totally three stages for an eight-point FFT algorithm. The number of stage M should satisfy $M = \log_2 N$, where N is the data length.

In Figure 5.13, there are $N = 8$ horizontal lines. There are N lines for an N-point FFT algorithm.

(3) The butterfly unit and in-place computation

As the fundamental calculation unit, the butterfly unit is shown in Figure 5.10. The calculation in the m stage butterfly unit is shown in eq. (5.5.87).

$$X_{m+1}(k) = X_m(k) + W_N^r X_m(j)$$
$$X_{m+1}(j) = X_m(k) - W_N^r X_m(j) \tag{5.87}$$

where k and j denote the lines in which the operation takes place. We can see from the above equations that the calculation in both k and j lines does not relate to other lines. Such operation style is referred to as the in-place computation. Thus, only one complex array of N storage registers is physically necessary to implement the complete computation if $X_{m+1}(k)$ and $X_{m+1}(p)$ are stored in the same registers as $X_m(k)$ and $X_m(p)$ for an N-point FFT.

(4) The distance between two nodes in a butterfly unit

We see from Figure 5.13 that the distance between two nodes in the same butterfly unit is 1 if $m = 0$. The distance is 2 in the stage of $m = 1$, the distance becomes 4 if

Table 5.5: The distribution of factor W^r of FFT in different stages.

m stage	W^r	r
$m = 0$	W_2^r	$r = 0$
$m = 1$	W_4^r	$r = 0, 1$
$m = 2$	W_8^r	$r = 0, 1, 2, 3$
$\vdots$	$\vdots$	$\vdots$
m	$W_{2^{m+1}}^r$	$r = 0, 1, \ldots, 2^m - 1$
$\vdots$	$\vdots$	$\vdots$
$m = M - 1$	W_N^r	$r = 0, 1, \ldots, N/2 - 1$

$m = 2$, and so on. For an $N = 2^M$-point FFT, the distance between two nodes in the m stage satisfies $j - k = 2^m$.

(5) The distribution of factor W^r

The distribution of factor W^r of FFT in different stages is listed in Table 5.5.

(6) The bit-revised order

In order that the computation may be done in place as just discussed, the input sequence must be stored (or accessed) in a nonsequential order. In fact, the order in which the input data are stored and accessed is referred to as the bit-revised order. The rule of the bit-revised order for an eight-point FFT is shown in Figure 5.14.

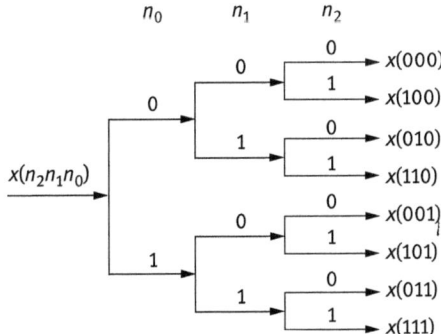

Figure 5.14: The tree diagram depicting the bit reversed storing.

5.5.4 Decimation-in-Frequency Radix-2 FFT Algorithm

1. Algorithm principle

Similar to but different from the decimation-in-time radix-2 FFT algorithm, the basic idea of the decimation-in-frequency radix-2 FFT algorithm is to form smaller and smaller subsequences of the output sequence $X(k)$. FFT algorithms based on this procedure are commonly called decimation-in-frequency algorithms.

We still suppose the sequence length of $x(n)$ satisfies $N = 2^M$, where M is a positive integer. Before $X(k)$ is divided into two subsequences based on its even and odd orders, the input sequence $x(n)$ is divided into two parts: one is the first half of the sequence in the order of n, and the other is the last part of the sequence. Thus,

$$X(k) = \sum_{n=0}^{N-1} x(n) W_N^{nk} = \sum_{n=0}^{N/2-1} x(n) W_N^{nk} \sum_{n=N/2}^{N-1} x(n) W_N^{nk} = \sum_{n=0}^{N/2-1} x(n) W_N^{nk}$$

$$+ \sum_{n=0}^{N/2-1} x(n + \tfrac{N}{2}) W_N^{nk} W_N^{Nk/2} = \sum_{n=0}^{N/2-1} [x(n) + W_N^{Nk/2} x(n + \tfrac{N}{2})] W_N^{nk}$$

(5.88)

where $W_N^{Nk/2} = (-1)^k$. If we suppose $\left.\begin{matrix} k = 2r \\ k = 2r+1 \end{matrix}\right\}$, $r = 0, 1, \ldots, N/2 - 1$, then

$$X(2r) = \sum_{n=0}^{N/2-1} \left[x(n) + x(n + \frac{N}{2}) \right] W_{N/2}^{nr}, \quad r = 0, 1, \ldots, N/2 - 1$$

(5.89)

$$X(2r+1) = \sum_{n=0}^{N/2-1} \left[x(n) - x(n + \frac{N}{2}) \right] W_{N/2}^{nr} W_N^n, \quad r = 0, 1, \ldots, N/2 - 1$$

(5.90)

If we let

$$\left.\begin{matrix} x_1(n) = x(n) + x(n + N/2) \\ x_2(n) = [x(n) - x(n + N/2)] W_N^n \end{matrix}\right\}, \quad n = 0, 1, \ldots, N/2 - 1$$

(5.91)

then

$$X(2r) = \sum_{n=0}^{N/2-1} x_1(n) W_{N/2}^{nr}$$
$$\left.\begin{matrix} \\ \\ \end{matrix}\right\}, \quad r = 0, 1, \ldots, N/2 - 1$$
$$X(2r+1) = \sum_{n=0}^{N/2-1} x_2(n) W_{N/2}^{nr}$$

(5.92)

The above relations can be expressed as a butterfly computation as shown in Figure 5.15.

Thus, $X(k)$ is divided into two $N/2$-point DFTs based on the order of k. For a more general case, such decomposition can be conducted further until two-point DFT is obtained. A signal flow graph of $N = 8$-point decimation-in-frequency radix-2 FFT is shown in Figure 5.16.

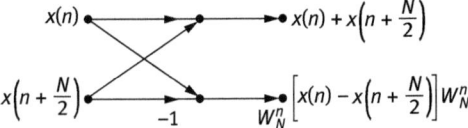

Figure 5.15: The butterfly unit of the decimation-in-frequency radix-2 FFT algorithm.

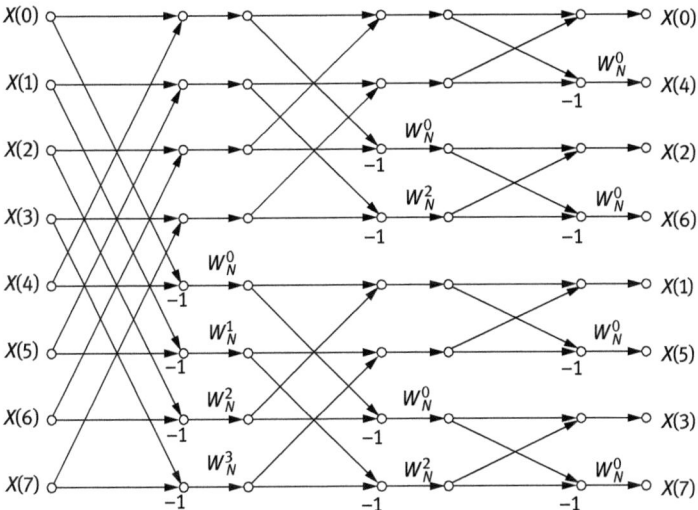

Figure 5.16: The signal graph of eight-point decimation-in-frequency radix-2 FFT.

We see from the figure that $x(n)$ in the decimation-in-frequency radix-2 FFT algorithm is in the normal sorting, while its output $X(k)$ is in the bit-revised sorting.

2. The features of the algorithm

Comparing the decimation-in-frequency radix-2 FFT algorithm and the decimation-in-time radix-2 FFT algorithm, we see that both algorithms have similar or even the same features, such as their in-place computation and the distance between two nodes in a butterfly unit.

3. Similarities and differences with the decimation-in-time radix-2 FFT algorithm

There are some similarities and differences between the decimation-in-frequency algorithm and the decimation-in-time algorithm. For example, the calculating quantity for both algorithms is the same. For a sequence with data length $N = 2^M$, the number of complex multiplication is $(N/2)\log_2 N$ for both algorithms. For the decimation-in-time algorithm, the input is in a bit-reversed sorting,

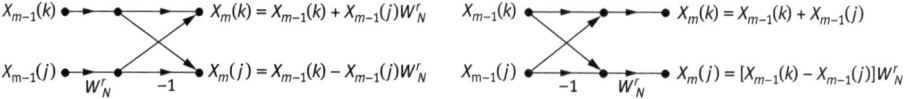

Figure 5.17: Butterfly structures for both algorithms: (a) the decimation-in-time algorithm and (b) the decimation-in-frequency algorithm.

while for the decimation-in-frequency algorithm, the output is in the bit-reversed sorting. Furthermore, the butterfly structure for both algorithms is different, as shown in Figure 5.17.

As a matter of fact, we can get the decimation-in-frequency structure by taking a transpose to the decimation-in-time structure. The transpose operation is to exchange input and output of the signal graph and to reverse all of the branches.

Example 5.9: Calculate FFT and then the inverse FFT of the given signal with the MATLAB programming. The signal is composed of two sinusoidal signals and a white noise.
Solution: Both sinusoidal signals are given as $x_1(n) = 5\sin(0.2\pi n)$ and $x_2(n) = 3\sin(0.4\pi n)$. The MATLAB program is listed as follows.

```
clear all;
% Generating the Gaussian white noise;
N=256; f1=0.1; f2=0.2; fs=1; a1=5; a2=3; w=2*pi/fs;
x=a1*sin(w*f1*(0:N-1))+a2*sin(w*f2*(0:N-1))+randn(1,N);
% Calculating the spectrum with FFT, and then returning to the original
signal;
subplot(3,1,1); plot(x(1:N/4)); f=-0.5:1/N:0.5-1/N; X=fft(x);
y=ifft(X);
title('Original signal'); ylabel('Amplitude'); xlabel('n');
subplot(3,1,2); plot(f,fftshift(abs(X))); title('Spectrum');
ylabel('Mugnitude'); xlabel('f');
subplot(3,1,3); plot(real(x(1:N/4))); title('Resumed signal');
ylabel('Amplitude'); xlabel('n');
```

Figure 5.18 shows the waveform of the original signal and its spectrum, and the waveform of the resumed signal.

5.5.5 Chirp Z-Transform (CZT)

We know that the DFT is actually the frequency sampling of the DTFT, and it is the discrete sampling on the unit circle in the z-plane. A different algorithm called the

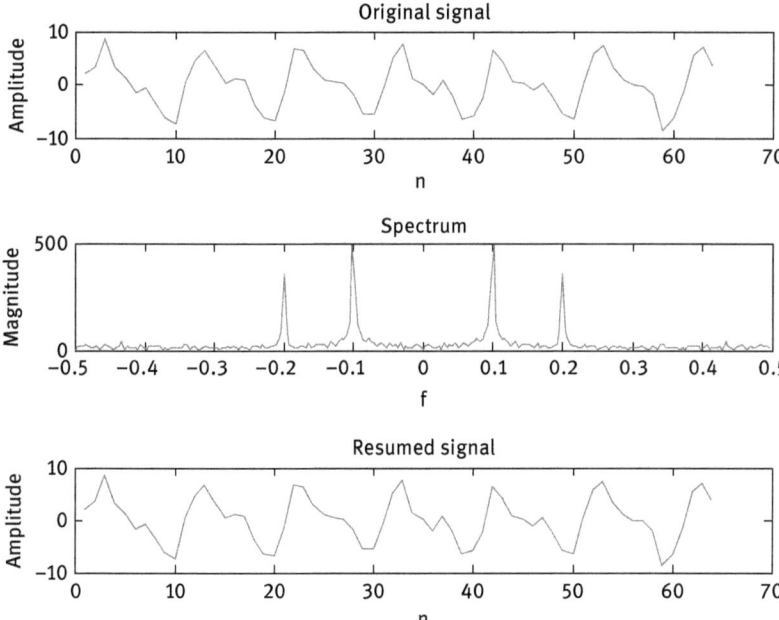

Figure 5.18: The waveform of the original signal and its spectrum, and the waveform of the resumed signal.

chirp z-transform (CZT) can be used to calculate the Fourier transform of an arbitrary section of an arc or a spiral on or not on the unit circle.

1. The algorithm principle

Set a sequence $x(n)$ with finite length N, whose unilateral z-transform is

$$X(z) = \sum_{n=0}^{N-1} x(n)z^{-n} \tag{5.93}$$

where

$$z = e^{sT} = e^{(\sigma + j\Omega)T} = e^{\sigma T}e^{j\Omega T} = Ae^{j\omega} \tag{5.94}$$

in which s is the complex frequency of the Laplace transform, $\omega = \Omega T$ is the digital radian frequency, and T is the sampling interval. If we let

$$z_k = AW^{-k}, \quad k = 0, 1, \ldots, M-1 \tag{5.95}$$

then z_k is actually the sampling points on a general spiral in the z-plane. In the above equation, M is the number of frequency points, which may not be the same as N. Since

$$A = A_0 e^{j\theta_0} \tag{5.96}$$

$$W = W_0 e^{-j\varphi_0} \tag{5.97}$$

z_k can be rewritten as

$$z_k = A_0 e^{j\theta_0} W_0^{-k} e^{-j\varphi_0 k} = A_0 W_0^{-k} e^{j(\theta_0 + k\varphi_0)} \tag{5.98}$$

From eq. (5.5.98) we have

$$z_0 = A_0 e^{j\theta_0}, \quad z_1 = A_0 W_0^{-1} e^{j(\theta_0 + \varphi_0)}, \quad \cdots, \quad z_k = A_0 W_0^{-k} e^{j(\theta_0 + k\varphi_0)}, \quad z_{M-1}$$
$$= A_0 W_0^{-(M-1)} e^{j[\theta_0 + (M-1)\varphi_0]} \tag{5.99}$$

The positions of these sampling points in the z-plane along the spiral position are shown in Figure 5.19.

Substituting eq. (5.95) into eq. (5.93), we obtain

$$X(z_k) = \sum_{n=0}^{N-1} x(n) z_k^{-n} = \sum_{n=0}^{N-1} x(n) A^{-n} W^{nk}, \quad k = 0, 1, \ldots, M-1 \tag{5.100}$$

Equation (5.100) is the computation equation for CZT. The direct calculation is similar to the calculation of DFT, with a rather large calculating complexity. If we take the Bluestein equation, the computation can be transformed to the convolution. As a result, the FFT can be used to reduce the calculating complexity significantly.

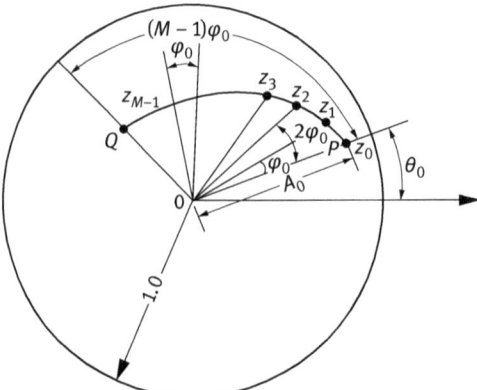

Figure 5.19: The spiral curve and the sampling positions of CZT.

2. The calculation of CZT

The calculation of CZT can be summarized as follows. First, we transform the calculation of the z-transform of samples for an arbitrary spiral on the z-plane into a convolution operation. Then we calculate the convolution with FFT to obtain the result of CZT.

Example 5.10: Combine three generated signals, and then find its CZT and FFT with MATLAB programming.

Solution: Generate three sinusoidal signals whose frequencies are $f_1 = 8$Hz, $f_2 = 8.22$Hz, and $f_3 = 9$Hz. Combine them together. The MATLAB program is as follows:

```
clear all;
% Combine 3 sinusoidal signals;
N=128; f1=8; f2=8.22; f3=9; fs=40; stepf=fs/N;
n=0:N-1; t=2*pi*n/fs; n1=0:stepf:fs/2-stepf;
x=sin(f1*t)+sin(f2*t)+sin(f3*t); M=N; W=exp(-j*2*pi/M);
% The CZT when A = 1;
A=1; Y1=czt(x,M,W,A); subplot(311); plot(n1,abs(Y1(1:N/2))); grid on;
title('CZT spectrum'); ylabel('Magnitude'); xlabel('k');
% Calculating FFT
Y2=abs(fft(x)); subplot(312); plot(n1,abs(Y2(1:N/2))); grid on;
   title('FFT spectrum');
ylabel('Magnitude'); xlabel('k');
% The CZT operation after the construction of A in detail;
M=60; f0=7.2; DELf=0.05; A=exp(j*2*pi*f0/fs); W=exp(-j*2*pi*DELf/fs);
   Y3=czt(x,M,W,A);
n2=f0:DELf:f0+(M-1)*DELf; subplot(313);plot(n2,abs(Y3));grid on;
   title('CZT spectrum');
ylabel('Magnitude'); xlabel('k');
```

Figure 5.20 shows a comparison of spectra obtained from CZT and FFT, respectively. It is obvious that the CZT spectrum has a better spectrum construction.

As a summary, we can get the characteristics of the CZT as follows:

First, similar to the DFT (or FFT), the CZT can be used to get the CZT spectrum from a discrete-time signal. Second, different from the DFT (or FFT), the data length of the input and the output of CZT may be different. Third, the CZT can be used to refine its spectrum, and it is usually used in the situation when data length is short.

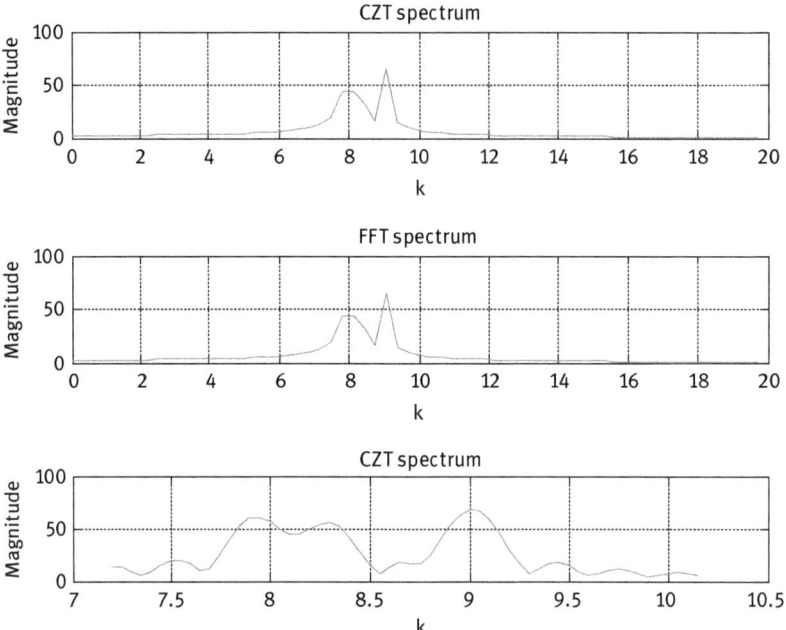

Figure 5.20: A comparison of spectra between CZT and FFT.

5.6 Applications of FFT

5.6.1 Calculating the Linear Convolution Based on FFT

1. Linear convolution and circular convolution

In Chapter 1, we introduced the fundamental concepts and methods of time-domain analysis and convolution operation for linear time-invariant systems. For a finite impulse response system, the output $y(n)$ is the linear convolution between its input signal $x(n)$ and its impulse response $h(n)$, which is shown as

$$y(n) = x(n) \star h(n) = \sum_{m=-\infty}^{\infty} h(m)x(n-m) \tag{5.101}$$

If the signal length is N and the system length is M, then the above equation can be rewritten as

$$y(n) = \sum_{m=0}^{M-1} h(m)x(n-m) \tag{5.102}$$

The data length of the output signal is $L = N + M - 1$.

The fast convolution is actually to substitute the linear convolution with a circular convolution. The circular convolution for $x_1(n)$ and $x_2(n)$ (both with data length N) is expressed as

$$
\begin{aligned}
y(n) = \text{IDFT}[Y(k)] &= \left[\sum_{m=0}^{N-1} x_1(m) x_2((n-m))_N \right] R_N(n) \\
&= \left[\sum_{m=0}^{N-1} x_2(m) x_1((n-m))_N \right] R_N(n)
\end{aligned}
\tag{5.103}
$$

Obviously, the output length of the circular convolution is still N, different from that of the linear convolution. Thus, the circular convolution cannot be directly used to calculate the linear convolution.

2. Calculating linear convolution with FFT

In order to avoid the alias when calculating the linear convolution with FFT and to meet the requirement of the output length, we have to conduct the zero-padding to both $x(n)$ and $h(n)$, having their data length at least equal to the length of the output of the linear convolution as

$$
x(n) = \begin{cases} x(n), & n = 0, 1, \ldots, N-1 \\ 0, & n = N, N+1, \ldots, L-1 \end{cases}
\qquad
h(n) = \begin{cases} h(n), & n = 0, 1, \ldots, M-1 \\ 0, & n = M, M+1, \ldots, L-1 \end{cases}
\tag{5.104}
$$

Thus, the circular convolution can be used to substitute the linear convolution. The calculating steps are listed as follows:

1. Calculate $H(k) = \text{FFT}[h(n)]$, $\quad n = 0, 1, \ldots, L-1$; $\quad k = 0, 1, \ldots, L-1$.
2. Calculate $X(k) = \text{FFT}[x(n)]$, $\quad n = 0, 1, \ldots, L-1$; $\quad k = 0, 1, \ldots, L-1$.
3. Calculate $Y(k) = X(k)H(k)$.
4. Calculate the inverse FFT $\quad y(n) = \text{IFFT}[Y(k)]$, $\quad n = 0, 1, \ldots, L-1$; $k = 0, 1, \ldots, L-1$.

3. The computation analysis

If the length of the input signal is N and the length of $h(n)$ is M, then the total computation of the linear convolution is

$$
m_{\text{d}} = NM
\tag{5.105}
$$

For a linear phase system, if the system satisfies $h(n) = \pm h(M-1-n)$, then the multiplication can be reduced to

$$
m_{\text{d}} = \frac{NM}{2}
\tag{5.106}
$$

For the same input signal and system unit impulse response as above, the total multiplication is

$$m_F = \frac{3}{2}L \log_2 L + L = L\left(1 + \frac{3}{2}\log_2 L\right) \tag{5.107}$$

Comparing the computation for both eqs. (5.106) and (5.107), we obtain

$$K_m = \frac{m_d}{m_F} = \frac{NM}{2L\left(1 + \frac{3}{2}\log_2 L\right)} = \frac{NM}{2(M+N-1)\left[1 + \frac{3}{2}\log_2(M+N-1)\right]} \tag{5.108}$$

If $N = M = 64$ is satisfied, then computation with FFT is much faster than direct computation of the linear convolution.

If $N \gg M$, the circular convolution with FFT is usually realized by the overlap–add method or the overlap–retain method.

4. Partition convolution with the overlap–add method
Suppose that the input signal $x(n)$ is a long sequence, and the length of the unit impulse response $h(n)$ is M. We usually divide the long signal sequence into various segments $x_i(n)$, each with data length N. Thus, the convolution between $x(n)$ and $h(n)$ is the summation of linear convolution between each segment signal $x_i(n)$ and $h(n)$. Since the length of $x_i(n) \star h(n)$ is $L = N + M - 1$, we need to conduct the zero-padding for both $x_i(n)$ and $h(n)$ to length L before the circular convolution is computed.

Since the length of $x_i(n)$ is N and the length of the convolution output is $L = N + M - 1$, there is an overlap for $M - 1$ data points in the output of two adjacent segments. This overlapped part should be added to the nonoverlapping parts to form the overall output $y(n)$. Thus, a method is obtained which is now called the overlap–add method.

5. Partition convolution with the overlap–retain method
The overlap–retain method is a little different from the overlap–add method as the overlap–retain method divides $x(n)$ into various segments. The data length of each segment is $N = L - M + 1$, and fills $M - 1$ data samples of the input sequence retained from the last segment, constructing $x_i(n)$ with data length of $L = N + M - 1$. If $L = N + M - 1 < 2^m$ is satisfied, then we conduct the zero-padding at the end of each segment. The data length then is 2^m, greater than the positive integer power of L. If the convolution between $x_i(n)$ and $h(n)$ is completed by FFT, then we need to give up the first $M - 1$ values of the circular convolution since they are not equal to the values of the linear convolution.

Example 5.11: Calculate the linear convolution between a long data sequence $x(n)$ and a short system $h(n)$ with the overlap–add method by MATLAB programming.

Solution: Suppose that the signal is a combination of a sinusoidal signal and a Gaussian white noise. The MATLAB program is shown as follows. The result of the convolution is shown in Figure 5.21.

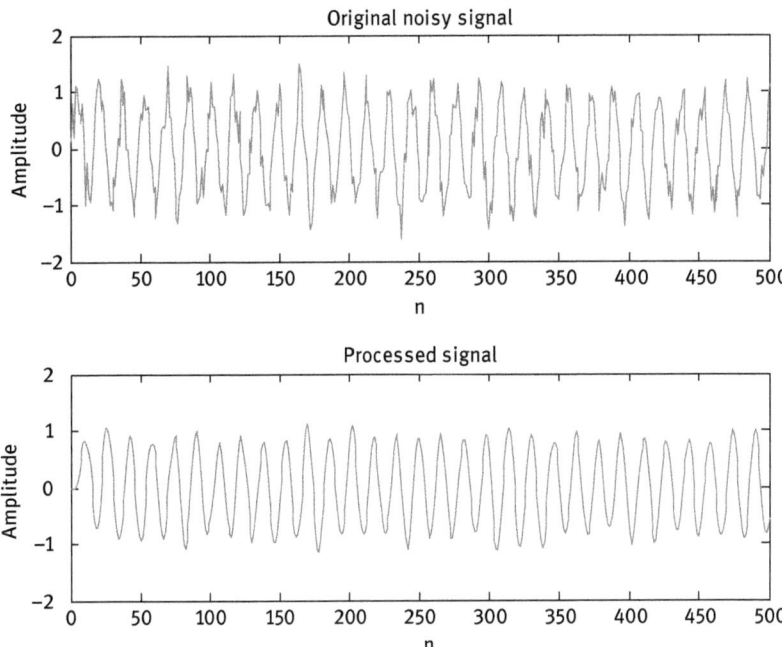

Figure 5.21: Noise redundancy with a linear convolution.

```
clear all;
h=fir1(10,0.3,hanning(11)); % h(n) is an FIR system. It is used to conduct
the low-pass filtering to get rid of noise;
N=500;p=0.05;f=1/16; u=randn(1,N)*sqrt(p); s=sin(2*pi*f*[0:N-1]);
x=u(1:N)+s; y=fftfilt(h,x);
subplot(211); plot(x); title('Original noisy signal'); xlabel('n');
ylabel('Amplitude');
subplot(212); plot(y); title('Processed signal'); xlabel('n'); ylabel
('Amplitude');
```

5.6.2 Calculating the Linear Correlation Based on FFT

The correlation operation is an important method to compare the similarity of signals, and it plays an important role in statistical signal processing. Set $x_1(n)$ and $x_2(n)$ as two discrete-time signals; the correlation between them is defined as

$$r_{x_1x_2}(m) = E[x_1(n)x_2^*(n+m)]$$

(5.109)

where $E[\cdot]$ denotes the mathematical expectation and "*" denotes the complex conjugation.

Similar to the operation for the calculation of linear convolution by FFT, the operation for the linear correlation by FFT is in fact substituting the linear correlation with the circular correlation. Set $x_1(n)$ and $x_2(n)$ as two finite-length sequences, whose data lengths are N and M, respectively. The calculating steps are listed as follows:

(1) Find $L = N + M - 1$-point FFT $X_1(k) = \text{FFT}[x_1(n)]$.
(2) Find L-point FFT $X_2(k) = \text{FFT}[x_2(n)]$.
(3) Find the product $R_{x_1 x_2}(k) = X_1(k)X_2{}^*(k)$.
(4) Find the L-point inverse FFT $r_{x_1 x_2}(m) = \text{IFFT}[R_{x_1 x_2}(k)]$.

ℹ️ Exercises

5.1 Summarize and compare the principles and relations of various types of Fourier transforms.

5.2 Describe the similarities and differences between DFT and DTFT.

5.3 Describe the relations between the z-transform and the DTFT.

5.4 Describe the relations between the z-transform and the DFT.

5.5 Summarize the properties of DFT.

5.6 Describe the difference between the circular convolution and the linear convolution.

5.7 What is frequency aliasing? How to avoid frequency aliasing?

5.8 What is frequency leakage? How to reduce frequency leakage?

5.9 Describe the concept of the fence effect.

5.10 Describe the concept of frequency resolution. Compare the frequency resolution of FT, DTFT, and DFT.

5.11 Try to analyze the effects of zero-padding.

5.12 Try to analyze the problem of time duration and frequency band of signals.

5.13 Describe the principle and method of the decimation-in-time radix-2 FFT algorithm.

5.14 Try to understand the basic principle of butterfly construction and the principle of in-place computation.

5.15 Try to understand the basic principle of the chirp z-transform.

5.16 Try to understand the basic principle and algorithm of FFT-based fast linear convolution.

5.17 Try to understand the basic principle and algorithm of FFT-based fast linear correlation.

5.18 Try to understand the overlap–add and overlap–retain algorithms.

5.19 Try to calculate the DFT for a $2N$-point real-valued sequence with an N-point DFT algorithm, based on its conjugating symmetry property.

5.20 Suppose that $x(n)$ is an N-point sequence, where N is an even number. The DFT of $x(n)$ is $X(k)$. Let $y(n) = \{\, x(n/2), \quad n \text{ is even } 0, n \text{ is odd.}$ Try to express $Y(k)$ by $X(k)$.

5.21 Try to prove Parseval's theorem: $\sum\limits_{n=0}^{N-1} |x(n)|^2 = \frac{1}{N} \sum\limits_{k=0}^{N-1} |X(k)|^2.$

5.22 Suppose $x(n) = [1, 2, 3, 4]$, and the system is $h(n) = [4, 3, 2, 1]$. Try to
  (1) Find the output of the system $y(n) = x(n) \star h(n)$.
  (2) Calculate $y(n)$ by the circular convolution.
  (3) Describe the basic idea for calculating $y(n)$ by DFT.

5.23 Write the MATLAB program to realize the following functions:
  (1) The decimation-in-time radix-2 FFT algorithm.
  (2) The decimation-in-frequency radix-2 FFT algorithm.

5.24 Calculate the convolution between $x(n)$ and $h(n)$ by the MATLAB programming.

5.25 Calculate the convolution between $x(n)$ and $h(n)$ by the MATLAB programming, where $x(n)$ is a long sequence and $h(n)$ is a short system.

5.26 Calculate the correlation of two finite-length data sequences by the MATLAB programming.

6 Digital Filter and Digital Filter Design

6.1 Introduction

The so-called digital filter is a discrete-time system, in which the input sequence is transformed into the output sequence, according to a predetermined algorithm.

Compared with the analog filter, the digital filter has obvious advantages, such as high precision, high reliability, high flexibility, programmable control, and easy integration. So it has been widely used in speech, image, radar, sonar, industrial process control, biomedical signal processing, and many other fields.

Figure 6.1 shows a general block diagram of digital filtering:

Figure 6.1: A general block diagram of digital filtering.

This chapter introduces the principles of digital filters, including the infinite impulse response (IIR) digital filter and the finite impulse response (FIR) digital filter. This chapter also introduces the design principles and methods of both kinds of digital filters.

6.1.1 Classification of Digital Filters

1. Classification based on the frequency characteristics
If we classify digital filters based on their frequency characteristics, digital filters can be divided into five basic types, including the low-pass filter (LPF), the high-pass filter (HPF), the band-pass filter (BPF), the band-stop filter (BSF), and the all-pass filter (APF). Figure 6.2 demonstrates their frequency characteristics and compares them with corresponding analog filters.

2. Classification based on the impulse response properties
The system function of a digital filter can be expressed as

$$H(z) = \frac{Y(z)}{X(z)} = \frac{\sum_{k=0}^{M} b_k z^{-k}}{1 - \sum_{k=1}^{N} a_k z^{-k}} \tag{6.1}$$

where $X(z)$ and $Y(z)$ denote the z-transforms of input signal $x(n)$ and output signal $y(n)$ respectively, M and N denote orders of input term and output term, b_k and a_k are

https://doi.org/10.1515/9783110465082-006

the weight coefficients of input and output terms, respectively. The corresponding difference equation is shown as

$$y(n) = \sum_{k=1}^{N} a_k y(n-k) + \sum_{k=0}^{M} b_k x(n-k) \tag{6.2}$$

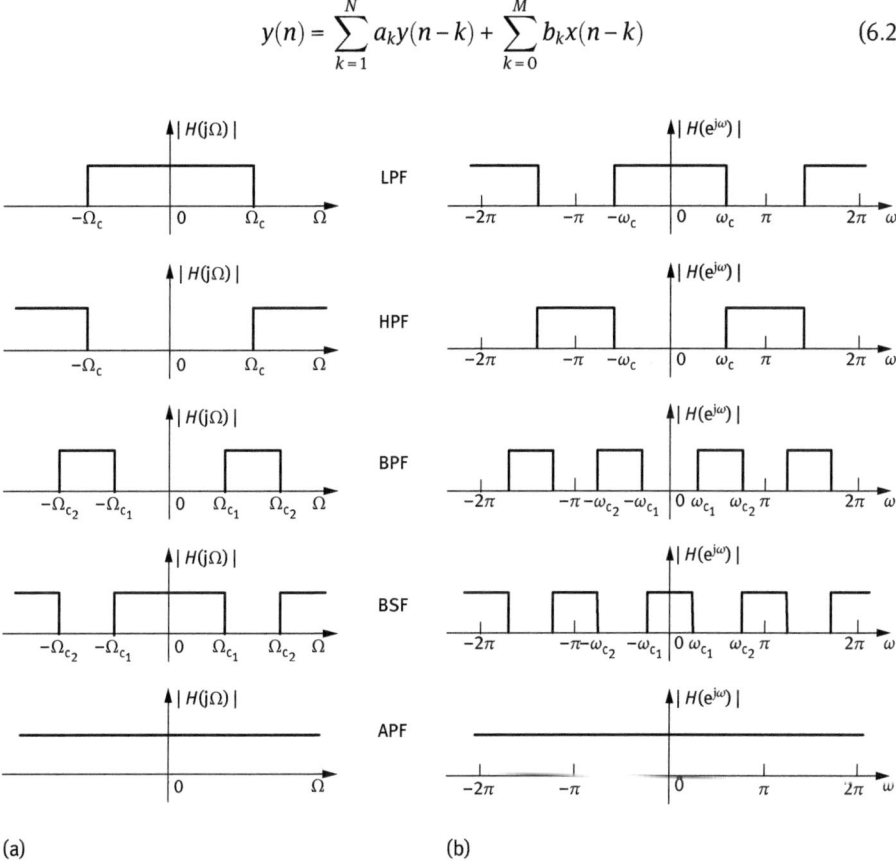

(a)

(b)

Figure 6.2: Five types of analog filters and digital filters: (a) ideal analog filters and (b) ideal digital filters.

If all a_k in eqs. (6.1) and (6.2) satisfy $a_k = 0$, $k = 1, 2, \ldots, N$, then the filter is referred to as an FIR filter. If any $a_k \neq 0$, $k = 1, 2, \ldots, N$ exists, then the filter is called an IIR filter.

6.1.2 Design of Digital Filters

The essence of the digital filter design is to approximate the performance requirements of the filter with a causal and stable discrete-time linear time-invariant system. Both IIR and FIR filters have their own design methods.

The most common design method for IIR digital filter is to employ the method used in analog filter design. Generally, we design or choose a suitable analog filter first based on the given technical specification, and then transform it into a digital filter by using the impulse response invariance, the step invariance, and the bilinear transform methods.

The design method for FIR digital filters is to design the filter directly according to the given characteristics. The usually used methods include the window method, the frequency sampling method, and the Chebyshev approximation method.

6.2 Structure of Digital Filters

There are multiple representation methods for digital or discrete-time filters, mainly including the difference equation in time domain, the system function in the z-domain, and the block diagram as well as the signal flow graph. These methods are introduced in this section.

6.2.1 Representation with Difference Equations

The difference equation is one of the most important expressions for the digital filter. In fact, eq. (6.2) demonstrates the general form of a difference equation.

For an IIR digital filter, the weight coefficients of the output term a_k, $k = 1, 2, \ldots, N$ cannot be all zeros. It means that the output of the IIR digital filter at present time is the linear combination of the input signal and the output at past time, which indicates the infinite time continuity of its impulse response.

Example 1.6 in Chapter 1 gives a typical example of the IIR digital filter, which is listed as follows:

$$y(n) = 1.01y(n-1) + x(n) \tag{6.3}$$

where $y(n)$ and $y(n-1)$ denote the balance of this month and last month, respectively, $x(n)$ denotes the net deposit of this month. $a_1 = 1.01$ denotes the interest rate.

For an FIR filter, the coefficients of its output term satisfy $a_k = 0$, $k = 1, 2, \ldots, N$. Usually, $a_0 = 1$ is taken. This shows that the current output signal of the FIR digital filter is only related to the input signal at the present time and past time, but not related to the output signal at past time. From eq. (6.2) we see that if $a_k = 0$, $k = 1, 2, \ldots, N$, then eq. (6.2) changes to

$$y(n) = \sum_{k=0}^{M} b_k x(n-k) \tag{6.4}$$

When comparing with the definition of the discrete-time linear convolution in Chapter 1, we find a finite causal linear system as follows:

$$y(n) = \sum_{k=0}^{M} h(k)x(n-k) \tag{6.5}$$

It is obvious that both eqs. (6.4) and (6.5) are the same, which means that the operation of an FIR digital filter is identical with the convolution of the discrete-time linear time-invariant system.

6.2.2 Representation with System Functions

The system function of a digital filter is given in eq. (6.1). In fact, if we take the z-transform to both sides of eq. (6.2), we can get the system function as shown in eq. (6.1).

For an IIR filter, since the denominator in eq. (6.1) is not a constant, the system is referred to as the all-zero (or pole–zero) system for the existence of its poles. For an FIR filter, since the denominator in eq. (6.1) is the constant 1, the system function degenerates as

$$H(z) = \sum_{k=0}^{M} b_k z^{-k} \tag{6.6}$$

The above equation is actually the same as that of the z-transform of the unit impulse response $h(n)$, which is shown as

$$H(z) = \sum_{n=0}^{M} h(n)z^{-n} \tag{6.7}$$

6.2.3 Representation with System Block Diagrams and Signal Flow Graphs

System block diagram and signal flow graph are important methods to represent digital filters, which is simple and clear. Figure 6.3 demonstrates the operation units of the block diagram and the signal flow graph.

The block diagram and the signal flow graph representations of a common second-order IIR digital filter is shown in Example 6.1.

Example 6.1: The difference equation of the second-order IIR digital filter is shown as follows. Try to represent the system by a block diagram and a signal flow graph, respectively:

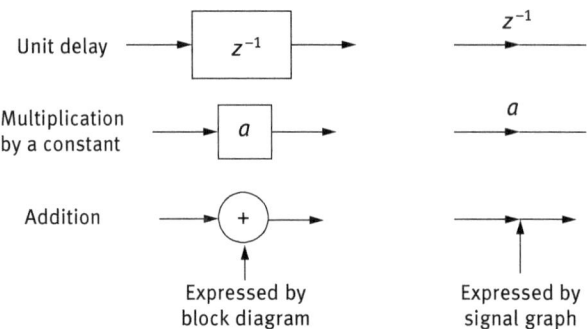

Unit delay

Multiplication by a constant

Addition

Expressed by block diagram

Expressed by signal graph

Figure 6.3: The basic operation units of the block diagram and the signal flow graph.

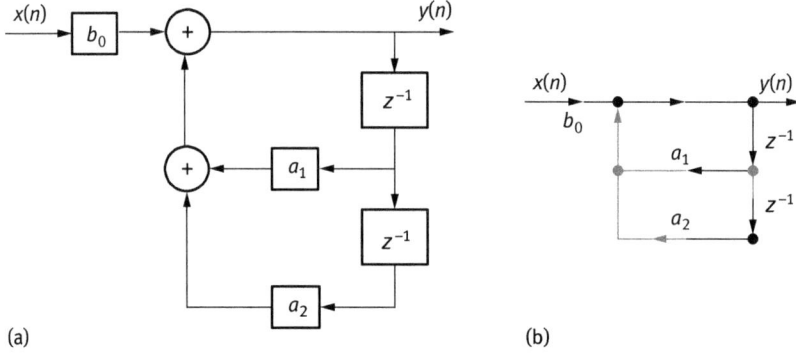

(a)

(b)

Figure 6.4: The block diagram and the signal flow graph: (a) the block diagram and (b) the signal flow graph.

$$y(n) = a_1 y(n-1) + a_2 y(n-2) + b_0 x(n)$$

Solution: The block diagram and the signal flow graph are shown in Figure 6.4.

6.3 IIR Digital Filter

The IIR digital filter has the following characteristics: first, its unit impulse response is infinite, which is the origin of the name of such filter; second, there exist poles in the system function of such kind of filters; third, the filter has a feedback in the structure from the output to the input, referred to as the recursive structure.

As the system function of a digital filter can be represented with a variety of forms, its realization has also different forms. This section will introduce four types of basic structures of the IIR digital filter, including direct form I, direct form II, cascade form, and parallel form.

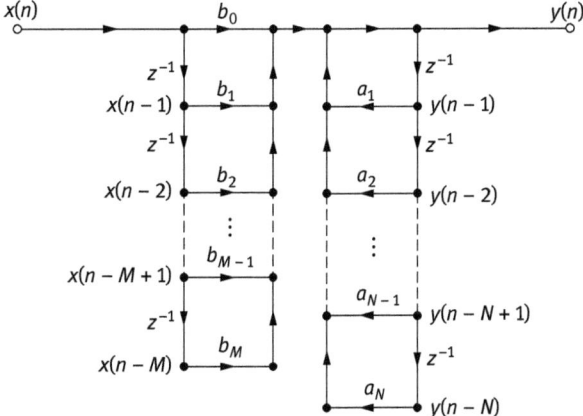

Figure 6.5: Block diagram representation for a general N th-order IIR digital filter with the direct form I.

6.3.1 The Structure of Direct Form I

We see from eq. (6.2) that the output signal $y(n)$ of the IIR filter is composed of two components, one is the input signal at present and past time, the other is the output at past time. In eq. (6.2), $\sum_{k=0}^{M} b_k x(n-k)$ represents a combination of the input signal at present and past time. It constructs M delayed network units, and forms a transversal filter structure. $\sum_{k=1}^{N} a_k y(n-k)$ in eq. (6.2) represents a weighted combination of the output signal at the previous time, constructing N delayed network units. Equation (6.2) constructs a feedback system since it contains the components of the delayed output. The direct realization of the IIR digital filter based on eq. (6.2) is called the direct form I, whose signal flow graph is shown in Figure 6.5.

We see from Figure 6.5 that the structure of direct form I is composed of two parts, where the left part is a forward network used to realize the zeros of the filter, and the right part is a feedback network, used to realize the poles of the filter. There are totally $N + M$ delayed units in the system.

6.3.2 The Structure of Direct Form II

A block diagram can be rearranged in a variety of ways without changing the overall system function. Since an LTI system obeys the commutative property, the order of the two parts in Figure 6.5 can be changed with each other, without changing the system function. Therefore, we can get another type of structure from the direct form I for the IIR filter, which is shown in Figure 6.6(a). Furthermore, we can get a new structure shown in Figure 6.6(b) by combining two delayed branches. Such structure is referred to as the direct form II.

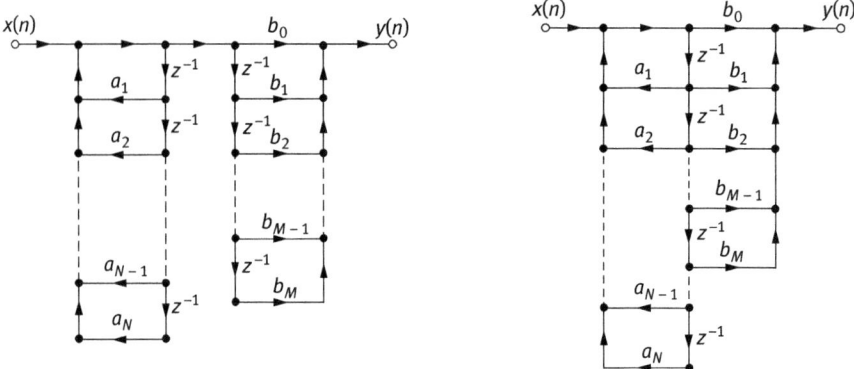

Figure 6.6: The direct form II of the IIR digital filter: (a) deformation of the direct form I and (b) direct form II.

The structure of the direct form II needs only N ($\geq M$) delay units, benefiting for its calculation. The significant drawback for the direct forms I or II is that their coefficients a_k and b_k cannot reflect their relations with the system poles. On the other hand, such structures may be unstable for their sensitivity to the change of their coefficients.

6.3.3 Cascade Structure

If we conduct a factorization to eq. (6.1) according to zeros and poles of the system function, we get

$$H(z) = \frac{\sum_{k=0}^{M} b_k z^{-k}}{1 - \sum_{k=1}^{N} a_k z^{-k}} = A \frac{\prod_{k=1}^{M_1}(1 - p_k z^{-1}) \prod_{k=1}^{M_2}(1 - q_k z^{-1})(1 - q_k^* z^{-1})}{\prod_{k=1}^{N_1}(1 - c_k z^{-1}) \prod_{k=1}^{N_2}(1 - d_k z^{-1})(1 - d_k^* z^{-1})} \tag{6.8}$$

where $M = M_1 + 2M_2$, $N = N_1 + 2N_2$. In the above equation, the first-order terms represent the real-valued roots of the system, and both p_k and c_k are the poles and zeros, respectively; the second-order terms represent the complex roots of the system, and q_k, q_k^*, d_k, and d_k^* are pairs of conjugate poles and conjugate zeros, respectively. A in the equation denotes the system gain. Equation (6.8) demonstrates the general form of zero–pole expression, which is rewritten as

$$H(z) = A \frac{\prod_{k=1}^{M_1}(1 - p_k z^{-1}) \prod_{k=1}^{M_2}(1 + \beta_{1k} z^{-1} + \beta_{2k} z^{-2})}{\prod_{k=1}^{N_1}(1 - c_k z^{-1}) \prod_{k=1}^{N_2}(1 - \alpha_{1k} z^{-1} - \alpha_{2k} z^{-2})} \tag{6.9}$$

Equation (6.9) can be further changed as

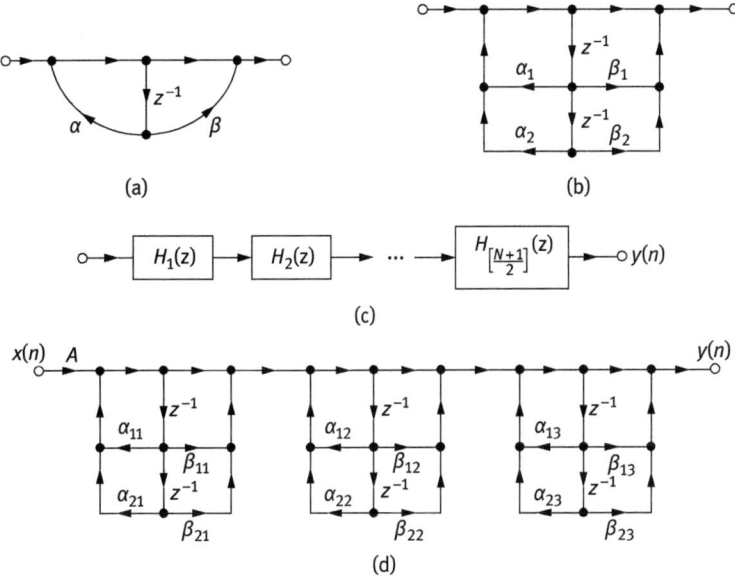

(a)

(b)

(c)

(d)

Figure 6.7: Structures of the IIR digital filters: (a) the first-order unit; (b) the second-order unit; (c) cascade structure; and (d) the cascade structure for a sixth-order IIR digital filter.

$$H(z) = A \prod_k \frac{1 + \beta_{1k}z^{-1} + \beta_{2k}z^{-2}}{1 - \alpha_{1k}z^{-1} - \alpha_{2k}z^{-2}} = A \prod_k H_k(z) \tag{6.10}$$

which can be realized by using a cascade structure, as shown in Figure 6.7.

We see from Figure 6.7(a) and (b) that every first- or second-order subsystem is referred to as a basic first-order unit or a basic second-order unit. The entire filter is in a cascade structure as shown in Figure 6.7(c). Figure 6.7(d) demonstrates the cascade structure of the sixth-order IIR digital filter, composed of three basic second-order units.

If we adjust coefficients β_{1k}, β_{2k} and α_{1k}, α_{2k} in the kth unit, then the poles and zeros in this unit can be adjusted, while the poles and zeros in other units are not influenced. Such structure is convenient for the filter adjustment.

6.3.4 Parallel Structure

The system function of the IIR digital filter shown in eq. (6.1) can be expanded into a partial fraction form as follows:

$$H(z) = \frac{\sum_{k=0}^{M} b_k z^{-k}}{1 - \sum_{k=1}^{N} a_k z^{-k}} = \sum_{k=1}^{N_1} \frac{A_k}{1 - c_k z^{-1}} + \sum_{k=1}^{N_2} \frac{B_k(1 - g_k z^{-1})}{(1 - d_k z^{-1})(1 - d_k^* z^{-1})} + \sum_{k=0}^{M-N} G_k z^{-k} \tag{6.11}$$

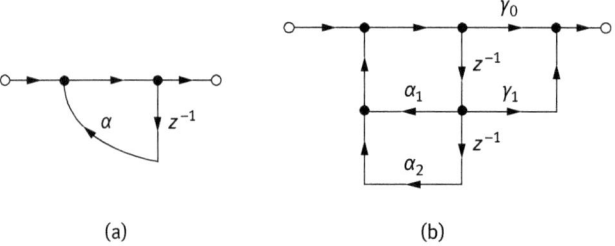

(a)   (b)

Figure 6.8: The parallel structure of the basic (a) first- and (b) second-order units.

where $N = N_1 + 2N_2$. Coefficients A_k, B_k, g_k, c_k, G_k are all integers and d_k^* is the conjugate of d_k. Generally, an IIR filter satisfies $M \le N$. The system function can be expressed as

$$H(z) = G_0 + \sum_{k=1}^{N_1} \frac{A_k}{1 - c_k z^{-1}} + \sum_{k=1}^{N_2} \frac{\gamma_{0k} + \gamma_{1k} z^{-1}}{1 - \alpha_{1k} z^{-1} - \alpha_{2k} z^{-2}} \tag{6.12}$$

In which, the basic first- and second-order units are shown in Figure 6.8.

The parallel structure can adjust a pair of poles by adjusting coefficients α_{1k} and α_{2k}. However, it cannot be used to adjust the position of zeros. Otherwise, the errors in each unit do not affect each other, and the error in this structure is less than that in the cascade structure.

In addition to the direct form, cascade and parallel structures, there are still some other implementations for the IIR digital filter, which are not introduced in this section.

6.4 FIR Digital Filter

The FIR digital filter is finite in length, and thus it is a stable system. The FIR filter can guarantee a strict linear phase characteristic under various arbitrary amplitude conditions. Therefore, FIR filter is widely used in communication, signal and image processing, pattern recognition, and other fields.

The FIR digital filter has following characteristics: first, its unit impulse response has a finite number of nonzero samples; second, for a causal FIR filter, the system function $H(z)$ in the entire z-plane (except $z = 0$) does not have any pole; third, the implementation of this kind of filter mainly uses nonrecursive structure.

An FIR filter is usually represented as

$$H(z) = \sum_{n=0}^{N-1} h(n) z^{-n} \tag{6.13}$$

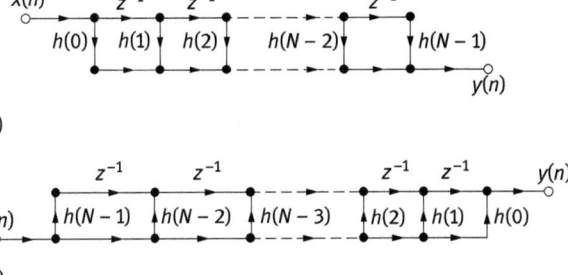

(a)

(b)

Figure 6.9: The structure of a transverse FIR filter: (a) the transverse structure and (b) the transpose of (a).

where $h(n)$ is the unit impulse response of the filter. There are various kinds of structures for the FIR filter, including the transverse structure, the cascade structure, the frequency sampling structure, the fast convolution structure, the linear phase structure, and the lattice structure .

6.4.1 Transverse Structure

The difference equation or the convolution of the system is usually written as

$$y(n) = \sum_{m=0}^{N-1} h(m)x(n-m) \tag{6.14}$$

where $x(n)$ and $y(n)$ are the input and output signals, respectively. This equation expresses the transverse structure of an FIR filter, as shown in Figure 6.9.

The transpose structure shown in Figure 6.9(b) reverses all the directions in the original structure in Figure 6.9(a), and exchanges all the input and output signals. The system function $H(z)$ remains unchanged.

6.4.2 Cascade Structure

The system function $H(z)$ can be rewritten as follows:

$$H(z) = \sum_{n=0}^{N-1} h(n)z^{-n} = \prod_{k=1}^{[N/2]} (\beta_{0k} + \beta_{1k}z^{-1} + \beta_{2k}z^{-2}) \tag{6.15}$$

where $[N/2]$ denotes the integer part of $N/2$. Figure 6.10 demonstrates the cascade structure of the FIR filter when N is odd.

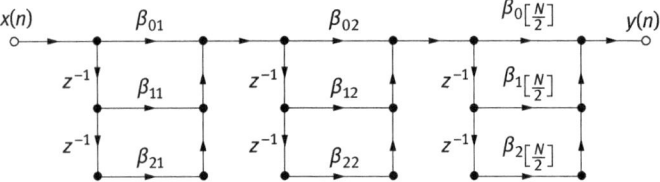

Figure 6.10: The cascade structure of an FIR filter when N is odd.

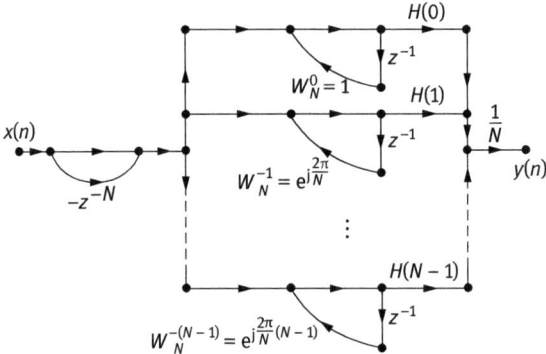

Figure 6.11: The frequency sampling structure of an FIR filter.

6.4.3 Frequency Sampling Structure

1. The frequency sampling structure

The general form of the frequency sampling structure is shown in Figure 6.11.

The frequency sampling structure of an FIR filter can be divided into two parts: the first part corresponds to the left half side in Figure 6.11, while the second part corresponds to the right half side in the same figure.

The signal flow graph and the frequency response of the first part are shown in Figure 6.12, which is in fact a comb filter.

The system function of the comb filter is shown as

$$H_c(z) = 1 - z^{-N} \tag{6.16}$$

The right side half of Figure 6.11 is a parallel connection of N first-order oscillators, cascaded with the comb filter. Each oscillator can be expressed by its subsystem function as

$$H'_k(z) = \frac{H(k)}{1 - W_N^{-k}z^{-1}}, \quad k = 0, 1, \ldots, N-1 \tag{6.17}$$

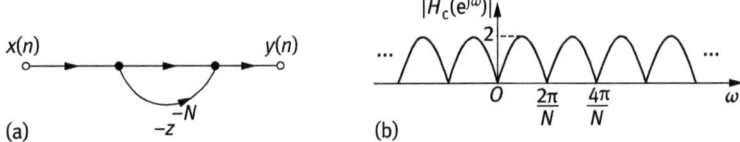

x(n)　　　　　　　　y(n)

$-z^{-N}$

(a)

$|H_c(e^{j\omega})|$

(b)

Figure 6.12: The frequency response (a) and structure (b) of a comb filter.

where $H(k)$ is the discrete Fourier transform of the unit impulse response $h(n)$ of the whole system.

2. The characteristics of the frequency sampling structure

There are following characteristics for the frequency sampling structure of the FIR filters. First, it is convenient to control the frequency response since the response at $\omega = (2\pi/N)k$ is just the same as the value of $H(k)$. Second, the coefficients in the structure are all complex numbers, and all poles are on the unit circle. Third, since W_N^{-k} is determined by poles, the errors may occur when the coefficients are quantified. Thus the stability of the system may be affected.

3. Modified structure

In order to avoid the system instability caused by the quantization of coefficients, we can consider moving all zeros and poles of the system to a circle of radius $r < 1$ inside the unit circle. Thus

$$H(z) = \frac{1 - r^N z^{-N}}{N} \sum_{k=0}^{N-1} \frac{H_r(k)}{1 - rW_N^{-k}z^{-1}} \tag{6.18}$$

where $H_r(k) \approx H(k)$ for $r \approx 1$. Then

$$H(z) \approx \frac{1 - r^N z^{-N}}{N} \sum_{k=0}^{N-1} \frac{H(k)}{1 - rW_N^{-k}z^{-1}} \tag{6.19}$$

The modified frequency sampling structure, if N is even, is shown as

$$H(z) = (1 - r^N z^{-N}) \frac{1}{N} \left[\frac{H(0)}{1 - rz^{-1}} + \frac{H(N/2)}{1 + rz^{-1}} + \sum_{k=1}^{\frac{N}{2}-1} \frac{\beta_{0k} + \beta_{1k}z^{-1}}{1 - z^{-1}2r\cos\left(\frac{2\pi}{N}k\right) + r^2 z^{-2}} \right] \tag{6.20}$$

$$= (1 - r^N z^{-N}) \frac{1}{N} \left[H_0(z) + H_{N/2}(z) + \sum_{k=1}^{\frac{N}{2}-1} H_k(z) \right]$$

and if N is odd

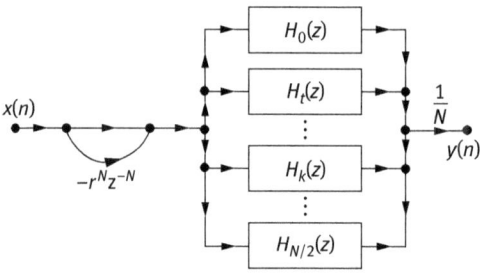

x(n)

y(n)

Figure 6.13: The modified frequency sampling structure of the FIR filter.

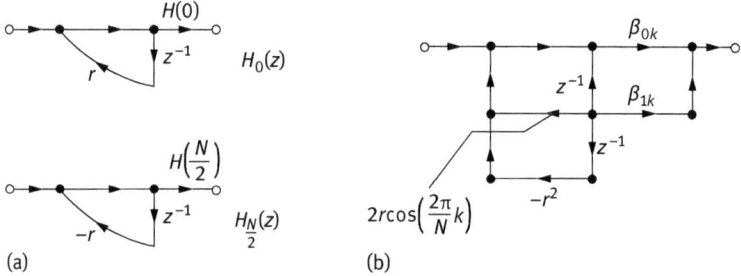

(a)   (b)

Figure 6.14: The signal flow graphs of the modified frequency sampling structure: (a) the first-order system and (b) the second-order system.

$$H(z) = (1 - r^N z^{-N})\frac{1}{N}\left[\frac{H(0)}{1 - rz^{-1}} + \sum_{k=1}^{(N-1)/2}\frac{\beta_{0k} + \beta_{1k}z^{-1}}{1 - z^{-1}2r\cos\left(\frac{2\pi}{N}k\right) + r^2 z^{-2}}\right]$$

$$= (1 - r^N z^{-N})\frac{1}{N}\left[H_0(z) + \sum_{k=1}^{(N-1)/2}H_k(z)\right]$$

(6.21)

4. The entire structure after the modification

The modified frequency sampling structure is shown in Figure 6.13.

If N is odd, then $H_{N/2}(z)$ in Figure 6.13 does not exist. The signal flow graphs for the first-order and the second-order systems are shown in Figure 6.14.

6.4.4 Fast Convolution Structure

The so-called fast convolution structure for the FIR digital filter employs FFT or DFT to solve the linear convolution in the discrete frequency domain. It should be noted that we have to conduct the zero padding to both signal sequence and impulse response to satisfy $L \geq N_1 + N_2 - 1$, where N_1 and N_2 are data length of $x(n)$ and $h(n)$, respectively.

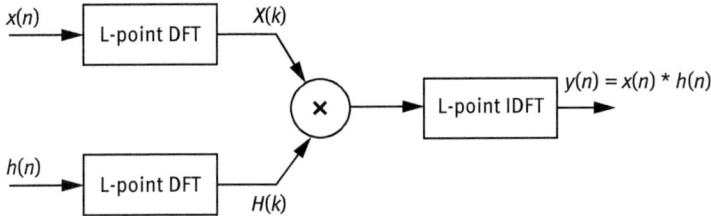

Figure 6.15: The block diagram of the fast convolution structure of the FIR filter.

Figure 6.15 shows the block diagram of the fast convolution structure of the FIR filter.

6.4.5 The Linear Phase FIR Digital Filter and the Minimum Phase Structure

1. Linear phase conditions
The linear phase is a very important feature in many digital filter applications. It is possible for an FIR filter to achieve a strict linear phase, since its impulse response is finite. In fact, if the FIR digital filter satisfies the following conditions, the filter has a strict linear phase:

$$h(n) = h(N - 1 - n) \tag{6.22}$$

or

$$h(n) = -h(N - 1 - n) \tag{6.23}$$

where eq. (6.22) represents the even symmetric structure, while eq. (6.23) represents the odd symmetric structure. The symmetry center is $n = (N - 1)/2$.

2. The direct form of a linear phase FIR filter
Suppose that the FIR digital filter is designed to meet the linear phase condition, then the system function $H(z)$ of the filter can be obtained when the length N of the filter sequence is odd or even, respectively:

$$H(z) = \sum_{n=0}^{\frac{N-1}{2}-1} h(n)[z^{-n} \pm z^{-(N-1-n)}] + h\left(\frac{N-1}{2}\right)z^{-\frac{N-1}{2}} \tag{6.24}$$

or

$$H(z) = \sum_{n=0}^{(N/2)-1} h(n)[z^{-n} \pm z^{-(N-1-n)}] \tag{6.25}$$

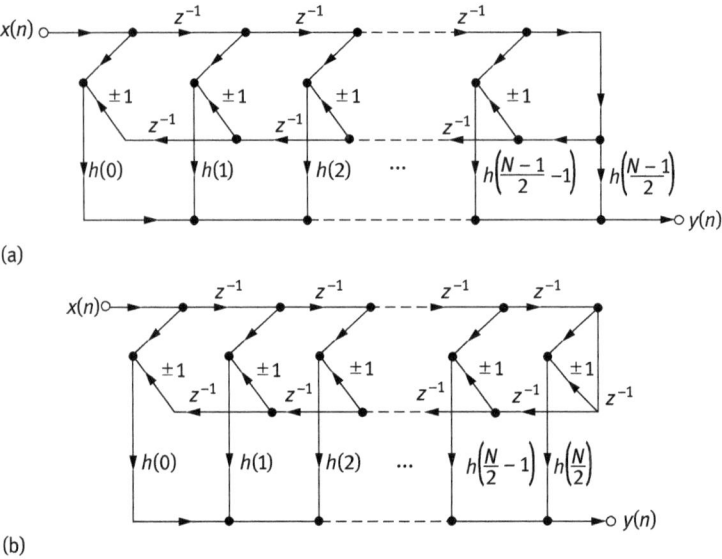

(a)

(b)

Figure 6.16: The structures of linear phase FIR filters: (a) N is odd and (b) N is even.

where symbols "+" and " – " denote even symmetry and odd symmetry, respectively. The structures of linear phase FIR filters are given in Figure 6.16.

3. Phase features of the linear phase FIR filters

We here investigate the phase features of the linear phase FIR filters. Suppose $h(n) = h(N-1-n)$ and N is even. From eq. (6.25) we have

$$
\begin{aligned}
H(e^{j\omega}) &= \sum_{n=0}^{(N/2)-1} h(n)e^{-j\omega n} + \sum_{n=N/2}^{N-1} h(n)e^{-j\omega n} \\
&= e^{-j(N-1)\omega/2} \sum_{n=0}^{(N/2)-1} h(n) \left[e^{j(\frac{N-1}{2}-n)\omega} + e^{-j(\frac{N-1}{2}-n)\omega} \right] \\
&= e^{-j(N-1)\omega/2} \sum_{n=0}^{(N/2)-1} 2h(n) \cos\left[\left(\tfrac{N-1}{2} - n \right)\omega \right]
\end{aligned}
\tag{6.26}
$$

Set $m = N/2 - n$, and change the variable m to n, we have

$$
H(e^{j\omega}) = e^{-j(N-1)\omega/2} \sum_{n=1}^{N/2} 2h\left(\frac{N}{2} - n \right) \cos[(n-1/2)\omega]
\tag{6.27}
$$

Obviously, $H(e^{j\omega})$ has a linear phase as

$$
\arg\left[H(e^{j\omega}) \right] = \varphi(\omega) = -(N-1)\omega/2
\tag{6.28}
$$

In the same way, we can get phase features for other linear phase structures of FIR filters.

4. The concept of the minimum phase system

The minimum phase system requires that all the poles and zeros of its system function for a causal and stable system are inside the unit circle. In contrast, if all the zeros of the system function are outside the unit circle, such system is called the maximum phase system. If some zeros are inside and others are outside of the unit circle, such system is called the mixed phase system.

The minimum phase system is widely used in signal processing. The main properties of the minimum phase system are listed as follows:

Property 6.1: The minimum phase filter has the minimal phase shift in a group of stable filters which have the same magnitude response.

Property 6.2: The energy of the impulse response for a minimum phase system is concentrated in a small range where the value of n is small, and the system has the minimal delay.

Property 6.3: In the systems with the same magnitude characteristics, there is only one minimal phase system, and the minimum phase system has the minimal group delay.

Property 6.4: Only if a causal and stable system is a minimum phase system, its inverse system is causal and stable.

Property 6.5: The system function for an arbitrary nonminimum phase system can be composed of a cascade of a minimum phase system and an all-pass system.

Example 6.2: Given $H_1(z) = \dfrac{1 - bz^{-1}}{1 - az^{-1}}$, $|a| < 1$, $|b| < 1$ and $H_2(z) = \dfrac{b - z^{-1}}{1 - az^{-1}}$, $|a| < 1$, $|b| < 1$.

(1) Compare the magnitude response of $H_1(z)$ and $H_2(z)$.
(2) Whether $H_1(z)$ and $H_2(z)$ are the minimum phase systems, the maximum phase systems, or the mixed phase systems?
(3) Plot the magnitude spectrum and phase spectrum of $H_1(z)$ and $H_2(z)$, and plot their unit impulse responses.

Solution: (1) Since $H_1(z)H_1(z^{-1}) = \dfrac{z - b}{z - a} \cdot \dfrac{z^{-1} - b}{z^{-1} - a} = \dfrac{1 - b(z + z^{-1}) + b^2}{1 - a(z + z^{-1}) + a^2}$, we get

$H_2(z)H_2(z^{-1}) = \dfrac{bz - 1}{z - a} \cdot \dfrac{bz^{-1} - 1}{z^{-1} - a} = \dfrac{1 - b(z + z^{-1}) + b^2}{1 - a(z + z^{-1}) + a^2}$. It is obvious that both $H_1(z)$ and $H_2(z)$ have the same magnitude response.

(2) $H_1(z)$ is a minimum phase system since its poles and zeros are all inside the unit circle. $H_2(z)$ is a maximum phase system since its zeros are all outside the unit circle.

(3) The magnitude and phase spectra of $H_1(z)$ and $H_2(z)$ are shown in Figure 6.17, as well as their unit impulse responses.

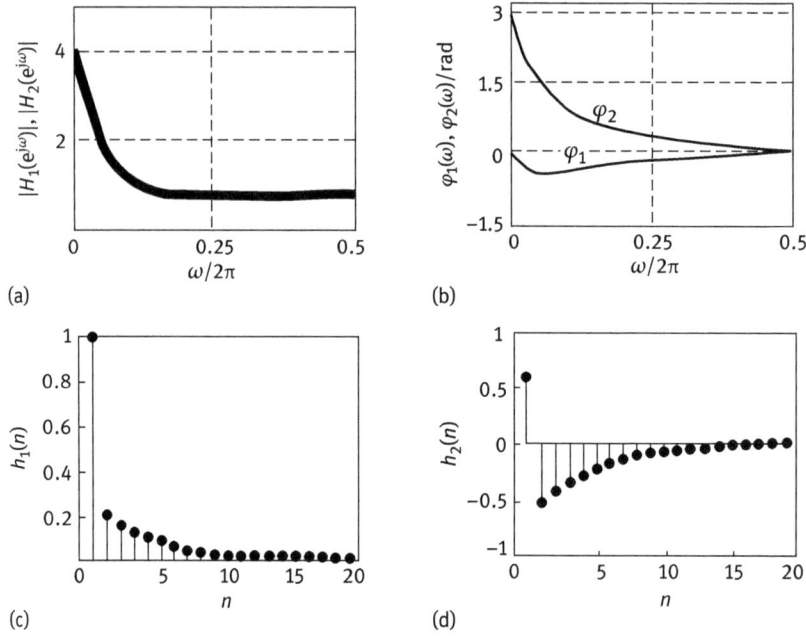

Figure 6.17: The magnitude and phase spectra of $H_1(z)$ and $H_2(z)$, as well as their unit impulse responses: (a) magnitude spectrum; (b) phase spectrum; (c) $h_1(n)$; and (d) $h_2(n)$.

Obviously, the phase shift of $H_1(z)$ is smaller than that of $H_2(z)$ if both have the same magnitude response, which means that the energy of $h_1(n)$ is more concentrated than $h_2(n)$.

6.5 Lattice Structure of Digital Filters

The lattice structure of the digital filter is a very useful modular structure. It is widely used in various areas since it is easy to achieve a high speed and parallel processing, including power spectrum estimation, speech processing, and adaptive filtering.

6.5.1 All-Zero Lattice FIR Systems

1. Lattice structure and basic transfer unit of FIR filters

Suppose an M th-order transversal FIR filter is given as

$$H(z) = B(z) = \sum_{i=0}^{M} h(i)z^{-i} = 1 + \sum_{i=1}^{M} b_i^{(M)} z^{-i} \tag{6.29}$$

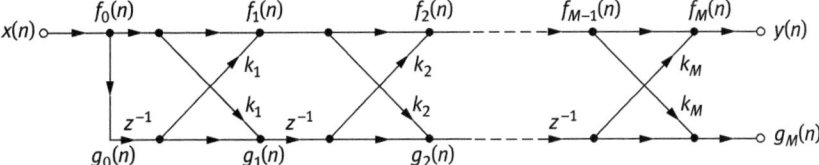

Figure 6.18: All-zero lattice structure for FIR filter.

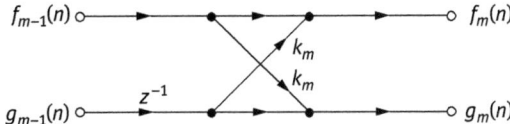

Figure 6.19: The basic transfer unit of the all-zero lattice FIR filter.

where $b_i^{(M)}$ denotes the i th coefficient of the filter, and $h(0) = 1$ is assumed. The lattice structure of the FIR filter is shown in Figure 6.18.

In Figure 6.18, k_i, $i = 1, 2, k, M$, represents reflection coefficients. The basic transfer unit of the lattice filter is shown in Figure 6.19.

The basic transfer unit of the lattice structure shown in Figure 6.19 can be expressed as

$$f_m(n) = f_{m-1}(n) + k_m g_{m-1}(n-1), \quad m = 1, 2, \ldots, M$$
$$g_m(n) = k_m f_{m-1}(n) + g_{m-1}(n-1), \quad m = 1, 2, \ldots, M \tag{6.30}$$

We see from Figure 6.18, $f_0(n) = g_0(n) = x(n)$, $f_M(n) = y(n)$.

Defining $B_m(z)$ and $\bar{B}_m(z)$ as the system functions at the upper and lower ends of the m th basic transfer unit, we have

$$B_m(z) = \frac{F_m(z)}{F_0(z)} = 1 + \sum_{i=1}^{m} b_i^{(m)} z^{-i}, \quad m = 1, 2, \ldots, M$$
$$\bar{B}_m(z) = \frac{G_m(z)}{G_0(z)}, \quad m = 1, 2, \ldots, M \tag{6.31}$$

The above equations reflect the modular structure of the lattice structure. The $B_m(z)$ in this stage is formed by the cascade connection between the $B_{m-1}(z)$ in the previous stage and the basic transfer unit in this stage.

2. The recursive of the lattice structure
From eq. (6.30) we have

$$F_m(z) = F_{m-1}(z) + k_m z^{-1} G_{m-1}(z)$$
$$G_m(z) = k_m F_{m-1}(z) + z^{-1} G_{m-1}(z) \tag{6.32}$$

Based on eq. (6.31), the above equation can be rewritten as

$$B_m(z) = B_{m-1}(z) + k_m z^{-1} \bar{B}_{m-1}(z)$$
$$\bar{B}_m(z) = k_m B_{m-1}(z) + z^{-1} \bar{B}_{m-1}(z) \tag{6.33}$$

or

$$B_{m-1}(z) = \frac{1}{1-k_m^2} [B_m(z) - k_m \bar{B}_m(z)]$$
$$\bar{B}_{m-1}(z) = \frac{1}{1-k_m^2} [-z k_m B_m(z) + z \bar{B}_m(z)] \tag{6.34}$$

Equations (6.33) and (6.34) represent the recursive relation between the previous stage and the present stage. By using the relation of $\bar{B}_m(z) = z^{-m} B_m(z^{-1})$, we have

$$B_m(z) = B_{m-1}(z) + k_m z^{-m} B_{m-1}(z^{-1}) \tag{6.35}$$

and

$$B_{m-1}(z) = \frac{1}{1-k_m^2} [B_m(z) - k_m z^{-m} B_m(z^{-1})] \tag{6.36}$$

Equations (6.35) and (6.36) show the recursive relations of the lattice filter, from the lower stage to the higher stage, or conversely.

3. The relations between the reflect coefficients and the transversal coefficients

Consider eqs. (6.31) and (6.35) jointly, we get both recursive relations as

$$\begin{cases} b_m^{(m)} = k_m \\ b_i^{(m)} = b_i^{(m-1)} + k_m b_{m-i}^{(m-1)} \end{cases}, \quad i = 1, 2, \ldots, (m-1); \quad m = 2, \ldots, M \tag{6.37}$$

and

$$\begin{cases} k_m = b_m^{(m)} \\ b_i^{(m-1)} = \frac{1}{1-k_m^2} [b_i^{(m)} - k_m b_{m-i}^{(m)}] \end{cases}, \quad i = 1, 2, \ldots, (m-1); \quad m = 2, \ldots, M \tag{6.38}$$

4. Solving for reflect coefficients

If $H(z) = B(z) = B_M(z)$ is satisfied , the reflect coefficients k_i, $i = 1, 2, \ldots, M$, can be obtained as the following steps:

(1) Find $k_M = b_M^{(M)}$ according to eq. (6.38).
(2) Find the coefficients $b_1^{(M-1)}$, $b_2^{(M-1)}$, $\ldots$, $b_{M-1}^{(M-1)} = k_{M-1}$ of $B_{M-1}(z)$ based on k_M and $b_1^{(M)}$, $b_2^{(M)}$, $\ldots$, $b_M^{(M)}$, according to eq. (6.38). We can also directly find $B_{M-1}(z)$ according to eq. (6.36), and then $k_{M-1} = b_{M-1}^{(M-1)}$.
(3) We can find k_M, k_{M-1}, $\ldots$, k_1 and $B_{M-1}(z)$, $\ldots$, $B_1(z)$ by repeating step (2).

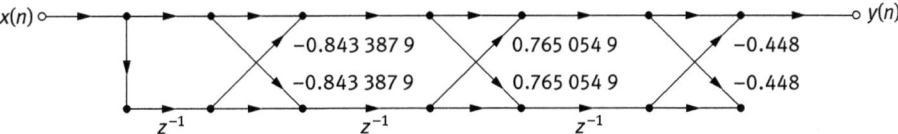

Figure 6.20: The structure of a lattice FIR filter.

Example 6.3: The system function of an FIR filter is shown as follows:

$$H(z) = \left(1 - 0.8e^{j\pi/4}z^{-1}\right)\left(1 - 0.8e^{-j\pi/4}z^{-1}\right)\left(1 - 0.7z^{-1}\right)$$

Try to find its lattice structure.

Solution: Rearranging the system function, we get

$$H(z) = 1 - 1.8313708z^{-1} + 1.4319595z^{-2} - 0.448z^{-3}$$

and $b_1^{(3)} = -1.8313708$, $b_2^{(3)} = 1.4319595$, $b_3^{(3)} = -0.448$, $k_3 = b_3^{(3)} = -0.448$. From eq. (6.38), we have $b_1^{(3)} = \frac{1}{1-k_3^2}[b_1^{(3)} - k_3b_2^{(3)}] = -1.4886262$, $b_2^{(3)} = \frac{1}{1-k_3^2}$

$[b_2^{(3)} - b_1^{(3)}] = 0.7650549$, $k_2 = b_2^{(2)} = 0.7650549$. Thus, $B_2(z) = 1 - 1.4886262z^{-1}$ $+ 0.7650549z^{-2}$. In the same way, we can get

$$b_1^{(1)} = \frac{1}{1-k_2^2}[b_1^{(2)} - k_2b_1^{(2)}] = -0.8433879, \qquad k_1 = -0.8433879$$

Thus, we get $B_1(z) = 1 - 0.8433879z^{-1}$. We can draw the lattice structure as shown in Figure 6.20.

6.5.2 All-Pole Lattice IIR Systems

The system function of an all-pole IIR digital filter can be expressed as

$$H(z) = \frac{1}{A(z)} = \frac{1}{1 + \sum_{i=1}^{M} a_i^{(M)}z^{-i}} \qquad (6.39)$$

where $a_i^{(M)}$ represents the ith coefficient of the all-pole IIR filter of order M. Its lattice structure is shown in Figure 6.21.

The basic transfer unit can be expressed as follows:

$$\begin{aligned} f_{m-1}(n) &= f_m(n) - k_m g_{m-1}(n-1), & m = 1, 2, \ldots, M \\ g_m(n) &= k_m f_{m-1}(n) + g_{m-1}(n-1), & m = 1, 2, \ldots, M \end{aligned} \qquad (6.40)$$

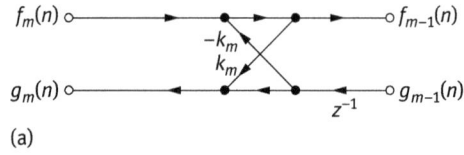

(a)

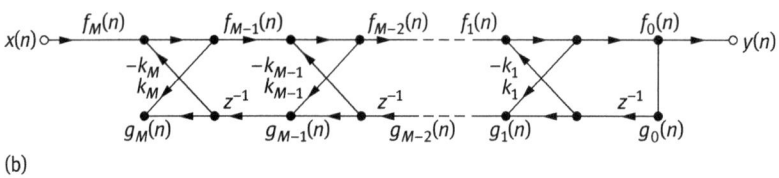

(b)

Figure 6.21: Lattice structure of an all-pole IIR filter: (a) the basic transfer unit and (b) the lattice structure.

Since eq. (6.40) is the same as eq. (6.30), the calculation for the reflect coefficients k_1, k_2, ..., k_M and $a_i^{(M)}$, $i = 1, 2, ..., M$; $m = 1, 2, ..., M$, is the same as that in the lattice FIR filter, unless $b_i^{(M)}$ in the all-zero system is substituted by $a_i^{(M)}$.

6.5.3 Pole–Zero Lattice IIR Systems

The IIR system that has both zeros and poles on a finite z-plane is expressed as

$$H(z) = \frac{B(z)}{A(z)} = \frac{\sum_{i=1}^{N} b_i^{(N)} z^{-i}}{1 + \sum_{i=1}^{N} a_i^{(N)} z^{-i}} \tag{6.41}$$

Its lattice structure is shown in Figure 6.22.

We see from Figure 6.22:

(1) If coefficients $k_1 = k_2 = \cdots = k_N = 0$, then the structure in Figure 6.22 changes to an FIR transversal structure of order N.

(2) If $c_1 = c_2 = \cdots = c_N = 0$, then the structure in Figure 6.22 changes to an all-pole IIR structure.

(3) Therefore, the upper part of Figure 6.22 corresponds to an all-pole structure, while its lower part corresponds to an all-zero structure. The coefficients k_1, k_2, ..., k_N can be solved with the same method as that used in the all-pole system, since the lower part of the figure does not affect the upper part. However, since the upper part affects the lower part, the solution method for c_i is different from that of b_i. The recursive equation is shown as follows:

$$\begin{cases} c_k = b_k^{(N)} - \sum_{m=k+1}^{N} c_m a_{m-k}^{(m)}, & k = 0, 1, ..., N-1 \\ c_N = b_N^{(N)} \end{cases} \tag{6.42}$$

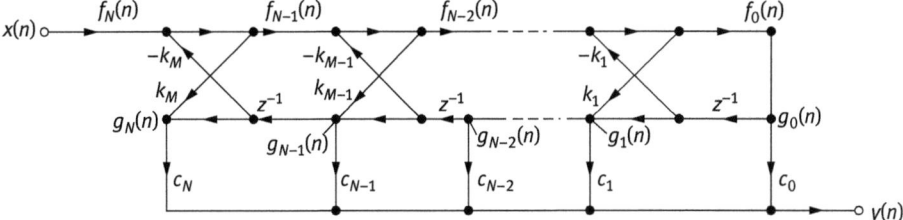

Figure 6.22: Pole–zero structure of a lattice IIR digital filter.

6.6 IIR Digital Filter Design

Digital filters can be divided into different types based on their frequency characteristics, including LPF, HPF, BPF, BSF, and APF. Due to the discretization of the unit impulse response of the digital filter, the frequency response of the digital filter is periodic. The frequency responses of different types of digital filters are shown in Figure 6.2. Usually, we used to represent the analog frequency with Ω in analog filters (in fact, it should be the analog angular frequency or the radian frequency, abbreviated as analog frequency). We used to represent the digital frequency with ω in digital filters. The relationship between the analog frequency and the digital frequency is shown as

$$\omega = \Omega T \tag{6.43}$$

where T is the sampling interval or period. Since $T = 1/F_s$, where F_s is the sampling frequency, we get $\omega = \Omega/F_s$.

By Fourier analysis theory, the frequency characteristics of the digital filter is periodic, whose period is 2π. Since it keeps symmetric in each period, we usually concern the range of $\omega < \omega_s/2 = \pi$, where $\omega_s - \Omega_s T - 2\pi F_s T - 2\pi$ is the sampling frequency in the digital frequency domain, while $\omega_s/2$ is referred to as the folding frequency.

Generally, the design of a digital filter contains the following steps:
(1) Determine the performance based on the requirement of the design.
(2) Approximate the performance by using a causal and stable discrete-time LTI system.
(3) Realize the system function with a finite precision algorithm.
(4) Implement the digital filter by using the software, the dedicated digital filter hardware, or the digital signal processor.

6.6.1 Technical Requirement of Filters and Design Brief for Analog Filters

1. The main technical requirements of filters
The frequency responses of various analog and digital filters are shown in Figure 6.2. However, these filters are generally not physically implemented, since their

frequency responses exist a discontinuity in the frequency domain. In order to physically achieve or approximate the filters, we usually need to arrange a transition zone in the frequency response, while the strict flatness in the passband and stopband is not necessarily required. Figure 6.23 shows the magnitude response of a typical nonideal LPF.

We see from the figure that the entire frequency response is divided into three parts: the passband, the stopband, and the transition band. Inside the passband, the magnitude response has an error tolerance α_1 as follows:

$$1 - \alpha_1 \le \left| H(e^{j\omega}) \right| \le 1, \quad |\omega| \le \omega_c \tag{6.44}$$

where ω_c is the cutoff frequency in the passband. Inside the stopband, the error tolerance α_2 is

$$\left| H(e^{j\omega}) \right| \le \alpha_2, \quad \omega_{st} \le |\omega| \le \pi \tag{6.45}$$

where ω_{st} is the cutoff frequency in the stopband. The range of the transition band is $\omega_c < \omega < \omega_{st}$.

In practical applications, the allowable maximum attenuation of passband δ_1 and minimum attenuation of stopband δ_2 should be satisfied:

$$\begin{aligned}
\delta_1 &= 20\log_{10} \frac{\left| H(e^{j0}) \right|}{\left| H(e^{j\omega_c}) \right|} = -20\log_{10} \left| H(e^{j\omega_c}) \right| = -20\log_{10}(1 - \alpha_1) \\
\delta_2 &= 20\log_{10} \frac{\left| H(e^{j0}) \right|}{\left| H(e^{j\omega_{st}}) \right|} = -20\log_{10} \left| H(e^{j\omega_{st}}) \right| = -20\log_{10} \alpha_2
\end{aligned} \tag{6.46}$$

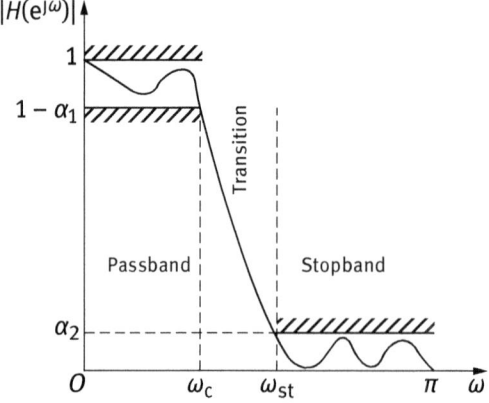

Figure 6.23: Magnitude response of a typical nonideal low-pass filter.

where $|H(e^{j0})| = 1$, and the 3 dB bandwidth $|H(e^{j\omega_c})| = 0.707$ and $|H(e^{j\omega_{st}})| = 0.001$ are supposed. Thus, the system gain decreases $\delta_1 = 3$ dB at ω_c, and $\delta_2 = 60$ dB at ω_{st}.

2. Three important properties of digital filters

(1) Magnitude-squared characteristics

The magnitude-squared characteristics are defined as

$$|H(e^{j\omega})|^2 = H(e^{j\omega})H^*(e^{j\omega}) = H(e^{j\omega})H(e^{-j\omega}) = \left[H(z)H(z^{-1})\right]_{z=e^{j\omega}} \tag{6.47}$$

The analysis shows that the poles of $H(z)H(z^{-1})$ are either conjugate symmetry or mirror symmetry. In order to have $H(z)$ as a realizable system, we always take the poles inside the unit circle as poles of $H(z)$, and take the poles outside the unit circle as poles of $H(z^{-1})$.

(2) Phase characteristics

The phase characteristics of a digital filter are shown as

$$\varphi(e^{j\omega}) = \arctan\left\{\frac{\text{Im}[H(e^{j\omega})]}{\text{Re}[H(e^{j\omega})]}\right\} \tag{6.48}$$

(3) Group delay characteristics

The group delay is a measure to the average delay of filters. The group delay of a digital filter is defined as

$$\tau(e^{j\omega}) = -\frac{d\varphi(e^{j\omega})}{d\omega} \tag{6.49}$$

3. A design summary for analog filters

(1) General design method for analog filters

We introduce the general design method of the analog filter by using a low-pass analog filter as an example. We need to determine the technical specifications, including the cut-off frequencies Ω_c and Ω_{st}, the allowable maximum attenuation of pass-band δ_1 and minimum attenuation of stop-band δ_2 in dB. Suppose the system function of the analog filter to be designed as

$$G(s) = \frac{d_0 + d_1 s + \cdots + d_{N-1}s^{N-1} + d_N s^N}{c_0 + c_1 s + \cdots + c_{N-1}s^{N-1} + c_N s^N} \tag{6.50}$$

The logarithmic magnitude-squared response $10\log_{10}|G(j\Omega)|^2$ should satisfy the requirements on δ_1 and δ_2 at both cut-off frequencies Ω_c and Ω_{st}.

Define the attenuation function as

$$\delta(\Omega) = 10\log_{10}\left|\frac{X(j\Omega)}{Y(j\Omega)}\right|^2 = 10\log_{10}\frac{1}{|G(j\Omega)|^2} \tag{6.51}$$

We get

$$\begin{aligned}\delta_1 = \delta(\Omega_c) &= -10\log_{10}|G(j\Omega_c)|^2\\ \delta_2 = \delta(\Omega_{st}) &= -10\log_{10}|G(j\Omega_{st})|^2\end{aligned} \tag{6.52}$$

On the other hand, since the unit impulse response of the designed filter is generally the real valued, we have

$$G(s)G^*(s) = G(s)G(-s)|_{s=j\Omega} = |G(j\Omega)|^2 \tag{6.53}$$

If we can find $|G(j\Omega)|^2$ based on the given technical specifications, then we can get the system function $G(s)$ of the designed filter.

(2) Some commonly used analog filters
① Butterworth filter

$$|G(j\Omega)|^2 = \frac{1}{1+C^2\Omega^{2N}} \tag{6.54}$$

where C is an undetermined constant and N is the order of the filter.
② Chebyshev type I filter

$$|G(j\Omega)|^2 = \frac{1}{1+\varepsilon^2 C_n^2(\Omega)} \tag{6.55}$$

where $C_n^2(\Omega) = \cos^2(n\arccos\Omega)$.
③ Chebyshev type II filter

$$|G(j\Omega)|^2 = \frac{1}{1+\varepsilon^2\left[\dfrac{C_n^2(\Omega_{st})}{C_n^2(\Omega_{st}/\Omega)}\right]^2} \tag{6.56}$$

④ Elliptic filter

$$|G(j\Omega)|^2 = \frac{1}{1+\varepsilon^2 U_n^2(\Omega)} \tag{6.57}$$

where $U_n^2(\Omega)$ is the Jacobian elliptic function.

6.6.2 Digital Filter Design Based on the Analog Filter

Major steps for designing an IIR digital filter based on the analog filter are listed as follows:

(1) Change the determined technical specifications of the digital filter into those of the corresponding analog filter based on some particular transform rule.

(2) Transform the specifications of HPF, BPF and stopband filters into those of the LPF, in order to use the existed materials of analog LPFs.

(3) Design the system function $G(s)$ of the analog filter by using some approximation algorithm, according to the specifications of the analog filter.

(4) Based on the rule used in steps (1) and (2), change the system function $G(s)$ into $H(z)$ of the digital filter.

6.6.3 Digital Filter Design by the Impulse Invariance Procedure

1. Principle

The impulse invariance is a basic method for designing the IIR digital filter from the corresponding analog filter. The principle of this method is to have the impulse response of the digital filter equals to the samples of the impulse response of the continuous-time filter as

$$h(n) = g(nT) \tag{6.58}$$

where T is the sampling period. The system functions of $h(n)$ and $g(t)$ are $H(z)$ and $G(s)$, respectively. Based on the relation between z-transform and Laplace transform

$$z = e^{sT} \tag{6.59}$$

we have

$$H(z)|_{z=e^{sT}} = \frac{1}{T} \sum_{k=-\infty}^{\infty} G\left(s - j\frac{2\pi}{T}k\right) \tag{6.60}$$

It can be seen from the above formula that the impulse response invariance procedure converts the s-plane of the analog filter into the z-plane of the digital filter, and $z = e^{sT}$ is the mapping relation of the transform.

In fact, the above map from the s-plane to the z-plane is not a simple algebraic mapping. Figure 6.24 shows the corresponding relationship of this kind of mapping. We see from Figure 6.24 that every horizontal stripe with the width of $2\pi/T$ on the s-plane is ae: overlappingly mapped to the entire z-plane, reflecting the periodic extension relation between $H(z)$ and $G(s)$. Furthermore, the left side half s-plane is mapped

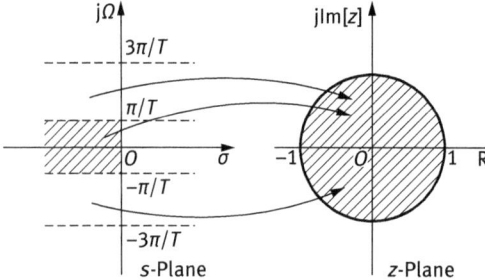

Figure 6.24: The mapping relation of the impulse invariant method.

inside the unit circle on the z-plane, while the right side half s-plane is mapped outside the unit circle. $j\Omega$ is mapped onto the unit circle.

2. Aliasing
From eq. (6.60) we have

$$H(e^{j\omega}) = \frac{1}{T} \sum_{k=-\infty}^{\infty} G\left[\frac{j(\omega - 2\pi k)}{T}\right] \tag{6.61}$$

which means that the frequency response of a digital filter is the periodic extension of the corresponding analog filter. Therefore, the analog filter should satisfy the band-limited condition, and its passband should be inside the range of $\Omega_s/2$, which yields

$$H(e^{j\omega}) = \frac{1}{T} G\left(j\frac{\omega}{T}\right), \quad |\omega| < \pi \tag{6.62}$$

3. Design method and examples
Without loss of generality, the system function $G(s)$ of a causal and stable analog filter of order N can be divided into a form of partial fraction, constructing N first-order systems as

$$G(s) = \sum_{i=1}^{N} \frac{A_i}{s + s_i} \tag{6.63}$$

Taking the inverse Laplace transform, we get

$$g(t) = \sum_{i=1}^{N} A_i e^{-s_i t} u(t) \tag{6.64}$$

According to the impulse invariance method, we can get the impulse response of the digital filter from eq. (6.58) as follows:

$$h(n) = g(nT) = \sum_{i=1}^{N} A_i e^{-s_i nT} u(n) = \sum_{i=1}^{N} A_i (e^{-s_i T})^n u(n) \tag{6.65}$$

Taking the z-transform to both sides of the above equation, we get the system function of the digital filter as

$$H(z) = \sum_{n=-\infty}^{\infty} h(n) z^{-n} = \sum_{n=0}^{\infty} A_i \sum_{i=1}^{N} (e^{-s_i T} z^{-1})^n = \sum_{i=1}^{N} \frac{A_i}{1 - e^{-s_i T} z^{-1}} \tag{6.66}$$

It is obvious if eq. (6.58) is satisfied, and the system function of the digital filter $H(z)$ can be obtained from the system function of the analog filter $G(s)$:

$$G(s) = \sum_{i=1}^{N} \frac{A_i}{s + s_i} \Rightarrow \sum_{i=1}^{N} \frac{A_i}{1 - e^{-s_i T} z^{-1}} = H(z) \tag{6.67}$$

We see from eq. (6.67) that the mapping from the analog filter to the digital filter by the impulse invariance method is a pole mapping method. If the poles of $G(s)$ are all in the left half of the s-plane, then the poles of $H(z)$ are all mapped inside the unit circle on the z-plane. If $G(s)$ is stable, then $H(z)$ is also stable.

But in fact, the impulse invariance method does not guarantee the similar relations for zero mapping. And on the other hand, eq. (6.58) is usually modified as follows to guarantee that the frequency response of the digital filter does not change with the sampling frequency:

$$h(n) = Tg(nT) \tag{6.68}$$

Example 6.4: An RC circuit is shown in Figure 6.25, where $\alpha = 1/RC$. Design a digital filter based on the corresponding analog filter, and draw its signal flow graph.
Solution: We can find the system $G(s)$ and the unit impulse response $g(t)$ as
$G(s) = \frac{Y(s)}{X(s)} = \frac{\alpha}{s + \alpha}$ and $g(t) = \alpha e^{-\alpha t}$, based on the given circuit. From eqs. (6.67) and (6.68), we can get the system function $H(z)$, the frequency response $H(e^{j\omega})$, and the unit impulse response $h(n)$ of the digital filter as follows:

$$H(z) = \frac{\alpha T}{1 - e^{-\alpha T} z^{-1}}, \quad H(e^{j\omega}) = \frac{\alpha T}{1 - e^{-\alpha T} e^{-j\omega}}, \quad h(n) = \alpha T e^{-\alpha Tn}$$

Figure 6.26 demonstrates $g(t)$, $h(n)$, and the signal flow graph of the digital filter.

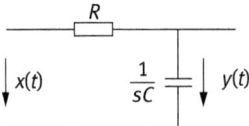

Figure 6.25: The given first-order RC circuit.

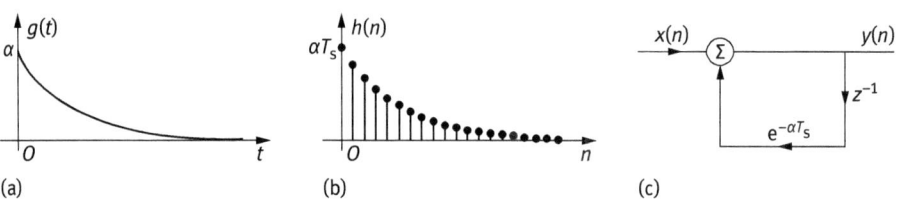

(a)   (b)   (c)

Figure 6.26: The unit impulse responses of the analog and digital filters and the signal flow graph: (a) the impulse response of the given circuit; (b) the impulse response of the corresponding digital filter; and (c) signal flow graph.

Further, we can also analyze the frequency characteristics of $H(e^{j\omega})$ and $G(j\Omega)$ for both digital and analog filters. The analysis shows that if the sampling frequency of the digital system is high enough, that is, the sampling period T is small enough, the frequency characteristics of the digital system are consistent with the frequency characteristics of the analog system. But if T is large, the error between the two filters becomes significant.

Example 6.5: Suppose the system function of an analog filter is $G(s) = \frac{2}{s^2 + 4s + 3}$. Design an IIR digital filter by using the impulse invariance method.
Solution: Taking the partial fraction decomposition to the system function, we get

$$G(s) = \frac{2}{s^2 + 4s + 3} = \frac{1}{s+1} - \frac{1}{s+3}$$

Based on eqs. (6.67) and (6.68), we get

$$H(z) = \frac{T}{1 - e^{-T}z^{-1}} - \frac{T}{1 - e^{-3T}z^{-1}} = \frac{Tz^{-1}(e^{-T} - e^{-3T})}{1 - z^{-1}(e^{-T} - e^{-3T}) + e^{-4T}z^{-2}}$$

If $T = 1$ is set, then

$$H(z) = \frac{0.318z^{-1}}{1 - 0.4177z^{-1} + 0.01831z^{-2}}$$

Thus the frequency response of the given analog filter and the designed digital filter is

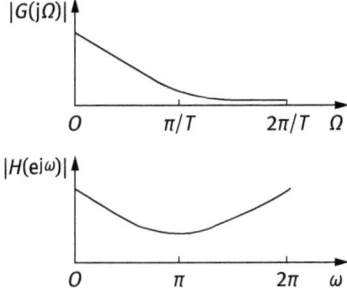

Figure 6.27: The magnitude spectra of the analog filter and corresponding digital filter.

$$G(j\Omega) = \frac{2}{(3-\Omega^2)+j4\Omega}, \quad H(e^{j\omega}) = \frac{0.318e^{-j\omega}}{1-0.4177e^{-j\omega}+0.01831e^{-j2\omega}}$$

Figure 6.27 demonstrates the curves of $|G(j\Omega)|$ and $|H(e^{j\omega})|$. We see from the figure that $H(e^{j\omega})$ has a significant aliasing since $G(j\Omega)$ is not sufficiently band limited.

6.6.4 Digital Filter Design by the Bilinear Transform

The basic idea of the impulse invariance method to design the digital filter introduced in Section 6.6.3 is to approximate the characteristics of the analog filter in the time domain. The advantage of this method is it has a good characteristic in time domain. However, since the multivalue mapping from s-plane to z-plane, the frequency aliasing may exist. The bilinear transform introduced in this section overcomes such shortcoming.

1. Principle
The bilinear transformation method can be divided into two steps. In the first step, the entire s-plane is compressed into a horizontal strip of s_1-plane, with a range from $-\pi/T$ to π/T. In the second step, the horizontal strip is mapped into the entire z-plane. In this way, the one-to-one correspondence relation between s-plane and z-plane can be guaranteed, and the multivalued mapping can also be eliminated. Figure 6.28 shows such transform steps.

The mapping from s-plane to s_1-plane is shown as

$$s = \tanh\left[\frac{s_1 T}{2}\right] = \frac{1-e^{-s_1 T}}{1+e^{-s_1 T}} \tag{6.69}$$

The mapping from s_1-plane to z-plane is

$$z = e^{s_1 T} \tag{6.70}$$

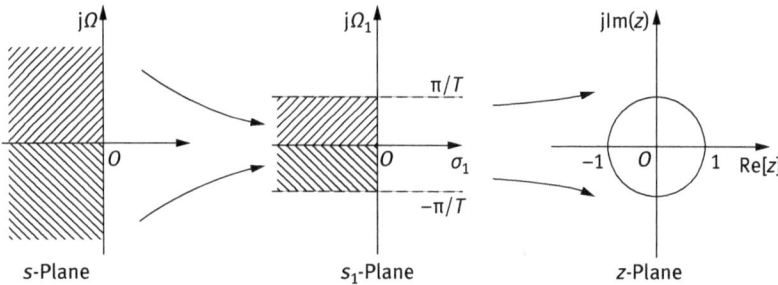

Figure 6.28: The transforms of the bilinear transform.

Thus we can get the single-valued mapping relations between the s-plane and the z-plane as

$$s = \frac{1-z^{-1}}{1+z^{-1}}, \qquad z = \frac{1+s}{1-s} \tag{6.71a}$$

An undetermined constant c can be introduced to construct a frequency relation between the analog filter and the digital filter. Thus eq. (6.71a) changes as

$$s = c\frac{1-z^{-1}}{1+z^{-1}}, \qquad z = \frac{c+s}{c-s} \tag{6.71b}$$

It can be proved that after the bilinear transform, the digital filter is still causal and stable if the original analog filter is causal and stable. The transformation from the analog filter $G(s)$ to the corresponding digital filter $H(z)$ can be realized as follows:

$$H(z) = G(s)\big|_{s=c\frac{1-z^{-1}}{1+z^{-1}}} = G\left(c\frac{1-z^{-1}}{1+z^{-1}}\right) \tag{6.72}$$

2. The selection of constant c

There are two methods to choose the undetermined constant c:

Method 1: Making both analog filter and digital filter to have a rather exact correspondence at low frequency band:

$$c = 2/T \tag{6.73}$$

is usually selected.

Method 2: Making a specific frequency of the digital filter to be strictly corresponding to a specific frequency of the analog filter, we have

$$c = \Omega_c \cot(\omega_c/2) \tag{6.74}$$

where Ω_c and ω_c represent the cutoff frequencies of the analog filter and the digital filter, respectively.

3. Nonlinear problem and predistortion

The significant advantage of the bilinear transform is it can avoid the frequency aliasing problem. However, the relationship between the analog frequency and the digital frequency is nonlinear as

$$\Omega = c \tan(\omega/2) \tag{6.75}$$

When Ω is far from 0, the nonlinear of the above equation becomes serious, resulting in a nonlinear phase and the distortion in the magnitude characteristics of the digital filter.

The predistortion technique is one of the effective methods to solve such problem. Figure 6.29 shows the predistortion technique.

In Figure 6.29, ω_1, ω_2, ω_3, ω_4 are digital frequencies in the digital filter to be designed. When the predistortion is conducted, the above four frequencies are primarily transformed into the corresponding analog frequencies Ω_1, Ω_2, Ω_3, Ω_4 based on $\Omega = c \tan(\omega/2)$, and then the digital filter is designed by the bilinear transform method. The digital frequencies obtained are just the same as those of original required ones.

Example 6.6: Given $f_c = 100$ Hz, $f_{st} = 300$ Hz, $\delta_1 = 3$ dB, $\delta_2 = 20$ dB, and the sampling frequency $F_s = 1000$Hz. Try to design a low-pass digital filter by the bilinear transform method.

Solution: (1) Change the given specifications into the form of radian frequency as $\omega_c = 0.2\pi$, $\omega_{st} = 0.6\pi$.

(2) Change the specifications of the digital filter into the form of the analog filter, shown as

$$\Omega_c = \tan(\omega_c/2) = 0.3249, \quad \Omega_{st} = \tan(\omega_{st}/2) = 1.37638$$

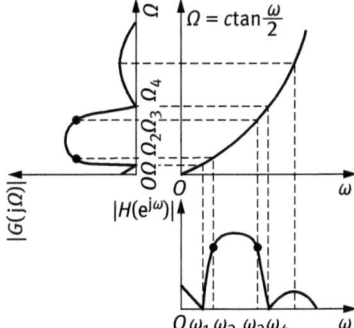

Figure 6.29: The predistortion of the bilinear transform.

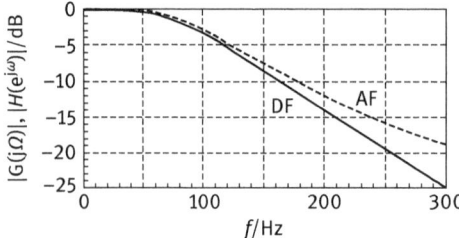

Figure 6.30: Frequency responses of the analog filter (AF, the dashed line) and the designed digital filter (DF, the solid line).

(3) Design the analog LPF as $G(s) = \frac{0.3249^2}{s^2 + 0.4595s + 0.3249^2}$.

(4) Find $H(z)$ from $G(s)$ by using the bilinear transform as

$$H(z) = G(s)\big|_{s=\frac{z-1}{z+1}} = \frac{0.06745 + 0.1349z^{-1} + 0.06745z^{-2}}{1 - 1.143z^{-1} + 0.4128z^{-2}}$$

Figure 6.30 demonstrates the frequency responses of the analog filter and the designed digital filter.

From the figure we see that the frequency characteristic of the digital filter designed by the bilinear transform method reaches the design requirements.

6.6.5 The Design Idea for High-Pass, Band-Pass, and Band-Stop Digital Filters

We introduced the basic method of designing a digital LPF based on the analog LPF in the above parts. In fact, digital filters have different forms, including high-pass, band-pass, and band-stop. This section gives the principle of design ideas and processes for such filters. Please refer to the relevant literature for the specific design method.

Figure 6.31 shows the basic method of designing various kinds of digital filters based on an analog low-pass prototype filter.

We see from Figure 6.31 that the design methods can be roughly divided into two types:

(1)   Conduct the analog frequency transform before the digitizing. The schematic diagram of this method is shown in the upper part of Figure 6.31.

(2)   Conduct the digitizing before the digital-to-digital frequency transform. The schematic diagram of this method is shown in the lower part of Figure 6.31.

Both methods are usually referred to as the frequency transform method of digital filter design due to the transform from an LPF to HPF, BPF, and BSF.

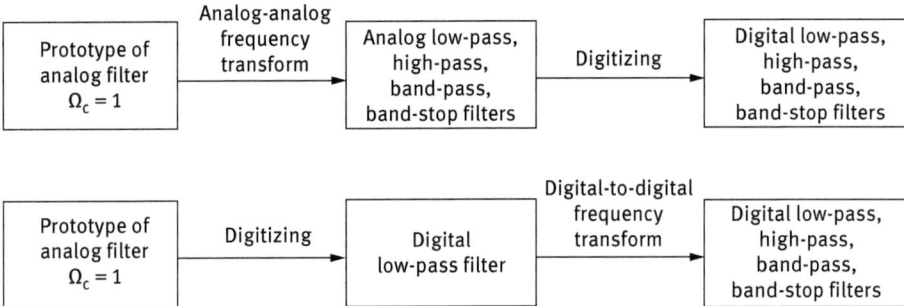

Figure 6.31: The block diagram for designing various types of digital filters.

6.6.6 IIR Digital Filter Design with MATLAB Programming

In practical applications of signal processing, we usually use MATLAB to design the filter and carry out the corresponding operation. This section gives a brief introduction to the functions of MATLAB on the filter design and demonstrates an example of the IIR filter design. Table 6.1 shows some commonly used MATLAB functions for the filter design.

Example 6.7: Design a digital filter with MATLAB 'programming. The given conditions are the same as those in Example 6.6.
Solution: The MATLAB program is shown as follows:

```
clear all; wp=.2*pi; ws=.6*pi; Fs=1000; rp=3, rs=20;
wap=2*Fs*tan(wp/2); was=2*Fs*tan(ws/2); [n,wn]=buttord(wap,was,rp,
rs,'s');
[z,p,k]=buttap(n); [bp,ap]=zp2tf(z,p,k); [bs,as]=lp2lp(bp,ap,wap)
w1=[0:499]*2*pi; h1=freqs(bs,as,w1); [bz,az]=bilinear(bs,as,Fs)
[h2,w2]=freqz(bz,az,500,Fs); plot(w1/2/pi,abs(h1),'-.',w2,abs(h2),'k');
grid on; xlabel('Frequency/Hz'); ylabel('Normalized magnitude');
```

The results of the design are $Bz = [0.0675, 0.1349, 0.06745]$, $Az = [1, -1.143, 0.4128]$. The system function of the digital filter can be obtained as

$$H(z) = \frac{0.06745 + 0.1349z^{-1} + 0.06745z^{-2}}{1 - 1.143z^{-1} + 0.4128z^{-2}}$$

The magnitude response of the digital filter, as well as the corresponding analog filter, is shown in Figure 6.32.

Table 6.1: Some commonly used MATLAB functions for the filter design.

Order	MATLAB function	Invocation style	Function	Explain
1	buttord.m	[N, Wn]=buttord (Wp, Ws, Rp, Rs)	Determine the order of the filter.	Wp and Ws are cutoff frequencies; Rp and Rs are the attenuations of the passband and stopband. N is the order of the filter, and Wn is the 3 dB frequency.
2	buttap.m	[z, p, k]=buttap (N)	Design the prototype analog low-pass filter.	z, p, and k are the zero, pole, and gain of the prototype analog low-pass filter, respectively.
3	lp2lp.m	[B, A]=lp2lp(b, a, Wo)	Transform the prototype analog filter into the low-pass, high-pass, band-pass, and band-stop filters.	b and a are the numerator and denominator coefficient vectors of the analog low-pass filter. B and A are the coefficient vectors of the transformed filter. Wo is the cutoff frequency of the low-pass and high-pass filters, or it is also the central frequency of the band-pass and band-stop filters. Bw is the bandwidth of the filter.
	lp2hp.m	[B, A]=lp2hp(b, a, Wo)		
	lp2bp.m	[B, A]=lp2bp(b, a, Wo, Bw)		
	lp2bs.m	[B, A]=lp2bs(b, a, Wo, Bw)		
4	bilinear.m	[Bz, Az]=bilinear (B, A, Fs)	Bilinear transform	B and A are the numerator and denominator coefficient vectors of the analog filter, respectively. Bz and Az are the coefficient vectors of the digital filter. Fs is the sampling frequency.
5	butter.m	[Bz, Az]=butter (N, Wn)	Butterworth BPF	Bz and Az are the coefficient vectors of the digital filter. Wn is the passband cutoff frequency.
		[Bz, Az]=butter (N, Wn, "high")	Butterworth HPF	
		[Bz, Az]=butter (N, Wn, "stop")	Butterworth BSF	
		[Bz, Az]=butter (N, Wn, "s")	Analog filter	
6	cheb1ord.m	[N, Wp] = cheb1ord(Wp, Ws, Rp, Rs)	Find the order of Chebyshev filter.	Consulting Butterworth filter
7	cheb1ap.m	[z, p, k] = cheb1ap(N, Rp)	Design Chebyshev type I analog filter.	Consulting Butterworth filter
8	cheby1.m	[B, A] = cheby11 (N, R, Wp)	Design Chebyshev type I digital filter.	Consulting Butterworth filter
9	impinvar.m	[Bz, Az] = impinvar(B, A, Fs)	Complete the transform from s to z by the impulse invariance procedure.	Consulting Butterworth filter

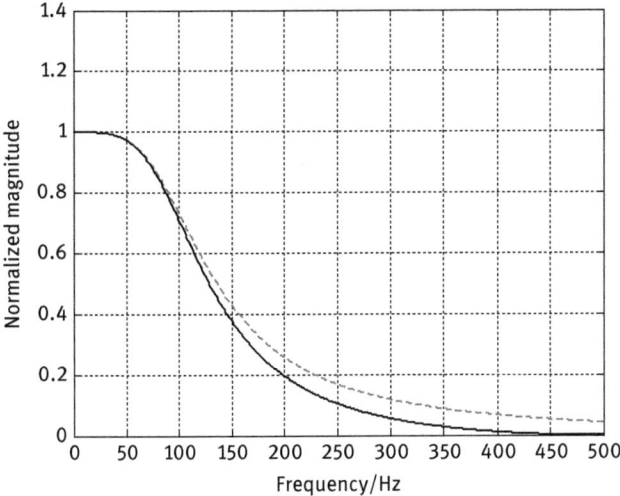

Figure 6.32: The frequency responses of the designed digital filter (the solid line) and corresponding analog filter (the dot-dashed line).

6.7 FIR Digital Filter Design

The FIR digital filter introduced in this section has following advantages and characteristics. First, the FIR digital filter can be designed to have not only a strict linear phase, but also a required amplitude characteristic. Second, since the unit impulse response of the FIR digital filter is finite, and it is the all-zero structure, the filter is definitely stable. Third, the FIR digital filter can be designed as a causal system since any noncausal finite length system can be transformed into a causal one, as long as a certain delay is conducted. Fourth, FFT can be used in the operation of the FIR digital filter to significantly improve the computational efficiency.

This section mainly introduces the window function design method and the frequency sampling design method for the FIR digital filter.

6.7.1 FIR Digital Filter Design by Windowing

1. Basic idea

The basic idea of the FIR digital filter design is to design an FIR filter $h(n)$ under the premise of a given ideal digital filter $H_d(e^{j\omega}) = \sum_{n=-\infty}^{+\infty} h_d(n)e^{-j\omega n}$, making its frequency response $H(e^{j\omega}) = \sum_{n=0}^{N-1} h(n)e^{-j\omega n}$ close to the ideal filter $H_d(e^{j\omega})$, that is, making $h(n)$ close to $h_d(n)$. The whole design is carried out in the time domain. The key technology is the windowing for the unit impulse response. In the process of the design, the conversion between time domain and frequency domain is also needed.

Therefore the FIR digital filter design with a window function is also known as the Fourier series method.

2. Design method
The design procedure of the FIR filter with the window function is as follows:
   (1) Find the impulse response of the ideal digital filter from the given frequency response of the ideal digital filter
   Suppose the frequency response of the desired ideal digital filter is as follows:

$$H_d(e^{j\omega}) = \sum_{n=-\infty}^{+\infty} h_d(n)e^{-j\omega n} \tag{6.76}$$

where the impulse response of the ideal filter $h_d(n)$ is shown as

$$h_d(n) = \frac{1}{2\pi} \int_{-\pi}^{\pi} H_d(e^{j\omega})e^{j\omega n} d\omega \tag{6.77}$$

(2) Truncate the ideal impulse response with a window function
Since the ideal digital filter has the rectangular frequency characteristics, it must be a noncausal sequence with infinite length. Because the impulse response of the FIR filter to be designed is $h(n)$, it is needed to truncate $h_d(n)$ with a window function, whose time duration is finite, which is shown as

$$h(n) = w(n)h_d(n) \tag{6.78}$$

Thus $h(n)$ becomes a causal filter with the FIR.
(3) Find the frequency characteristics for the windowed impulse response
Suppose that the window function $w(n) = R_N(n)$ is a rectangular sequence. The frequency characteristics of the ideal LPF with linear phase are as follows:

$$H_d(e^{j\omega}) = \begin{cases} e^{-j\omega\alpha}, & -\omega_c \le \omega \le \omega_c \\ 0, & -\pi < \omega < -\omega_c, \quad \omega_c < \omega \le \pi \end{cases} \tag{6.79}$$

where $-\omega\alpha$ represents the linear phase. After the inverse Fourier transform we get

$$h_d(n) = \frac{1}{2\pi} \int_{-\omega_c}^{\omega_c} e^{-j\omega\alpha} e^{j\omega n} d\omega = \frac{\omega_c}{\pi} \frac{\sin[\omega_c(n-\alpha)]}{\omega_c(n-\alpha)} \tag{6.80}$$

Figure 6.33 demonstrates the frequency response and impulse response of the ideal LPF. It also shows the frequency domain and time domain features of the rectangular window.

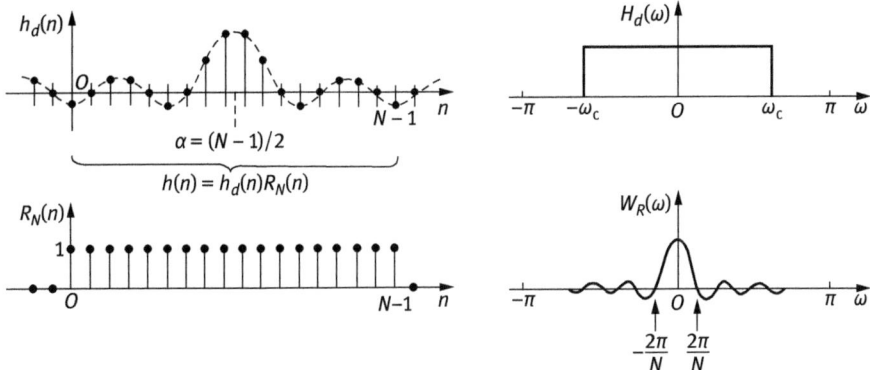

Figure 6.33: The frequency response and impulse response of the ideal LPF, and the frequency domain and time domain features of the rectangular window.

Consider the linear phase characteristics of an FIR digital filter, whose impulse response $h(n)$ is even symmetric. Therefore

$$h(n) = w(n)h_d(n) = \begin{cases} h_d(n), & 0 \le n \le N-1 \\ 0, & \text{others} \end{cases} \tag{6.81}$$

From the convolution property of the Fourier transform, we have

$$H(e^{j\omega}) = \frac{1}{2\pi} \int_{-\pi}^{\pi} H_d(e^{j\theta}) W(e^{j(\omega-\theta)}) d\theta \tag{6.82}$$

The approximation degree from $H(e^{j\omega})$ to $H_d(e^{j\omega})$ depends entirely on the frequency characteristics of the window function. The time domain sequence of the window $w(n)$ is a rectangular sequence, so its frequency response is a sinc function in the frequency domain, as shown in the right-bottom of Figure 6.33.

(4) Analysis

Figure 6.34 demonstrates the operation of eq. (6.82); it also gives a frequency comparison between the designed FIR digital filter and the ideal filter.

In Figure 6.34, $H_d(\theta)$ and $H(\omega)$ represent the frequency response of the ideal LPF and the designed LPF, respectively. $W_R(\omega)$ is the frequency response of the rectangular window. It is obvious that the windowed filter is of significant fluctuations in its passband and stopband, forming a certain spectrum leakage.

Example 6.8: Design an FIR low-pass digital filter. The desired frequency response is as

$$H_d(e^{j\omega}) = \begin{cases} e^{-jM\omega/2}, & 0 \le |\omega| \le 0.25\pi \\ 0, & 0.25\pi < |\omega| \le \pi \end{cases}$$

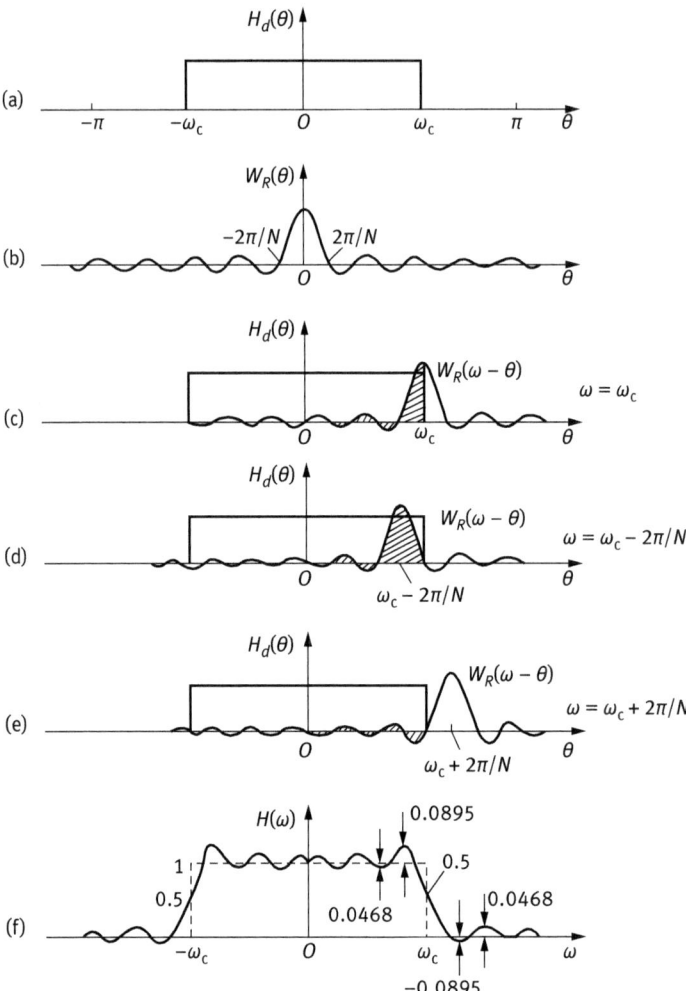

Figure 6.34: A comparison between the designed FIR digital filter and the ideal filter.(a) the frequency response of the ideal LPF; (b) the frequency response of the rectangular window; (c)~(e) the convolution operation of eq. (6.82) (f) the frequency response of the designed LPF.

M is taken as $M = 10,\ 20,\ 40$. Please investigate its amplitude response.

Solution: Taking the inverse Fourier transform to the given spectrum, we have

$$h_d(n) = \frac{\sin[(n - M/2) \times 0.25\pi]}{\pi(n - M/2)}$$

If $M = 10$, then

$h(0) = h(10) = -0.045,\ \ h(1) = h(9) = 0,\ \ h(2) = h(8) = 0.075,\ \ h(3) = h(7) = 0.1592,$
$h(4) = h(6) = 0.2551,\ \ h(5) = 0.25$

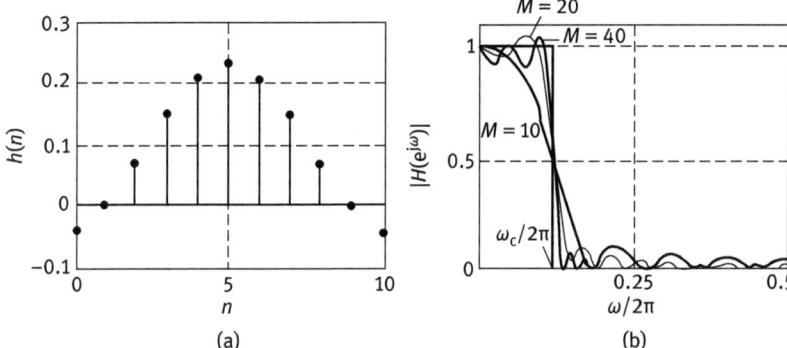

Figure 6.35: The designed results of Example 6.8: (a) normalized impulse response and (b) magnitude response.

Obviously, it satisfies the symmetric relation. Figure 6.35 shows the normalized $h(n)$ when $M = 10$. It also gives the magnitude response of $H(e^{j\omega})$ when $M = 10,\ 20,\ 40$.

We see from Figure 6.35 when M increases, the fluctuation will become serious, which is referred to as the Gibbs phenomenon.

6.7.2 The Concept of Window Function and Some Commonly Used Windows

1. Problems caused by windowing
The purpose of truncation to the impulse response of an ideal filter is to modify the infinite noncausal impulse response of an ideal filter to a causal and realizable FIR digital filter. On the other hand, the windowing operation will inevitably lead to Gibbs phenomenon and spectrum leakage, which may affect the performance of the FIR filter.

2. General criteria for selection of window function
The general criteria for the selection of window function are as follows. First, the main lobe (or called the first lobe) of the spectrum of the window function is required to be as narrow as possible, so that it can obtain a more abrupt transition zone. Second, the magnitude of the maximum side lobe is required to be as small as possible, so that the energy of the window function is concentrated in its main lobe, which can reduce the acromion and ripples of the spectrum.

However, both requirements cannot usually be satisfied at the same time. In fact, the purpose of different window functions is to make the FIR digital filter get a flat passband characteristics and a small stopband ripple.

3. Commonly used windows
Commonly used window functions in signal processing contain the rectangular window, triangular window, Hanning window, Hamming window, Blackman window, and so on.

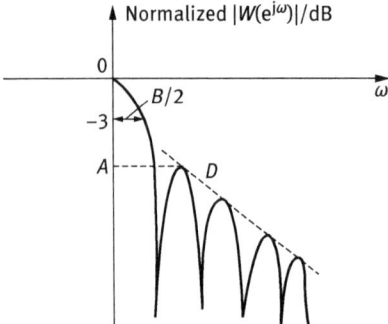

Figure 6.36: Parameters in window functions.

Figure 6.36 demonstrates some spectrum parameters for the window function, which can be used to evaluate the performance of a window. In the figure, parameter A is the maximum side lobe value in dB, and parameter B is the 3 dB bandwidth of the window. $B_0 = 4\pi/N$ represents the width of the main lobe for the rectangular window between two zero crossings, where N is the data length. In this way, $\Delta\omega = 2\pi/N$ can be used to express parameter B. Parameter D denotes the attenuation rate of the side lobes in dB/oct.

An ideal window function should satisfy the following conditions: with minimal B and A, and maximal D. In addition, $w(n)$ should be a real-valued, even symmetric, and nonnegative sequence. $W(e^{j\omega})$ should be as positive as possible. $w(0) = 1$ should also be satisfied.

The definition and characteristics of commonly used windows are listed in Table 6.2.

In Table 6.2, $W_R(\omega) = \frac{\sin(N\omega/2)}{\sin(\omega/2)}$ represents the magnitude spectrum of the rectangular window. Figure 6.37 demonstrates curves in the time domain and spectra in the frequency domain for the commonly used windows.

6.7.3 FIR Digital Filter Design with the Frequency Sampling Method

1. Design idea and method
Different from the windowing method, the design for the FIR filter by using the frequency sampling method is conducted in the frequency domain, by sampling the given ideal filter $H_d(e^{j\omega})$, which is shown as

$$H_d(k) = H_d(e^{j\omega})\big|_{\omega = \frac{2\pi}{N}k} \tag{6.83}$$

Let the discrete frequency response of the designed digital filter be

Table 6.2: The definition and characteristics of commonly used windows.

Order	Window functions	Expression in time/frequency domains	Parameters
1	Rectangular	$w(n) = R_N(n)$ $W(e^{j\omega}) = W_R(\omega)e^{-j\left(\frac{N-1}{2}\right)\omega}$ $W_R(\omega) = \sin\left(\frac{N\omega}{2}\right)/\sin\left(\frac{\omega}{2}\right)$	$B = 0.89\Delta\omega$, $B_0 = 4\pi/N$, $A = -13$ dB, $D = -6$ dB/oct
2	Triangular (Bartlett)	$w(n) = \begin{cases} \dfrac{2n}{N-1}, & 0 \le n \le \dfrac{N-1}{2} \\ 2 - \dfrac{2n}{N-1}, & \dfrac{N-1}{2} < n \le N-1 \end{cases}$ $W(e^{j\omega}) \approx \frac{2}{N}\left(\frac{\sin(N\omega/4)}{\sin(\omega/2)}\right)^2 e^{-j\left(\frac{N-1}{2}\right)\omega}$	$B = 1.28\Delta\omega$, $B_0 = 8\pi/N$, $A = -27$ dB, $D = -12$ dB/oct
3	Hanning	$w(n) = \frac{1}{2}\left[1 - \cos\left(\frac{2\pi n}{N-1}\right)\right] R_N(n)$ $W(e^{j\omega}) = W(\omega)e^{-j\left(\frac{N-1}{2}\right)\omega}$ $W(\omega) \approx 0.5 W_R(\omega)$ $\quad + 0.25\left[W_R\left(\omega - \frac{2\pi}{N}\right) + W_P\left(\omega + \frac{2\pi}{N}\right)\right]$	$B = 1.44\Delta\omega$, $B_0 = 8\pi/N$, $A = -32$ dB, $D = -18$ dB/oct
4	Hamming	$w(n) = \left[0.54 - 0.46\cos\left(\frac{4\pi n}{N-1}\right)\right] R_N(n)$ $W(\omega) = 0.54 W_R(\omega)$ $\quad + 0.23\left[W_R\left(\omega - \frac{2\pi}{N}\right) + W_R\left(\omega + \frac{2\pi}{N}\right)\right]$	$B = 1.3\Delta\omega$, $B_0 = 8\pi/N$, $A = -43$ dB, $D = -6$ dB/oct
5	Blackman	$w(n) = \left[0.42 - 0.5\cos\left(\frac{2\pi n}{N-1}\right) + 0.08\cos\left(\frac{4\pi n}{N-1}\right)\right] R_N(n)$ $W(\omega) = 0.42 W_R(\omega) + 0.25\left[W_R\left(\omega - \frac{2\pi}{N-1}\right) + W_R\left(\omega + \frac{2\pi}{N-1}\right)\right]$ $\quad + 0.04\left[W_R\left(\omega - \frac{4\pi}{N-1}\right) + W_R\left(\omega + \frac{4\pi}{N-1}\right)\right]$	$B = 1.68\Delta\omega$, $B_0 = 12\pi/N$, $A = -58$ dB, $D = -18$ dB/oct

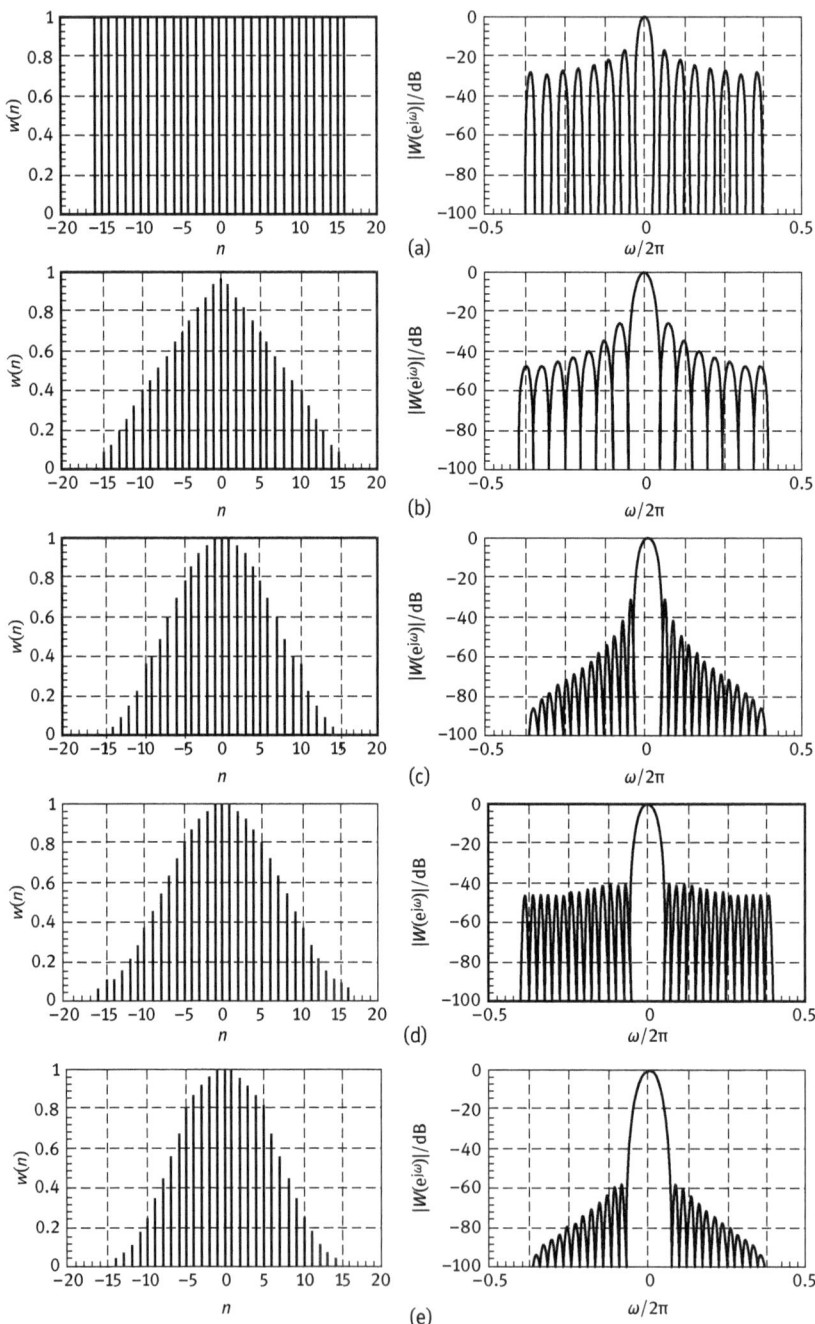

Figure 6.37: The spectra and curves in the time domain for the commonly used windows: (a) rectangular window; (b) triangular window; (c) Hanning window; (d) Hamming window; and (e) Blackman window.

$$H(k) = H_d(k) = H_d(e^{j\omega})|_{\omega = \frac{2\pi}{N}k}, \quad k = 0, 1, \ldots, N-1 \tag{6.84}$$

Taking the inverse DFT (IDFT), we can get the finite sequence $h(n)$ from $H(k)$. $h(n)$ is the unit impulse response of the designed FIR filter.

2. Analysis

From the theory of the frequency sampling, we can get the system function $H(z)$ and frequency response $H(e^{j\omega})$ of the FIR digital filter by using N samples of $H(k)$, and can further analyze the approximation degree between $H(z)$ (or $H(e^{j\omega})$) and $H_d(z)$ (or $H_d(e^{j\omega})$).

Based on the theory of the frequency sampling, the interpolation formula from $H(k)$ to $H(z)$ and $H(e^{j\omega})$ are as follows:

$$H(z) = \frac{1 - z^{-N}}{N} \sum_{k=0}^{N-1} \frac{H(k)}{1 - W_N^{-k} z^{-1}} \tag{6.85}$$

and

$$H(e^{j\omega}) = \sum_{k=0}^{N-1} H(k) \Phi\left(\omega - \frac{2\pi}{N} k\right) \tag{6.86}$$

where the interpolation function $\Phi(\omega)$ is shown as

$$\Phi(\omega) = \frac{1}{N} \frac{\sin(\omega N/2)}{\sin(\omega/2)} e^{-j\omega(\frac{N-1}{2})} \tag{6.87}$$

Substituting eq. (6.87) into eq. (6.86) and taking a simplification, we get

$$H(e^{j\omega}) = e^{-j\omega(\frac{N-1}{2})} \sum_{k=0}^{N-1} H(k) \cdot \frac{1}{N} e^{j\frac{k\pi}{N}(N-1)} \cdot \frac{\sin[N(\omega/2 - k\pi/N)]}{\sin(\omega/2 - k\pi/N)} \tag{6.88}$$

Compared with the ideal filter $H_d(e^{j\omega})$, it can be seen that the frequency response of the filter $H(e^{j\omega})$ is equal to the frequency response of the ideal filter at each sampling point, which is shown as

$$H\left(\exp\left(j\frac{2\pi}{N}k\right)\right) = H(k) = H_d(k) = H_d\left(\exp\left(j\frac{2\pi}{N}k\right)\right), \quad k = 0, 1, \ldots, N-1 \tag{6.89}$$

However, the frequency response out of the frequency sampling points has a certain approximation error compared with the ideal filter. Figure 6.38 shows the comparison of the frequency responses between the designed filter and the ideal filter.

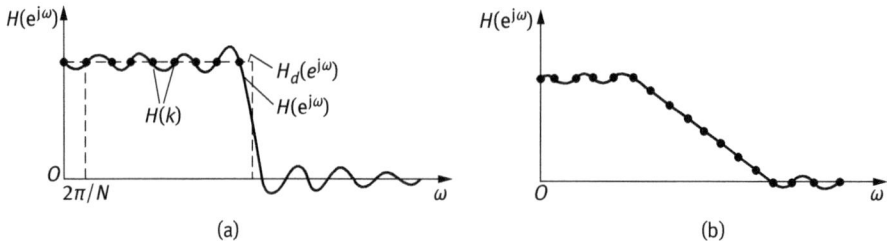

Figure 6.38: A comparison of the frequency responses between the designed filter and the ideal filter.

We see from the figure that when the frequency characteristics of the ideal filter changes slowly, the approximation error is small (see Figure 6.38(b)), and when the frequency characteristics of the ideal filter changes relatively steep, the approximation error is significant (see Figure 6.38(a)).

3. The determination of $H_d(k)$ and the design steps

The general criteria for determining $H_d(k)$ are as follows:

(1) Let $|H_d(k)| = 1$ in the passband and $|H_d(k)| = 0$ in the stopband. Set a phase function for $H_d(k)$ in the passband.

(2) Guarantee that the unit impulse response $h(n)$ obtained from IDFT is real valued.

(3) Guarantee that $H(e^{j\omega})$ obtained from $h(n)$ has a linear phase.

If N is an even number, we have

$$H_d(k) = \begin{cases} e^{-j(N-1)k\pi/N}, & k = 0, 1, \ldots, N/2-1 \\ 0, & k = N/2 \\ -e^{-j(N-1)k\pi/N}, & k = 0, 1, \ldots, N/2-1 \end{cases} \tag{6.90}$$

If N is an odd number, we have

$$H_d(k) = e^{-j(N-1)k\pi/N}, \quad k = 0, 1, \ldots, N-1 \tag{6.91}$$

Thus, the design steps for the FIR digital filter of the frequency sampling method are listed as follows:

(1) Based on the requirements of the passband and the stopband of the filter, determine $H_d(k)$ according to eq. (6.90) or eq. (6.91). In the stopband, set $H_d(k) = 0$.

(2) Find the system function $H(z)$ from $H_d(k)$, and find the frequency response of the designed filter $H(e^{j\omega})$ further.

Example 6.9: Design a low-pass FIR digital filter by using the frequency sampling method. The cutoff frequency is 1/10 of the sampling frequency, and the order of the filter is $N = 20$.

Solution: Since N is even, we can get only two nonzero frequency bins when we conduct the frequency sampling to $H_d(e^{j\omega})$, which is shown as

$$H_d(0) = 1, \quad H_d(1) = e^{-j\frac{19\pi}{20}}, \quad H_d(k) = H_d(N-k) = 0, \quad k = 2, 3, \ldots, 10$$
$$H_d(19) = H_d(20-1) = e^{j\frac{19\pi}{20}} = H_d^*(1)$$

After the inverse Fourier transform, we get the samples of $h(n)$ as

$$h(0) = h(19) = -0.04877, \quad h(1) = h(18) = -0.0391, \quad h(2) = h(17) = -0.0207,$$
$$h(3) = h(16) = 0.0046, \quad h(4) = h(15) = 0.03436, \quad h(5) = h(14) = 0.0656,$$
$$h(6) = h(13) = 0.0954, \quad h(7) = h(12) = 0.12071, \quad h(8) = h(11) = 0.1391,$$
$$h(9) = h(10) = 0.14877$$

Furthermore, we can get $H(e^{j\omega})$, and it can be compared with $H_d(e^{j\omega})$.

6.7.4 Some Simple Digital Filters

1. The average filter
The average filter is one of the most simple and commonly used digital filters, whose unit impulse response is

$$h(n) = \begin{cases} 1/N, & n = 0, 1, \ldots, N-1 \\ 0, & \text{others} \end{cases} \tag{6.92}$$

The system function of the filter is given as

$$H(z) = \frac{1}{N} \sum_{n=0}^{N-1} z^{-n} = \frac{1}{N} \frac{1-z^{-N}}{1-z^{-1}} \tag{6.93}$$

The corresponding difference equation is shown as

$$y(n) = \frac{1}{N} \sum_{k=0}^{N-1} x(n-k) \tag{6.94}$$

Figure 6.39 shows the processing of a three-point moving average filter to a signal.

In Figure 6.39, the upper figure demonstrates the input signal, while the lower figure shows the processed result. It is obvious that the processed signal becomes more smoothed.

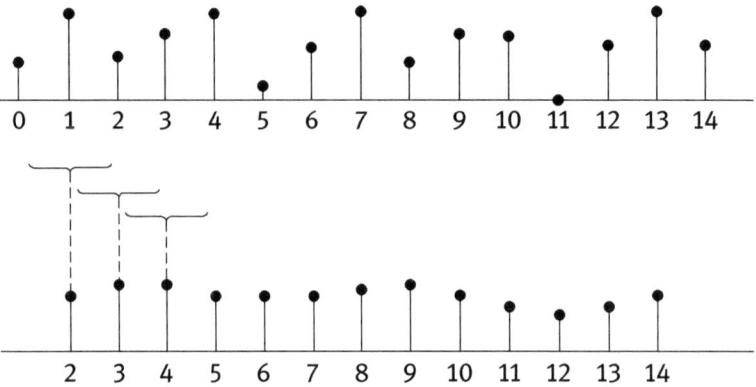

Figure 6.39: The processing of a three-point moving average filter to a signal.

2. The smoothing filter

A simple LPF is called the Savitzky–Golay smoothing filter. The basic idea of such filter is to construct a p th-order polynomial to fit a group of data $x(i)$, $i = -M, \ldots, 0, \ldots, M$, in the input signal $x(n)$, which is shown as

$$f_i = a_0 + a_1 i + a_2 i^2 + \cdots + a_p i^p = \sum_{k=0}^{p} a_k i^k, \quad p \leq 2M \tag{6.95}$$

By defining a fitting error and taking the first-order partial derivative, we can get parameters a_k, $k = 0, 1, \ldots, p$, of the polynomial. In practice, the parameters can also be obtained as follows:

$$f_i|_{i=0} = a_0, \quad \frac{df_i}{di}\Big|_{i=0} = a_1, \quad \ldots, \quad \frac{d^p f_i}{di^p}\Big|_{i=0} = p! a_p \tag{6.96}$$

Example 6.10: The walking data recorded by a double-axis inclination sensor is plotted in Figure 6.40(a). Obviously, there are rather serious noises in the signal, which may affect the subsequent processing. Try to process the signal by using the Savitzky–Golay smoothing filter.

Solution: Set the order of the filter as $p = 7$. The coefficients a_0, a_1, $\ldots$, a_7 of the filter can be obtained by eq. (6.96), where a_0 is the value of f_i when $i = 0$. Parameters a_1, $\ldots$, a_7 are $-2/21$, $3/21$, $6/21$, $7/21$, $6/21$, $3/21$, $-2/21$, respectively. Figure 6.40(b) gives the filtered signal. The noises in the signal are reduced effectively.

3. The amplitude clamping filter

Generally, the sudden changes happened in a signal are often caused by external interferences. The amplitude clamping filter assumes the maximum deviation value between adjacent two samples in a signal is limited. If the deviation exceeds a

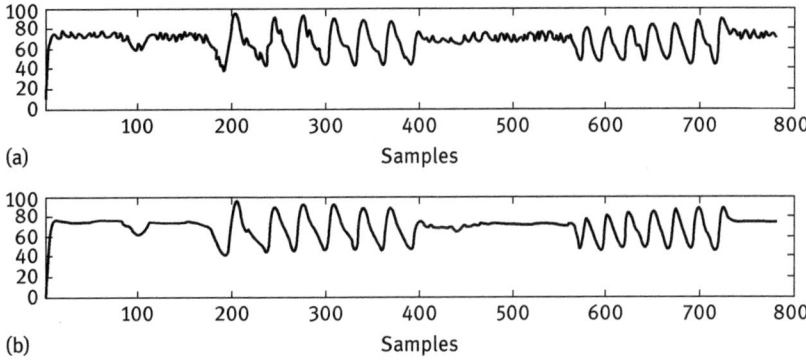

(a)

(b)

Figure 6.40: Noise reduction by using Savitzky–Golay smooth filter: (a) original noisy signal and (b) processed signal.

reasonable value, then the input signal may have been impacted by an outside interference, which should be eliminated.

Suppose the maximum error between adjacent two samples is C, then the criterion to determine the effective signal samples is as follows:

$$x(n) = \begin{cases} x(n), & \text{if } \Delta x(n) \le C \\ x(n-1), & \text{if } \Delta x(n) > C \end{cases} \qquad (6.97)$$

where $\Delta x(n) = |x(n) - x(n-1)|$.

Example 6.11: Suppose a group of measured temperature data is

$$T = [25.40,\ 25.50,\ 25.38,\ 25.48,\ 25.42,\ 25.46,\ 25.45,\ 25.43,\ 25.51]$$

Try to process the signal with an amplitude clamping filter, if $C = 0.1$ is set. *Solution:* The original signal and filtered result by the amplitude clamping filter are shown in Figure 6.41, respectively.

4. The median filter
The median filter is a nonlinear digital filter, commonly used in the removal of noise in signals and images. The basic idea of the median filter is to construct a sliding window on the signal sequence, which is composed of a number of adjacent samples arranged by their values in ascending (or descending) order. The sample value at the intermediate position of the sliding window is replaced by the intermediate value inside the sliding window. Its function is to eliminate the influence of the impulse noise.

Example 6.12: Suppose the temperature changes as

$$T = [25.40,\ 25.50,\ 25.68,\ 25.48,\ 25.42,\ 25.46,\ 25.45,\ 25.43,\ 25.51]$$

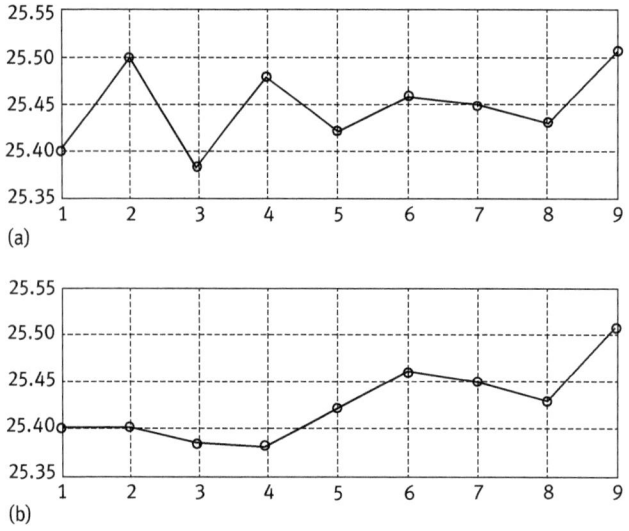

(a)

(b)

Figure 6.41: The signal filtered result by an amplitude clamping filter: (a) original signal and (b) processed signal.

Try to process the signal with the median filter by MATLAB programming.
Solution: The MATLAB program is shown as follows:

```
clear; clc;
T1=[25.4, 25.5, 25.68, 25.48, 25.42, 25.46, 25.45, 25.43, 25.51];
L=length(T1); N=3; k=0; m=0;
for i=1:L
    if i+N-1>L
        break
    else
        for j=i:i+N-1
            m=m+1; W(m)=T1(j);
        end
        k=k+1; T(k)=median(W); m=0;
    end
end
figure(1); subplot(211); plot(T1,'-o'); grid; axis([1 7 25.3 25.7]);
title('Original temperature data'); xlabel('Measurement Times');
    ylabel('Measured temperature');
subplot(212); plot(T,'-o'); grid; axis([1 7 25.3 25.7]);
title('Median filtered result'); xlabel('Measurement Times');
    ylabel('Filtered result');
```

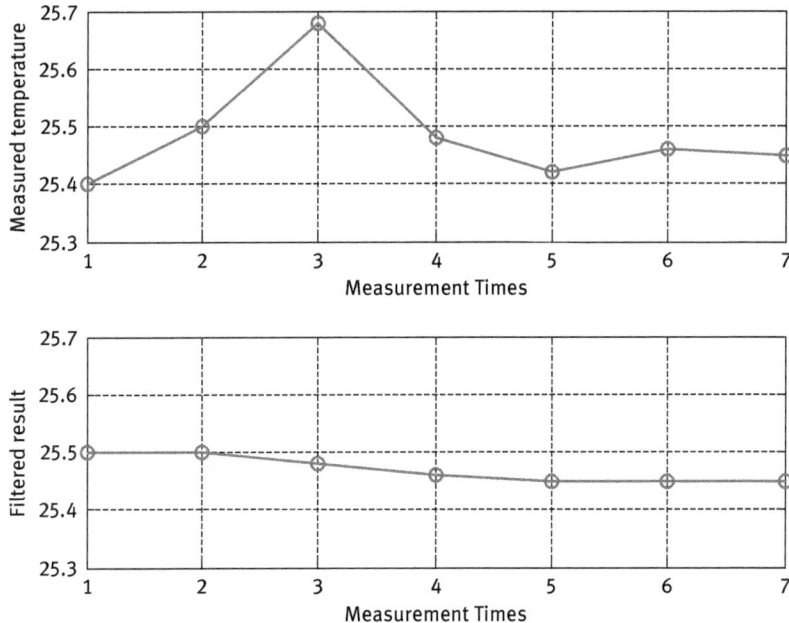

Figure 6.42: The original measurement and the median filtered result.

Figure 6.42 demonstrates the data processing result with the median filter as well as the original data. We see from the results that the peak caused by the measurement errors is eliminated.

Exercises

6.1 Describe the concept and characteristics of digital filters.

6.2 Summarize the classification of digital filters.

6.3 Describe the advantages and disadvantages of IIR and FIR filters.

6.4 Summarize the characteristics of the direct type IIR digital filter.

6.5 Construct the cascade structure of an IIR digital filter based on its system function.

6.6 Construct the parallel structure of an IIR digital filter based on its system function.

6.7 Describe the characteristics of the transversal and cascade structures of the digital filter.

6.8 Describe the concept and characteristics of the frequency sampling structure of a digital filter.

6.9   Describe the concept and characteristics of the fast convolution structure of a digital filter.

6.10 Describe the linear phase condition and characteristics of an FIR filter.

6.11 Describe the concept and characteristics of the minimum phase system.

6.12 What is the lattice structure of a digital filter? What are their characteristics?

6.13 Describe the basic steps for designing a digital filter.

6.14 What are the basic specifications for designing a digital filter?

6.15 Describe the commonly used analog filters.

6.16 Describe the basic idea and method for designing an IIR digital filter with the impulse invariance method.

6.17 Describe the basic idea and method for designing an IIR digital filter with the bilinear transform method.

6.18 Describe the basic idea and method for designing an FIR digital filter with the window function method.

6.19 Summarize the characteristics of the main window functions.

6.20 Describe the basic idea and method for designing an FIR digital filter with the frequency sampling method.

6.21 Implement the direct type I structure of the IIR digital filter with system function
$H(z) = \frac{3 + 4.2z^{-1} + 0.8z^{-2}}{2 + 0.6z^{-1} - 0.4z^{-2}}$.

6.22 Implement the cascade structure of the IIR filter with system function
$H(z) = \frac{4(z+1)(z^2 - 1.4z + 1)}{(z - 0.5)(z^2 + 0.9z + 0.8)}$.

6.23 Implement the transversal structure of the FIR digital filter with system function $H(z) = \left(1 - \frac{1}{2}z^{-1}\right)\left(1 + 6z^{-1}\right)\left(1 - 2z^{-1}\right)\left(1 + \frac{1}{6}z^{-1}\right)\left(1 - z^{-1}\right)$.

6.24 Implement the frequency sampling structure of the FIR filter with system function $H(z) = \frac{5 - 2z^{-3} - 3z^{-6}}{1 - z^{-1}}$, where the sampling number is $N = 6$, and the correction radius is $r = 0.9$.

6.25 Given the system function of the FIR filter as $H(z) = \frac{1}{5}(1 + 3z^{-1} + 5z^{-2} + 3z^{-3} + z^{-4})$. Try to depict its linear phase structure.

6.26 Transform $G(s)$ into $H(z)$ by using the impulse invariance method. The sampling period is set as T.
(1) $G(s) = \frac{s+a}{(s+a)^2 + b^2}$; (2) $G(s) = \frac{A}{(s - s_0)^{n_0}}$.
where n_0 is an arbitrary positive integer.

6.27 The system function of an analog filter is given as $G(s) = \frac{1}{s^2 + s + 1}$. If the sampling period is $T = 2$, try to transform it into a digital filter $H(z)$ using the bilinear transform method.

6.28 Any nonminimal phase system $H(z)$ can be expressed as a cascade between a minimal phase system $H_{min}(z)$ and an all-pass system $H_{ap}(z)$, that is, $H(z) = H_{min}(z)H_{ap}(z)$. Let $\Phi(\omega) = \arg[H(e^{j\omega})]$ and $\Phi_{min}(\omega) = \arg[H_{min}(e^{j\omega})]$. Prove $-\frac{d\Phi(\omega)}{d\omega} > -\frac{d\Phi_{min}(\omega)}{d\omega}$ for all ω. This inequality means that the minimal phase system has the minimal group delay.

6.29 Design a linear phase FIR LPF with the rectangular window. Given $\omega_c = 0.5\pi$, $N = 21$, find $h(n)$ and plot the curve of $20\log_{10}\left|H(e^{j\omega})\right|$.

6.30 Design a linear phase FIR LPF with the triangular window. Given $\omega_c = 0.5\pi$, $N = 51$, find $h(n)$ and plot the curve of $20\log_{10}\left|H(e^{j\omega})\right|$.

6.31 Design a linear phase FIR LPF with the frequency sampling method. Given $\omega_c = 0.5\pi$, $N = 51$, find $h(n)$ and plot the curve of $20\log_{10}\left|H(e^{j\omega})\right|$.

7 Finite-Precision Numerical Effects in Digital Signal Processing

7.1 Introduction

All the theories and methods of discrete-time signals, discrete-time systems, and digital signal processing introduced in previous chapters assumed that the data and parameters have an infinite precision. But in fact, when discrete-time signals, discrete-time systems, and digital filters are implemented in software and hardware, data and parameters of signals and systems need to be stored in the storage unit of the finite word length. Thus, the accuracy of the signals and the filter coefficients are no longer unlimited, resulting in errors. This chapter studies errors in digital signal processing caused by the finite-precision numerical effects and the possible methods of improvements.

There are a variety of finite-precision numerical effects in digital signal processing. This chapter mainly studies the following finite-precision numerical effects and the caused errors.

(1) The quantization error caused by an analog to digital (A/D) converter when it converts an analog signal to a digital signal.
(2) The error caused by the finite-precision numerical effects of the coefficients of digital filters.
(3) The error caused by the operation process in digital signal processing.
(4) The finite-precision numerical effects in discrete Fourier transform.

The objective to study the finite-precision numerical effects is to understand the reliability of digital signal processing and to choose a suitable word length for the trade-off between the required accuracy and the equipment cost.

7.2 Quantization Effects in Analog-to-Digital (A/D) Conversion

7.2.1 The Basic Concept and the Principle of A/D Conversion

The A/D converter is an electronic component used to change an analog signal into a digital signal, which is also known as an analog-to-digital converter. Resolution is one of the most important indicators of the A/D converter.

The so-called resolution is the output number of discrete signal values for an allowed range of the analog signal. These signal values are usually stored in common binary numbers; hence, the resolution often uses "bit" as its unit, and the number of discrete values is a power of 2.

The process of A/D conversion can be divided into two parts, namely, sampling and quantization. Sampling is the process of converting a continuous-time signal (or

https://doi.org/10.1515/9783110465082-007

analog signal) $x(t)$ to a discrete-time signal $x(n)$. Here, the value of the discrete-time signal $x(n)$ at discrete-time points is equal to the value of $x(t)$ at that time; that is, it has an infinite precision. The basic theory and the method of sampling have been introduced in detail in Chapter 4. Therefore, we will only briefly introduce the concept of quantification of A/D conversion.

The so-called quantization is to quantize the amplitude of the discrete-time signal with infinite precision to obtain a digital signal, so that it is suitable for the requirement of the digital signal processing system with limited precision. The basic methods of quantization include truncating and rounding. Both methods will map each sample value $x(n)$ into the nearest quantization level so as to obtain a digital signal $x_q(n)$. Figure 7.1 demonstrates a process in which an analog signal $x(t)$ is converted to a discrete-time signal $x(n)$ by sampling, and then $x(n)$ is converted to a digital signal $x_q(n)$.

$x(t), x(n), x_q(n)$

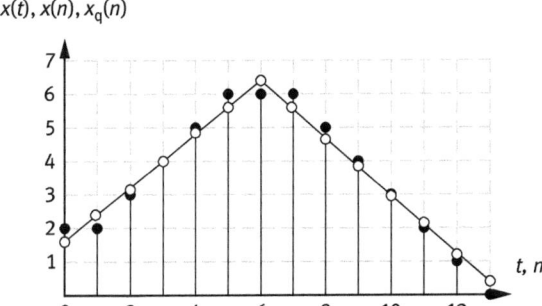

Figure 7.1: A sketch map of an analog signal (solid line) and its corresponding discrete-time signal (open circle) and digital signal (solid circle).

In Figure 7.1, the solid line represents the analog signal $x(t)$; the open circle denotes the discrete-time signal $x(n)$, which is quantized in time but is continuous in its amplitude; and the solid circle is the digital signal $x_q(n)$, which is quantized in both time and amplitude.

7.2.2 Quantization Effects in A/D Conversion and Error Analysis

If the word length of the A/D converter is b, then the quantization step size is defined as $q = 2^{-b}$. When $b = 8$, the quantization step size is $q = 2^{-8} = 1/256$. Without loss of generality, if we suppose that the value of a discrete-time signal $x(n)$ in time n is positive, then $x(n)$ can be expressed as

$$x(n) = \sum_{i=1}^{+\infty} a_i 2^{-i}, \quad a_i = 0, 1 \tag{7.1}$$

The above equation is the fixed-point notation for a binary positive number. For the truncated quantization, $x(n)$ can be expressed as

$$x_T(n) = \sum_{i=1}^{b} \alpha_i 2^{-i}, \quad \alpha_i = 0, 1 \tag{7.2}$$

For the rounding quantization, we obtain

$$x_R(n) = \sum_{i=1}^{b} \alpha_i 2^{-i} + \alpha_{b+1} 2^{-b}, \quad \alpha_i = 0, 1 \tag{7.3}$$

If $\alpha_{b+1} = 1$, a "1" is added at the bth place; otherwise, a "0" is added. The errors caused by both truncating quantization and rounding quantization are shown as

$$e_T = x_T(n) - x(n) = - \sum_{i=b+1}^{+\infty} \alpha_i 2^{-i}, \quad \alpha_i = 0, 1 \tag{7.4}$$

and

$$e_R = x_R(n) - x(n) = \alpha_{b+1} 2^{-(b+1)} - \sum_{i=b+2}^{+\infty} \alpha_i 2^{-i}, \quad \alpha_i = 0, 1 \tag{7.5}$$

The analysis shows that the truncation quantization error meets the following relation:

$$-q < e_T \le 0 \tag{7.6}$$

However, the rounding quantization error meets

$$-q/2 < e_R \le q/2 \tag{7.7}$$

Both truncating and rounding errors are represented as $e(n)$. Thus, the quantization model can be expressed as

$$x_q(n) = x(n) + e(n) \tag{7.8}$$

It means that the quantized digital signal $x_q(n)$ is equal to the infinite accuracy of discrete-time signal $x(n)$ plus a quantization error $e(n)$. Since the truncating quantization error has the nonzero-mean characteristics and the rounding quantization error has the zero-mean characteristics, the rounding quantization method is used widely in practical applications.

The signal-to-noise ratio (SNR) is defined as

$$\text{SNR}_{\text{dB}} = 10\log_{10}\left(\frac{\sigma_x^2}{\sigma_e^2}\right) = 10\log_{10}\left(\frac{\sigma_x^2}{2^{-2b}/12}\right) = 6.02b + 10.79 + 10\log_{10}\sigma_x^2 \quad (\text{dB}) \qquad (7.9)$$

where σ_x^2 and σ_e^2 represent the powers of the infinite accuracy signal and the quantization error, respectively. It can be seen from the above formula that the greater the signal power, the higher the SNR. On the other hand, the longer the word length b of the A/D converter, the higher the quantization SNR, too. If the length of an A/D converter increases by 1 bit, then the quantization SNR will increase by about 6 dB.

The root mean square error (RMSE) in eq. (7.10) is often used to represent the quantization error in practice as

$$e_{\text{rms}} = \sqrt{\sigma_e^2} = q/\sqrt{12} \qquad (7.10)$$

Example 7.1: In digital audio applications, the dynamic range of an A/D converter is 0–10 V. If the RMSE of the quantization error is less than 50 μV, determine the word length of the A/D converter.
Solution: From eq. (7.10), we obtain $e_{\text{rms}} = q/\sqrt{12} < 50\mu$V. From $q = 10/2^b$, we find $b > 15.82$. Then, $b = 16$ is taken.

7.3 Quantization Effects of Digital Filters

7.3.1 Coefficient Quantization Effects of IIR Digital Filter Coefficients

The system function of an IIR digital filter is shown as

$$H(z) = \frac{B(z)}{A(z)} = \frac{\sum\limits_{k=0}^{M} b_k z^{-k}}{1 - \sum\limits_{k=1}^{N} a_k z^{-k}} \qquad (7.11)$$

where the coefficients a_k and b_k are all in the infinite precision. In practice, a_k and b_k have to be quantized, forming finite-precision coefficients $\hat{a}_k = a_k + \Delta a_k$ and $\hat{b}_k = b_k + \Delta b_k$, where Δa_k and Δb_k represent the coefficient errors caused by quantization. Thus, eq. (7.11) becomes

$$\hat{H}(z) = \frac{\hat{B}(z)}{\hat{A}(z)} = \frac{\sum\limits_{k=0}^{M} \hat{b}_k z^{-k}}{1 - \sum\limits_{k=1}^{N} \hat{a}_k z^{-k}} \qquad (7.12)$$

For an IIR digital filter, what we are actually most concerned with is the impact of quantization on the poles of the system. $A(z)$ can be expressed as

$$A(z) = 1 - \sum_{k=1}^{N} a_k z^{-k} = \prod_{i=1}^{N}(1 - p_i z^{-1}) \tag{7.13}$$

where p_i, $i = 1, 2, \cdots, N$, denote the poles of the system, while the poles after the coefficient quantization corresponding to $\hat{A}(z)$ may be expressed as

$$\hat{p}_i = p_i + \Delta p_i, \quad i = 1, 2, \ldots, N \tag{7.14}$$

Thus, the error caused by the coefficient quantization is given as

$$\Delta p_i = \sum_{k=1}^{N} \frac{\partial p_i}{\partial a_k} \Delta a_k, \quad i = 1, 2, \ldots, N \tag{7.15}$$

It can be seen that the quantization error for each pole of the closed system relates to the quantization of all the coefficients in the denominator polynomial. We can further obtain

$$\Delta p_i = -\sum_{i=1}^{N} \frac{p_i^{N-k}}{\prod_{\substack{l=1 \\ l \neq i}}^{N}(p_i - p_l)} \Delta a_k, \quad i = 1, 2, \ldots, N \tag{7.16}$$

We see from the above equation that if certain two poles are very closed, the value of $p_i - p_l$ will be very small, which may produce a very large Δp_i, even causing the system to be unstable since the poles may move out of the unit circle under an extreme condition.

Example 7.2: The coefficient vector of an IIR digital filter is given as

$$\begin{aligned} a &= [1.0000 \quad -2.2188 \quad 3.0019 \quad -2.4511 \quad 1.2330 \quad -0.3109] \\ b &= [0.0079 \quad 0.0397 \quad 0.0794 \quad 0.0794 \quad 0.0397 \quad 0.0079] \end{aligned}$$

Take the truncating quantization to both vectors with 4-bit word length and 5-bit word length, respectively, and evaluate the impact of the quantization effects.
Solution: (1) Taking the 4-bit quantization to the coefficients in both vectors, we obtain

$$\begin{aligned} \hat{a}_{4\text{bit}} &= [1.0000 \quad -2.0000 \quad 3.0000 \quad -2.2500 \quad 1.1250 \quad -0.2500] \\ \hat{b}_{4\text{bit}} &= [0 \quad 0 \quad 0.0625 \quad 0.0625 \quad 0 \quad 0] \end{aligned}$$

Thus, the poles before and after quantization are as follows:

$$\boldsymbol{p}=[0.2896+j0.8711 \quad 0.2896-j0.8711 \quad 0.6542 \quad 0.4936+j0.5675 \quad 0.4936-j0.5675]$$

and

$$\hat{\boldsymbol{p}}_{4bit}=[0.4569+j1.0239 \quad 0.4569-j1.0239 \quad 0.4234 \quad 0.3314+j0.5999 \quad 0.3314-j0.5999]$$

Figure 7.2 shows the distribution of poles before and after quantization, and it also shows the frequency response of the system.

(2) Then the 5-bit quantization is used to quantify the coefficient vectors. The pole distribution map and the system frequency response are shown in Figure 7.3.

It is obvious that a pair of poles is out of the unit circle after the 4-bit quantization is done, resulting in an unstable system. Moreover, the frequency response distortion is serious.

If the 5-bit quantization is conducted, the system becomes stable and its frequency response is also improved.

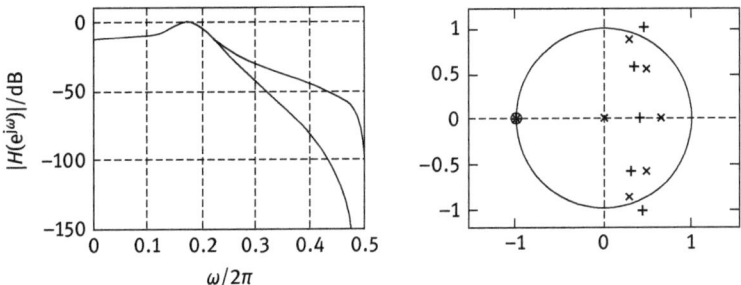

Figure 7.2: The distribution of poles before and after quantization and the frequency response of the system with 4-bit quantization: (a) the frequency response of the system and (b) the distribution map of poles before (×) and after (+) quantization.

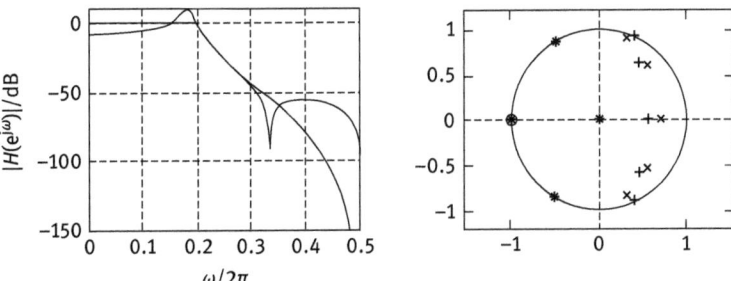

Figure 7.3: The distribution of poles before and after quantization and the frequency response of the system with 5-bit quantization: (a) the frequency response of the system and (b) the distribution map of poles before (×) and after (+) quantization.

In fact, a higher order system is usually implemented by the cascade or parallel structures in practice. In such structures, the order is not greater than the second order in each stage, in order to reduce the negative effects caused by quantization.

7.3.2 Coefficient Quantization Effects of FIR Digital Filter Coefficients

The system function of an FIR digital filter is as follows:

$$H(z) = \sum_{n=0}^{M} h(n)z^{-n} \tag{7.17}$$

where $h(n)$ denotes the unit impulse response of the system with infinite precision. When we quantify the coefficients of $h(n)$, the quantization error may yield. Thus, the quantized impulse response is shown as

$$\hat{h}(n) = h(n) + e(n) \tag{7.18}$$

After quantization, the system becomes

$$\hat{H}(z) = \sum_{n=0}^{M} h(n)z^{-n} + \sum_{n=0}^{M} e(n)z^{-n} = H(z) + E(z) \tag{7.19}$$

The frequency characteristics of the quantization error are

$$E(e^{j\omega}) = \sum_{n=0}^{M} e(n)e^{-j\omega n} \tag{7.20}$$

Furthermore, we have

$$\left| E(e^{j\omega}) \right| = \left| \sum_{n=0}^{M} e(n)e^{-j\omega n} \right| \leq \sum_{n=0}^{M} |e(n)| \left| e^{-j\omega n} \right| \leq \sum_{n=0}^{M} |e(n)| \tag{7.21}$$

It means that the magnitude response at any frequency bin is not greater than the summation of the absolute values of the error sequence. If the rounding quantization is taken, we obtain

$$\left| E(e^{j\omega}) \right| \leq \sum_{n=0}^{M} |e(n)| \leq \frac{(M+1)q}{2} \tag{7.22}$$

Example 7.23: Suppose the order of an FIR filter is $M = 30$. Determine the word length of the filter if the error of the frequency response caused by quantization is less than 0.001 or -60 dB.

Solution: Based on the given conditions and eq. (7.22), we have $(30+1)2^{-b}/2 \leq 0.001$. The word length can be solved as $b = 13.92$, i.e., $b = 14$.

7.4 Quantization Effects in the Operation of Digital Filters

In digital signal processing, the basic operation of the digital filter includes the unit delay, the multiplication by a coefficient, and the addition. We know that the coefficients of the digital filter and data in the operations are all stored in registers (or memories), and their word lengths are limited. For example, if two data of b-bit length are multiplied, the product will be $2b$ bit in its word length, beyond the length of the register. Therefore, we have to conduct rounding or truncating operation to meet the requirement of the word length. This section mainly discusses the effect of multiplication by a coefficient in a digital filter on the performance of the system.

7.4.1 The Limit Cycle Oscillation in IIR Digital Filters

The limit cycle oscillation is an oscillation phenomenon of the IIR filter caused by setting of poles. The reason is that the poles of the system function are set on the unit circle. In the IIR digital filter, if the word-length problem cannot be solved properly, it is possible to generate the oscillation phenomenon so that the system becomes unstable.

Example 7.4: Given the difference equation of a first-order IIR digital filter as $y(n) = ay(n-1) + x(n)$ and if the input signal is $x(n) = 0.875\delta(n)$, $a = 0.5$, and the initial condition $y(-1) = 0$,

(1) Find the output signal of the system.

(2) If the word length of the system is 4-bit and the first bit is the sign bit, find the output of the system again.

Solution: (1) Taking the z-transform to both sides of the equation, we obtain $Y(z) = \frac{1}{1-az^{-1}}X(z)$. Since $x(n) = 0.875\delta(n)$, we obtain $X(z) = 0.875$. Thus, $Y(z) = \frac{1}{1-az^{-1}}X(z) = \frac{0.875}{1-0.5z^{-1}}$. Taking the inverse z-transform, we obtain $y(n) = 0.875a^n u(n) = 0.875 \times (0.5)^n u(n)$.

(3) If the word length of the system is 4-bit and the first bit is the sign bit, both signal and the system can be expressed in the binary form as $x(n) = 0.111_B \delta(n)$, $a = 0.100_B$, where "B" in subscript represents the binary number. Taking the rounding process expressed as $[\cdot]_R$ to the product, we obtain

$$n = 0, \quad \hat{y}(0) = x(0) = 0.111_B$$

$$n = 1, \quad \hat{y}(1) = [a\hat{y}(0)]_R = [0.100_B \times 0.111_B]_R = [0.011100_B]_R = 0.100_B$$

$$n = 2, \quad \hat{y}(2) = [a\hat{y}(1)]_R = [0.100_B \times 0.100_B]_R = [0.001110_B]_R = 0.010_B$$

$$n = 3, \quad \hat{y}(3) = [a\hat{y}(2)]_R = [0.100_B \times 0.010_B]_R = [0.000111_B]_R = 0.001_B$$

$$n = 4, \quad \hat{y}(4) = [a\hat{y}(3)]_R = [0.100_B \times 0.001_B]_R = [0.000111_B]_R = 0.001_B$$

If we take the recursive operation continuously, we find $\hat{y}(5) = \hat{y}(6) = \cdots = \hat{y}(n) = 0.001_B$, $n \to \infty$. It means that $\hat{y}(n)$ becomes a constant sequence, no longer a monotonic decreasing one. This result demonstrates a limit cycle oscillation phenomenon caused by rounding to the multiplication operation. Such rounding moves the poles onto the unit circle.

7.4.2 Finite Word-Length Effects in Multiplication Operation of IIR Digital Filters

In fixed-point arithmetic, after each multiplication such as $y(n) = ax(n)$, a rounding (or a truncating) operation is needed. Suppose the rounding error can be expressed as

$$e(n) = [ax(n)]_R - ax(n) = [y(n)]_R - y(n) \tag{7.23}$$

If the error sequence is a uniformly distributed stationary white sequence, and the error signal, the input signal as well as the intermediate result are not correlated, then the rounding error is uniformly distributed within $(-2^{-b}/2, \ +2^{-b}/2]$, and the mean m_e and the variance σ_e^2 of the error sequence can be expressed as

$$m_e = E[e(n)] = 0$$

$$\sigma_e^2 = E[e^2(n)] = \frac{q^2}{12}, \quad q = 2^{-b} \tag{7.24}$$

Thus, the total rounding error $e_f(n)$ caused by each rounding error $e(n)$ is expressed as

$$e_f(n) = \hat{y}(n) - y(n) \tag{7.25}$$

The mean and variance for each round noise are expressed as

$$\sigma_f^2 = \sigma_e^2 \sum_{n=-\infty}^{+\infty} h_e^2(n)$$

$$m_f = m_e \sum_{n=-\infty}^{+\infty} h_e(n) \tag{7.26}$$

where $h_e(n)$ represents the unit impulse response from the node of $e(n)$ entered to the output node. Under the assumptions of uniformly distributed white noise and uncorrelated assumption, the total variance of the output noise is equal to the sum of the variance of each output noise.

Suppose the system function of an IIR digital filter is $H(z) = \frac{0.2}{(1-0.7z^{-1})(1-0.6z^{-1})}$. Table 7.1 shows the rounding errors for the direct type, the cascade, and the parallel structures under the fixed-point arithmetic condition.

We can see from Table 7.1 that the rounding error of the direct type is most significant. This is because all rounding errors of the structure are accumulated by the feedback of the network. Since the rounding error of each parallel subsystem affects only itself, the effect of accumulated error for a parallel structure is minimal. The error of the cascade system is generally between the direct structure and the parallel structure.

7.4.3 Finite Word-Length Effects in Multiplication Operation of FIR Digital Filters

The difference equation of an FIR digital filter of order M is shown as

$$y(n) = \sum_{k=0}^{M} b_k x(n-k) \tag{7.27}$$

For any n, the number of multiplication is $M+1$, yielding $M+1$ rounding errors. Suppose such errors are not correlated; we have

$$\hat{y}(n) = y(n) + \sum_{k=0}^{M} e_k(n) \tag{7.28}$$

The noise variance caused by the rounding errors is

$$\sigma_f^2 = \frac{M+1}{12} q^2 \tag{7.29}$$

Obviously, the rounding error caused by the multiplication in the FIR system directly appears in the output, and has no relation with the filter coefficients. On the other hand, the higher the filter order, the more significant the rounding noise. The longer the word length, the smaller the rounding error, too.

7.5 Finite Word-Length Effects in Discrete Fourier Transform

The DFT of an N-point sequence $x(n)$ is defined as

$$X(k) = \sum_{n=0}^{N-1} x(n) W_N^{nk}, \quad k = 0, 1, \ldots, N-1 \tag{7.30}$$

Table 7.1: A rounding error comparison among three different structures.

Structure	System function	System structure	Rounding error analysis
Direct type	$H(z) = \dfrac{0.2}{(1-0.7z^{-1})(1-0.6z^{-1})}$ $= \dfrac{0.2}{1-1.3z^{-1}+0.42z^{-2}}$		$\sigma_f^2 = 3\sigma_e^2 \times 7.5008$ $= 1.8752q^2$
Cascade	$H(z) = \dfrac{0.2}{(1-0.7z^{-1})(1-0.6z^{-1})}$ $= 0.2 \cdot \dfrac{1}{A_1(z)} \cdot \dfrac{1}{A_2(z)}$ $A_1(z) = 1-0.7z^{-1}$ $A_1(z) = 1-0.6z^{-1}$		$\sigma_f^2 = 16.5641\sigma_e^2$ $= 1.3803q^2$
Parallel	$H(z) = \dfrac{1.4}{1-0.7z^{-1}} + \dfrac{-1.2}{1-0.6z^{-1}}$		$\sigma_f^2 = 7.0466\sigma_e^2$ $= 0.5872q^2$

For every frequency bin k, there are N complex multiplications when calculating $X(k)$, which contains $4N$ real multiplications. If the rounding error for each multiplication is e_i, $i = 1, 2, \ldots, 4N$, and suppose the errors obey the uniform distribution and are uncorrelated with each other and with $x(n)$, then the total variance reflected to a particular $X(k)$ of the $4N$ errors is

$$\sigma_f^2 = 4N\sigma_e^2 = 4N\frac{q^2}{12} = \frac{2^{-2b}N}{3} \tag{7.31}$$

Furthermore, taking the absolute values for both sides of eq. (7.30), we obtain

$$|X(k)| \le \sum_{n=0}^{N-1} |x(n)| \tag{7.32}$$

If $x(n)$ is supposed to obey the uniform distribution in the region $\left(-\frac{1}{N}, \frac{1}{N}\right)$, then

$$\sigma_x^2 = \frac{(2/N)^2}{12} = \frac{1}{3N^2} \tag{7.33}$$

Thus, the variance of $X(k)$ is $\sigma_X^2 = N\sigma_x^2 = \frac{1}{3N}$. The signal-to-noise ratio between the output of DFT and the rounding error is shown as

$$\text{SNR}_{\text{dB}} = 10\log_{10}\left(\frac{\sigma_X^2}{\sigma_f^2}\right) = 10\log_{10}\left(\frac{2^{2b}}{N^2}\right) = 6.02b - 20\log_{10}N \ \ (\text{dB}) \tag{7.34}$$

It means that the SNR is inversely proportional to N^2. If $N = 1024 = 2^{10}$ and $\text{SNR}_{\text{dB}} = 30\text{dB}$, we can find the word length as $b = 15\text{bit}$.

Exercises

7.1 Describe the concept of the A/D converter and its quantization effects.

7.2 What are the truncating quantization and its error? What are the rounding quantization and its error?

7.3 Describe the quantization effects of the IIR digital filter.

7.4 Describe the quantization effects of the FIR digital filter.

7.5 Describe the finite word-length effects of the IIR digital filter.

7.6 Describe the finite word-length effects of the FIR digital filter.

7.7 What is the limit cycle oscillation of the IIR digital filter? What is its cause?

7.8 Describe the finite word-length effects of the multiplication operation in the IIR digital filter.

7.9 Describe the finite word-length effects of the multiplication operation in the FIR digital filter.

7.10 Describe the finite word-length effects of the DFT.

7.11 The system function of an IIR digital filter is $H(z) = \frac{0.01722133z^{-1}}{1-1.7235682z^{-1}+0.74081822z^{-2}}$. If the coefficients are stored with 8-bit word length (rounding quantization is used regardless of the sign bit), find the practical system function $\hat{H}(z)$.

7.12 The word length of an A/D converter is b. A discrete-time system is connected to the output end of the A/D converter, whose unit impulse response is $h(n) = [a^n + (-a)^n]u(n)$. Find the quantization noise variance of the system.

7.13 The difference equation of a second-order IIR filter is given as $y(n) = y(n-1) - ay(n-2) + x(n)$. Suppose that the fixed-point operation of $b = 3$ is adopted and the rounding quantization is processed.

(1) If the coefficient is $a = 0.75$ and $x(n) = 0$ with the initial conditions $\hat{y}(-2) = 0$ and $\hat{y}(-1) = 0.5$, find the output $\hat{y}(n)$ when $0 \le n \le 9$.

(2) Prove that the limit cycle oscillation will happen when $[a\hat{y}(n-2)]_R = \hat{y}(n-2)$ and explain it.

7.14 Given $X(k) = \sum_{n=0}^{N-1} x(n)W_N^{nk}$, $k = 0, 1, \cdots, N-1$, where $W_N = e^{-j2\pi/N}$, suppose $x(n)$ is a zero-mean stationary white noise sequence, and $E[x(n)x(m)] = \sigma_x^2\delta(n-m)$ and $E[x(n) = 0]$ are satisfied.

(1) Determine the variance of $|X(k)|^2$.

(2) Determine $E[X(k)X^*(r)]$ and represent it as a function of k and r.

7.15 The difference equation of a first-order all-pass system is $y(n) = ay(n-1) + x(n) - a^{-1}x(n-1)$. If the coefficient is taken as $a = 0.9$ and $a = 0.98$, respectively, check whether the system is still an all-pass system, when the quantization word lengths are 4-bit and 8-bit, respectively.

7.16 The system function of an IIR digital filter is $H(z) = \frac{1-0.5z^{-1}}{(1-0.8z^{-1})(1-0.9z^{-1})}$.

(1) Find the output power of the quantization noise of the input signal, if the quantization word length is b-bit.

(2) Find and plot the system structures with direct, cascade, and parallel types.

(3) Find the output power of the rounding error of the above three different structures.

8 Data Error Analysis and Signal Preprocessing

8.1 Introduction

Many signals in a signal processing area are from a variety of measurements or tests of certain physical quantities, such as the measurements of length, height, distance, and other geometrical quantities, and the measurements of light, sound, temperature, humidity, electricity, and other physical quantities. From this point of view, signal detection and measurement are closely related to signal processing.

An unavoidable problem of measurement is the error problem. The existence of error will affect the accuracy of the measurement; even it will result in a mistake. Therefore, it is necessary to make a systematic theoretical analysis and the actual effective treatment of the measurement error.

According to the cause of the error, the measurement error can be roughly divided into two types: the random error and the systematic error. The characteristics and causes of the two types of errors are different. We need to make a compensation and processing to different types of errors, including the elimination of gross errors and the removal of outliers and trends in the data, so as to improve the measurement accuracy.

The methods for elimination or reduction of errors are similar to the methods of signal preprocessing in signal processing. In addition, the least squares method and the regression analysis for the measurement of data are the basic means and important contents of data analysis and processing, which will be introduced in this chapter.

This chapter will briefly introduce the concept of error and discuss the uncertainty assessment. It will also introduce the least squares method and regression analysis. The judgment and processing of the gross error, and the signal trend and outlier removal methods will also be included. Finally, this chapter will give some application examples of the error analysis and data processing.

8.2 The Basic Concept and the Principle of Errors

8.2.1 The Basic Concept of Errors

The measurement error (abbreviated as "error") refers to the difference between the measurement result and the truth value of the measured quantity. Errors can be divided into many types, such as the absolute error, the relative error, the quoted error, etc.

1. The absolute error
The absolute error Δ_θ is defined as the difference between the measured value θ and its truth value θ_0, which is shown as

https://doi.org/10.1515/9783110465082-008

$$\Delta_\theta = \theta - \theta_0 \tag{8.1}$$

The absolute error is not the absolute value of the error, which can be either positive or negative. The absolute error indicates the deviation degree of the measured value from its truth value. The unit of the absolute error is the same as that of the measured value. In addition, because the truth value cannot be obtained in many situations, the arithmetic mean is generally used to replace the truth value. In this way, the absolute error is converted into the residual error (abbreviated as "residual") as follows:

$$v_\theta = \theta - \bar{\theta} \tag{8.2}$$

where v_θ denotes the residual error and $\bar{\theta}$ is the arithmetic mean of multimeasurements.

2. The relative error
The relative error δ_θ is defined as the ratio of the absolute error to the truth value, which is generally expressed as a percentage:

$$\delta_\theta = \left|\frac{\Delta_\theta}{\theta_0}\right| \times 100\% \tag{8.3}$$

3. The quoted error
The so-called quoted error is defined as the ratio of the error from the measurement instrument to the specific value of the instrument. That is,

$$Y_a = \frac{D}{B} \times 100\% \tag{8.4}$$

where D represents the error from the measurement instrument and B is the specific value (or quoted value) of the instrument. The quoted error can be used to express the accuracy of the measurement instrument.

8.2.2 Stochastic Error

1. The basic concept of the stochastic error
The stochastic error (or random error), which is also known as the "accidental error" or "not determined error, refers to the unpredictable changes in the measurement results of multiple measurements under the same conditions.

Although the appearance of a stochastic error is not regular, it is possible to find certain statistical rules if a large number of repeated experiments are carried out. In this way, we can estimate the general trend and distribution of random errors by using probabilistic and statistical methods, and take corresponding measures to reduce its impact.

2. The concept of the arithmetic mean and its application

The arithmetic mean is defined as the arithmetic average of repeated measurements, which is shown as

$$\bar{\theta} = \frac{1}{N}\sum_{i=1}^{N}\theta_i \tag{8.5}$$

where $\bar{\theta}$ is the average value of N measurement results and θ_i is the ith measured value.

Since the arithmetic mean is usually close to the truth value, if the number of measurements tends to infinity, the absolute error of a measurement can be written as

$$\Delta_i = \theta_i - \theta_0 = (\theta_i - E[\theta_i]) + (E[\theta_i] - \theta_0) \tag{8.6}$$

where $E[\theta_i]$ is the mathematical expectation of the measured results. From eq. (8.6), it can be seen that the measurement error can be divided into two parts: $(\theta_i - E[\theta_i])$ is the deviation between the measurement result and the expectation, commonly known as the stochastic error or the random error; while $(E[\theta_i] - \theta_0)$ represents the deviation between the expectation and the truth value, often referred to as the systematic error (or system error).

3. The standard deviation of measurements

The standard deviation is a measure of the dispersion of the measurement. It is a key indicator to measure the random error. The standard deviation of a single measurement is defined as

$$\sigma = \sqrt{\frac{\sum_{i=1}^{N}\Delta_i^2}{N}} \tag{8.7}$$

Since the absolute error Δ_i cannot be easily obtained in practice, it is usually replaced by the residual error v_i. As a result, eq. (8.7) is rewritten as

$$\sigma = \sqrt{\frac{\sum_{i=1}^{N}v_i^2}{N-1}} \tag{8.8}$$

The standard deviation of the arithmetic mean $\bar{\theta}$ is defined as

$$\sigma_{\bar{\theta}} = \frac{\sigma}{\sqrt{N}} \tag{8.9}$$

Obviously, the standard deviation of the arithmetic mean is $1/\sqrt{N}$ of the standard deviation of a single measurement. When the number of measurement increases, the arithmetic mean is more close to the truth value.

8.2.3 Systematic Error

1. The concept of the systematic error
The systematic error refers to the error whose size and sign remain unchanged in multiple repeated measurements. Or when the measurement condition changes, the measurement results alter according to a certain rule.

The variation law and the reason of the systematic error can be determined by experiment or analysis. If the measuring condition or the measuring method is improved before the measurement, the systematic error can also be reduced or eliminated. However, it cannot be reduced or eliminated by increasing the number of measurements.

2. The source of the systematic error
The source of the systematic error mainly includes the following aspects:
(1) The error introduced by the measurement instrument, which refers to the imperfect instrument structure or the error and deviation of the parts.
(2) The error introduced by the instrument adjustment, which refers to the error introduced by the improper installation or adjustment to the correct state of the instrument.
(3) The error caused by the habit of users, which refers to the error introduced by the measurement habit, such as the advance or lag in measurement time.
(4) The error introduced by the measuring condition, which refers to the error caused by the change in temperature, pressure, air current, vibration, etc., during the measurement.
(5) The error introduced by the measuring method, which refers to the error introduced by inappropriate measurement or data processing methods.

3. Ways for finding, reducing, and eliminating errors
The most important way to improve the measurement accuracy is to find the systematic error. Due to the complexity in the formation of the systematic error, there is no general error detection method for a variety of systematic errors. Generally, a method is needed to conduct a comprehensive analysis of the measurement process and measuring instruments. Furthermore, methods to find a systematic error include the experimental comparison, theoretical analysis, and data analysis. As soon as the systematic error is found, we need to reduce or eliminate such error. Commonly used methods include the elimination of the cause of error, the introduction of correction term, and the employment of the systematic error-free measurement.

8.2.4 Gross Error

The so-called gross error is the error which significantly exceeds the expected value. The main causes of gross errors include the mistaken reading of the indication, the use of defective measuring instruments, and the electromagnetic and vibration interference to the instruments.

1. Decision criteria for gross errors

(1) 3σ criterion

This criterion supposes that only the stochastic error exists in the measured data. The decision for the gross error with 3σ criterion is as follows:

$$|v_d| = |\theta_d - \bar{\theta}| > 3\sigma \tag{8.10}$$

where θ_d is the suspicious datum, $\bar{\theta}$ is the average value of the measured data, v_d is the residual error, and σ is the standard deviation of the residual error. If the residual error of the suspicious datum is greater than 3σ, the datum should be eliminated.

(2) Romanovski criterion

To determine the gross error, the Romanovski criterion (also known as the t distribution inspection criterion) removes a suspicious measurement value first and then determines whether the measured values include gross errors in accordance with the t distribution. The measurement data obtained from independent measurement with equal precision are θ_1, θ_2, $\ldots$, θ_N. If θ_d is a suspicious datum, then the arithmetic mean after eliminating θ_d is

$$\bar{\theta} = \frac{1}{N-1} \sum_{\substack{i=1 \\ i \neq d}}^{N} \theta_i \tag{8.11}$$

Calculating the standard deviation excluding $v_d = \theta_d - \bar{\theta}$, we obtain

$$\sigma = \sqrt{\frac{\sum_{i=1}^{N-1} v_i^2}{N-2}} \tag{8.12}$$

According to the number of measurements N and the significance α, we can obtain the coefficient $K(N, \alpha)$ of the t distribution test. If the following inequality is satisfied

$$|\theta_d - \bar{\theta}| \geq K(N, \alpha)\sigma \tag{8.13}$$

then θ_d contains the gross error, which should be eliminated. Otherwise, the datum should be retained.

2. Elimination methods of the gross error
There are mainly two principles for the elimination of the gross error: first, the decision criterion of the gross error should be properly selected; second, a stepwise elimination method should be taken. Generally, the maximal gross error should be eliminated first.

8.2.5 Combination of Errors

1. Combination of random errors
Suppose q single random errors are in the measurement process. Their standard deviations are σ_1, σ_2, ..., σ_q, and the corresponding error propagation coefficients are a_1, a_2, ..., a_q, respectively. These error propagation coefficients can be obtained by the indirect measured function as follows:

$$a_i = \frac{\partial y}{\partial \theta_i}, \quad i = 1, 2, \ldots, q \tag{8.14}$$

where $y = f(\theta_1, \theta_2, \ldots, \theta_q)$ is the indirect measurement function.
Therefore, the combined standard deviation of the random errors is shown as

$$\sigma = \sqrt{\sum_{i=1}^{q} (a_i\sigma_i)^2 + 2\sum_{1 \le i < j} \rho_{ij} a_i a_j \sigma_i \sigma_j} \tag{8.15}$$

If various measurement errors are uncorrelated, then the correlation coefficient is $\rho_{ij} = 0$. The above equation is simplified as

$$\sigma = \sqrt{\sum_{i=1}^{q} (a_i\sigma_i)^2} \tag{8.16}$$

If all of the propagation coefficients satisfy $a_i = 1$, $i = 1, 2, \ldots, q$, then

$$\sigma = \sqrt{\sum_{i=1}^{q} \sigma_i^2} \tag{8.17}$$

2. Synthesis of the systematic error

(1) Synthesis of the fixed systematic error

The fixed systematic error is an error whose size and sign are determined. If N single fixed systematic errors are given, then the total fixed systematic error is

$$\Delta y = a_1 \Delta_1 + a_2 \Delta_2 + \cdots + a_N \Delta_N = \sum_{i=1}^{N} a_i \Delta_i \tag{8.18}$$

where Δ_i, $i = 1, 2, \ldots, N$, is the single fixed systematic error. The fixed systematic error should be eliminated or corrected before or after the error synthesis.

(2) Synthesis of the uncertainty systematic error

The uncertainty systematic error is an error whose exact numerical value is unknown but its limit range is known. The uncertainty systematic error has a certain randomness and can be synthesized by the method of random error synthesis.

Suppose that the number of single random errors is q and the number of single uncertainty systematic errors is r. The standard deviations of both errors are $\sigma_1, \sigma_2, \ldots, \sigma_q$ and $s_1, s_2, \ldots, s_r$, respectively. Without loss of generality, if each error propagation coefficient is set as 1, then the total measurement standard deviation is

$$\sigma = \sqrt{\sum_{i=1}^{q} \sigma_i^2 + \sum_{j=1}^{r} s_j^2 + R} \tag{8.19}$$

where R denotes the summation of various covariance of errors. If various errors are uncorrelated, then

$$\sigma = \sqrt{\sum_{i=1}^{q} \sigma_i^2 + \sum_{j=1}^{r} s_j^2} \tag{8.20}$$

8.2.6 Distribution of Errors

Error distribution is the inverse process of error synthesis. It is the problem of how to reasonably determine the individual error under the constraint of allowed total error. The general principle of the error distribution is shown as follows:

$$\sqrt{\sigma_{y_1}^2 + \sigma_{y_2}^2 + \cdots + \sigma_{y_N}^2} \leq \sigma_y \tag{8.21}$$

where σ_y is the total error and $\sigma_{y_i} = a_i \sigma_i$, $i = 1, 2, \ldots, N$, represents each individual error.

1. Accepting or rejecting a small error

The standard deviation of a measurement result is given as

$$\sigma_y = \sqrt{\sigma_{y_1}^2 + \sigma_{y_2}^2 + \cdots \sigma_{y_{k-1}}^2 + \sigma_{y_k}^2 + \sigma_{y_{k+1}}^2 + \cdots + \sigma_{y_N}^2} \tag{8.22}$$

If the new standard deviation satisfies $\sigma'_y = \sqrt{\sigma_{y_1}^2 + \sigma_{y_2}^2 + \cdots \sigma_{y_{k-1}}^2 + \sigma_{y_{k+1}}^2 + \cdots + \sigma_{y_N}^2}$ $\approx \sigma_y$ after σ_{y_k} is eliminated, the error corresponding to σ_{y_k} is called a small error.

For the general accuracy of the measurement, the significant digit of the measurement error is usually taken as only one bit. If the following relation is satisfied:

$$\sigma_{y_k} \le (0.3 \sim 0.4)\sigma_y \tag{8.23}$$

then the error term corresponding to σ_{y_k} can be eliminated. For the precision measurement, the significant digit of the measurement error is usually taken as two bits. If the following relation is satisfied:

$$\sigma_{y_k} \le (0.10 \sim 0.14)\sigma_y \tag{8.24}$$

then the error term corresponding to σ_{y_k} can be given up.

2. The equal effect principle of the error distribution

The so-called equal effect principle is to control an equal influence of each error component on the total variance which is shown as follows:

$$\sigma_{y_1} = \sigma_{y_2} = \cdots = \sigma_{y_N} = \sigma_y/\sqrt{N} \tag{8.25}$$

Thus, we obtain

$$\sigma_i = \frac{\sigma_y}{\sqrt{N}} \cdot \frac{1}{a_i} \tag{8.26}$$

3. The possibility principle of the error adjustment

The basic idea of the possibility principle for the error adjustment is to give a proper adjustment under the basis of equal effect error distribution, in accordance with the specific circumstances. We usually expand the distribution of the error which is difficult to achieve and reduce the distribution of the error which is easy to achieve. The result of such adjustments is to avoid the use of more expensive instruments and to increase experiment times.

8.3 The Assessment and Estimation of Measurement Uncertainty

Since the measurement error exists, we need to know the degree of confidence or uncertainty of the measurement results. The quality of the measurement results is often the key factor in scientific research and experiments, and it is also the important basis for many applications.

8.3.1 The Basic Concept of Measurement Uncertainty

1. The definition of measurement uncertainty
Measurement uncertainty is a concept based on the error theory; it can be understood as the credibility of the measurement results or as the degree of uncertainty. There are many different forms of measurement uncertainty, such as a standard deviation or a half-width of a given confidence interval. The measurement uncertainty is often composed of various components: some of which can be estimated by the statistical distribution of a series of measurement, and other components can be estimated based on experience or other information.

According to the concept of measurement uncertainty, a complete measurement result should contain two parts: the estimated value and the dispersion parameter.

For example, the measurement result is $y \pm \Delta$, where y is the estimation of the quantity and Δ is the uncertainty. The uncertainty represented by the standard deviation is called the standard uncertainty, which is often denoted as u. If the uncertainty is represented by the multiple of the standard deviation, it is referred to as the expanded uncertainty, denoted as U.

2. Measurement uncertainty and error
Error and uncertainty are two concepts that are interrelated but different from each other.

On the one hand, both error and uncertainty are caused by the system effect and the random effect of the measurement. Because of the existence of these effects, the measured value may have some error and uncertainty.

On the other hand, there is a significant difference between the error and the uncertainty. The error is in terms of its truth value: it is the difference between the measured value and the truth value. The error cannot be accurate in many cases because the truth value may be unknowable. Uncertainty is actually a representation of the undetermined degree of measurement results due to the existence of system effects and random effects, which indicates the possible range of the measured values. In addition, the error represents a difference value, whose sign may be positive or negative; while the uncertainty of measurement represents a range, whose value is always positive.

8.3.2 The Assessment of Standard Uncertainty

1. Class A assessment of standard uncertainty

Class A assessment of standard uncertainty uses statistical analysis methods. In the standard uncertainty, its measurement uncertainty u is equal to the standard deviation σ of the measurement.

In the measurement, the arithmetic mean of the independent observation is generally used as the measurement result. The standard uncertainty is shown as follows:

$$u(\bar{\theta}) = s(\bar{\theta}) = \frac{s(\theta_i)}{\sqrt{N}} \tag{8.27}$$

where $s(\theta_i)$ is the sample standard deviation (or referred to as the experiment standard deviation), which represents an unbiased estimate of the standard deviation σ of the measured value θ_i. In fact, $s(\theta_i)$ characterizes the dispersion around the arithmetic mean value $\bar{\theta}$ of θ_i. $s(\bar{\theta})$ is the averaged experiment standard deviation and N is the measurement times.

Only when the number of measurements is sufficiently large, Class A assessment of the uncertainty is reliable. Generally, $N > 5$ should be satisfied.

2. Class B assessment of standard uncertainty

If the measured value θ_i is estimated from experience, experiment, or other information, a different assessment of the measurement should be taken, which is called Class B assessment of the standard uncertainty.

If the estimates are taken from manufacturing specifications, verification or calibration certificates, or other sources, and its uncertainty $U(\theta_i)$ is the k_i times of the standard deviation, then the standard uncertainty $u(\theta_i)$ is shown as

$$u(\theta_i) = \frac{U(\theta_i)}{k_i} \tag{8.28}$$

Suppose that the measured values are subject to the normal distribution. If the extended uncertainty of θ_i is given in accordance with the half-width a of the confidence interval at probability P, then the standard uncertainty of θ_i is

$$u(\theta_i) = \frac{a}{k_P} \tag{8.29}$$

where k_P is the confidence factor, whose value can be obtained from the normal distribution integral table.

8.3.3 The Combination of Measurement Uncertainty

If the measurement result is affected by many factors, which results in uncertainty in several components, then the combined standard uncertainty u_c can be used to express the standard uncertainty of the measurement results. In order to obtain u_c, we need to analyze the relationship between various factors and measurement results.

If the influence of various factors on the measurement results is indirect, then the value of each uncertainty component should be determined according to the Class A assessment or Class B assessment before the combination. The combined uncertainty $u_c(y)$ of the measurement $y = f(x_1, x_2, \cdots, x_N)$ is defined as

$$u_c(y) = \sqrt{\sum_{i=1}^{N} \left(\frac{\partial f}{\partial x_i}\right)^2 (u_{x_i})^2 + 2 \sum_{1 \le i < j}^{N} \frac{\partial f}{\partial x_i} \frac{\partial f}{\partial x_j} \rho_{ij} u_{x_i} u_{x_j}} \tag{8.30}$$

where ρ_{ij} represents the correlation coefficient between the i th measured value and the j th measured value. Both u_{x_i} and u_{x_j} are the standard uncertainties of the directly measured values.

If the uncertainties of x_i and x_j are independent of each other, then we obtain $\rho_{ij} = 0$. Thus, the above equation is changed as follows:

$$u_c(y) = \sqrt{\sum_{i=1}^{N} \left(\frac{\partial f}{\partial x_i}\right)^2 (u_{x_i})^2} \tag{8.31}$$

If the influence of the measurement results of the N components which cause the uncertainty is direct, then

$$u_c(y) = \sqrt{\sum_{i=1}^{N} (u_{x_i})^2} \tag{8.32}$$

8.4 The Least Squares (LS) Method for Data Processing

The least squares (LS) method is a mathematical optimization method, which can be used to find the optimal matching function of data by minimizing the square of the error. This method is widely concerned and applied in science and technology.

8.4.1 The Principle of the Least Squares Method

1. Problem description

In order to determine the estimation x_1, x_2, ..., x_t of t unknown quantities X_1, X_2, ..., X_t which cannot be directly measured, we can directly measure Y with N times and obtain the measured data l_1, l_2, ..., l_N, in which Y is a function of X_1, X_2, ..., X_t. Define the functional relation between Y and X_1, X_2, ..., X_t as

$$\begin{cases} Y_1 = f_1(X_1, X_2, \ldots, X_t) \\ Y_2 = f_2(X_1, X_2, \ldots, X_t) \\ \quad\vdots \\ Y_N = f_N(X_1, X_2, \ldots, X_t) \end{cases} \tag{8.33}$$

If $N = t$ is set, we can obtain x_1, x_2, ..., x_t from eq. (8.33). In practice, $N > t$ is usually satisfied; thus, eq. (8.33) becomes an overdetermined equation set, with a nonunique solution. How to obtain the most reliable solution from the measured data is the problem of the least squares method which needs to be solved.

2. The principle of the LS method

Suppose the estimates of directly measured data Y_1, Y_2, ..., Y_N are y_1, y_2, ..., y_N. We have

$$\begin{cases} y_1 = f_1(x_1, x_2, \ldots, x_t) \\ y_2 = f_2(x_1, x_2, \ldots, x_t) \\ \quad\vdots \\ y_N = f_N(x_1, x_2, \ldots, x_t) \end{cases} \tag{8.34}$$

The residual errors of the measured data l_1, l_2, ..., l_N are

$$v_i = l_i - y_i, \quad i = 1, 2, \ldots, N \tag{8.35}$$

or

$$v_i = l_i - f_i(x_1, x_2, \ldots, x_t), \quad i = 1, 2, \ldots, N \tag{8.36}$$

Equations (8.35) and (8.36) are residual equations. If the measurement of data l_1, l_2, ..., l_N is unbiased and independent, obeying the normal distribution with standard deviations as σ_1, σ_2, ..., σ_N, then the probability of each measurement value appearing in the neighborhood $d\delta_1$, $d\delta_2$, ..., $d\delta_N$ of its truth value is

$$P_i = \frac{1}{\sigma_i \sqrt{2\pi}} e^{-\frac{\delta_i^2}{2\sigma_i^2}} d\delta_i, \quad i = 1, 2, \ldots, N \tag{8.37}$$

By the multiplicative property of the probability theory, the probability P of each measurement value appearing in the corresponding area $d\delta_1$, $d\delta_2$, $\ldots$, $d\delta_N$ is

$$P = \prod_{i=1}^{N} P_i = \frac{1}{\sigma_1 \sigma_2 \cdots \sigma_N \left(\sqrt{2\pi}\right)^N} e^{-\sum_{i=1}^{N} \frac{\delta_i^2}{2\sigma_i^2}} d\delta_1 d\delta_2 \cdots d\delta_N \tag{8.38}$$

In an equal precision measurement, $\sigma_1 = \sigma_2 = \cdots = \sigma_N$. We know from the analysis of eq. (8.38) that the most reliable value of the measurement results will be obtained when the square of the residue reaches its minimum, which is the principle of the least squares method:

$$v_1^2 + v_2^2 + \cdots + v_N^2 = \sum_{i=1}^{N} v_i^2 = \min \tag{8.39}$$

The estimated value according to the LS method is customarily said the most probable value, also known as the most reliable value, and it is unbiased and the most reliable. It should be explained that, although the LS principle is derived under the conditions of unbiased, normal distribution but independent of each other, it is often used in the case of not strictly obeying the normal distribution.

3. The linear parameter LS principle of equal precision measurement

For the measurement of linear parameters, the function $f(\cdot)$ in eq. (8.34) is set as a linear function:

$$\begin{cases} y_1 = a_{11}x_1 + a_{12}x_2 + \cdots + a_{1t}x_t \\ y_2 = a_{21}x_1 + a_{22}x_2 + \cdots + a_{2t}x_t \\ \quad\vdots \\ y_N = a_{N1}x_1 + a_{N2}x_2 + \cdots + a_{Nt}x_t \end{cases} \tag{8.40}$$

whose residual equations is as follows:

$$\begin{cases} v_1 = l_1 - (a_{11}x_1 + a_{12}x_2 + \cdots + a_{1t}x_t) \\ v_2 = l_2 - (a_{21}x_1 + a_{22}x_2 + \cdots + a_{2t}x_t) \\ \quad\vdots \\ v_N = l_N - (a_{N1}x_1 + a_{N2}x_2 + \cdots + a_{Nt}x_t) \end{cases} \tag{8.41}$$

The above equation can be rewritten in a matrix form as

$$V = L - A\hat{X} \tag{8.42}$$

where

$$L = [l_1 \quad l_2 \quad \cdots \quad l_N]^T, \quad \hat{X} = [x_1 \quad x_2 \quad \cdots \quad x_N]^T, \quad V = [v_1 \quad v_2 \quad \cdots \quad v_N]^T$$

$$A = \begin{bmatrix} a_{11} & a_{12} & \cdots & a_{1t} \\ a_{21} & a_{22} & \cdots & a_{2t} \\ \vdots & \vdots & \ddots & \vdots \\ a_{N1} & a_{N2} & \cdots & a_{Nt} \end{bmatrix} \tag{8.43}$$

When the equal precision measurement is implemented, the minimum of the square summation of the residual equation is expressed as

$$V^T V = \min \tag{8.44}$$

or

$$(L - A\hat{X})^T (L - A\hat{X}) = \min \tag{8.45}$$

4. The linear parameter LS principle of unequal precision measurement
For the unequal precision measurement of linear parameters, the LS matrix form is given as

$$(L - A\hat{X})^T \Gamma (L - A\hat{X}) = \min \tag{8.46}$$

where Γ is an $N \times N$ weighting matrix which is given as

$$\Gamma = \begin{bmatrix} \gamma_1 & 0 & \cdots & 0 \\ 0 & \gamma_2 & \cdots & 0 \\ \vdots & \vdots & \ddots & \vdots \\ 0 & 0 & \cdots & \gamma_N \end{bmatrix} = \begin{bmatrix} \sigma^2/\sigma_1^2 & 0 & \cdots & 0 \\ 0 & \sigma^2/\sigma_2^2 & \cdots & 0 \\ \vdots & \vdots & \ddots & \vdots \\ 0 & 0 & \cdots & \sigma^2/\sigma_N^2 \end{bmatrix} \tag{8.47}$$

in which $\gamma_i = \frac{\sigma^2}{\sigma_i^2}$, $i = 1, 2, \ldots, N$, are the weighting factors of the measured data, σ^2 is the unit variance, and σ_i^2, $i = 1, 2, \ldots, N$, are the variances of the measured data $l_1, l_2, \ldots, l_N$. The unequal precision measurement of linear parameters can be transformed into the equal precision form. Thus, it can be processed by the LS method. To this end, the residual equation (8.41) should be multiplied by $\sqrt{\gamma_i}$ on both sides, and then it is changed into the form of equal precision residual equation as follows:

$$(\tilde{L} - \tilde{A}\hat{X})^T (\tilde{L} - \tilde{A}\hat{X}) = \min \tag{8.48}$$

where

$$\tilde{L} = \left[\sqrt{\gamma_1}l_1 \quad \sqrt{\gamma_2}l_2 \quad \cdots \quad \sqrt{\gamma_N}l_N \right]^T$$

$$\tilde{A} = \begin{bmatrix} \sqrt{\gamma_1}a_{11} & \sqrt{\gamma_1}a_{12} & \cdots & \sqrt{\gamma_1}a_{1t} \\ \sqrt{\gamma_2}a_{21} & \sqrt{\gamma_2}a_{22} & \cdots & \sqrt{\gamma_2}a_{2t} \\ \vdots & \vdots & \ddots & \vdots \\ \sqrt{\gamma_N}a_{N1} & \sqrt{\gamma_N}a_{N2} & \cdots & \sqrt{\gamma_N}a_{Nt} \end{bmatrix}^T \tag{8.49}$$

8.4.2 Normal Equation: The Basic Method of the LS Processing

The so-called normal equation is a set of algebraic equations with definite solutions. For the purpose of reducing the error, the number of measurement times N is usually greater than the number of unknown parameters t, so that the number of residues in eq. (8.41) is more than the number of unknowns. Such equation cannot obtain the unique solution, and the LS method can be used to transform the residual equation into the algebraic equation with the definite solution, i.e., the normal equation, so as to obtain the unknown parameters.

The procedures of the LS method for the linear parametric processing can be divided down as follows. First, list the residual equation according to the specific problems. Second, transform the residual equation into the normal equation based on the LS principle. Third, find the estimated values by solving the normal equation. And finally, demonstrate the accuracy of the estimation. For nonlinear parameters, it should be linearized first, and then it should be processed by the method of the above linear parameters.

1. The linear parameter LS processing of equal precision measurement
The residual equation of linear parameters is shown in eq. (8.41), and the equal precision LS condition is given in eq. (8.44) or eq. (8.45). In order to realize $V^T V = \min$, we usually use the extremum method to obtain

$$\begin{cases} \sum_{i=1}^{N} a_{i1}l_i = \sum_{i=1}^{N} a_{i1}a_{i1}x_1 + \sum_{i=1}^{N} a_{i1}a_{i2}x_2 + \cdots + \sum_{i=1}^{N} a_{i1}a_{it}x_t \\ \sum_{i=1}^{N} a_{i2}l_i = \sum_{i=1}^{N} a_{i2}a_{i1}x_1 + \sum_{i=1}^{N} a_{i2}a_{i2}x_2 + \cdots + \sum_{i=1}^{N} a_{i2}a_{it}x_t \\ \qquad\qquad\qquad\qquad\qquad \vdots \\ \sum_{i=1}^{N} a_{it}l_i = \sum_{i=1}^{N} a_{it}a_{i1}x_1 + \sum_{i=1}^{N} a_{it}a_{i2}x_2 + \cdots + \sum_{i=1}^{N} a_{it}a_{it}x_t \end{cases} \tag{8.50}$$

The above equation set is the LS normal equation of the equal precision linear parameter measurement. Obviously, it is an equation with t unknowns. It has the unique solution when its coefficient determinant is not equal to zero. Equation (8.50) can be rewritten as

$$\begin{cases} a_{11}V_1 + a_{21}V_2 + \cdots + a_{N1}V_N = 0 \\ a_{12}V_1 + a_{22}V_2 + \cdots + a_{N2}V_N = 0 \\ \quad\quad\quad\quad\vdots \\ a_{1t}V_1 + a_{2t}V_2 + \cdots + a_{Nt}V_N = 0 \end{cases} \tag{8.51}$$

Writing the equation into the matrix form as

$$\boldsymbol{A}^\mathrm{T}\boldsymbol{V} = \boldsymbol{0} \tag{8.52}$$

where both $\boldsymbol{A}$ and $\boldsymbol{V}$ are given in eq. (8.43), vector $\boldsymbol{0}$ is defined as

$$\boldsymbol{0} = [0 \quad 0 \quad \cdots \quad 0]^\mathrm{T} \tag{8.53}$$

Since $\boldsymbol{V} = \boldsymbol{L} - \boldsymbol{A}\hat{\boldsymbol{X}}$, the normal equation can be written as

$$(\boldsymbol{A}^\mathrm{T}\boldsymbol{A})\hat{\boldsymbol{X}} = \boldsymbol{A}^\mathrm{T}\boldsymbol{L} \tag{8.54}$$

If we let $\boldsymbol{C} = \boldsymbol{A}^\mathrm{T}\boldsymbol{A}$, then the normal equation can also be written as

$$\boldsymbol{C}\hat{\boldsymbol{X}} = \boldsymbol{A}^\mathrm{T}\boldsymbol{L} \tag{8.55}$$

If the rank of $\boldsymbol{A}$ is t, then $\boldsymbol{C}$ is a full-rank matrix, whose determinant satisfies $|\boldsymbol{C}| \neq 0$. As a result, $\hat{\boldsymbol{X}}$ has a unique solution. Such solution can be expressed as

$$\hat{\boldsymbol{X}} = \boldsymbol{C}^{-1}\boldsymbol{A}^\mathrm{T}\boldsymbol{L} \tag{8.56}$$

Taking the expectation of the above equation, we obtain

$$E[\hat{\boldsymbol{X}}] = E[\boldsymbol{C}^{-1}\boldsymbol{A}^\mathrm{T}\boldsymbol{L}] = \boldsymbol{C}^{-1}\boldsymbol{A}^\mathrm{T}E[\boldsymbol{L}] = \boldsymbol{C}^{-1}\boldsymbol{A}^\mathrm{T}\boldsymbol{Y} = \boldsymbol{C}^{-1}\boldsymbol{A}^\mathrm{T}\boldsymbol{A}\boldsymbol{X} = \boldsymbol{X} \tag{8.57}$$

where $\boldsymbol{Y} = [Y_1 \quad Y_2 \quad \cdots \quad Y_N]^\mathrm{T}$, whose elements are the truth values of the measured values. The elements of $\boldsymbol{X} = [X_1 \quad X_2 \quad \cdots \quad X_t]^\mathrm{T}$ are the truth values of unknowns. Obviously, $\hat{\boldsymbol{X}}$ is the unbiased estimation of $\boldsymbol{X}$.

2. The linear parameter LS processing of unequal precision measurement

For the linear parameter LS processing of the unequal precision measurement, the least squares sum of the weighted residuals is required:

$$A^T \Gamma V = 0 \tag{8.58}$$

where the weighting coefficient matrix is defined as

$$\Gamma = \begin{bmatrix} \gamma_1 & 0 & \cdots & 0 \\ 0 & \gamma_2 & \cdots & 0 \\ \vdots & \vdots & \ddots & \vdots \\ 0 & 0 & \cdots & \gamma_N \end{bmatrix} \tag{8.59}$$

Substituting $V = L - A\hat{X}$ into eq. (8.58), we have

$$A^T \Gamma L - A^T \Gamma A \hat{X} = 0 \tag{8.60}$$

or

$$A^T \Gamma A \hat{X} = A^T \Gamma L \tag{8.61}$$

After arrangement, we obtain

$$\hat{X} = (A^T \Gamma A)^{-1} A^T \Gamma L \tag{8.62}$$

If we let $\tilde{C} = A^T \Gamma A$, then we obtain

$$\hat{X} = \tilde{C}^{-1} A^T \Gamma L \tag{8.63}$$

Taking the expectation, we have $E[\hat{X}] = E\{\tilde{C}^{-1} A^T \Gamma L\} = \tilde{C}^{-1} A^T \Gamma E\{L\} = \tilde{C}^{-1} A^T \Gamma A X = X$. It is obvious that $\hat{X}$ is the unbiased estimation of X.

3. Nonlinear parameter LS processing

Generally, $y_i = f_i(x_1, x_2, \cdots, x_t)$, $i = 1, 2, \ldots, N$, are nonlinear functions, and the residual equations $v_i = l_i - f_i(x_1, x_2, \cdots, x_t)$, $i = 1, 2, \ldots, N$, are also nonlinear functions. Under this condition, we usually take a linearization method to change the nonlinear functions into linear functions, and then process them as that in the linear parameters.

Take $x_{10}, x_{20}, \ldots, x_{t0}$ as the approximate values of $x_1, x_2, \ldots, x_t$ and set

$$x_j = x_{j0} + \delta_j, \quad j = 1, 2, \ldots, t \tag{8.64}$$

where δ_j, $j = 1, 2, \ldots, t$, are the differences between the estimates and their approximate values. If δ_j, $j = 1, 2, \ldots, t$, can be solved, then $x_1, x_2, \ldots, x_t$ can be found from eq. (8.64). Taking the Taylor expansion of the nonlinear functions

$y_i = f_i(x_1, x_2, \ldots, x_t)$, $i = 1, 2, \ldots, N$, at x_{10}, x_{20}, $\cdots$, x_{t0}, and retaining the first-order terms, we have

$$
\begin{aligned}
y_i &= f_i(x_1, x_2, \ldots, x_t) \\
&= f_i(x_{10}, x_{20}, \ldots, x_{t0}) + \left(\frac{\partial f_i}{\partial x_1}\right)_0 \delta_1 + \left(\frac{\partial f_i}{\partial x_2}\right)_0 \delta_2 + \cdots + \left(\frac{\partial f_i}{\partial x_t}\right)_0 \delta_t, \quad i = 1, 2, \ldots, N
\end{aligned}
\tag{8.65}
$$

where $\left(\frac{\partial f_i}{\partial x_j}\right)_0$, $j = 1, 2, \ldots, t$, represent values of the partial derivative f_i with respect to x_j at x_{j0}.

Taking the expansion to the above equation and substituting it into the residual equation, we let

$$
l'_i = l_i - f_i(x_{10}, x_{20}, \ldots, x_{t0}), \quad i = 1, 2, \ldots, N
\tag{8.66}
$$

and let $a_{ij} = \left(\frac{\partial f_i}{\partial x_j}\right)_0$, $j = 1, 2, \ldots, t$; the nonlinear residual equations are transformed into the linear equations as

$$
\begin{cases}
v_1 = l'_1 - (a_{11}\delta_1 + a_{12}\delta_2 + \cdots + a_{1t}\delta_t) \\
v_2 = l'_2 - (a_{21}\delta_1 + a_{22}\delta_2 + \cdots + a_{2t}\delta_t) \\
\quad \vdots \\
v_N = l'_N - (a_{N1}\delta_1 + a_{N2}\delta_2 + \cdots + a_{Nt}\delta_t)
\end{cases}
\tag{8.67}
$$

The normal equations can be listed in the case of linear parameters, and then the solutions δ_1, δ_2, $\ldots$, δ_t can be obtained. Furthermore, the estimates x_1, x_2, $\ldots$, x_t can also be solved by eq. (8.64).

4. The relationship between the LS method and the arithmetic mean

Suppose the measurement data l_1, l_2, $\ldots$, l_N are obtained for X by N measurements. The corresponding weighting factors are γ_1, γ_2, $\ldots$, γ_N, and the residual equations are

$$
v_i = l_i - x, \quad i = 1, 2, \ldots, N
\tag{8.68}
$$

Based on the LS principle, we have

$$
x = \frac{\sum\limits_{i=1}^{N} \gamma_i l_i}{\sum\limits_{i=1}^{N} \gamma_i}
\tag{8.69}
$$

When the measurement is in the equal precision, the estimator is

$$x = \frac{1}{N} \sum_{i=1}^{N} l_i \tag{8.70}$$

Comparing the above two results, we see that the LS principle is the generalized arithmetic mean, while the arithmetic mean is a special case of the LS method.

8.4.3 The Estimation Accuracy of the LS Processing

1. Estimation accuracy of the directly measured data

The accuracy of the measured data is usually expressed by its standard deviation σ. However, because the truth value of σ is not easy to obtain, we usually use its estimation $\hat{\sigma}$. In this way, the estimation of the accuracy is changed into the estimation of the standard deviation.

Suppose an LS problem is the estimation to t unknowns. For the equal precision measurement, we have

$$\hat{\sigma}^2 = \frac{\sum_{i=1}^{N} v_i^2}{N-t} \tag{8.71}$$

Thus, the estimator of the standard deviation is

$$\hat{\sigma} = \sqrt{\frac{\sum_{i=1}^{N} v_i^2}{N-t}} \tag{8.72}$$

It is also written as

$$\sigma = \sqrt{\frac{\sum_{i=1}^{N} v_i^2}{N-t}} \tag{8.73}$$

For the unequal precision measurement, eq. (8.73) becomes

$$\sigma = \sqrt{\frac{\sum_{i=1}^{N} \gamma_i v_i^2}{N-t}} \tag{8.74}$$

2. The precision estimation of the LS estimator

In the case of the equal precision measurement, the basic idea of the precision estimation of the LS estimator is to multiply a group of constrained indefinite numbers d_{11}, d_{12}, ..., d_{1t}; d_{21}, d_{22}, ..., d_{2t}; d_{t1}, d_{t2}, ..., d_{tt} by the normal equations. It yields

$$\begin{cases} \sigma_{x_1} = \sigma\sqrt{d_{11}} \\ \sigma_{x_2} = \sigma\sqrt{d_{22}} \\ \vdots \\ \sigma_{x_t} = \sigma\sqrt{d_{tt}} \end{cases} \tag{8.75}$$

where σ is the standard deviation of the measured data. Equation (8.75) transforms the precision estimation into the problem of solving indefinite numbers. We can further obtain

$$D = \begin{bmatrix} d_{11} & d_{21} & \cdots & d_{N1} \\ d_{12} & d_{22} & \cdots & d_{N2} \\ \vdots & \vdots & \ddots & \vdots \\ d_{1t} & d_{2t} & \cdots & d_{Nt} \end{bmatrix} = (A^T A)^{-1} \tag{8.76}$$

where

$$A = \begin{bmatrix} a_{11} & a_{12} & \cdots & a_{1t} \\ a_{21} & a_{22} & \cdots & a_{2t} \\ \vdots & \vdots & \ddots & \vdots \\ a_{N1} & a_{N2} & \cdots & a_{Nt} \end{bmatrix} \tag{8.77}$$

is the coefficient matrix of the normal equations.

In the case of unequal precision measurement, we need to multiply weighting factors γ_i on both sides of the normal equations and then take the same method as that of the equal precision measurement:

$$\begin{cases} \sigma_{x_1} = \sigma\sqrt{d_{11}} \\ \sigma_{x_2} = \sigma\sqrt{d_{22}} \\ \vdots \\ \sigma_{x_t} = \sigma\sqrt{d_{tt}} \end{cases} \tag{8.78}$$

where the indefinite numbers d_{11}, d_{22}, ..., d_{tt} are given by eq. (8.79) as

$$D = \begin{bmatrix} d_{11} & d_{21} & \cdots & d_{N1} \\ d_{12} & d_{22} & \cdots & d_{N2} \\ \vdots & \vdots & \ddots & \vdots \\ d_{1t} & d_{2t} & \cdots & d_{Nt} \end{bmatrix} = (A^T \Gamma A)^{-1} \tag{8.79}$$

8.4.4 The LS Processing for the Combined Measurement

The combined measurement is a direct measure of a set of different combinations of measured values to determine the best estimate of the measured values. This method can effectively reduce the random errors of the measurement, and it also contributes to the reduction of systematic errors. Therefore, it is widely used in the precision measurements.

Figure 8.1 gives an example of the spacing calibration measurement among the four given lines.

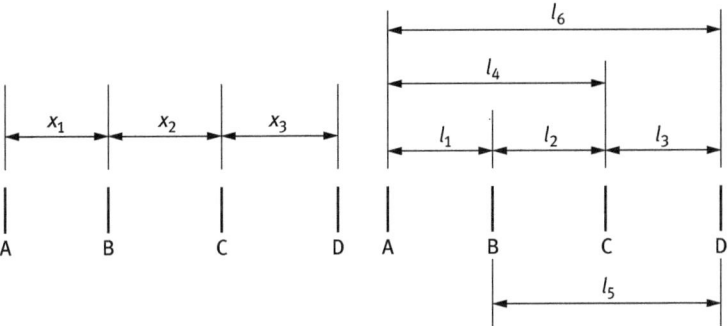

Figure 8.1: An example of the spacing calibration measurement among the four given lines.

In the figure, x_1, x_2, and x_3 are the spacing among lines A, B, C, and D, respectively. The combined direct measurement values are $l_1 = 1.015$, $l_2 = 0.985$, $l_3 = 1.020$, $l_4 = 2.016$, $l_5 = 1.981$, $l_6 = 3.032$ in mm. The residual equations are

$$
\begin{cases}
v_1 = l_1 - x_1 \\
v_2 = l_2 - x_2 \\
v_3 = l_3 - x_3 \\
v_4 = l_4 - (x_1 + x_2) \\
v_5 = l_5 - (x_2 + x_3) \\
v_6 = l_6 - (x_1 + x_2 + x_3).
\end{cases}
$$

From the above equations, we obtain

$$
\boldsymbol{L} = [l_1 \ l_2 \ l_3 \ l_4 \ l_5 \ l_6]^{\mathrm{T}} = [1.015 \ 0.985 \ 1.020 \ 2.016 \ 1.918 \ 3.032]^{\mathrm{T}}
$$
$$
\hat{\boldsymbol{X}} = [x_1 \ x_2 \ x_3]^{\mathrm{T}}
$$
$$
\boldsymbol{A} =
\begin{bmatrix}
1 & 0 & 0 & 1 & 0 & 1 \\
0 & 1 & 0 & 1 & 1 & 1 \\
0 & 0 & 1 & 0 & 1 & 1
\end{bmatrix}^{\mathrm{T}}
$$

Thus, we have

$$\hat{X} = [x_1 \ x_2 \ x_3]^T = C^{-1}A^T L = (A^T A)^{-1} A^T L$$

and

$$\hat{X} = \begin{bmatrix} x_1 \\ x_2 \\ x_3 \end{bmatrix} = \begin{bmatrix} 1.028 \\ 0.983 \\ 1.013 \end{bmatrix}$$

The above results are the optimal estimations of the spacing among the four given lines. Substituting the estimators into the residual errors, we have

$$\begin{cases} v_1 = l_1 - x_1 = -0.013 \\ v_2 = l_2 - x_2 = 0.002 \\ v_3 = l_3 - x_3 = 0.007 \\ v_4 = l_4 - (x_1 + x_2) = 0.005 \\ v_5 = l_5 - (x_2 + x_3) = -0.015 \\ v_6 = l_6 - (x_1 + x_2 + x_3) = 0.008 \end{cases}$$

Then,

$$\sum_{i=1}^{6} v_i^2 = 0.000536 \, (\text{mm}^2)$$

The standard deviation of the measured data is obtained as

$$\sigma = \sqrt{\frac{\sum_{i=1}^{N} v_i^2}{N - t}} = 0.013 \, (\text{mm})$$

From $D = C^{-1} = (A^T A)^{-1}$, we can obtain $d_{11} = 0.5$, $d_{22} = 0.5$, and $d_{33} = 0.5$. Finally, the standard deviations of the LS estimators x_1, x_2, x_3 are as follows:

$$\begin{cases} \sigma_{x_1} = \sigma \sqrt{d_{11}} = 0.009 \ \text{mm} \\ \sigma_{x_2} = \sigma \sqrt{d_{22}} = 0.009 \ \text{mm} \\ \sigma_{x_3} = \sigma \sqrt{d_{33}} = 0.009 \ \text{mm} \end{cases}$$

Those are the precision estimations of the optimal estimators.

8.5 The Regression Analysis

The regression analysis is a statistical analysis method used to determine the quantitative interdependent relations between two variables or among more variables, and to obtain mathematical expressions in accordance with the internal law of things.

8.5.1 Univariate Linear Regression

1. Determination of the regression equation

The linear regression is one of the simplest models that describe the linear relationship between two variables. In fact, it is the linear fitting of the data. In this section, the basic principle and the method of linear regression analysis are introduced by an example of the wire resistance value y at a certain temperature x. Table 8.1 demonstrates the data of x and y.

Table 8.1: The data of wire resistance values at certain temperatures.

$x/°C$	19.1	25.0	30.1	36.0	40.0	46.5	50.0
y/Ω	76.30	77.80	79.75	80.80	82.35	83.90	85. 10

Figure 8.2 shows the curve based on the data in Table 8.1.

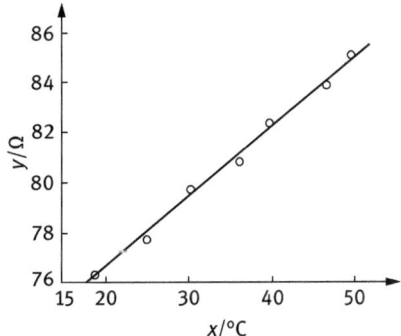

Figure 8.2: The scatter plot of the data in Table 8.1.

We see from the figure that the relation between resistance values and temperatures is roughly a linear relation. We suppose the input and output have the following relation:

$$y_i = \beta_0 + \beta x_i + \varepsilon_i, \quad i = 1, 2, \ldots, N \tag{8.80}$$

where ε_i, $i = 1, 2, \ldots, N$, denote the effects of random factors on the resistance measurement. We take the LS method to find the solution of eq. (8.80).

Suppose b_0 and b are the LS estimators of β_0 and β, respectively. The unitary linear regression equation is obtained as

$$\hat{y} = b_0 + bx \tag{8.81}$$

For each x_i, the residual equations between the measured values y_i and the regression values $\hat{y}_i$ are as follows:

$$v_i = y_i - \hat{y}_i = y_i - b_0 - bx_i, \quad i = 1, 2, \ldots, N \tag{8.82}$$

Its matrix form is given as

$$V = Y - X\hat{b} \tag{8.83}$$

where

$$V = \begin{bmatrix} v_1 \\ v_2 \\ \vdots \\ v_N \end{bmatrix}, \quad Y = \begin{bmatrix} y_1 \\ y_2 \\ \vdots \\ y_N \end{bmatrix}, \quad X = \begin{bmatrix} 1 & x_1 \\ 1 & x_2 \\ \vdots & \vdots \\ 1 & x_N \end{bmatrix}, \quad \hat{b} = \begin{bmatrix} b_0 \\ b \end{bmatrix} \tag{8.84}$$

If y_i is in an equal precision, then

$$\hat{b} = (X^\mathsf{T} X)^{-1} X^\mathsf{T} Y \tag{8.85}$$

From the above, we obtain

$$b = \frac{l_{xy}}{l_{xx}}, \quad b_0 = \bar{y} - b\bar{x} \tag{8.86}$$

where

$$\bar{x} = \frac{1}{N} \sum_{i=1}^{N} x_i$$

$$\bar{y} = \frac{1}{N} \sum_{i=1}^{N} y_i$$

$$l_{xx} = \sum_{i=1}^{N} (x_i - \bar{x})^2 = \sum_{i=1}^{N} x_i^2 - \frac{1}{N} \left(\sum_{i=1}^{N} x_i \right)^2 \tag{8.87}$$

$$l_{xy} = \sum_{i=1}^{N} (x_i - \bar{x})(y_i - \bar{y}) = \sum_{i=1}^{N} x_i y_i - \frac{1}{N} \left(\sum_{i=1}^{N} x_i \right) \left(\sum_{i=1}^{N} y_i \right)$$

$$l_{yy} = \sum_{i=1}^{N} (y_i - \bar{y})^2 = \sum_{i=1}^{N} y_i^2 - \frac{1}{N} \left(\sum_{i=1}^{N} y_i \right)^2$$

Substituting (8.86) into (8.81), we get another form of regression linear equation as

$$\hat{y} - \bar{y} = b(x - \bar{x}) \tag{8.88}$$

2. The group average method for establishing the linear regression equation

Since the establishment of the linear regression equations is on the basis of the LS operation, its calculation is complicated. In order to reduce the computation complexity, a group average method for establishing the linear regression equations is introduced.

In the establishment of the regression equations, N variable data are sequenced in the ascending order, and then they are divided into M groups with equal or similar numbers of data. For example, N values are divided into two groups. The first group contains x_1, x_2, ..., x_K, and the second group contains x_{K+1}, x_{K+2}, ..., x_N. Two observation equations are established as follows:

$$
\begin{cases} y_1 = b_0 + bx_1 \\ \quad \vdots \\ y_K = b_0 + bx_K \end{cases}
\qquad
\begin{cases} y_{K+1} = b_0 + bx_{K+1} \\ \quad \vdots \\ y_N = b_0 + bx_N \end{cases}
\tag{8.89}
$$

We can obtain two equations about b_0 and b by adding both the above equations, respectively:

$$
\begin{cases} \displaystyle\sum_{i=1}^{K} y_i = Kb_0 + b\sum_{i=1}^{K} x_i \\ \displaystyle\sum_{i=K+1}^{N} y_i = (N-K)b_0 + b\sum_{i=K+1}^{N} x_i \end{cases}
\tag{8.90}
$$

And then, we can obtain the solutions of b_0 and b.

8.5.2 Univariate Nonlinear Regression

In scientific research and other applications, the relationship between two variables is more likely to be nonlinear. In this way, it is necessary to carry out nonlinear regression analysis on the data.

Nonlinear regression analysis needs to solve two basic problems, namely, the choice of the regression function and the solution of unknown parameters. This section briefly introduces both issues, which may make readers have a preliminary understanding of the nonlinear regression analysis.

1. Selection and test of the type of regression function

The selection of nonlinear regression function generally can be done in two ways, that is, the direct judgment method and the curve observation method. The former is based on the professional knowledge and the experience to determine the type of function; the latter is to plot the observation data first and then compare the data with the typical curve to determine which type it belongs to. Figure 8.3 shows the typical curves of nonlinear functions.

After selecting the type of nonlinear regression function, it is often required to test the function by the linear test method or other methods.

2. Solution of unknown parameters
In order to obtain the unknown parameters in the regression equation, the basic idea is to convert the nonlinear regression equation to a linear one, and then to solve the linear regression equation. Generally speaking, such nonlinear regression equation which can be tested by the linear inspection can be converted into a linear equation by variable substitution.

Example 8.1: Suppose the regression equation of a group of measurement data is $y = ax^b$. Transform it into a linear regression equation.

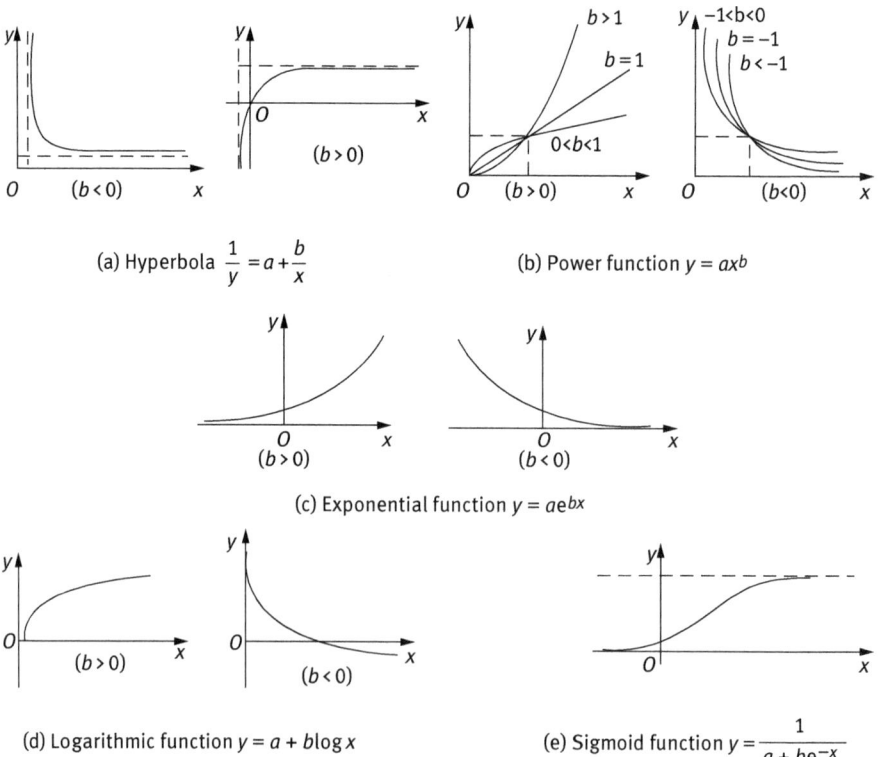

(a) Hyperbola $\dfrac{1}{y} = a + \dfrac{b}{x}$ (b) Power function $y = ax^b$

(c) Exponential function $y = ae^{bx}$

(d) Logarithmic function $y = a + b\log x$ (e) Sigmoid function $y = \dfrac{1}{a + be^{-x}}$

Figure 8.3: Typical curves of nonlinear functions.

Solution: Taking the logarithm on both sides of the nonlinear function, we have

$$\ln y = \ln a + b \ln x$$

If we let $y' = \ln y$, $x' = \ln x$, and $b_0 = \ln a$, then we obtain

$$y' = b_0 + bx'$$

Thus, the original nonlinear regression function is converted into a linear function.

8.5.3 Multiple Linear Regression

1. Multiple linear regression equations

Suppose there is an intrinsic linear relation between dependent variable y and M variables $x_1, x_2, \ldots, x_M$. The N groups of measured data are $x_{i1}, x_{i2}, \ldots, x_{iM}$; y_i, $i = 1, 2, \ldots, N$, whose multiple linear regression equations are as follows:

$$y_i = \beta_0 + \beta_1 x_{i1} + \cdots + \beta_M x_{iM} + \varepsilon_i, \quad i = 1, 2, \ldots, N \tag{8.91}$$

where $\beta_0, \beta_1, \ldots, \beta_M$ are $M + 1$ parameters to be determined and $\varepsilon_1, \varepsilon_2, \ldots, \varepsilon_N$ are N independent random variables, obeying the normal distribution. Equation (8.91) can be rewritten as

$$Y = X\beta + \varepsilon \tag{8.92}$$

where

$$Y = \begin{bmatrix} y_1 \\ y_2 \\ \vdots \\ y_N \end{bmatrix}, \quad X = \begin{bmatrix} 1 & x_{11} & \cdots & x_{1M} \\ 1 & x_{21} & \cdots & x_{2M} \\ \vdots & \vdots & \ddots & \vdots \\ 1 & x_{N1} & \cdots & x_{NM} \end{bmatrix}, \quad \beta = \begin{bmatrix} \beta_1 \\ \beta_2 \\ \vdots \\ \beta_M \end{bmatrix}, \quad \varepsilon = \begin{bmatrix} \varepsilon_1 \\ \varepsilon_2 \\ \vdots \\ \varepsilon_N \end{bmatrix} \tag{8.93}$$

If we estimate β with the LS method and suppose $b_0, b_1, \ldots, b_M$ are the LS estimators of $\beta_0, \beta_1, \ldots, \beta_M$, then the regression equation is given as

$$\hat{y} = b_0 + b_1 x_1 + b_2 x_2 + \cdots + b_M x_M \tag{8.94}$$

The matrix form of the normal equation can be obtained as

$$(X^T X)b = X^T Y \tag{8.95}$$

where $b = [b_0 \quad b_1 \quad \cdots \quad b_M]^T$. If we define $A = X^T X$ and $B = X^T Y$, then eq. (8.95) is written as

$$Ab = B \tag{8.96}$$

Since A is usually a full-rank matrix,

$$b = CB = A^{-1}B = (X^{T}X)^{-1}X^{T}Y \tag{8.97}$$

Then the regression coefficients b_0, b_1, $\ldots$, b_M can be obtained. Generally, A is called the coefficient matrix (or the information matrix) of the normal equation. C is called the correlation matrix. The column vector B is called the constant matrix of the normal equation.

2. Solution of the multiple linear regression equations
In the multiple linear regression, the commonly used regression model is

$$\hat{y} = \mu_0 + b_1(x_1 - \bar{x}_1) + b_2(x_2 - \bar{x}_2) + \cdots + b_M(x_M - \bar{x}_M) \tag{8.98}$$

where $\bar{x}_j = \frac{1}{N}\sum\limits_{i=1}^{N} x_{ij}$, $j = 1, 2, \ldots, M$. The structure matrix is shown as

$$Ab = B \tag{8.99}$$

where

$$A = X^{T}X, \quad B = X^{T}Y \tag{8.100}$$

The matrices A and B are represented as

$$A = \begin{bmatrix} N & 0 & 0 & \cdots & 0 \\ 0 & l_{11} & l_{12} & \cdots & l_{1M} \\ \vdots & \vdots & \vdots & \ddots & \vdots \\ 0 & l_{M1} & l_{M2} & \cdots & l_{MM} \end{bmatrix}, \quad B = \begin{bmatrix} \sum\limits_{i=1}^{N} y_i \\ l_{1y} \\ \vdots \\ l_{My} \end{bmatrix} \tag{8.101}$$

where

$$l_{jk} = \sum_{i=1}^{N}(x_{ij} - \bar{x}_j)(x_{ik} - \bar{x}_k) = \sum_{i=1}^{N} x_{ij}x_{ik} - \frac{1}{N}\left(\sum_{i=1}^{N} x_{ij}\right)\left(\sum_{i=1}^{N} x_{ik}\right), \quad j, k = 1, 2, \ldots, M$$

$$l_{ky} = \sum_{i=1}^{N}(x_{ik} - \bar{x}_k)y_i = \sum_{i=1}^{N} x_{ik}y_i - \frac{1}{N}\left(\sum_{i=1}^{N} x_{ik}\right)\left(\sum_{i=1}^{N} y_i\right), \quad j, k = 1, 2, \ldots, M$$

$$\tag{8.102}$$

It should be noted that matrices A and B defined by eq. (8.101) are different from the matrices A and B shown in eq. (8.97). The inverse matrix C of the matrix A is shown as

$$C = \begin{bmatrix} 1/N & 0 \\ 0 & L^{-1} \end{bmatrix} \tag{8.103}$$

where L is the partitioned matrix of A, which is defined as

$$L = \begin{bmatrix} l_{11} & l_{12} & \cdots & l_{1M} \\ \vdots & \vdots & \ddots & \vdots \\ l_{M1} & l_{M2} & \cdots & l_{MM} \end{bmatrix} \tag{8.104}$$

Thus, the regression coefficient vector of the regression equation is shown as

$$b = CB \tag{8.105}$$

where

$$b = \begin{bmatrix} \mu_0 & b_1 & \cdots & b_M \end{bmatrix}^T \tag{8.106}$$

in which

$$\mu_0 = \frac{1}{N} \sum_{i=1}^{N} y_i = \bar{y} \tag{8.107}$$
$$\begin{bmatrix} b_1 & b_2 & \cdots & b_M \end{bmatrix}^T = L^{-1} \begin{bmatrix} l_{1y} & l_{2y} & \cdots & l_{My} \end{bmatrix}^T$$

8.6 Removal of Trends and Outliers in Signals

In signal processing, it is very important to preprocess the signals. These preprocessing may include the removal of signal trends, the recognition of outliers, and the reduction of noise and interference in signals. This section introduces the method of removing signal trend and the recognition and treatment of outliers in the signal.

8.6.1 Removal of Trends in Signals

1. The meaning of the signal trend
The so-called signal trend term generally refers to the function expressed by the non-zero or nonconstant mean. Such trend function may be the linear function, the power function, and the exponential function which stably grows or decays with time. It may also have the periodic property of sine and cosine functions.

The variation of the signal mean with time is a typical nonstationary phenomenon. The issue of the nonstationary random signal processing will be introduced in later chapters.

In practical applications, the signal trend has a variety of causes, including the imperfect contact of the measurement electrodes, the temperature drift, the instability in the sensors, and so on. These factors may cause the signal measured to deviate from the baseline; even the deviation may change with time. The whole process of the signal deviated from the baseline is the trend of the signal.

2. Removal of the signal trend

There are many ways to remove the signal trend items, including the autoregressive integrated moving average (ARIMA) model, the seasonal model, and other nonstationary random signal processing methods. The empirical mode decomposition (EMD) method developed in recent years can give better results. In this section, a simple and practical method to remove the signal trend is presented from the point of view of engineering applications.

To eliminate the signal trend, we often need to analyze the characteristics and physical model of the signal, and give the boundary conditions, initial conditions, and statistical characteristics. The correction function obtained from the analysis is usually in the polynomial form. Once the polynomial and its coefficients are determined, the correction function of the trend term can be subtracted from the original signal.

The original signal with a trend term is provided as $x(t)$, and the corresponding discrete-time signal is $x(n)$. A pth-order polynomial is constructed with the LS method as

$$y(t) = a_0 + a_1 t + a_2 t^2 + \cdots + a_p t^p = \sum_{k=0}^{p} a_k t^k \tag{8.108}$$

where p is a positive integer, representing the order of the polynomial. If the signal trend is linear, then $p = 1$ is selected. Thus, the original signal $x(t)$ is subtracted from $y(t)$, and we can obtain the signal in which the trend is removed, as shown in

$$\hat{y}(t) = x(t) - y(t) \tag{8.109}$$

Example 8.2: In the ECG examination, due to the movement of the body and other factors, the baseline drift sometimes occurs, as shown in Figure 8.4(a). Try to correct the baseline.

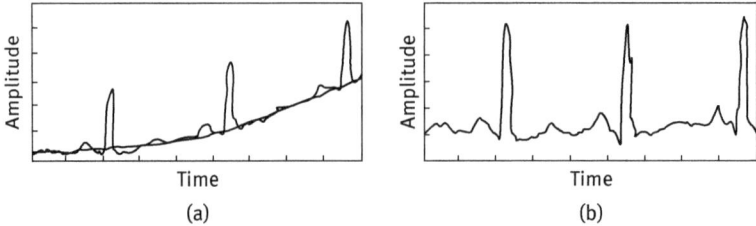

Figure 8.4: (a) The original signal and (b) the trend removed signal.

Solution: Construct a second-order polynomial with the LS method, reflecting the signal trend. Subtracting the constructed trend from the original signal, we can obtain the signal in which the trend is removed, as shown in Figure 8.4(b).

8.6.2 Recognition and Processing of Outlines in Signals

The "outliers" in signals generally refer to the abnormal values which are far from other data sample values. For the measurement data, an outlier usually represents a large measurement error, similar to the gross error introduced in Section 8.2. We will introduce how to identify and determine outliers dynamically in the process of measurement and signal processing applications.

In the outlier detection, we usually establish the LS polynomial based on the previous observations to extrapolate the estimate for the next time and, furthermore, to determine whether the difference between the estimated and the measured values exceeds a given threshold. If the threshold exceeds, then the observation datum is known as an outlier. Otherwise, it is a normal one.

Suppose that four consecutive data samples $x(n-4)$, $x(n-3)$, $x(n-2)$, $x(n-1)$ in the signal $x(n)$ are obtained from the process of measurement or signal processing. The data sample at present time can be estimated by the LS extrapolation as

$$\hat{x}(n) = x(n-1) + \frac{1}{2}x(n-2) - \frac{1}{2}x(n-4) \tag{8.110}$$

Taking the absolute value of the error between $\hat{x}(n)$ and $x(n)$ and comparing it with the threshold δ, we have

$$|x(n) - \hat{x}(n)| \le \delta \tag{8.111}$$

If the above in equation is satisfied, it means that $x(n)$ is a normal sample value. Otherwise, $x(n)$ is an outlier, which should be canceled or substituted by $\hat{x}(n)$. The 3σ criterion and the Romanovski criterion introduced in Section 8.2 can be used to determine the threshold.

Example 8.3: There are some outliers in the measurement of a sinusoidal signal as shown in Figure 8.5. Try to modify the measurement data with MATLAB programming.
Solution: The MATLAB program is shown as follows:

```
% Main program
close all; clear all; clc;
Fs =1000;N =1024; t =557:0.05:559.3; tt =t/Fs;
yy =25 *sin(2 *pi *Fs *tt-pi/2);
yy(5) =-99; yy(14) =90; yy(15) =90; yy(27) =90; yy(28) =80;
```

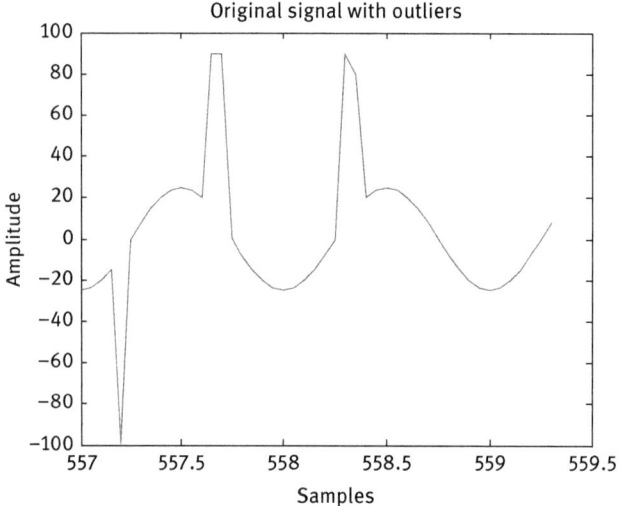

Figure 8.5: The measurement data with some outliers.

```
y =yy; L =length(y);
i =5;
while i <L;
   [u,delta,i] =pand(y,i,L);
   if i = =L
   else
      j =i-1; w =delta; [u,delta,i] =pand1(y,i,L);
      k =i-j; y =guji(k,j,y,w); i =k +j;
   end
end
plot(t,yy, 'b',t,y, ':'); xlabel( 'Samples'); ylabel( 'Amplitude');
legend( 'Original data', 'Modified data'); title( 'Example for eliminate
   outliers');
% Function pand
function[u,delta,i] =pand(y,i,L)
u =y(i-1) +0.5 *y(i-2)-0.5 *y(i-4); delta =3 *std(y(i-4:i-1));
while (abs(u-y(i)) ≤delta) & (i <L);
    i =i +1; u =y(i-1) +0.5 *y(i-2)-0.5 *y(i-4); delta =3 *std(y(i-4:i-1));
end
% Function pand1
function[u,delta,i] =pand1(y,i,L)
u =y(i-1) +0.5 *y(i-2)-0.5 *y(i-4); delta =3 *std(y(i-4:i-1));
while (abs(u-y(i)) >delta) &(i ≤L);
   i =i +1;
```

```
if y(i) = =y(i-1)
    i =i +1;
   else
    u =y(i-1) +0.5 *y(i-2)-0.5 *y(i-4); delta =3 *std(y(i-4:i-1));
   end
end
% Function guji
function y =guji(k,j,y,w)
for i =1:4
   a(i,1) =1; a(i,2) =k +5-i; a(i,3) =(k +5-i) ^2; l(i,1) =y(j +k +5-i);
end
for i =5:8
   a(i,1) =1; a(i,2) =5-i; a(i,3) =(5-i) ^2; l(i,1) =y(j +5-i);
end
for i =1:k
   b(i,1) =1; b(i,2) =k +1-i; b(i,3) =(k +1-i) ^2;
end
h =b *(a '*a) ^(-1) *a' *l;
for i =1:k
   if abs(y(j +k +1-i)-h(i)) >w
      y(j +k +1-i) =h(i);
   end
end
```

Figure 8.6 demonstrates the curves before and after eliminating the outliers.

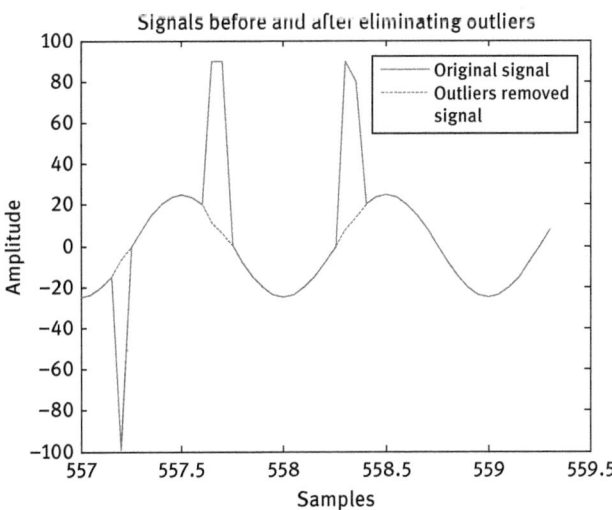

Figure 8.6: A comparison between the original data and the modified data.

8.7 Data Processing Examples of Temperature Measurements

This section introduces applications of the error theory and data processing in practical engineering using examples on temperature measurement and data processing.

8.7.1 Temperature and Its Measurement

Temperature is a commonly used basic physical quantity. It represents the degree of hotness and coldness macroscopically, and it is the average intensity which indicates the massive molecular heat movement from the microscopic sense.

Temperature can only be measured indirectly through the object with some characteristics. The temperature scale is used to represent the value of temperature. The commonly used temperature scales include the international thermodynamic temperature scale (K), Fahrenheit temperature scale (°F), and the centigrade temperature scale (°C).

The temperature measurement is a process in which the temperature of the object is measured by an instrument. In general, the temperature measuring instrument can be divided into two types: contact type and noncontact type. For the contact-type instrument, the sensor of the instrument is directly contacted with the measured medium, which has the characteristics of simple, reliable, and high precision. Unlike contact type, the sensor in the noncontact temperature measurement does not need to be directly contacted with the measured medium, so the temperature of the moving object can be measured. The optical pyrometer, radiation thermometer, and colorimetric thermometer are all noncontact type of temperature measurement instruments, which measure the heat radiation of objects. Generally speaking, a temperature measuring instrument has two parts: the detection part and the display (or recording) part.

8.7.2 An Engineering Example of Temperature Measurement with Platinum Resistance

A PT1000 platinum resistor is selected as the temperature sensor. The basic principle is that the resistance of the platinum resistor changes with the change in temperature. The PT1000 platinum resistor has many advantages, including good long-term stability, wide measurement range, high measurement accuracy, and high linearity. Its temperature measurement accuracy can reach 0.01°C.

Figure 8.7 shows the basic principle of the temperature measurement with PT1000, in which the amplifier is used to amplify the signal obtained from the sensor, the filter is used to suppress the noise in the signal, the A/D converter is used to convert the analog signal into the digital signal, and the computer is used to process the digital signal and then send the result to the display.

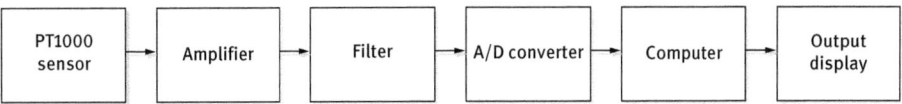

Figure 8.7: The basic principle of the temperature measurement with PT1000.

8.7.3 Data Analysis and Processing of the Temperature Measurement

The data obtained from a high-precision temperature measurement are shown in Table 8.2.

Table 8.2: The measured temperature data and corresponding resistances and voltages.

Temperature/°C	0.00	10.00	25.00	50.00	65.00	70.00	85.00	100.00
Resistance/Ω	1,000	1,039.025	1,097.347	1,193.971	1,251.6	1,270.751	1,328.033	1,385.055
Voltage/V	0.2541	0.2612	0.2759	0.3002	0.3147	0.3195	0.3339	0.3482

Let the voltage $V = x$ and the temperature $C = y$. The relationship between x and y is $y = \beta_0 + \beta_1 x$. The estimates of β_0 and β_1 using the LS method are given as

$$\hat{\beta}_1 = \frac{l_{xy}}{l_{xx}}, \qquad \hat{\beta}_0 = \bar{y} - \hat{\beta}_1 \bar{x} \qquad (8.112)$$

where

$$\bar{x} = \frac{1}{N}\sum_{i=1}^{N} x_i, \qquad \bar{y} = \frac{1}{N}\sum_{i=1}^{N} y_i$$

$$l_{xx} = \sum_{i=1}^{N} (x_i - \bar{x})^2, \qquad l_{xy} = \sum_{i=1}^{N} (x_i - \bar{x})(y_i - \bar{y})$$

Substituting the data in Table 8.2 into eq. (8.112), we obtain

$$\bar{x} = 0.300625, \quad \bar{y} = 50.625, \quad l_{xx} = 0.008574, \quad l_{xy} = 8.7885$$

Furthermore, we have

$$\hat{\beta}_1 = 1,025.02, \qquad \hat{\beta}_0 = -257.521$$

The linear regression equation is obtained as follows:

$$y = 1,025.02x - 257.521 \tag{8.113}$$

Solving for the correlation coefficient $r = \dfrac{l_{xy}}{\sqrt{l_{xx}l_{yy}}}$, where $l_{yy} = \sum_{i=1}^{N}(y_i - \bar{y})^2 = 29,254.6387$, we obtain

$$r = 0.996 \tag{8.114}$$

Obviously, the correlation coefficient between y and x is very close to 1, resulting in a very good linearity. Figure 8.8 shows the relations between the output voltages and the temperature values. In the figure, the solid circles are the measured data, and the line represents the regressed linear function. The error analysis can also be carried out based on the results.

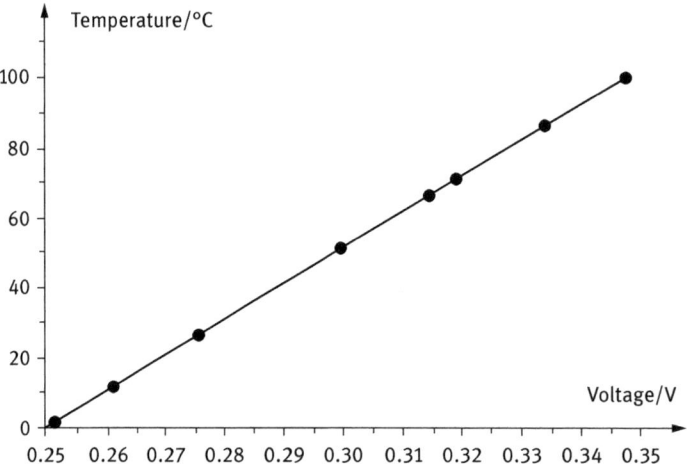

Figure 8.8: The relation between output voltages and the temperature values.

Exercises

8.1 Describe the concept of the error.

8.2 What is the absolute error? What is the relative error? What is the quoted error?

8.3 Describe the concept of the random error.

8.4 Describe the concept of the systematic error.

8.5 Describe the concepts of the arithmetic average and the standard deviation.

8.6 What is the gross error? How to determine the gross error?

8.7 Describe the methods for eliminating a gross error.

8.8 Describe the combination method of the random errors.

8.9 Describe the combination method of the systematic errors.

8.10 What is the fixed systematic error? What is the uncertainty systematic error?

8.11 Describe the basic method for error distribution.

8.12 Describe the concept of the measurement uncertainty.

8.13 How to assess the standard uncertainty?

8.14 Describe the combination of the measurement uncertainty?

8.15 Describe the basic principle of the least squares method.

8.16 Try to understand the establishment of the normal equation.

8.17 How to analyze the precision of the LS processing?

8.18 Describe the method of the univariate linear regression.

8.19 Describe the method of the univariate nonlinear regression.

8.20 Describe the concepts of the outlier and the trend in a signal. How can we remove such outlier and trend?

8.21 Ten measurement data for a length are given as: 802.40, 802.50, 802.38, 802.48, 802.42, 802.46, 802.39, 802.47, 802.43, and 802.44 (in mm). Calculate its arithmetic average and standard deviation.

8.22 Fifteen measured data for a value are given as: 20.42, 20.43, 20.40, 20.43, 20.42, 20.43, 20.39, 20.30, 20.40, 20.43, 20.42, 20.41, 20.39, 20.39, and 20.40. Suppose the systematic error has already been removed. Determine whether the gross error is included in the measurement.

8.23 The power rating of an electrical product is estimated between 56 W and 64 W, with a probability of 50%. Determine its standard uncertainty.

8.24 The relation between the measured value and the temperature is linear as $y = y_0(1 + \alpha t)$. The measured data at different temperatures are listed in the following table. Try to determine the most reliable values of the length y_0 and the coefficient α at 0°C.

i	1	2	3	4	5	6
t_i /°C	10	20	25	30	40	45
l_i /mm	2,000.36	2,000.72	2,000.80	2,001.07	2,001.48	2,001.60

8.25 Find the measurement precision in the last question.

8.26 The measurement values for a resistor at different temperatures are listed as follows:

i	1	2	3	4	5	6	7
x_i /°C	19.1	25.0	30.1	36.0	40.0	46.5	50.0
y_i /Ω	76.30	77.80	79.75	80.80	82.35	83.90	85.10

Please determine the relation between the resistor and the temperature.

8.27 Find the regression equation for the last question.

9 Fundamentals of Random Signal Processing

9.1 Introduction

Random signal processing or statistical signal processing is an important part of signal processing. The random signal processing mentioned here refers to the processing for random signals. Such signals cannot be described by a mathematical function. The statistical signal processing mentioned here means that the signal processing method is based on the theory of probability and statistics. Therefore, the concepts for both random signal processing and statistical signal processing are consistent.

Different from the deterministic signal, the random signal can neither be described by a deterministic function nor be accurately reproduced. Signals we encounter in practice, such as speech, music, seismic, radar and sonar signals, measured temperature signals, detected biomedical signals, and mobile communication signals are all random signals. These signals have to be described and studied using statistical methods. Therefore, the statistics of random signals plays a very important role in the analysis of random signals. The most commonly used statistics are the mean (the first-order statistics), the correlation function, and the power spectrum density (the second-order statistics). In addition, higher-order statistics, such as the third- or fourth-order moments, and the higher-order cumulants are also widely used in signal processing. The theory of fractional lower-order statistics and its signal processing methods developed in recent years have also been broadly concerned and studied.

On the other hand, a signal is inevitable to be affected by noise and interference in the signal detection, transformation, transmission, and processing. These noises and interferences are often random and can be referred to as random noises. One task of random signal processing is to extract or recover the pure signal from noises and interferences according to the statistical characteristics of signals and noises or to estimate parameters of the signals.

This chapter introduces the basic concepts and principles for random signals, and it also introduces both classical and modern random signal processing methods.

9.2 The Concept of Random Variable and Its Properties

9.2.1 The Concept of Random Variable

The experiments or observations under certain conditions may give different results, that is, there are more than one possible test results. Generally, when no one can predict what will be the outcome before the test, such a phenomenon is called the

https://doi.org/10.1515/9783110465082-009

random phenomenon. In a randomized trial, some event may or may not occur. However, such event has some regularity in a large number of repeated trials, which is called a random event.

Random variables are the variables that represent the various outcomes of random phenomena. For example, flipping a coin at random; there may be either a head or a tail showing. If X is defined as the number of times of head, then X is a random variable. The value of X is 1 when it is a head showing. Otherwise, the value of X is 0. Another example is for flipping a dice; all the possible outcomes are 1, 2, 3, 4, 5, and 6 points. If X is defined as the number of points that appears, then X is a random variable. In addition, the number of passengers waiting for a bus at a certain period of time and the number of calls received in a certain period of time are all examples of random variables.

If the value of the variable X is determined based on the outcome of the random test, then X is called a random variable. In fact, the random variable is a variable that depends on the random test. Strictly speaking, if E is a random test, then its sample space is $S = \{e_i\}$. If for each $e_i \in S$, there is a corresponding real number $X(e_i)$, then a real single-valued function $X = X(e)$ defined on S is obtained, where $X(e)$ (abbreviated as X) is known as the random variable. In general, a random variable is represented by a capital letter, such as X, Y, and Z. And the corresponding possible value of a random variable is represented by a lower-case letter, such x, y, and z. This book also uses this notation. If the value of the random variable X is continuous, it is called a continuous random variable; if all the number of possible values is countable or infinite countable, it is called the discrete random variable. In addition, there is a mixture of random variables.

In some practical problems, the results of some randomized trials may be described by two or more random variables. For example, if the random variable X can be used to express the voltage amplitude of a random signal, then X is a one-dimensional random variable. If we need to describe the amplitude and phase of a random signal, we have to use two random variables, such as X and Y. Both X and Y construct a two-dimensional random vector (X, Y), also referred to as the two-dimensional random variable. For more complex randomized trials, it may be possible to use more random variables (i.e., multivariate random variables) to describe.

9.2.2 The Distribution of Random Variables

For random variable X, the probability distribution function, the probability density function (PDF), and the numerical characteristics are usually used to describe the properties of the random variable.

1. Probability distribution function
The probability distribution function or the cumulative distribution function is defined as

$$F(x) = P(X \le x) \tag{9.1}$$

where $F(x)$ is the probability distribution function, and $P(\cdot)$ represents the probability. The concept of probability distribution function is not only suitable for continuous random variables, but also suitable for discrete random variables. Its main properties can be described as follows:

(1) $F(x)$ is a monotonically nondecreasing function. That is, $F(x_2) \ge F(x_1)$ if $x_2 > x_1$.
(2) $F(x)$ is nonnegative, satisfying $0 \le F(x) \le 1$. At the two ends of $(-\infty, \infty)$, we have $F(-\infty) = 0$ and $F(\infty) = 1$.
(3) The probability of the random variable X in the range of (x_1, x_2) is $P(x_1 < X \le x_2) = F(x_2) - F(x_1)$.
(4) $F(x)$ is right continuous, satisfying $F(x+0) = F(x)$. For the discrete random variables, the probability may have the step form. The step height is equal to the probability of the random variable at the point, that is, $F(x) = \sum_{i=1}^{+\infty} P(X = x_i)u(x - x_i) = \sum_{i=1}^{+\infty} P_i u(x - x_i)$, where $u(x)$ is the unit step function, and P_i is the probability when $X = x_i$.

2. Probability density function

The PDF of the random variable X is defined as the derivative of the probability distribution function $F(x)$ respect to x:

$$f(x) = \frac{dF(x)}{dx} \tag{9.2}$$

The main properties of the PDF are listed as follows:
(1) $f(x)$ is nonnegative, that is, $f(x) \ge 0$ for all x.
(2) The integration of $f(x)$ in $(-\infty, +\infty)$ is 1, that is, $\int_{-\infty}^{+\infty} f(x)dx = 1$.
(3) The integration of $f(x)$ in $(x_1, x_2]$ is the probability in the interval, that is,

$$P(x_1 < X \le x_2) = \int_{x_1}^{x_2} f(x)dx \tag{9.3}$$

(4) The probability of the discrete random variable is given as

$$f(x) = \sum_{i=1}^{+\infty} P(X = x_i)\delta(x - x_i) = \sum_{i=1}^{+\infty} P_i \delta(x - x_i) \tag{9.4}$$

where $\delta(x)$ is the unit impulse function.

Some examples of probability distribution functions and PDFs are shown in Figure 9.1.

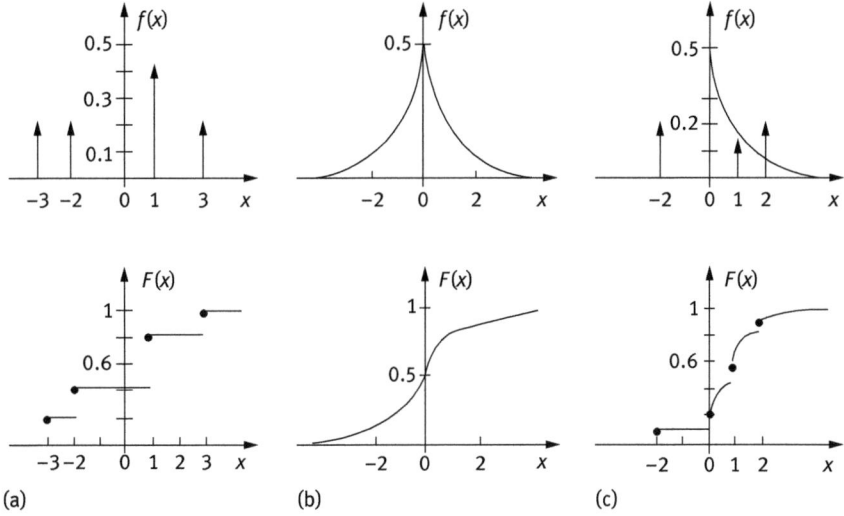

Figure 9.1: Some examples of probability distribution functions and PDFs: (a) discrete random variable; (b) continuous random variable; and (c) mixed random variable.

3. Multidimensional random variable and its distribution

Two-dimensional random variable (X, Y) can be considered as a random point on a two-dimensional plane, and the n-dimension random variable is a random point in the n-dimensional space. The nature of the n-dimensional random variable is not only related to $X_1, X_2, \ldots, X_n$, but also related to the relations among the n random variables. The joint distribution function of the two-dimensional random variable (X, Y) is defined as

$$F(x, y) = P((X \le x) \cap (Y \le y)) = P(X \le x, \ Y \le y) \tag{9.5}$$

where x, y are arbitrary real values. The joint PDF of the two-dimensional random variable (X, Y) is defined as

$$f(x, y) = \frac{\partial^2 F(x, y)}{\partial x \partial y} \tag{9.6}$$

The main properties of the two-dimensional joint PDF are as follows:
(1) The two-dimensional joint PDF is a nonnegative function, that is, $f(x, y) \ge 0$.
(2) The integration of the two-dimensional joint PDF in the whole interval is 1, that is, $\int_{-\infty}^{+\infty} \int_{-\infty}^{+\infty} f(x, y)\mathrm{d}x\mathrm{d}y = 1$.
(3) The integration of the two-dimensional joint PDF in an interval gives the value of the probability in that interval, that is,

$$P(x_1 < X \le x_2, \; y_1 < X \le y_2) = \int_{x_1}^{x_2} \int_{y_1}^{y_2} f(x,y)\mathrm{d}x\mathrm{d}y \qquad (9.7)$$

(4) The integration of the two-dimensional joint PDF in the whole interval of one random variable gives the PDF of the other random variable, that is, $f_X(x) = \int_{-\infty}^{+\infty} f(x,y)\mathrm{d}y$ or $f_Y(y) = \int_{-\infty}^{+\infty} f(x,y)\mathrm{d}x$. Both $f_X(x)$ and $f_Y(y)$ are referred to as the marginal PDFs. The corresponding $F_X(x)$ and $F_Y(y)$ are called marginal probability distribution functions.

Under the condition of $X \le x$, the conditional probability distribution function and conditional PDF of the random variable Y are expressed as $F_Y(y|x) = \frac{F(x,y)}{F_X(x)}$ and $f_Y(y|x) = \frac{f(x,y)}{f_X(x)}$, respectively. The statistically independent conditions of both random variables X and Y are $f_X(x|y) = f_X(x)$ and $f_Y(y|x) = f_Y(y)$. The sufficient and necessary condition of the statistical independence for both random variables X and Y is that the two-dimensional joint PDF equals to the product of both marginal PDFs, that is,

$$f(x,y) = f_X(x)f_Y(y) \qquad (9.8)$$

The probability distribution function and the PDF of the n-dimensional random variable are defined as

$$F(x_1, x_2, \ldots, x_n) = P(X_1 \le x_1, \; X_2 \le x_2, \; \ldots, \; X_n \le x_n) \qquad (9.9)$$

$$f(x_1, x_2, \ldots, x_n) = \frac{\partial^n F(x_1, x_2, \ldots, x_n)}{\partial x_1 \partial x_2 \ldots \partial x_n} \qquad (9.10)$$

The sufficient and necessary condition of the statistical independence for the n-dimensional random variable is that all $x_1, x_2, \ldots, x_n$ satisfy

$$f(x_1, x_2, \ldots, x_n) = f_{X_1}(x_1)f_{X_2}(x_2)\cdots f_{X_n}(x_n) = \prod_{i=1}^{n} f_{X_i}(x_i) \qquad (9.11)$$

9.2.3 Numerical Characteristics of Random Variables

The probability distribution function and the PDF can completely describe the statistical characteristics of random variables. However, in many practical problems, it is not necessary to find out such probability functions by a large number of tests. We may only need to know some of main statistical features of random variables. These features are called numerical characteristics of random variables, including the mathematical expectation, variance, correlation coefficient, and moments, which are mainly used to describe the statistical characteristics of random variables.

1. Mathematical expectation

The mathematical expectation, also known as the statistical mean or mean, is used to describe the ensemble characteristics of random variables. For continuous random variable X, the mathematical expectation is defined as

$$E[X] = \int_{-\infty}^{+\infty} xf(x)dx \tag{9.12}$$

For discrete random variable X, the mathematical expectation is

$$E[X] = \sum_{i=1}^{+\infty} x_i P(X = x_i) = \sum_{i=1}^{+\infty} x_i P_i \tag{9.13}$$

The mathematical expectation of random variable X are usually denoted by μ_X or m_X. The mathematical expectation has following main properties:

(1) The mathematical expectation of constant c is itself, that is $E[c] = \int_{-\infty}^{+\infty} cf(x)dx = c \int_{-\infty}^{+\infty} f(x)dx = c$.

(2) The mathematical expectation of a linear combination of random variables is equal to the linear combination of the mathematical expectations of the various random variables, that is, $E[\sum_{i=1}^{n} a_i X_i] = \sum_{i=1}^{n} a_i E[X_i]$.

(3) If X is a nonnegative random variable, then $E[X] \geq 0$.

(4) If and only if $P(X = 0) = 1$, then $E[X^2] = 0$.

(5) For random variable X, $|E[X]| \leq E[|X|]$ holds.

(6) If $X_1, X_2, \ldots, X_n$ are independent random variables, then $E[X_1, X_2, \ldots, X_n] = \prod_{i=1}^{n} E[X_i]$.

2. Variance

Suppose X is a random variable. If $E[(X - E[X])^2]$ holds, then $E[(X - E[X])^2]$ is referred to as the variance of X, denoted as $\mathrm{Var}[X]$.

$$\mathrm{Var}[X] = E[(X - E[X])^2] \tag{9.14}$$

The variance of random variable X is used to measure the deviation degree between the random variable and its mathematical expectation. Its calculating equation is

$$\mathrm{Var}[X] = E[X^2] - (E[X])^2 \tag{9.15}$$

In practice, we usually use $\sqrt{\mathrm{Var}[X]}$ as the standard deviation, denoted as σ_X. The main properties of the variance are listed as follows:

(1) The variance of a constant c is zero, that is, $\mathrm{Var}[c] = 0$.

(2) If X is a random variable, and c is a constant, then $\mathrm{Var}[cX] = c^2\mathrm{Var}[X]$.

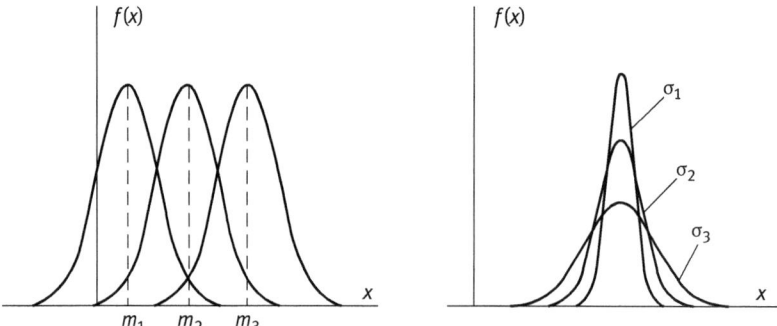

Figure 9.2: The PDF curves for different mathematical expectations and variances.

(3)   If X and Y are two independent random variables, then $\text{Var}[X+Y] = \text{Var}[X] + \text{Var}[Y]$. This property can be extended to multiple independent variables.

(4)   The sufficient and necessary condition of $\text{Var}[X] = 0$ is that X takes the value of constant c with the probability of 1.

The mathematical expectation and the variance are both important numerical characteristics of random variables. Figure 9.2 demonstrates the PDF curves for different mathematical expectations and variances.

3. Covariance and correlation coefficient

The covariance of both random variables X and Y is defined as

$$\text{Cov}[X, Y] = E[(X - E[X])(Y - E[Y])] \tag{9.16}$$

The correlation coefficient of X and Y is defined as

$$\rho_{XY} = \frac{\text{Cov}[X, Y]}{\sqrt{\text{Var}[X]}\sqrt{\text{Var}[Y]}} \tag{9.17}$$

If $X = Y$ is satisfied, then $\rho_{XY} = 1$. If X and Y are statistically independent, then $\rho_{XY} = 0$, which is referred to as uncorrelated.

The main properties of the covariance $\text{Cov}[X, Y]$ are as follows:

(1)   Symmetry property, that is, $\text{Cov}[X, Y] = \text{Cov}[Y, X]$.

(2)   If a and b are constants, then $\text{Cov}[aX, bY] = ab\text{Cov}[X, Y]$.

(3)   The covariance satisfies $\text{Cov}[X_1 + X_2, Y] = \text{Cov}[X_1, Y] + \text{Cov}[X_2, Y]$.

The correlation coefficient satisfies $|\rho_{XY}| \leq 1$. The sufficient and necessary condition for $|\rho_{XY}| = 1$ is $P(Y = a + bX) = 1$ if constants a and b are existent.

4. Moment and covariance matrix

Set X and Y as random variables. If $E[X^k]$, $k = 1, 2, \ldots$, exist, then they are called the kth-order origin moments of X, or abbreviated as the kth moments. If $E[(X - E[X])^k]$, $k = 1, 2, \ldots$, exist, then they are referred to as the kth-order central moments of X. If $E[X^k Y^l]$, $k, l = 1, 2, \ldots$, exist, then they are called the $k + l$th-order mixed moments of X and Y. If $E[(X - E[X])^k (Y - E[Y])^l]$, $k, l = 1, 2, \ldots$, exist, then they are called the $k + l$th-order mixed central moments of X and Y. Obviously, mathematical expectation $E[X]$ is the first-order origin moment of X, and the variance $\text{Var}[X]$ is the second-order central moment of X.

The covariance matrix of the n-dimensional random variable $(X_1, X_2, \ldots, X_n)$ is given as

$$
C = \begin{bmatrix}
c_{11} & c_{12} & \cdots & c_{1n} \\
c_{21} & c_{22} & \cdots & c_{2n} \\
\vdots & \vdots & \vdots & \vdots \\
c_{n1} & c_{n2} & \cdots & c_{nn}
\end{bmatrix}
\tag{9.18}
$$

where $c_{ij} = \text{Cov}[X_i, X_j] = E[(X_i - E[X_i])(X_j - E[X_j])]$, $i, j = 1, 2, \ldots, n$. (It is supposed that both moments exist.) Since $c_{ij} = c_{ji}$, the covariance matrix C is symmetric.

9.2.4 Characteristic Functions of Random Variables

1. Definition of the characteristic function

Suppose X is a continuous random variable whose PDF is $f(x)$. The characteristic function of X is defined as

$$
\Phi_X(u) = E[e^{jux}] = \int_{-\infty}^{+\infty} f(x) \exp[jux] dx
\tag{9.19}
$$

If X is a discrete random variable, then its characteristic function is defined as

$$
\Phi_X(u) = \sum_{i=0}^{+\infty} P_i \exp[jux_i]
\tag{9.20}
$$

The second characteristic function of X is defined as

$$
\Psi_X(u) = \ln[\Phi_X(u)]
\tag{9.21}
$$

2. The relationship between the characteristic function and PDF as well as moment function

We see from eqs. (9.19) and (9.20) that the characteristic function $\Phi_X(u)$ of X is a mathematical transformation of its PDF $f(x)$ similar to the Fourier transform. On the other hand, the characteristic function is corresponding to the moment function one by one. Therefore, the characteristic function is also referred to as the moment generating function, satisfying the following relations:

$$E[X] = \int_{-\infty}^{+\infty} xf(x)dx = -j\frac{d\Phi_X(u)}{du}\bigg|_{u=0} \tag{9.23}$$

$$E[X^n] = \int_{-\infty}^{+\infty} x^n f(x)dx = (-j)^n \frac{d^n\Phi_X(u)}{du^n}\bigg|_{u=0} \tag{9.24}$$

9.3 Random Processes and Random Signals

9.3.1 The Statistical Distribution of Random Processes and Random Signals

1. Random process and random signal
The value of a random variable can be used to indicate the possible result of the random test. In many cases, these random variables will change with some parameters, or they are the function of some parameters. Such kind of random variables is usually called random function. In mathematics, it is known as the stochastic process or random process.

In engineering area, we usually use the concept of stochastic signal or random signal. The so-called random signal is the signal in which at least one parameter (such as amplitude) is a random function. Thermal noise in the instrument, for example, is a typical random signal.

Definition 9.1: Suppose the sample space of a random test is $S = \{e_i\}$. Each sample $e_i \in S$ in the space is always corresponding to a time function $X(t, e_i)$, $t \in T$. Thus for all of the samples $e \in S$ in the space S, there is a family of functions $X(t, e)$ corresponding to them. The family of time functions is defined as the random process.

The random process is a family of time functions; each sample function in the random process is a deterministic time function $x(t)$, while the whole family of time functions is a random process, denoted as $X(t)$. On the other hand, at a particular time t_1, the random process is a random variable $X(t_1)$. This shows that the random process and random variable are both different and closely connected. In this book, we use upper case letters $X(t)$, $Y(t)$ (or $\{x(t)\}$, $\{y(t)\}$) to represent random processes, and we use lower case letters $x(t)$ and $y(t)$ to represent the sample function of a random process. We also use $x(t)$ and $y(t)$ to represent random signals under the premise without causing a confusion. Figure 9.3 shows an example of the random signal.

As shown in Figure 9.3, if we take the observation on the temperature drift of a transistor amplifier as a random test, then we can get a sample function $x_i(t)$ for each trail. The assemblage of all sample functions $x_i(t)$, $i = 1, 2, \ldots, N$, construct the

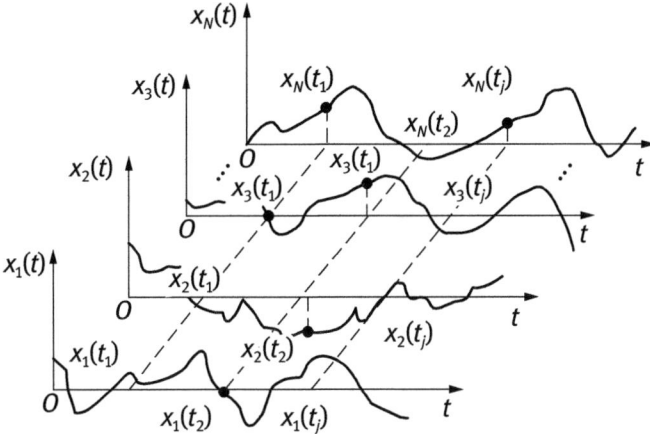

Figure 9.3: An example of random signal – the temperature drift of a transistor amplifier.

whole process of the temperature drift when $N \rightarrow \infty$, that is, the random process $X(t)$. On the other hand, sample values $x_1(t_1)$, $x_2(t_1)$, $\ldots$, $x_N(t_1)$ at a given moment $t = t_1$ constitute a random variable, which is equivalent to measuring the output value of N same amplifiers at the same time.

2. Probability distribution and PDF of random process or random signal
One-dimensional distribution function and PDF of a random process $X(t)$ are defined as

$$F_X(x_1, t_1) = P[X(t_1) \leq x_1] \tag{9.24}$$

$$f_X(x_1, t_1) = \frac{\partial F_X(x_1, t_1)}{\partial x_1} \tag{9.25}$$

Similarly, the probability distribution and PDF of the n th-dimensional random process can be defined as

$$F_X(x_1, x_2, \ldots, x_n; t_1, t_2, \ldots, t_n) = P[X(t_1) \leq x_1, X(t_2) \leq x_2, \ldots, X(t_n) \leq x_n] \tag{9.26}$$

$$f_X(x_1, x_2, \ldots, x_n; t_1, t_2, \ldots, t_n) = \frac{\partial^n F_X(x_1, x_2, \ldots, x_n; t_1, t_2, \ldots, t_n)}{\partial x_1 \partial x_2 \ldots \partial x_n} \tag{9.27}$$

3. The numerical characteristics of random process or random signal
Although the distribution function can completely describe the statistical properties of the random process, it is often difficult to determine the distribution function based on the information obtained in the practical application. Therefore, the concept of the numerical characteristics of the random process is introduced, including

the mathematical expectation, the variance, and the correlation function of the random process.

The mathematical expectation of a random process is the statistical average of the random process at the moment t, defined as a deterministic time function

$$\mu_X(t) = E[X(t)] = \int_{-\infty}^{+\infty} x f_X(x, t) dx \tag{9.28}$$

The variance of the random process is defined as the degree of dispersion of all the sample functions with respect to the mathematical expectation $\mu_X(t)$ as

$$\sigma_X^2(t) = D[X(t)] = \int_{-\infty}^{+\infty} [x - \mu_X(t)]^2 f_X(x, t) dx \tag{9.29}$$

where $\sigma_X(t)$ is the standard deviation of the random process. For two arbitrary moments t_1 and t_2, the autocorrelation function of the real random process is defined as

$$R_{XX}(t_1, t_2) = E[X(t_1)X(t_2)] = \int_{-\infty}^{+\infty} \int_{-\infty}^{+\infty} x_1 x_2 f_X(x_1, x_2; t_1, t_2) dx_1 dx_2 \tag{9.30}$$

The cross-correlation function between $X(t)$ and $Y(t)$ is defined as

$$R_{XY}(t_1, t_2) = E[X(t_1)Y(t_2)] = \int_{-\infty}^{+\infty} \int_{-\infty}^{+\infty} xy f_{XY}(x, y; t_1, t_2) dx dy \tag{9.31}$$

The concept of correlation represents the association degree of the random process between two different moments. Similarly, the autocovariance function and the cross-covariance function of the random process are defined as

$$\begin{aligned} C_{XX}(t_1, t_2) &= E\{[X(t_1) - \mu_X(t_1)][X(t_2) - \mu_X(t_2)]\} \\ &= \int_{-\infty}^{+\infty} \int_{-\infty}^{+\infty} [x_1 - \mu_X(t_1)][x_2 - \mu_X(t_2)] f_X(x_1, x_2; t_1, t_2) dx_1 dx_2 \end{aligned} \tag{9.32}$$

$$\begin{aligned} C_{XY}(t_1, t_2) &= E\{[X(t_1) - \mu_X(t_1)][Y(t_2) - \mu_Y(t_2)]\} \\ &= \int_{-\infty}^{+\infty} \int_{-\infty}^{+\infty} [x - \mu_X(t_1)][y - \mu_Y(t_2)] f_{XY}(x, y; t_1, t_2) dx dy \end{aligned} \tag{9.33}$$

We can see that the correlation function and the covariance function meet the following relationship:

$$C_{XX}(t_1, t_2) = R_{XX}(t_1, t_2) - \mu_X(t_1)\mu_X(t_2) \tag{9.34}$$

$$C_{XY}(t_1, t_2) = R_{XY}(t_1, t_2) - \mu_X(t_1)\mu_Y(t_2) \tag{9.35}$$

For the cross-correlation function and the cross-covariance function, if $R_{XY}(t_1, t_2) = 0$ holds for any two moments t_1 and t_2, then $X(t)$ and $Y(t)$ are called the orthogonal processes. If $C_{XY}(t_1, t_2) = 0$ holds for any two moments t_1 and t_2, then $X(t)$ and $Y(t)$ are called uncorrelated. If $X(t)$ and $Y(t)$ are independent, then they both are uncorrelated.

9.3.2 The Stationary Random Process

Generally, random signals can be divided into stationary random signals and non-stationary random signals according to their statistical characteristics. Furthermore, the stationary random signal can be divided into the strict-sense stationary random signal and the wide-sense stationary random signal.

The stationary random signal is a kind of very important random signal. In practical applications, many random signals are stationary or nearly stationary. Because the analysis and processing of stationary random signals are much simpler than those of the general random signals, it is possible to take various random signals as approximately stationary both in theory and practice.

Definition 9.2 (strict-sense stationary): If the n-dimensional distribution function of the random process (or random signal) $X(t)$ does not change with time for arbitrary n moments $t_1, t_2, \ldots, t_n$ and an arbitrary real number τ, that is, satisfying

$$F_X(x_1, x_2, \ldots, x_n; t_1, t_2, \ldots, t_n) = F_X(x_1, x_2, \ldots, x_n; t_1 + \tau, t_2 + \tau, \ldots, t_n + \tau) \quad (9.36)$$

then $X(t)$ is called the strict-sense stationary random process (or random signal).

If arbitrary $n + m$-dimensional joint probability distribution of the two random signals $X(t)$ and $Y(t)$ does not change with time, that is, satisfying

$$F_{XY}(x_1, x_2, \ldots, x_n, y_1 y_2 \ldots y_m; t_1, t_2, \ldots, t_n, t'_1, t'_2, \ldots, t'_m)$$
$$= F_{XY}(x_1, x_2, \ldots, x_n, y_1 y_2 \ldots y_m; t_1 + \tau, t_2 + \tau, \ldots, t_n + \tau, t'_1 + \tau, t'_2 + \tau, \ldots, t'_n + \tau)$$
$$(9.37)$$

then random signals $X(t)$ and $Y(t)$ are referred to as the joint stationary random signals.

Definition 9.3 (wide-sense stationary): If the random process (or random signal) $X(t)$ satisfies the following conditions

$$E[X(t)] = \mu_X(t) = \mu_X$$
$$E[X^2(t)] < \infty \quad (9.38)$$
$$R_{XX}(\tau) = E[X(t)X(t + \tau)] = E[X(t + t_1)X(t + t_1 + \tau)]$$

then $X(t)$ is called the wide-sense stationary random process (or random signal).

Because the definition of the strict-sense stationary is too strict, we usually use the concept of the wide-sense stationary in practice. In the following parts of this book, if it is not specified, the wide-sense stationary is generally concerned. The second-order stationary signal must be a wide-sense stationary signal as long as the mean square value is finite. But the opposite is not necessarily the case.

Suppose both $X(t)$ and $Y(t)$ are wide-sense stationary signals. If their cross-correlation function satisfies

$$R_{XY}(\tau) = E[X(t)Y(t+\tau)] \tag{9.39}$$

then $X(t)$ and $Y(t)$ are jointly wide-sense stationary.

The discrete-time random signal $X(n)$ has following properties:

$$E[X(n)] = \mu_X(n) = \mu_X \tag{9.40}$$

$$E[X(n)X(n+m)] = R_{XX}(n_1, n_2) = R_{XX}(m), \quad m = n_2 - n_1 \tag{9.41}$$

$$E[(X(n) - \mu_X)^2] = \sigma_X^2(n) = \sigma_X^2 \tag{9.42}$$

$$E[(X(n) - \mu_X)(X(n+m) - \mu_X)] = C_{XX}(n_1, n_2) = C_{XX}(m) \tag{9.43}$$

If both $X(n)$ and $Y(n)$ are wide-sense stationary signals, then

$$E[X(n)Y(n+m)] = R_{XY}(m) \tag{9.44}$$

$$E[(X(n) - \mu_X)(Y(n+m) - \mu_Y)] = C_{XY}(m) \tag{9.45}$$

9.3.3 Ergodicity

For stationary random signal $X(t)$ or $X(n)$, if the first- and second-order statistics of all its sample functions in a fixed moment are consistent with the statistics of the single sample function at a long time, then $X(t)$ or $X(n)$ is referred to as the ergodic signal. Suppose $x(t)$ is a sample function of an ergodic signal $X(t)$, then

$$E[X(t)] = \overline{x(t)} = \mu_X = \lim_{T \to +\infty} \frac{1}{2T} \int_{-T}^{T} x(t)dt = \mu_x \tag{9.46}$$

$$E[X(t)X(t+\tau)] = R_{XX}(\tau) = \lim_{T \to +\infty} \frac{1}{2T} \int_{-T}^{T} x(t)x(t+\tau)dt = R_{xx}(\tau) \tag{9.47}$$

where $E[\cdot]$ represents the assemble average, and $\overline{x(t)}$ denotes the time average of $x(t)$. Similarly, we have

$$E[X(n)] = \overline{x(n)} = \mu_X = \lim_{M \to +\infty} \frac{1}{2M+1} \sum_{n=-M}^{M} x(n) = \mu_x \qquad (9.48)$$

$$E[X(n)X(n+m)] = R_{XX}(m) = \lim_{M \to +\infty} \frac{1}{2M+1} \sum_{n=-M}^{M} x(n)x(n+m) = R_{xx}(m) \qquad (9.49)$$

9.3.4 Power Spectral Density of Random Signals

The Fourier transform is a powerful tool for the spectrum analysis of deterministic signals. However, the Dirichlet conditions need to be satisfied. For a random signal, since its energy is generally infinite, it does not meet such conditions. On the other hand, the average power of a random signal is finite, and we can still use Fourier transform to analyze random signals.

The power spectral density (PSD) function of the random process is defined as

$$S_{XX}(\Omega) = E\left[\lim_{T \to +\infty} \frac{1}{2T} |X_T(j\Omega)|^2 \right] = \lim_{T \to +\infty} \frac{1}{2T} E\left[|X_T(j\Omega)|^2 \right] \qquad (9.50)$$

where $X_T(j\Omega)$ represents Fourier transform of $X_T(t)$, which is the truncation of the random signal $X(t)$. From Wiener–Khintchine theorem, the PSD function of a random signal is the Fourier transform of its autocorrelation function, shown as

$$S_{XX}(\Omega) = \int_{-\infty}^{+\infty} R_{XX}(\tau) e^{-j\Omega\tau} d\tau \qquad (9.51)$$

The cross-power spectral density (cross-PSD) function of $X(t)$ and $Y(t)$ is defined as

$$S_{XY}(\Omega) = \int_{-\infty}^{+\infty} R_{XY}(\tau) e^{-j\Omega\tau} d\tau, \qquad S_{YX}(\Omega) = \int_{-\infty}^{+\infty} R_{YX}(\tau) e^{-j\Omega\tau} d\tau \qquad (9.52)$$

where $R_{XY}(\tau)$ and $R_{YX}(\tau)$ represent the cross-correlation functions between $X(t)$ and $Y(t)$, respectively. The PSD shown in eq. (9.50) is also called the auto-PSD function.

9.3.5 Nonstationary Random Signals

A random signal that belongs to neither strict-sense stationary nor wide-sense stationary random signals is known as a nonstationary random signal. From the point of view of statistics, if the statistics of random signal changes with time, it is a nonstationary random signal. The most commonly used nonstationary random signal is such a signal whose mean, variance, and correlation function change with time.

The PDF of the nonstationary random signal is the function of time. When $t = t_i$, its PDF is shown as

$$f(x, t_i) = \lim_{\Delta x \to 0} \frac{P[x < x(t_i) < x + \Delta x]}{\Delta x} \qquad (9.53)$$

and

$$\int_{-\infty}^{+\infty} f(x, t_i) dx = 1 \qquad (9.54)$$

where P denotes the probability. On the basis of the probability, the numerical characteristics of the nonstationary random signal is defined as follows:

Mean: $\qquad\qquad\qquad m_x(t) = E[x(t)] = \int_{-\infty}^{+\infty} x f(x, t) dx \qquad (9.55)$

Mean square value: $\qquad D_x(t) = E[x^2(t)] = \int_{-\infty}^{+\infty} x^2 f(x, t) dx \qquad (9.56)$

Variance: $\qquad\qquad\quad \sigma_x^2(t) = D_x(t) - m_x^2(t) \qquad (9.57)$

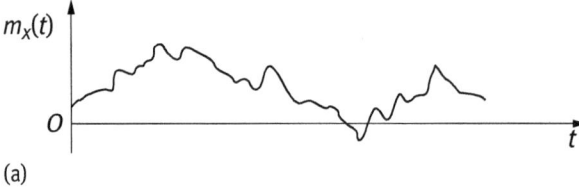

(a)

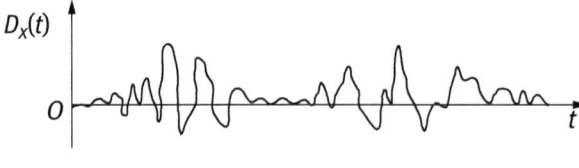

(b)

Figure 9.4: A nonstationary random signal: (a) time-varying mean and (b) time-varying mean square value.

They are all functions of time. Figure 9.4 shows an example of time-varying mean and mean square value of a nonstationary random signal.

9.4 Commonly Used Random Signals and Random Noises

9.4.1 Gaussian (Normal) Distributed Random Signals

Suppose a random process $X(t)$. For an arbitrary finite moment t_i $(i = 1, 2, \ldots, n)$, if the probability distribution of the n-dimensional random variable $X_i = X(t_i)$ is Gaussian distributed, then the random process is referred to as Gaussian process.

The n-dimensional PDF and the n-dimensional characteristic function are defined as

$$f_X(x_1, x_2, \ldots, x_n; t_1, t_2, \ldots, t_n) = \frac{1}{(2\pi)^{n/2}|C|^{1/2}} \exp\left[-\frac{1}{2|C|} \sum_{i=1}^{n} \sum_{j=1}^{n} |C|_{ij}(x_i - \mu_{X_i})(x_j - \mu_{X_j})\right]$$

$$\Phi_X(u_1, u_2, \ldots, u_n; t_1, t_2, \ldots, t_n) = \exp\left(j \sum_{i=1}^{n} u_i \mu_{X_i} - \frac{1}{2} \sum_{i=1}^{n} \sum_{j=1}^{n} C_{ij} u_i u_j\right)$$

$$(9.58)$$

where $X_i = X(t_i)$, and $\mu_{X_i} = E[X(t_i)]$. $|C|_{ij}$ is the cofactor of element C_{ij} in determinant $|C|$. $C_{ij} = E[(X_i - \mu_{X_i})(X_j - \mu_{X_j})]$ compose the following determinant:

$$|C| = \begin{vmatrix} C_{11} & C_{12} & \cdots & C_{1n} \\ C_{21} & C_{22} & \cdots & C_{2n} \\ \vdots & \cdots & \cdots & \vdots \\ C_{n1} & C_{n2} & \cdots & C_{nn} \end{vmatrix}$$

$$(9.59)$$

We also have $C_{ij} = C_{ji}$, and $C_{ii} = \sigma_{X_i}^2$.

The PDF and the characteristic function of the Gaussian distributed one-dimensional wide-sense stationary random process $X(t)$ are shown in the following equation:

$$f_X(x, t) = \frac{1}{\sqrt{2\pi}\sigma} e^{-\frac{(x-\mu)^2}{2\sigma^2}}$$
$$\Phi_X(u, t) = e^{(j\mu u - \sigma^2 u^2/2)}$$

$$(9.60)$$

Similarly, it is convenient to give the n-dimensional PDF and the n-dimension characteristic function of the Gaussian process. Set $X = [X_1\ X_2\ \cdots\ X_n]^T$, its mean vector is shown as

$$E[\boldsymbol{X}] = \boldsymbol{\mu} = [E[X_1] \ E[X_2] \ \cdots \ E[X_n]]^{\mathrm{T}} = [\mu_{X_1} \ \mu_{X_2} \ \cdots \ \mu_{X_n}]^{\mathrm{T}} \tag{9.61}$$

The covariance matrix is

$$\boldsymbol{C} = \begin{bmatrix} E[(X_1 - \mu_{X_1})^2] & \cdots & E[(X_1 - \mu_{X_1})(X_n - \mu_{X_n})] \\ \vdots & \ddots & \vdots \\ E[(X_n - \mu_{X_n})(X_1 - \mu_{X_1})] & \cdots & E[(X_n - \mu_{X_n})^2] \end{bmatrix} \tag{9.62}$$

The n-dimensional PDF of Gaussian process is

$$f_X(\boldsymbol{X}) = \frac{1}{(2\pi)^{n/2}\sqrt{|\boldsymbol{C}|}} \exp[-\frac{1}{2}(\boldsymbol{X} - \boldsymbol{\mu})^{\mathrm{T}} \boldsymbol{C}^{-1}(\boldsymbol{X} - \boldsymbol{\mu})] \tag{9.63}$$

The n-dimensional characteristic function of Gaussian process is

$$\Phi_X(\boldsymbol{u}) = \exp[\mathrm{j}\boldsymbol{\mu}^{\mathrm{T}}\boldsymbol{u} - \boldsymbol{u}^{\mathrm{T}}\boldsymbol{C}\boldsymbol{u}/2] \tag{9.64}$$

where $\boldsymbol{u} = [u_1 \ u_2 \ \cdots \ u_n]^{\mathrm{T}}$.

Gaussian signal is one of the most commonly used random signal models. From the above description, we see as long as we know the mean vector $E[\boldsymbol{X}] = \boldsymbol{\mu}$ and the covariance matrix $\boldsymbol{C}$ of the signal, the PDF of any order can be analytically expressed. If the Gaussian process is a wide-sense stationary signal, then it must be a strict-sense stationary signal. If the random variables of the Gaussian process are uncorrelated, then the process must be statistically independent. In addition, the Gaussian process is still Gaussian distributed after a linear operation.

9.4.2 White Noise and Band-Limited White Noise

1. White noise
White noise is defined as a stochastic process with equal power at all frequencies, whose PSD is $S_{ww}(\Omega) = N_0/2$. Since the PSD of the white noise is the Fourier transform of its autocorrelation function, the corresponding autocorrelation function is shown as

$$R_{ww}(\tau) = (N_0/2)\delta(\tau) \tag{9.65}$$

$R_{ww}(\tau)$ has nonzero value only when $\tau = 0$. If $\tau \neq 0$, $R_{ww}(\tau) \equiv 0$ holds, which means the white noise is uncorrelated at different moments. On the other hand, the average power of the white noise tends to be infinite, so the white noise is not physically realizable. However, in the practical applications, the white noise as a random signal model is very important and meaningful for simplifying the analysis of signals and systems.

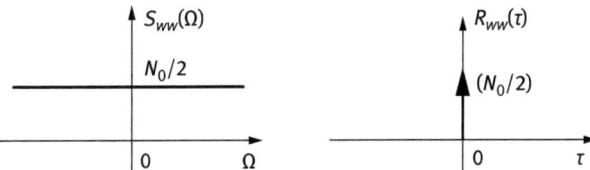

Figure 9.5: Curves of PSD function and autocorrelation function of the white noise: (a) PSD function and (b) autocorrelation function.

Figure 9.5 shows curves of PSD function and autocorrelation function of the white noise.

In many practical problems, the white noise of Gaussian distribution is usually used. Therefore, Gaussian white noise is usually used to represent the white random process with the Gaussian distribution.

2. Band-limited white noise
From the definition of white noise, we see that the pure white noise is only a theoretical concept, because it needs infinite power to cover an infinite wide range of frequencies. On the other hand, all discrete-time signals must be of finite bandwidth, and its highest frequency component must be less than or equal to the half of the sampling frequency. The power spectrum of the band-limited white noise with a bandwidth of WHz is defined as

$$S(\Omega) = \begin{cases} N_0/2, & |\Omega| \leq W \\ 0, & |\Omega| > W \end{cases} \tag{9.66}$$

Taking Fourier transform to both sides of eq. (9.66), we get the autocorrelation function of the band-limited white noise:

$$R(\tau) = \frac{WN_0}{2\pi} \cdot \frac{\sin W\tau}{W\tau} \tag{9.67}$$

It should be noticed that the autocorrelation function of the band-limited white noise has $R(\tau) = 0$ at $\tau = K\frac{\pi}{W}$. Therefore, if the sampling rate satisfies the Nyquist sampling rate π/W, the sampled data will be uncorrelated with each other.

9.4.3 The Gauss–Markov Process

The random process with the exponential type of autocorrelation function is called Gauss–Markov process, whose autocorrelation function and PSD function are as follows:

$$R(\tau) = \sigma^2 e^{-\beta|\tau|}$$
$$S(\Omega) = \frac{2\sigma^2\beta}{\Omega^2 + \beta^2} \qquad (9.68)$$

It can be seen from the above formulas, with the increase of τ, the sampling value of the signal tends to be uncorrelated. Such random process can be seen as an output of a Gaussian white noise passing through a first-order autoregressive (AR) system.

9.4.4 Other Common Noises

1. Colored noise
The so-called colored noise refers to the noise whose PSD in the entire frequency domain does not show the uniform distribution. Because of the different power in the frequency band, it is called the colored noise similar to the colored light. In practical applications, most of the noise in the audio frequency band, such as the vehicle noise, computer fan noise, and electric drill noise, are all colored noises.

2. Thermal noise
Thermal noise, also known as Johnson noise, is produced by the random movement of charged particles in the conductor. The spontaneous motion of electrons in a conductor constitutes the spontaneous current, which is the thermal noise. As the temperature increases, the free electrons jump to a higher energy level, and the thermal noise increases. The thermal noise has a flat power spectrum, so it belongs to the white noise. Thermal noise cannot be avoided by shielding and grounding the system.

3. Shot noise
Shot noise is a random noise caused by the motion of the discrete charge, and the noise intensity increases with the increase of the average current through the conductor. The name of shot noise comes from the random variation of electron emission from the cathode in a vacuum tube. The electron flow in semiconductors and the photoelectron stream emitted by the photosensitive diode will also form shot noises. The random variation of the shot noise can be described by Poisson probability distribution.

4. Electromagnetic noise
Electromagnetic noise refers to the noise caused by the alternation of the electromagnetic field in the environment. In practical applications, the sources of electromagnetic noise mainly include transformer, radio and television transmitters, mobile phones, microwave transmitters, AC power lines, motors and motor starters,

generators, fluorescent lamps, and electromagnetic storms. The electromagnetic noise is usually pulsating and random, and it may also be periodic.

9.4.5 The Generation of Random Signals and Noises

1. The generation of uniformly distributed random numbers

The PDF of the uniform distributed random sequence $\{u_i\}$, $i = 1, 2, \ldots$, is shown as

$$f(u_i) = \begin{cases} 1, & u_i \in [0, 1) \\ 0, & \text{others} \end{cases} \tag{9.69}$$

There are three basic methods for generating the uniformly distributed random numbers: the first one is the table method, in which the existed random numbers are stored into a table for subsequent applications; the second method is to generate the random numbers by electronic circuits; and the third method generates the random numbers by a mathematic algorithm, referred to as the pseudo-random numbers. Since the third method is easy to be used by a computer, it has been widely used.

The common method for generating the pseudo-random numbers in accordance with arithmetic formula is the linear congruence method, given as

$$\begin{aligned} y_0 &= 1, \\ y_n &= k y_{n-1} \pmod{N}, \\ u_n &= y_n / N, \end{aligned} \tag{9.70}$$

where y_0 is the initial value, k is a coefficient, and N is the modulus of the data, usually a very large number. Three groups of commonly used parameters are listed as follows:

(1) $N = 10^{10}$, $k = 7$, whose cycle period is about 5×10^7.
(2) $N = 2^{31}$, $k = 2^{16} + 3$, whose cycle period is about 5×10^8.
(3) $N = 2^{31} - 1$, $k = 7^5$, whose cycle period is about 2×10^9.

The random numbers generated by the above methods are uniformly distributed in $[0, 1)$.

In fact, pseudo-random numbers are not really random, which are periodic. If the parameter and calculation formula are chosen properly, the period is quite long. It can be regarded as a random process, and it can also pass through the random test of mathematical statistics.

2. Other distributions obtained by the transformation

Other distributed random numbers can be obtained by the transform method from the uniformly distributed random numbers.

Suppose that the distribution function $F(X)$ is strictly monotonic. Its inverse function $F^{-1}(\cdot)$ is used to transform the uniform distributed random variable U as

$$X = F^{-1}(U) \tag{9.71}$$

Thus, the distribution function of X is $F(X)$.

Example 9.1: Generate the exponential distributed random numbers based on the given uniformly distributed random numbers.

Solution: Suppose that U is a uniformly distributed random variable in $[0, 1)$, and the distribution function of the exponential distribution is shown as $F(x) = 1 - e^{-\lambda x}$, $x \geq 0$, where λ is the parameter. Then $X = F^{-1}(U) = -\frac{1}{\lambda}\ln(1 - U)$ where $1 - U$ is also the uniformly distributed random variable in $[0, 1)$. After getting the uniformly distributed random numbers $\{u_i\}$, the exponential distributed random numbers $\{x_i\}$ can be obtained from $x_i = -\frac{1}{\lambda}\ln u_i$.

3. The generation of normal distributed random numbers

(1) Cumulative approximation method

Based on the central limit theorem for independent identical distributed (i.i.d.) random variables, generate 12 independent uniformly distributed random variables u_1, u_2, $\ldots$, u_{12} first, and then calculate the normal distributed (approximately $N(0, 1)$) random numbers according to the following equation:

$$x_i = \sum_{k=1}^{12} u_k - 6 \tag{9.72}$$

x_i will approximately be the random numbers with normal distribution obeying $N(0, 1)$.

(2) Transform method

Generate two independent uniformly distributed random numbers first, and then calculate two standard normal distributed random numbers according to eq. (9.73):

$$x_1 = \sqrt{-\ln u_1} \cos 2\pi u_2, \qquad x_2 = \sqrt{-\ln u_1} \sin 2\pi u_2 \tag{9.73}$$

The random number y_i obeying $N(\mu, \sigma^2)$ can be obtained by taking a transform in eq. (9.74) from the random numbers obeying $N(0, 1)$

$$y_i = \sigma x_i + \mu \tag{9.74}$$

9.5 Linear Systems with Random Inputs

9.5.1 The Output of the Linear System and Its Probability Distribution

If a sample $x(t)$ of the random process $X(t)$ is inputted into a linear time-invariant (LTI) system, the output of the system can be expressed as

$$y(t) = \int_{-\infty}^{+\infty} h(\tau)x(t-\tau)d\tau \tag{9.75}$$

Since $x(t)$ is a deterministic signal, the output $y(t)$ is also a continuous-time deterministic signal. If each $x(t)$ in $X(t)$ converges in mean square sense, then the output of the system corresponding to $X(t)$ is

$$Y(t) = \int_{-\infty}^{+\infty} h(\tau)X(t-\tau)d\tau \tag{9.76}$$

For the discrete-time input signal and system, we have

$$Y(m) = \sum_{m=-\infty}^{+\infty} h(m)X(n-m) \tag{9.77}$$

Generally speaking, it is difficult to determine the distribution of the output of a linear system. However, if the input of the linear system is Gaussian distributed, then its output is also Gaussian distributed. Because of the time invariance of the system, if $X(t)$ and $X(t+\tau)$ have the same distribution, then $Y(t)$ and $Y(t+\tau)$ are also with the same distribution.

9.5.2 Numerical Characteristics of the Linear System

1. The mathematical expectation and autocorrelation of the system output
Suppose the input signal is a stationary random process, and its mathematical expectation is expressed as m_X. Taking the mathematical expectation to eq. (9.76), we have

$$E[Y(t)] = m_Y = E\left[\int_{-\infty}^{+\infty} h(\tau)X(t-\tau)d\tau\right] = m_X \int_{-\infty}^{+\infty} h(\tau)d\tau \tag{9.78}$$

If the system is causal, then the autocorrelation function of the output $Y(t)$ is

$$R_{YY}(\tau) = \int_{0}^{+\infty} \int_{0}^{+\infty} R_X(\tau+\lambda_1-\lambda_2)h(\lambda_1)h(\lambda_2)d\lambda_1 d\lambda_2 \tag{9.79}$$

If $X(t)$ is a stationary random process, then the output $Y(t)$ is also stationary. For a continuous-time system, eq. (9.79) can be written as

$$R_{YY}(\tau) = R_{XX}(\tau) \star h(-\tau) \star h(\tau) \tag{9.80}$$

The average power of the output process is

$$R_{YY}(0) = \int_0^{+\infty} \int_0^{+\infty} R_{XX}(\lambda_1 - \lambda_2)h(\lambda_1)h(\lambda_2)\mathrm{d}\lambda_1\mathrm{d}\lambda_2 \tag{9.81}$$

For the discrete-time random sequence, the results are similar.

2. The cross-correlation function between input and output

The cross-correlation function between input and output of a linear system is given as

$$R_{XY}(t, t+\tau) = E[X(t)Y(t+\tau)] = \int_{-\infty}^{\infty} E[X(t)X(t+\tau-\lambda)]h(\lambda)\mathrm{d}\lambda$$
$$= \int_{-\infty}^{+\infty} R_{XX}(t, t+\tau-\lambda)h(\lambda)\mathrm{d}\lambda \tag{9.82}$$

If $X(t)$ is a stationary process, then

$$R_{XY}(\tau) = \int_{-\infty}^{+\infty} R_{XX}(\tau-\lambda)h(\lambda)\mathrm{d}\lambda = R_{XX}(\tau) \star h(\tau) \tag{9.83}$$

It is indicated that the cross-correlation function between the input signal and the output signal can be obtained by the convolution of the autocorrelation function of the input signal and the unit impulse response of the system. Since $X(t)$ and $Y(t)$ are stationary and jointly stationary, the cross-correlation function between the output and input can be written as

$$R_{YX}(\tau) = \int_{-\infty}^{+\infty} R_{XX}(\tau-\lambda)h(-\lambda)\mathrm{d}\lambda = R_{XX}(\tau) \star h(-\tau) \tag{9.84}$$

Furthermore, we can get

$$R_{YY} = \int_{-\infty}^{+\infty} R_{XY}(\tau+\lambda)h(\lambda)\mathrm{d}\lambda = R_{XY}(\tau) \star h(-\tau)$$
$$R_{YY} = \int_{-\infty}^{+\infty} R_{YX}(\tau-\lambda)h(\lambda)\mathrm{d}\lambda = R_{YX}(\tau) \star h(\tau) \tag{9.85}$$

3. Output spectrum of the linear system

There are two ways to obtain the output PSD function of the system: one is through the autocorrelation of the system output, and the other is through the PSD of the system input. For the stationary random process $X(t)$, the output autocorrelation function is as given in eq. (9.79). Taking Fourier transform to both sides, we get

$$S_{YY}(\Omega) = \int_{-\infty}^{+\infty} R_{YY}(\tau)e^{-j\Omega\tau}d\tau = \int_{-\infty}^{+\infty} h(\lambda_1) \int_{-\infty}^{+\infty} h(\lambda_2) \int_{-\infty}^{+\infty} R_{XX}(\tau + \lambda_1 - \lambda_2)e^{-j\Omega\tau}d\tau d\lambda_1 d\lambda_2$$
$$= H^*(j\Omega)H(j\Omega)S_{XX}(\Omega) = S_{XX}(\Omega)|H(j\Omega)|^2$$

$$(9.86)$$

By using the PSD function $S_{XX}(\Omega)$ of the input signal, we get

$$S_{YY}(\Omega) = H(-j\Omega)H(j\Omega)S_{XX}(\Omega) = S_{XX}(\Omega)|H(j\Omega)|^2 \qquad (9.87)$$

Generally, $|H(j\Omega)|^2$ in above equation is referred to as the power transfer function.

4. Numerical characteristics of white noise through LTI systems

Set the unit impulse response of the LTI system as $h(t)$. If the output of the system whose input is a white noise is $Y(t)$, then the numerical characteristics of the output process are as follows:

Mean:	$m_Y(t) = 0$	(9.88)
Autocorrelation:	$R_{YY}(\tau) = \frac{N_0}{2}h(\tau) * h^*(-\tau) = \frac{N_0}{2}r_h(\tau)$	(9.89)
PSD function:	$S_{YY}(\Omega) = \frac{N_0}{2}|H(j\Omega)|^2$	(9.90)
Average power:	$R_{YY}(0) = \frac{N_0}{4\pi}\int_{-\infty}^{\infty}|H(j\Omega)|^2 d\Omega = \frac{N_0}{2}r_h(\tau)$	(9.91)

where $r_h(\tau)$ denotes the system correlation.

Example 9.2: Suppose the autocorrelation function of a white noise $X(t)$ is $R_{XX}(\tau) = \frac{N_0}{2}\delta(\tau)$. Calculate the autocorrelation function of the output $Y(t)$ when $X(t)$ passing through the RC circuit shown in Figure 9.6.

Solution: Calculating the PSD function $S_{XX}(\Omega)$ of $X(t)$ based on the given autocorrelation function, we have $S_{XX}(\Omega) = \int_{-\infty}^{+\infty} \frac{N_0}{2}\delta(\tau)e^{-j\Omega\tau}d\Omega = \frac{N_0}{2}$. Since the transfer function of the circuit is $H(j\Omega) = \frac{1}{1+j\Omega RC}$, the PSD of the output process is $S_{YY}(\Omega) = \frac{N_0}{2}|H(j\Omega)|^2 = \frac{N_0}{2} \cdot \frac{1}{1+(\Omega RC)^2}$. From the inverse Fourier transform, we get the autocorrelation function of the output process as

$$R_{YY}(\tau) = 12\pi \int_{-\infty}^{+\infty} S_{YY}(\Omega)e^{j\Omega\tau}d\Omega = \frac{N_0}{4RC}e^{-\frac{|\tau|}{RC}}$$

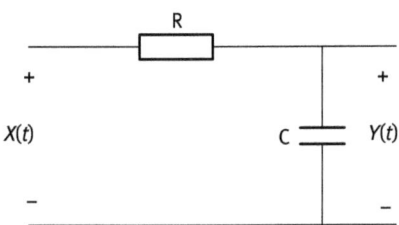

Figure 9.6: An RC integral circuit.

9.5.3 Equivalent Noise Bandwidth of System and Bandwidth of Random Signal

1. Equivalent noise bandwidth of the system

Generally, the bandwidth of the system refers to the 3 dB width of the system transfer function. For random signals, the transfer function of the system is replaced by the power transfer function of the system. For a random signal that passes through the linear system, we concern not only the frequency components of its output, but also the average power. Therefore, the bandwidth corresponding to the average power needs to be defined. The so-called system equivalent noise bandwidth is defined as the PSD of the white noise passing through a system. The essence is that the power transfer function of a practical system is equivalent to the power transfer function of the ideal system.

In general, after the white noise passes through a practical system, its power spectrum is often similar to the power transfer function of the system. Figure 9.7 shows the schematic diagram of the power transfer function and the equivalent noise bandwidth of the low-pass and the band-pass systems.

By Figure 9.7, the equivalent noise bandwidth is represented by $\Delta\Omega_e$ and is defined as the equivalent width of the system power transfer function $|H(j\Omega)|^2$, the area inside the rectangle in the figure. For a low-pass system, the equivalent noise bandwidth is represented as

$$|H_e(j\Omega)|^2 = \begin{cases} |H(0)|^2, & |\Omega| \le \Delta\Omega_e \\ 0, & |\Omega| > \Delta\Omega_e \end{cases} \tag{9.92}$$

For a band-pass system, the equivalent noise bandwidth is represented as

$$|H_e(j\Omega)|^2 = \begin{cases} |H(j\Omega_0)|^2, & \Omega_0 - \Delta\Omega_e/2 < |\Omega| < \Omega_0 + \Delta\Omega_e/2 \\ 0, & \text{others} \end{cases} \tag{9.93}$$

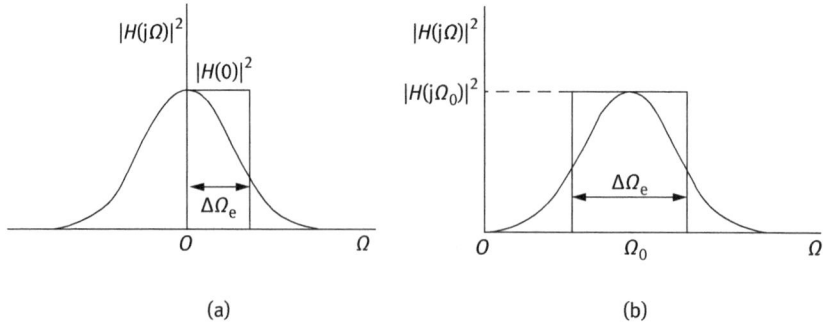

(a)   (b)

Figure 9.7: The schematic diagram of the power transfer function and the equivalent noise bandwidth of the low-pass system (a) and the band-pass system (b).

where Ω_0 represents the central frequency in the system power transfer function. The equivalent noise bandwidths of low-pass and band-pass systems can be calculated as follows, respectively:

$$\Delta\Omega_e = \int_0^{+\infty} \left|\frac{H(j\Omega)}{H(0)}\right|^2 d\Omega \tag{9.94}$$

$$\Delta\Omega_e = \int_0^{+\infty} \left|\frac{H(j\Omega)}{H(0)}\right|^2 d\Omega \tag{9.95}$$

2. Bandwidth of random signals

The rectangular equivalent bandwidth of a stationary random signal is defined as the equivalent noise bandwidth of the system, as shown in eq. (9.96).

$$B_{eq} = \frac{1}{2\pi} \int_0^{+\infty} \frac{S_{XX}(\Omega)}{S_{XX}(\Omega_0)} d\Omega = \frac{R_{XX}(0)}{2S_{XX}(\Omega_0)} \tag{9.96}$$

where $R_{XX}(0)$ is the average power of the random signal. For low-pass signals, $\Omega_0 = 0$, and for band-pass signals, Ω_0 is taken as the frequency of the maximum of $S_{XX}(\Omega)$.

The root mean square (RMS) bandwidth of a stationary random signal is a kind of commonly used bandwidth representation, which is defined as

$$B_{rms} = \frac{1}{2\pi} \left[\frac{\int_0^{+\infty} (\Omega - \Omega_0)^2 S_{XX}(\Omega) d\Omega}{\int_0^{+\infty} S_{XX}(\Omega) d\Omega}\right]^{1/2} \tag{9.97}$$

where $\Omega_0 = 0$ is taken for low-pass signals, and Ω_0 is taken as the center of gravity of $S_{XX}(\Omega)$ for band-pass signals:

$$\Omega_0 = \frac{\int_0^{+\infty} \Omega S_{XX}(\Omega) d\Omega}{\int_0^{+\infty} \Omega d\Omega} \tag{9.98}$$

9.6 Classical Analysis of Random Signals

9.6.1 The PDF of Common Random Signals

The PDFs of commonly used random signals are listed in Table 9.1.

Table 9.1: The PDFs of commonly used random signals.

Distribution	PDF curves	PDF expression	Numerical characteristics		Comments
			Mean	Variance	
Discrete		$f(x) = A\delta(x - a) + B\delta(x - b)$ $+ N\delta(x - n), A + B + N = 1$			
Uniform		$f(x) = \begin{cases} \frac{1}{b-a}, & a < x \le b \\ 0, & \text{others} \end{cases}$	$\frac{a+b}{2}$	$\frac{(b-a)^2}{12}$	The probability of the random variable takes the same value on $(a, b]$.
Gaussian		$f(x) = \frac{1}{\sqrt{2\pi}\sigma_x} e^{-\frac{(x-\mu_x)^2}{2\sigma_x^2}}$	μ_x	σ_x^2	Gaussian distribution is the most commonly used PDF. If the value of a random variable is affected by a variety of factors, and the degree of influence of each factor is similar, then the variable finally takes the Gaussian distribution.

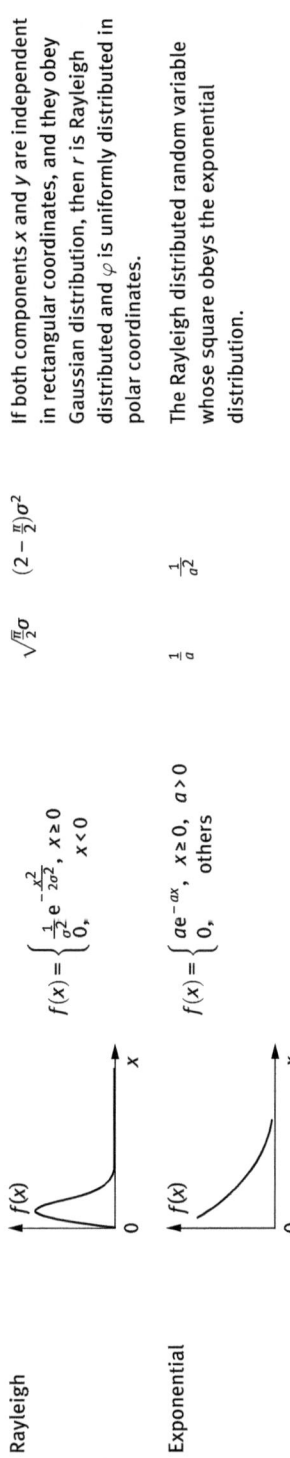

Rayleigh

$$f(x) = \begin{cases} \frac{1}{\sigma^2} e^{-\frac{x^2}{2\sigma^2}}, & x \geq 0 \\ 0, & x < 0 \end{cases}$$

$\sqrt{\frac{\pi}{2}}\sigma \qquad (2 - \frac{\pi}{2})\sigma^2$

If both components x and y are independent in rectangular coordinates, and they obey Gaussian distribution, then r is Rayleigh distributed and φ is uniformly distributed in polar coordinates.

Exponential

$$f(x) = \begin{cases} a e^{-ax}, & x \geq 0, \quad a > 0 \\ 0, & \text{others} \end{cases}$$

$\frac{1}{a} \qquad \frac{1}{a^2}$

The Rayleigh distributed random variable whose square obeys the exponential distribution.

9.6.2 The Calculation of the Numerical Characteristics for Random Signals

Suppose the stationary random sequence $X(n)$, also denoted as $\{x(n)\}$, is obtained by sampling the stationary signal $X(t)$. One sample sequence of $X(n)$ is denoted as $x(n)$. The purpose of this subsection is to calculate or estimate the numerical characteristics based on a sample sequence $x(n)$.

1. Sample mean $\hat{m}_x$

$$\hat{m}_x = \frac{1}{N}\sum_{n=1}^{N} x(n) \tag{9.99}$$

where N is the data length. The symbol "^" is used to express the estimated value.

2. Sample mean square value $E[\hat{m}_x^2]$

$$E[\hat{m}_x^2] = \frac{1}{N}\sum_{n=1}^{N} x^2(n) \tag{9.100}$$

3. Sample variance $\hat{\sigma}_x^2$

$$\hat{\sigma}_x^2 = \frac{1}{N}\sum_{n=1}^{N} (x(n) - \hat{m}_x)^2 \tag{9.101}$$

4. Sample covariance $\hat{C}_{xy}(m)$

$$\hat{C}_{xy}(m) = \frac{1}{N}\sum_{n=1}^{N} (x(n) - \hat{m}_x)(y(n+m) - \hat{m}_y) \tag{9.102}$$

where $\{y(n)\}$ represents another random sequence, whose sample mean is $\hat{m}_y$. If $\{x(n)\}$ and $\{y(n)\}$ are the same sequence, then the aforementioned covariance is called autocovariance.

5. Sample correlation $\hat{R}_{xy}(m)$

$$\hat{R}_{xy}(m) = \frac{1}{N}\sum_{n=1}^{N} x(n)y(n+m) \tag{9.103}$$

Example 9.3: Generate 100 Gaussian distributed random numbers with zero mean and $\sigma^2 = 1$, and calculate the sample numerical characteristics.
Solution: Figure 9.8 demonstrates the curve of the random sequence generated by the MATLAB program.

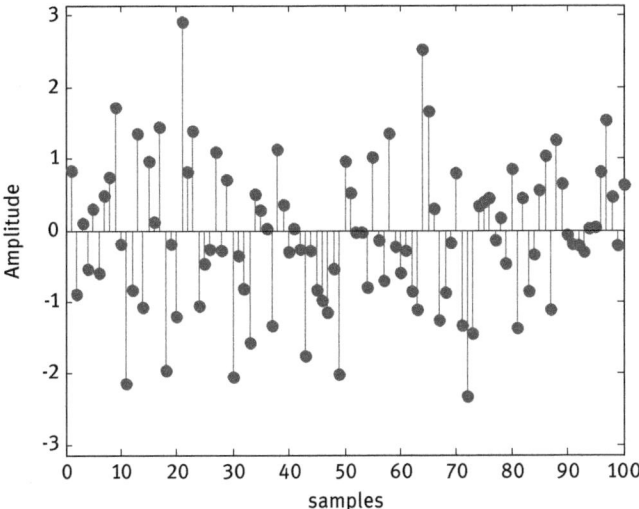

Figure 9.8: The Gaussian distributed random sequence.

According to the formulas for estimating the statistics, we have $\hat{m}_x = \frac{1}{N} \sum\limits_{n=1}^{N} x(n) =$ -0.0804, and $\hat{\sigma}_x^2 = \frac{1}{N} \sum\limits_{n=1}^{N} (x(n) - \hat{m}_x)^2 = 0.925$. We can also get other sample statistics. If the data length N is increased, then the estimated values will be more close to their true values.

9.7 Analysis for Random Signals with Modern Parameter Models

9.7.1 Wold Decomposition Theorem for Random Signals

Wold decomposition theorem for random signals is described as follows: the general stationary random signals $\{x(n)\}$ can be decomposed into a combination of predictable part $x_p(n)$ and unpredictable part $x_u(n)$, that is,

$$x(n) = x_p(n) + x_u(n) \tag{9.104}$$

and for arbitrary n_1 and n_2, both $x_p(n)$ and $x_u(n)$ satisfy $E[x_p(n_1)x_u(n_2)] \equiv 0$.

The predictable random sequence is such a sequence in which the future values can be accurately determined by its past values. The unpredictable random sequence is such a sequence whose future values cannot be determined by its past values.

Example 9.4: Suppose a random sequence $\{x(n)\}$ is represented as follows:

$$x(n) = ax(n-1) + w(n), \quad a < 1$$

where $w(n)$ denotes a Gaussian white noise with zero mean and unit variance. Determine whether it is predictable.

Solution: (1) If $w(n) = 0$, then $x(n) = ax(n-1)$. The future values of $x(n)$ is determined by its past values. Thus, $x(n)$ is a predictable sequence when $w(n) = 0$.

(2) If $w(n) \neq 0$, then $w(n)$ is undetermined. As a result, $x(n)$ is an unpredictable sequence.

9.7.2 Linear Parameter Models for Stationary Random Signals

Many stationary random signals $x(n)$ can be seen as a response of a particular linear system $h(n)$ when it is excited by a white noise $w(n)$. In this way, as long as the parameters of the white noise are determined, the investigation to random signals can be transformed into the study of the linear system that generates random signals. This is the parametric modeling method of random signal analysis. This method contains two aspects of basic issues: first, from the given sample sequence of the stationary random signal, we can establish the corresponding parameter model; and second, from the given white noise and the parameter model, we can produce the required random sequences. Figure 9.9 gives the diagram of the parameter model.

1. Moving average model
The moving average (MA) model of the wide-sense stationary random signal $x(n)$ satisfies the following equation:

$$x(n) = \sum_{k=0}^{q} b_k w(n - k) \tag{9.105}$$

where $w(n)$ is the white noise, b_k, $k = 0, 1, \ldots, q$, are the coefficients of MA model, and q is the model order. The above equation means that the random signal $x(n)$ is a weighted combination of q past values of $w(n)$ and its present value. Taking the z-transform, we get the system function of MA model as

$$H(z) = \frac{X(z)}{W(z)} = \sum_{k=0}^{q} b_k z^{-k} \tag{9.106}$$

The MA system is definitely a stable system since it does not have any pole. Thus, MA model is usually called the full zeros model, denoted as $MA(q)$.

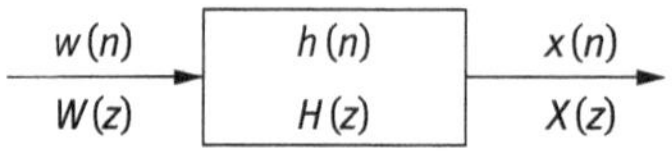

Figure 9.9: The diagram of the parameter model.

2. Autoregressive model

The AR model of the wide-sense stationary random signal $x(n)$ satisfies

$$x(n) = -\sum_{k=1}^{p} a_k x(n-k) + w(n) \tag{9.107}$$

where a_k, $k = 1, 2, \ldots, p$, are the model coefficients of the AR model, and p is the model order. The above equation means that the random signal $x(n)$ is composed of p past values of $x(n)$ and the present excitement $w(n)$. The system function of the AR system is given as

$$H(z) = \frac{1}{1 + \sum_{k=1}^{p} a_k z^{-k}} \tag{9.108}$$

There are only poles in AR system. So the model is called as an all poles model, denoted as $AR(p)$. Since the existence of poles, we need to concern the stability of the system.

3. ARMA model

The autoregressive moving average (ARMA) model of the wide-sense stationary random signal $x(n)$ satisfies

$$x(n) = \sum_{k=0}^{q} b_k w(n-k) - \sum_{k=1}^{p} a_k x(n-k) \tag{9.109}$$

Obviously, the ARMA model is a combination of AR and MA models. Its system function is

$$H(z) = \frac{\sum_{k=0}^{q} b_k z^{-k}}{1 + \sum_{k=1}^{p} a_k z^{-k}} \tag{9.110}$$

We see from the above equation, the ARMA model has not only zeros but also poles, referred to as the zero-pole model. It is denoted as $ARMA(p, q)$.

9.7.3 The Estimation of AR Parameters

The parameter estimation or modeling of random signals is widely used in many areas. On the basis of Wold decomposition theorem, AR, MA, and ARMA models can be transformed with each other. So here we only discuss the modeling problem of the AR model.

1. The autocorrelation function of AR(p)

Suppose the white noise $w(n)$ is of zero mean and variance σ_w^2, satisfying $n > p$. Taking the autocorrelation function to the AR model in eq. (9.109), and employing the symmetry property, we get

$$
\begin{aligned}
R_{xx}(m) &= E[x(n)x(n-m)] = E[w(n)x(n-m) - \sum_{k=1}^{p} a_k x(n-k)x(n-m)] \\
&= R_{xw}(m) - \sum_{k=1}^{p} a_k R_{xx}(m-k)
\end{aligned}
\tag{9.111}
$$

where

$$
\begin{aligned}
R_{xw}(m) &= E[x(n)w(n+m)] = \sum_{k=0}^{+\infty} h(k)E[w(n-k)w(n+m)] \\
&= \sum_{k=0}^{\infty} h(k)R_{ww}(m+k) = \sum_{k=0}^{+\infty} h(k)\sigma_w^2 \delta(m+k) = \sigma_w^2 h(-m)
\end{aligned}
\tag{9.112}
$$

where $h(n)$ is the unit impulse response of the system. After arrangement, we have

$$
R_{xw}(m) = \begin{cases} 0, & m > 0 \\ \sigma_w^2 h(-m), & m \le 0 \end{cases}
\tag{9.113}
$$

Substituting eq. (9.113) into eq. (9.111), we get

$$
R_{xx}(m) = \begin{cases} -\sum_{k=1}^{p} a_k R_{xx}(m-k), & m > 0 \\ -\sum_{k=1}^{p} a_k R_{xx}(m-k) + h(0)\sigma_w^2, & m = 0 \\ R_{xx}(-m), & m < 0 \end{cases}
\tag{9.114}
$$

For causal system $h(n)$, we have $h(0) = 1$. We see from the above equation, the autocorrelation function of AR(p) has a recursive property.

2. Yule–Walker equation

From eq. (9.114), we get

$$
\begin{aligned}
R_{xx}(0) + a_1 R_{xx}(-1) + \cdots + a_p R_{xx}(-p) &= \sigma_w^2 \\
R_{xx}(1) + a_1 R_{xx}(0) + \cdots + a_p R_{xx}(1-p) &= 0 \\
&\vdots \qquad\qquad\qquad\qquad \vdots \\
R_{xx}(p) + a_1 R_{xx}(p-1) + \cdots + a_p R_{xx}(0) &= 0
\end{aligned}
\tag{9.115}
$$

Equation (9.115) can be rewritten into a matrix form as

$$
\begin{bmatrix}
R_{xx}(0) & R_{xx}(-1) & \cdots & R_{xx}(-p) \\
R_{xx}(1) & R_{xx}(0) & \cdots & R_{xx}(1-p) \\
\vdots & \vdots & \ddots & \vdots \\
R_{xx}(p) & R_{xx}(p-1) & \cdots & R_{xx}(0)
\end{bmatrix}
\begin{bmatrix}
1 \\ a_1 \\ \vdots \\ a_p
\end{bmatrix}
=
\begin{bmatrix}
\sigma_w^2 \\ 0 \\ \vdots \\ 0
\end{bmatrix}
\tag{9.116}
$$

The above equation is the famous Yule–Walker equation. If the order p and the autocorrelation function $R_{xx}(m)$ are known, the AR modeling problem is transformed to a problem for solving Yule–Walker equation.

Example 9.5: Suppose $w(n)$ is a white noise sequence, whose variance is $\sigma_w^2 = 1$. $x(n)$ is an AR(3) random sequence, satisfying $x(n) = \frac{14}{24}x(n-1) + \frac{9}{24}x(n-2) - \frac{1}{24}x(n-3) + w(n)$.
(1) Find the values of the autocorrelation function $R_{xx}(m)$ when $m = 0$, 1, 2, 3, 4, 5.
(2) Find the estimates of $\hat{a}_k$, $k = 1$, 2, 3, by the estimated values of the autocorrelation obtained in (1), and find the estimation of $\hat{\sigma}_w^2$.
(3) Generate $N = 32$ samples of $x(n)$ based on the given model AR(3). Directly estimate $\hat{a}_k$, $k = 1$, 2, 3, by the observed values, and estimate the variance $\hat{\sigma}_w^2$.
Solution: (1) From the given conditions we know that $a_1 = -14/24$, $a_2 = -9/24$, $a_3 = 1/24$. Substituting a_k, $k = 1$, 2, 3, and $\sigma_w^2 = 1$ into eq. (9.116), we get

$$
R_{xx}(0) = 4.9377, \quad R_{xx}(1) = 4.3287, \quad R_{xx}(2) = 4.1964, \quad R_{xx}(3) = 3.8654
$$

According to eq. (9.114) we can further get the values of $R_{xx}(4)$ and $R_{xx}(5)$. If it is needed, we can get more values of the autocorrelation.

(2) On the basis of Yule–Walker equation, substituting the estimated values of the autocorrelation into eq. (9.116), we get $\hat{a}_1 = -14/24$, $\hat{a}_2 = -9/24$, $\hat{a}_3 = 1/24$, $\hat{\sigma}_w^2 = 1$ by solving the linear equations. The estimated values are the same as those of the truth values.

(3) Generate the observation values by a computer program and substitute them into Yule–Walker equation; we get the estimation values as $\hat{a}_1 = -0.6984$, $\hat{a}_2 = -0.2748$, $\hat{a}_3 = 0.0915$, and $\hat{\sigma}_w^2 = 0.4678$.

Comparing the results of (2) and (3), we see a significant error occurs in the result of (3). In fact, we use the truth value of the autocorrelation sequence for the estimation in (2), while in (3), we just use the estimated autocorrelation sequence to estimate the AR parameters. In addition, the estimated autocorrelation sequence used in (3) is calculated by only 32 sample points, which may result in a large error. If we increase the data length N, the parameter estimation will be improved.

3. The solution of Yule–Walker equation: Levinson–Durbin recursive algorithm

(1) Linear prediction error filter and AR model

Assume that a linear combination of past observations is used to represent the prediction of the current data $x(n)$; such method is called the forward predictor, that is,

$$\hat{x}(n) = - \sum_{k=1}^{m} a_m(k)x(n-k) \tag{9.117}$$

where $a_m(k)$, $k = 1, 2, \ldots, m$, represent prediction coefficients of the mth-order predictor. Obviously, the error exists between the predicted result and the true $x(n)$. Such error is referred to as the prediction error, represented as

$$e(n) = x(n) - \hat{x}(n) = x(n) + \sum_{k=1}^{m} a_m(k)x(n-k) \tag{9.118}$$

If $e(n)$ is considered as the output of a system and $x(n)$ is the system input, then the system function is shown as

$$\frac{E(z)}{X(z)} = 1 + \sum_{k=1}^{m} a_m(k)z^{-k} \tag{9.119}$$

If $m = p$, and the prediction coefficients are the same as AR parameters, then the prediction error system is the inverse of the system function of the AR model. Thus, the problem of solving AR model is transformed into the problem for solving the prediction coefficients of the prediction error filter. Taking the mean square value of the prediction error, we have

$$\begin{aligned} E[e^2(n)] &= E[(x(n) + \sum_{k=1}^{m} a_m(k)x(n-k))^2] \\ &= R_{xx}(0) + 2[\sum_{k=1}^{m} a_m(k)R_{xx}(k)] + \sum_{k=1}^{m}\sum_{l=1}^{m} a_m(l)a_m(k)R_{xx}(l-k) \end{aligned} \tag{9.120}$$

To minimize the mean square error, letting the partial derivative of the right-hand side of the above equation with respect to the prediction coefficients as zero, we get

$$R_{xx}(l) = - \sum_{k=1}^{m} a_m(k)R_{xx}(l-k), \quad l = 1, 2, \ldots, m \tag{9.121}$$

Substituting the above equation into eq. (9.120), we get the minimal mean square error as

$$E_m[e^2(n)] = E_p[e^2(n)] = R_{xx}(0) + \sum_{k=1}^{p} a_k R_{xx}(k) \tag{9.122}$$

where $a_k = a_m(k)$ and $m = p$ are supposed. It means that the prediction coefficients of the predictor equal the parameters of the pth-order AR model.

(2) Levinson–Durbin algorithm for solving AR model parameters

Levinson–Durbin recursive algorithm is a relatively fast algorithm for solving Yule–Walker equations, whose basic idea is to successively increase the order of the model to calculate the model parameters based on Yule–Walker equations and the recurrence of the autocorrelation sequence. For example, we usually calculate the prediction coefficient $a_m(k) = a_1(1)$ and σ_{w1}^2 first when $m = 1$. Then, we calculate $a_2(1)$, $a_2(2)$, and σ_{w2}^2 when $m = 2$. We conduct the calculation in the way mentioned above until we get $a_p(1)$, $a_p(2)$, $\ldots$, $a_p(p)$, and σ_{wp}^2 when $m = p$. The recursive calculation will be stopped if the precision is satisfied. Equation (9.123) shows the general form of the recursive equation for calculating the prediction coefficients and mean square error:

$$\begin{cases} a_m(k) = a_{m-1}k) + a_m(m)a_{m-1}(m-k) \\ a_m(m) = -\dfrac{R_{xx}(m) + \sum\limits_{k=1}^{m-1} a_{m-1}(k)R_{xx}(m-k)}{E_{m-1}} \\ E_m = \sigma_{wm}^2 = [1 - a_m^2(m)]E_{m-1} = R_{xx}(0) \prod\limits_{k=1}^{m} [1 - a_k^2(k)] \end{cases} \tag{9.123}$$

where $a_m(m)$ is the reflection coefficient. To carry out the Levinson–Durbin recursive operation, we need to know the autocorrelation function and initial values of $E_0 = R_{xx}(0)$ and $a_0(0) = 1$, as well as the order p of the AR model.

9.7.4 The Order Determination of the AR Model

The order estimation is very important for the models. In practice, the order of the model is always unknown, which needs to be estimated based on the observation data. If the order estimated is too low, then the mean square error of the parameter estimation will be large. If the order estimated is too high, then the calculation burden will be increased.

1. FPE criterion
The final prediction error (FPE) criterion is defined as

$$\mathrm{FPE}(m) = \sigma_{wm}^2 \left(\frac{N+m+1}{N-m-1} \right) \tag{9.124}$$

where N is the data length and σ_{wm}^2 is the variance of the residual error. With the increase of the fitting order, σ_{wm}^2 will decrease and $\frac{N+m+1}{N-m-1}$ will increase. At the order $m = p$, the FPE(m) will reach its minimal value, which is taken as the order of the model.

2. AIC criterion

This Akaike information (AIC) criterion determines the order of the AR model by means of the maximum logarithm likelihood function, given as

$$\text{AIC}(m) = \ln \sigma^2_{wm} + \frac{2m}{N} \tag{9.125}$$

When AIC(m) reach its minimal value, $m = p$ is taken as the optimal order of the model.

ℹ Exercises

9.1 Describe the concepts of random variable, random process, and random signal.

9.2 Describe the concepts of the means, variance, covariance function, and correlation function for the random process.

9.3 Describe the concepts of the probability density function and the probability distribution function for a random signal.

9.4 Describe the concepts of the stationary random signal and the nonstationary random signal.

9.5 Describe the concept of ergodicity of a random signal.

9.6 Describe the concept of the power spectrum density of a random signal.

9.7 What is the equivalent noise bandwidth of a system?

9.8 What are the rectangular equivalent bandwidth and the RMS equivalent bandwidth.

9.9 Describe the properties of the main random processes.

9.10 Describe the characteristics and differences of white noise and colored noise.

9.11 Describe the classical analysis method of the random signals.

9.12 Describe the modern analysis method of the random signals.

9.13 Describe the characteristics of AR, MA, and ARMA models.

9.14 Suppose a sinusoidal signal with a random phase is $x(t) = A \sin(\Omega_0 t + \varphi)$. The phase φ obeys the uniform distribution as $p(\varphi) = \begin{cases} \frac{1}{2\pi}, & 0 \le \varphi < 2\pi \\ 0, & \text{others} \end{cases}$. Find the PDF of $x(t)$, and determine whether it is correlated with time t.

9.15 Suppose $z(t) = \cos(\omega_0 t + \varphi)$ is a sinusoidal carrier with a random phase as $p(\phi) = \begin{cases} \frac{1}{2\pi}, & 0 \le \phi < 2\pi \\ 0, & \text{others} \end{cases}$. Modulating $z(t)$ with a signal $r(t)$, we get the AM signal $s(t) = r(t)z(t)$. If $r(t)$ obeys the Rayleigh distribution as $p(r) = \begin{cases} re^{-r^2/2}, & r \ge 0 \\ 0, & r < 0 \end{cases}$, prove that $s(t)$ obeys Gaussian distribution, and find the values of m_s and σ_s^2. Suppose that θ and $r(t)$ are independent of each other.

9.16 (1) Suppose $x(t) = a$ and a is a binarized random variable, whose probabilities for $+1$ and -1 are p and $q = 1 - p$, respectively. Find the time average, the ensemble average, the time autocorrelation, and the ensemble autocorrelation functions. Describe whether the process has the ergodic property.

(2) If a stationary random signal $x(t)$ is added with a deterministic signal $y(t)$ as $z(t) = x(t) + y(t)$, is $z(t)$ a stationary signal?

9.17 The autocorrelation function of a stationary random signal $x(t)$ is given as follows. Find the PSD and the mean square values, and describe whether the random process has the direct component and the periodic component.

(1) $R_{xx}(\tau) = 4e^{-|\tau|}\cos\pi\tau + \cos 3\pi\tau$;　　(2)$R_{xx}(\tau) = 25e^{(-4|\tau|)}\cos\Omega_0\tau + 16$

9.18 Determine which functions are correct expressions for the PSDs of real-valued signals based on the properties of PSD function. Describe their means and mean square values for correct functions.

(1) $S_1(\Omega) = \frac{\Omega^2 + 9}{(\Omega^2 + 4)(\Omega + 1)^2}$;　　　　(2) $S_2(\Omega) = \frac{\Omega^2 + 1}{\Omega^4 + 5\Omega^2 + 6}$;

(3) $S_3(\Omega) = \frac{\Omega^2 + 4}{\Omega^4 - 4\Omega^2 + 3}$;　　　　(4) $S_2(\Omega) = \frac{\Omega^2 + 1}{\Omega^4 + 5\Omega^2 + 6}$

9.19 An AM signal is given as $y(t) = a(t)\cos(\Omega_0 t + \varphi)$, where Ω_0 is a constant, φ is uniformly distributed in $[0, 2\pi]$, $a(t)$ is a random process, whose autocorrelation function is $R_{aa}(\tau)$, and PSD function is $S_{aa}(\Omega)$. $a(t)$ and θ are independent with each other. Find the correlation function and the PSD of $y(t)$, expressed by $R_{aa}(\tau)$ and $S_{aa}(\Omega)$.

9.20 A system is given in Figure E9.1, in which $x(t)$ is a stationary random process. Prove the PSD of $y(t)$ is $S_{yy}(\Omega) = 2S_{xx}(\Omega)(1 + \cos\Omega T)$.

Figure E9.1

9.21 The autocorrelation function of a random signal $x_1(t)$ is $R_1(\tau) = A_1 e^{-|\tau|}$. The autocorrelation function of $x_2(t)$ is $R_2(\tau) = A_2 e^{-|\tau|}$. Find the autocorrelation function $R_{xx}(\tau)$ of $x(t) = x_1(t) + x_2(t)$. Suppose that

(1) $x_1(t)$ and $x_2(t)$ are independent of each other.

(2) $x_1(t)$ and $x_2(t)$ are from the same signal source, satisfying $x_2(t) = Kx_1(t)$, $K \neq 1$ (constant).

9.22 (1) The complex random signal $z(t)$ is constructed by the joint real stationary processes $x(t)$ and $y(t)$, satisfying $z(t) = x(t) + jy(t)$. Prove $E[|z(t)|^2] = R_{xx}(0) + R_{yy}(0)$.

(2) Prove $E[|z(t + \tau) + z(t)|^2] = 2R_{xx}(0) + 2Re[R_{zz}(\tau)]$.

9.23 A signal is given as $s(t) = x(t)\cos\Omega_0 t - y(t)\sin\Omega_0 t$, where $x(t)$ and $y(t)$ are complex random processes. The autocorrelation and cross-correlation of $x(t)$ and $y(t)$ are all known, and Ω_0 is a fixed value.

(1) Find the autocorrelation function of $x(t)$.

(2) If $R_{xx}(\tau) = R_{yy}(\tau)$ and $R_{xy}(\tau) = 0$, prove $R_{ss}(\tau) = R_{xx}(\tau)\cos\Omega_0\tau$.

9.24 A random signal is given as $x(t) = A\sin(\Omega_0 t + \varphi)$, where A and Ω_0 are constants, and φ is a uniformly distributed random variable in $[0, 2\pi]$. $y(t) = x^2(t)$.

(1) Find the autocorrelation function $R_{yy}(\tau)$ of $y(t)$.

(2) Find the cross-correlation function $R_{xy}(\tau)$ between $x(t)$ and $y(t)$. Describe whether $x(t)$ and $y(t)$ are jointly stationary.

9.25 The zero mean statistically independent random processes $x(t)$ and $y(t)$ have the following autocorrelation functions as $R_{xx}(\tau) = e^{-|\tau|}$ and $R_{yy}(\tau) = \cos 2\pi\tau$. Construct new processes based on $x(t)$ and $y(t)$ as $w_1(t) = x(t) + y(t)$, and $w_2(t) = x(t) - y(t)$. Find $R_{w_1 w_2}(\tau)$.

9.26 The samples of a random sequence $x(n)$ are independent with each other, with the identical Gaussian distribution, whose means are all 3, and variances are all 4.

(1) If $y(n) = \frac{1}{2}(x(n) + x(n-1))$, find the mean, variance, and PDF of $y(n)$.

(2) If $z(n) = \frac{1}{2}(x(n) - x(n-1))$, find the mean, variance, and PDF of $z(n)$.

9.27 The PDF of an input signal $x(n)$ is $p[x(n)] = \begin{cases} 2e^{-2x(n)}, & x(n) \geq 0 \\ 0, & \text{others} \end{cases}$.

(1) Prove $E[x(n)] = 1/2$.

(2) If $y = 2x(1) + 4x(2)$, $x(1)$ and $x(2)$ all obey the above distribution, find $E(y)$.

9.28 The samples of a random sequence $x(n)$ are independent with each other, uniformly distributed in $[-1, +1]$. The signal processing is conducted based on the following relations:

(1) $y(n) = x(n) - x(n-1)$; (2) $z(n) = x(n) + 2x(n-1) + x(n-2)$;

(3) $w(n) = -\frac{1}{2}w(n-1) + x(n)$.

Find (1) the mean m_x and variance σ_x^2 of $x(n)$. (2) The autocorrelation functions and PDFs of $y(n)$, $z(n)$ and $w(n)$.

(Hint: You may try to find $R_{yy}(m) = ?$ when $|m| > 2$, and $R_{zz}(m) = ?$ when $|m| > 3$.)

9.29 The autocorrelation function is given as $r_{xx}(k) = \rho^k$, $k = 0, 1, 2, 3$. Solve for the parameters of AR(3) model by both Yule–Walker equation and Levinson–Durbin recursive method.

9.30 Suppose $N = 5$ and the data samples are as $x(0) = 1$, $x(1) = 2$, $x(2) = 3$, $x(3) = 4$, $x(4) = 5$. If the order of the AR model is 3, find the predicted value $\hat{x}(4)$ by using Levinson–Durbin recursive method.

10 Correlation Estimation and Power Spectral Density (PSD) Estimation of Random Signals

10.1 Introduction

The correlation function is a measure of the similarity between two deterministic or random signals. The power spectral density (PSD) function is the Fourier transform of the correlation function, which is the frequency representation of a random signal in the frequency domain.

The correlation function and the PSD function play important roles in signal processing, and the estimation of both functions is an important part in signal processing. On the basis of the basic concepts of the correlation function and the PSD function, this chapter introduces the basic theory and methods for the estimation of both correlation function and PSD function. This chapter also introduces the basic theory and applications of the cepstrum, and finally gives some typical applications of the spectral analysis.

10.1.1 Basic Tasks of Parameter Estimation

Parameter estimation is a method to determine the unknown parameters of the ensemble distribution according to the samples taken from the population. The task of signal parameter estimation is to determine a certain or some parameters of the signal from the noisy signal using the finite observation data. The most common estimation problem is to determine the main statistics based on samples of a given set of random variables. Suppose a set of data samples is given as x_i, $i = 1, 2, \ldots, N$; the estimations of the mean and variance from x_i are shown as

$$\hat{m}_x = \frac{1}{N} \sum_{i=1}^{N} x_i$$
$$\hat{\sigma}_x^2 = \frac{1}{N} \sum_{i=1}^{N} [x_i - E(x)]^2$$

(10.1)

Figure 10.1 shows the block diagram for estimating parameters from a noisy signal.

We see from the figure that the observation data x is composed of signal s and additive noise n, i.e., $x = s(\theta) + n$, where θ denotes the parameter of the signal which needs to be estimated. The parameter estimation problem requires obtaining the estimated value $\hat{\theta}$ of the true value θ and to have the relation between θ and $\hat{\theta}$ to achieve the optimal. For example, we generally use the minimum-mean square-error criterion to have $E[(\theta - \hat{\theta})^2]$ minimized.

https://doi.org/10.1515/9783110465082-010

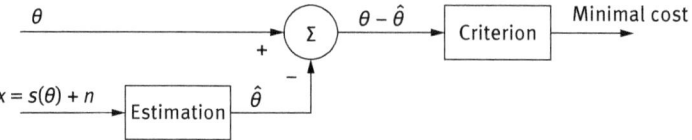

Figure 10.1: The block diagram for estimating parameters from a noisy signal.

10.1.2 Evaluation Criteria of Parameter Estimation

1. Unbiasedness of estimation

(1)　Deviation of estimation

The estimation deviation, also known as the estimation bias, is the difference between the estimated value $\hat{\theta}$ and the truth value θ or the mean value $E[\hat{\theta}]$, which can be used to measure the accuracy of the estimated results.

　　For the estimation of nonrandom parameters, if $E[\hat{\theta}] = \theta$, then $\hat{\theta}$ is referred to as the unbiased estimate of θ. If $E[\hat{\theta}] \neq \theta$, then $\hat{\theta}$ is the biased estimate of θ, whose bias is $b(\hat{\theta}) = E[\hat{\theta}] - \theta$. When data sample size N increases and if $\lim_{N \to \infty} b(\hat{\theta}) = 0$ is satisfied, then the estimation is called the asymptotically unbiased estimation.

　　In random parameter estimation, because the random parameter does not have the truth value, the truth value is substituted by its mean value. If $E[\hat{\theta}] = E[\theta]$, then $\hat{\theta}$ is called the unbiased estimation. If $E[\hat{\theta}] \neq E[\theta]$, then $\hat{\theta}$ is called the biased estimation, whose bias is $b(\hat{\theta}) = E[\hat{\theta}] - E[\theta]$. If $\lim_{N \to \infty} b(\hat{\theta}) = 0$ is satisfied, then the estimation is called the asymptotically unbiased estimation.

　　If θ is not a random quantity, then $E[\theta] = \theta$ holds. Thus, the nonrandom parameter estimation can be used to unify both the nonrandom parameter estimation and the random parameter estimation.

(2)　Variance of estimation

The estimation variance is used to measure the degree of dispersion between the estimated value $\hat{\theta}$ and its mathematical expectation $E[\hat{\theta}]$. That is,

$$\mathrm{Var}[\hat{\theta}] = \sigma_{\hat{\theta}}^2 = E[(\hat{\theta} - E[\hat{\theta}])^2] \tag{10.2}$$

The smaller the estimation variance, the smaller the estimation dispersion, which means that the parameter estimate is concentrated in a small neighborhood of its truth value or mean value.

(3)　Mean square error of the estimation

The mean square error of the estimator can be used to evaluate the quality of the estimation better, which combines the deviation and variance of estimation. It can be proved that the mean square error of the estimator is the summation of the square of the estimated bias and the variance of the estimator. That is,

$$E[(\theta - \hat{\theta})^2] = b^2(\hat{\theta}) + \sigma_{\hat{\theta}}^2 \qquad (10.3)$$

2. Effectiveness of estimation
If we need to compare the pros and cons of two unbiased estimates of the same parameter in the case of the same sample size N, we usually see which one of the estimates has the smaller variance. If

$$\mathrm{Var}[\hat{\theta}_1] < \mathrm{Var}[\hat{\theta}_2] \qquad (10.4)$$

then $\hat{\theta}_1$ is the efficient estimator of the parameter θ or $\hat{\theta}_1$ is more efficient than $\hat{\theta}_2$.

3. Consistency of estimation
If the sample size $N \to \infty$ and the parameter estimate is close to the truth value, i.e., the bias and the variance tend to be zero, then the estimation is a consistent estimation, which is shown as

$$\lim_{N \to \infty} E[(\hat{\theta} - \theta)^2] = 0 \qquad (10.5)$$

10.2 Correlation Functions and Power Spectral Density Functions

10.2.1 Correlation Functions

1. Definition of the correlation function
The concept of the correlation function has been given in Chapter 9. Here, we mainly introduce the properties of the correlation function of the wide-sense stationary random signals.

The autocorrelation function of a wide-sense stationary random signal $X(t)$ and the cross-correlation between $X(t)$ and $Y(t)$ are given as follows:

$$R_{XX}(\tau) = E[X(t)X^*(t+\tau)] = \int_{-\infty}^{+\infty} X(t)X^*(t+\tau)dt$$
$$R_{XY}(\tau) = E[X(t)Y^*(t+\tau)] = \int_{-\infty}^{+\infty} X(t)Y^*(t+\tau)dt \qquad (10.6)$$

where "*" denotes the complex conjugate. If the wide-sense stationary discrete random sequences $X(n)$ and $Y(n)$ are considered, the autocorrelation function and the cross-correlation function are defined as follows:

$$R_{XX}(m) = E[X(n)X^*(n+m)] = \sum_{n=-\infty}^{+\infty} X(n)X^*(n+m)$$

$$R_{XY}(m) = E[X(n)Y^*(n+m)] = \sum_{n=-\infty}^{+\infty} X(n)Y^*(n+m) \tag{10.7}$$

If the wide-sense stationary processes $X(n)$ and $Y(n)$ are ergodic, then the ensemble averages shown in eqs. (10.6) and (10.7) can be realized by the time averages of sample signals $x(n)$ and $y(n)$ as

$$R_{xx}(m) = \lim_{N\to\infty} \frac{1}{2N+1} \sum_{n=-N}^{N} x(n)x^*(n+m)$$

$$R_{xy}(m) = \lim_{N\to\infty} \frac{1}{2N+1} \sum_{n=-N}^{N} x(n)y^*(n+m) \tag{10.8}$$

where N is the data length. We will take $x(n)$ and $y(n)$ as the ergodic random signals in the following discussion.

2. Main properties of the correlation function

(1)   The autocorrelation function is even, which is shown as

$$R_{xx}(m) = R_{xx}(-m) \tag{10.9}$$

(2)   The autocorrelation function gets its maximal value at $m = 0$:

$$R_{xx}(m) \le R_{xx}(0) \tag{10.10}$$

(3)   The autocorrelation function of a periodic signal is still periodic with the same period, but it does not hold the phase information of the original signal.
(4)   The autocorrelation function of a random signal attenuates to zero quickly with the increase of $|m|$.
(5)   The cross-correlation function is neither even nor odd, but it satisfies

$$R_{xy}(-m) = R_{yx}(m) \tag{10.11}$$

(6)   The cross-correlation function of two periodic signals is still periodic with the same period, but it holds the phase difference of the original signals.

The above properties of the correlation function are important for determining the type of different signals.

3. Autocorrelation function of commonly used random signals

(1)   Autocorrelation function of the white noise
The autocorrelation function of a white noise $x(n) = w(n)$ is given as

$$R_{xx}^{WN}(m) = \frac{N_0}{2}\delta(m) \tag{10.12}$$

where N_0 is the power spectrum density of the white noise. If the bandwidth of the white noise is limited as W, then the white noise becomes a bandlimited white noise, whose autocorrelation function is

$$R_{xx}^{BLWN}(m) = \frac{WN_0}{2} \cdot \frac{\sin Wm}{Wm} \tag{10.13}$$

(2)   Autocorrelation function of an AR(p) signal
An AR(p) signal is shown as

$$x(n) = -\sum_{k=1}^{p} a_k x(n-k) + w(n) \tag{10.14}$$

where $w(n)$ denotes the white noise and p is the order of the AR model. The autocorrelation function is shown as

$$R_{xx}(m) = -\sum_{k=1}^{p} a_k R_{xx}(m-k) \tag{10.15}$$

(3)   Autocorrelation function of an MA(q) signal
An MA(q) signal is shown as

$$x(n) = \sum_{k=0}^{q} b_k w(n-k) \tag{10.16}$$

where q is the model order. The autocorrelation function of $x(n)$ is

$$R_{xx}(m) = \sigma^2 \sum_{k=1}^{q-m} b_k b_{k+m}, \quad 0 \le m \le q \tag{10.17}$$

(4)   Autocorrelation function of an ARMA(p, q) signal
An ARMA(p, q) signal is shown as

$$x(n) = \sum_{k=0}^{q} b_k w(n-k) - \sum_{k=1}^{p} a_k x(n-k) \tag{10.18}$$

The autocorrelation function of $x(n)$ is

$$R_{xx}(m) = - \sum_{k=1}^{p} a_k R_{xx}(m-k) + f_m(\boldsymbol{a}, \boldsymbol{b}), \quad m = 1, 2, \ldots, p \qquad (10.19)$$

where $f_m(\boldsymbol{a}, \boldsymbol{b}) = \sum_{k=0}^{q} b_k w(n-k) x(n+m), \quad m = 1, 2, \ldots, p$, is a nonlinear function of parameter vectors $\boldsymbol{a}$ and $\boldsymbol{b}$ of the ARMA (p, q) model.

10.2.2 Power Spectral Density Functions

1. Definition of the power spectral density function

The power spectral density (PSD) function is an important tool for analysis of random signals. The PSD function of the wide-sense stationary discrete random signal has two equivalent definitions, which are shown as follows:

$$P_{xx}(e^{j\omega}) = \sum_{m=-\infty}^{+\infty} R_{xx}(m) e^{-j\omega n} \qquad (10.20)$$

$$P_{xx}(e^{j\omega}) = \lim_{N \to \infty} E\left[\frac{1}{2N+1} \left| \sum_{n=-\infty}^{+\infty} x(n) e^{-j\omega n} \right|^2 \right] \qquad (10.21)$$

where N is the data length. It can be proved that both definitions are equivalent.

2. The relationship between correlation function and PSD function

The definition of the PSD function in eq. (10.20) is called the Wiener–Khintchine theorem, in which $R_{xx}(m)$ is the autocorrelation function of the signal $x(n)$. Obviously, the autocorrelation function and the PSD function of $x(n)$ are a pair of Fourier transform.

The cross-PSD function of $x(n)$ and $y(n)$ can also be obtained by

$$P_{xy}(e^{j\omega}) = \sum_{m=-\infty}^{+\infty} R_{xy}(m) e^{-j\omega n} \qquad (10.22)$$

10.3 Estimation for Correlation Sequences

10.3.1 Unbiased Estimation for Autocorrelation Sequences

Suppose the random signal $x(n)$ is ergodic, whose autocorrelation function is as follows:

$$R_{xx}(m) = E[x(n)x^*(n+m)] \qquad (10.23)$$

If the observation data length N is finite and the signal is denoted as $x_N(n)$, then $R_{xx}(m)$ can be obtained by the following estimation:

$$\hat{R}_{xx}(m) = \frac{1}{N} \sum_{n=0}^{N-1} x_N(n) x_N^*(n+m) \tag{10.24}$$

Since $x_N(n)$ has just a finite number of data samples, the number of usable samples is only $N-1-|m|$ for every lag m. We still denote $x_N(n) = x(n)$ and rewrite $\hat{R}_{xx}(m)$ as $\hat{R}_N^{(1)}(m)$. Thus, the unbiased estimation for the autocorrelation function in eq. (10.24) is written as

$$\hat{R}_N^{(1)}(m) = \frac{1}{N-|m|} \sum_{n=0}^{N-1-|m|} x(n) x^*(n+m), \quad |m| \leq N-1 \tag{10.25}$$

where the superscript "(1)" in $\hat{R}_N^{(1)}(m)$ represents the unbiased estimation. The length of $\hat{R}_N^{(1)}(m)$ is $2N-1$, and it is symmetric with respect to $m = 0$.

1. Properties of estimation

(1)   Bias analysis of estimation
Taking the mathematical expectation on both sides of eq. (10.25), we obtain

$$E[\hat{R}_N^{(1)}(m)] = \frac{1}{N-|m|} \sum_{n=0}^{N-1-|m|} E[x(n)x^*(n+m)] = \frac{1}{N-|m|} \sum_{n=0}^{N-1-|m|} R_{xx}(m) \tag{10.26}$$
$$= R_{xx}(m), \quad |m| \leq N-1$$

Obviously, its bias is zero, which is shown as

$$b[\hat{R}_N^{(1)}(m)] = E[\hat{R}_N^{(1)}(m)] - R_{xx}(m) = 0 \tag{10.27}$$

(2)   Mean square value analysis of estimation
The mean square value of $\hat{R}_N^{(1)}(m)$ is given as

$$E[(\hat{R}_N^{(1)}(m))^2] = \frac{1}{(N-|m|)^2} \sum_{n=0}^{N-1-|m|} \sum_{k=0}^{N-1-|m|} E[x(n)x^*(n+m)x(k)x^*(k+m)], \quad |m| \leq N-1 \tag{10.28}$$

If $x(n)$ is a zero-mean white Gaussian random signal, we have

$$E[x(k)x(l)x(m)x(n)] = E[x(k)x(l)]E[x(m)x(n)] + E[x(m)x(k)]E[x(l)x(n)] \\ + E[x(k)x(n)]E[x(l)x(m)] \tag{10.29}$$

Thus, eq. (10.28) can be rewritten as

$$E[(\hat{R}_N^{(1)}(m))^2] = \frac{1}{(N-|m|)^2} \sum_{n=0}^{N-1-|m|} \sum_{k=0}^{N-1-|m|} [R_{xx}^2(m) + R_{xx}^2(n-k) \tag{10.30}$$
$$+ R_{xx}(n-k-m)R_{xx}(n-k+m)], \quad |m| \le N-1$$

Taking further arrangement, we have

$$E[(\hat{R}_N^{(1)}(m))^2] = R_{xx}^2(m) + \frac{1}{(N-|m|)^2} \sum_{r=-(N-1-|m|)}^{N-1-|m|} (N-|m|-|r|)[R_{xx}^2(r) \tag{10.31}$$
$$+ R_{xx}(r-m)R_{xx}(r+m)]$$

where $r = n - k$. If $N \gg |m| - |r|$ is satisfied, the estimation variance of $\hat{R}_N^{(1)}(m)$ is given as

$$\text{Var}[\hat{R}_N^{(1)}(m)] = E\{[\hat{R}_N^{(1)}(m)]^2\} - \{E[\hat{R}_N^{(1)}(m)]\}^2 = E\{[\hat{R}_N^{(1)}(m)]^2\} - R_{xx}^2(m)$$
$$\approx \frac{N}{(N-|m|)^2} \sum_{r=-(N-1-|m|)}^{N-1-|m|} [R_{xx}^2(r) + R_{xx}(r-m)R_{xx}(r+m)] \tag{10.32}$$

When $N \to \infty$, we have

$$\lim_{N \to \infty} \text{Var}[\hat{R}_N^{(1)}(m)] = 0 \tag{10.33}$$

Therefore, $\hat{R}_N^{(1)}(m)$ is a consistent estimation.

It should be noticed that the variance of $\hat{R}_N^{(1)}(m)$ in eq. (10.32) is obtained under the assumption of $N \gg |m| - |r|$. If $|m|$ is close to N, the variance of $\hat{R}_N^{(1)}(m)$ will increase, since the data samples used for calculating $\hat{R}_N^{(1)}(m)$ are reduced.

10.3.2 Biased Estimation for Autocorrelation Sequences

1. Biased estimation of the autocorrelation sequence
If we change the coefficient $1/(N-|m|)$ in eq. (10.25) into $1/N$, we obtain the biased estimation of the autocorrelation sequence:

$$\hat{R}_N^{(2)}(m) = \frac{1}{N} \sum_{n=0}^{N-1-|m|} x(n)x^*(n+m), \quad |m| \le N-1 \tag{10.34}$$

Comparing eqs. (10.25) and (10.34), we see

$$\hat{R}_N^{(2)}(m) = \frac{N - |m|}{N} \hat{R}_N^{(1)}(m), \quad |m| \leq N - 1 \tag{10.35}$$

2. Properties of estimation

(1) Bias analysis of estimation

The mean value of $\hat{R}_N^{(2)}(m)$ is shown as

$$E[\hat{R}_N^{(2)}(m)] = \frac{N - |m|}{N} E[\hat{R}_N^{(1)}(m)] = \frac{N - |m|}{N} R_{xx}(m), \quad |m| \leq N - 1 \tag{10.36}$$

Thus, the bias of $\hat{R}_N^{(2)}(m)$ is

$$b[\hat{R}_N^{(2)}(m)] = R_{xx}(m) - E[\hat{R}_N^{(2)}(m)] = \frac{|m|}{N} R_{xx}(m), \quad |m| \leq N - 1 \tag{10.37}$$

We see from eq. (10.37) that $\hat{R}_N^{(2)}(m)$ is a biased estimate of $R_{xx}(m)$. However, due to

$$\lim_{N \to +\infty} b[R_N^{(2)}(m)] = 0 \tag{10.38}$$

$\hat{R}_N^{(2)}(m)$ is asymptotically unbiased.

(2) Variance analysis of estimation

The variance of the biased estimation for the autocorrelation function is given as

$$\begin{aligned}
\mathrm{Var}[\hat{R}_N^{(2)}(m)] &= \left(\frac{N - |m|}{N}\right)^2 \mathrm{Var}[\hat{R}_N^{(1)}(m)] \\
&\approx \frac{1}{N} \sum_{r = -(N-1-|m|)}^{N-1-|m|} [R_{xx}^2(r) + R_{xx}(r-m)R_{xx}(r+m)]
\end{aligned} \tag{10.39}$$

The above equation holds only when $N \gg |m| - |r|$ is satisfied. Obviously,

$$\lim_{N \to +\infty} \mathrm{Var}[\hat{R}_N^{(2)}(m)] = 0 \tag{10.40}$$

From eq. (10.40) and the asymptotical unbiasedness of $\hat{R}_N^{(2)}(m)$, we see that $\hat{R}_N^{(2)}(m)$ is the consistent estimate of $R_{xx}(m)$.

(3) Comparing with unbiased estimation

Comparing the unbiased estimation $\hat{R}_N^{(1)}(m)$ and the biased estimation $\hat{R}_N^{(2)}(m)$ of autocorrelations, we come to the following conclusions:

First, if the value of $|m|$ is close to N, the mean and bias of $\hat{R}_N^{(1)}(m)$ are not affected, while the mean of $\hat{R}_N^{(2)}(m)$ tends to zero, whose bias tends to $R_{xx}(m)$.

Second, the variance of $\hat{R}_N^{(2)}(m)$ is always less than or equal to that of $\hat{R}_N^{(1)}(m)$.

Third, if the value of $|m|$ is close to N, neither $\hat{R}_N^{(1)}(m)$ nor $\hat{R}_N^{(2)}(m)$ is a good estimate for the autocorrelation. However, for small values of $|m|$, both $\hat{R}_N^{(1)}(m)$ and $\hat{R}_N^{(2)}(m)$ are consistent estimates of $R_{xx}(m)$.

Finally, since the variance of $\hat{R}_N^{(2)}(m)$ is always less than that of $\hat{R}_N^{(1)}(m)$ and other reasons discussed later, the biased estimation $\hat{R}_N^{(2)}(m)$ is generally used in practical applications.

10.3.3 Fast Estimation for Autocorrelation Sequences

In practical applications, we generally use fast Fourier transform (FFT) to calculate the estimation of autocorrelation function. Suppose the data length is N and the observation signal is denoted as $x_N(n)$. Taking $\hat{R}_N^{(2)}(m)$ as an example, we obtain

$$\hat{R}_N^{(2)}(m) = \frac{1}{N}\sum_{n=0}^{N-1} x_N(n)x_N^*(n+m) \tag{10.41}$$

The data length of the correlation for two N-point sequences is $2N-1$. Similar to the fast convolution, the data length of the two input sequences should be padded to $2N-1$ with zeros, and then the FFT algorithm can be used:

$$x_{2N}(n) = \begin{cases} x_N(n), & 0 \le n \le N-1 \\ 0, & N \le n \le 2N-1 \end{cases} \tag{10.42}$$

Thus, eq. (10.41) can be rewritten as

$$\hat{R}_{2N}^{(2)}(m) = \frac{1}{N}\sum_{n=0}^{N-1} x_{2N}(n)x_{2N}^*(n+m) \tag{10.43}$$

The Fourier transform of $x_{2N}(n)$ is denoted as $X_{2N}(e^{j\omega})$. Taking Fourier transform to $\hat{R}_{2N}^{(2)}(m)$, we have

$$\sum_{m=-(N-1)}^{N-1} \hat{R}_{2N}^{(2)}(m)e^{-j\omega m} = \frac{1}{N}\sum_{n=0}^{2N-1} x_{2N}(n)e^{j\omega n}\sum_{l=0}^{2N-1} x_{2N}(l)e^{-j\omega l} = \frac{1}{N}|X_{2N}(e^{j\omega})|^2 \tag{10.44}$$

where $X_{2N}(e^{j\omega})$ can be calculated by FFT. The steps followed for the calculation are summarized as follows:

(1) Pad zeros for $x_N(n)$ to $2N-1$, forming $x_{2N}(n)$. Take FFT to $x_{2N}(n)$, getting $X_{2N}(k)$, $k=0,1,\ldots,2N-1$.

(2) Find the magnitude square of $X_{2N}(k)$ and we will obtain $\frac{1}{N}|X_{2N}(k)|^2$.

(3) Take the inverse Fourier transform to $\frac{1}{N}|X_{2N}(k)|^2$ and we will obtain $\hat{R}_0^{(2)}(m)$.

$\hat{R}_0^{(2)}(m)$ here is not simply the same as $\hat{R}_{2N}^{(2)}(m)$. In fact, it is a new sequence which is obtained by shifting the part of $-(N-1) \leq m < 0$ in $\hat{R}_{2N}^{(2)}(m)$ to the right by $2N$ samples. However, both PSDs of $\hat{R}_0^{(2)}(m)$ and $\hat{R}_{2N}^{(2)}(m)$ are the same.

Example 10.1: Generate a white noise $w(n)$ and a sinusoidal signal $s(n)$ with MATLAB programming, and then combine them into a noisy sinusoidal signal $x(n)$. Calculate the autocorrelation functions $R_{ww}(m)$ and $R_{xx}(m)$ of $w(n)$ and $x(n)$, respectively, and plot their curves.

Solution: The MATLAB program is shown as follows:

```
clear all; clc;
fs=1000; t=0:1/fs:(1-1/fs); nn=1:1000; maxlag=100; f=12;
w=zeros(1,1000); s=zeros(1,1000); w=randn(1,1000); s=sin(2*pi*f/fs.
*nn); x=0.5*w+s;
[cw,maxlags]=xcorr(w,maxlag); [cx,maxlags]=xcorr(x,maxlag);
figure(1)
subplot(221); plot(t,w); xlabel('n'); ylabel('w(n)'); title('White
noise'); axis([0 1 -4 4]);
subplot(222); plot(maxlags/fs,cw); xlabel('m'); ylabel('Rww(m)'); axis
([-0.1 0.1 -300 1200]);
title('Correlation function of white noise');
subplot(223); plot(t,x); xlabel('n'); ylabel('x(n)'); title('Noisy
sinusoidal signal'); axis([0 1 -4 4])
subplot(224); plot(maxlags/fs,cx); xlabel('m'); ylabel('Rxx(m)'); axis
([-0.1 0.1 -800 1000]);
title('Correlation function of noisy sinusoidal');
```

Figure 10.2 demonstrates the original signals and their correlation functions.

10.4 Classical Methods for PSD Estimation

10.4.1 The General Development of PSD Estimation

Power spectrum estimation is one of the main contents of signal processing. It mainly studies the characteristics of signals in the frequency domain, whose purpose is to extract signals and their parameters contaminated by noises from the limited observation data.

The earliest study of the spectrum can be traced back to the ancient times of human beings. At that time, people constructed the concepts of the year, month, and day based on experiences, and summed up the timing method and the calendar. In 600 BC, the Greek mathematician Pythagoras produced pure sinusoidal vibrations with a fixed string at both ends. The British scientist Newton first gave the concept of

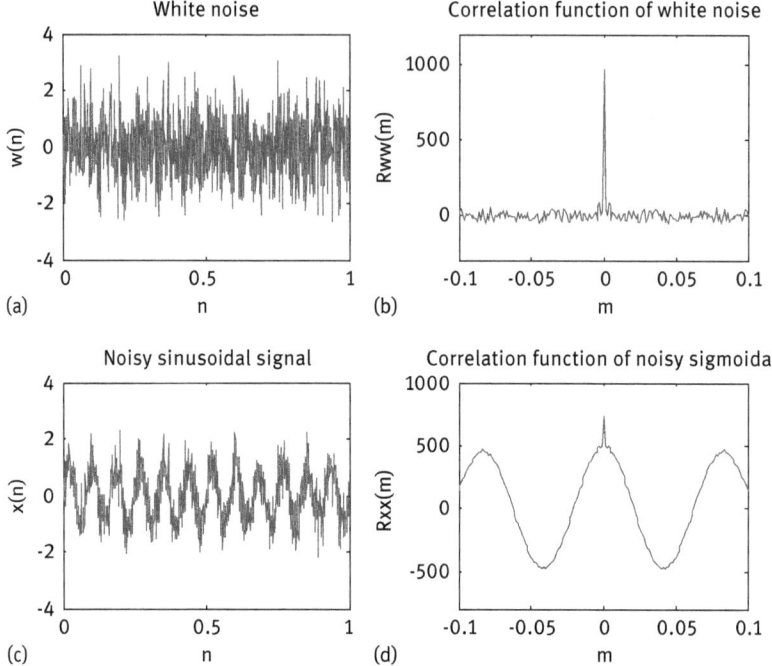

Figure 10.2: The original signals and their correlation functions: (a) white noise, (b) correlation function of white noise, (c) a noisy sinusoidal signal, and (d) correlation function of the noisy sigmoidal.

"spectrum," who analyzed a beam of sunlight into a rainbow of the spectrum by using a prism. In 1822, the French engineer Fourier proposed the famous harmonic analysis theory, known as Fourier series, which is still the theoretical basis for signal analysis and processing.

At the end of the nineteenth century, Schuster proposed the concept of "periodogram," which is the earliest formulation of the classical spectrum estimation, and is still in use today.

The variance of the periodogram estimation is poor, which urges people to study other methods. In 1927, Yule proposed a linear regression equation to model the time series, which constitutes an important method, the parameter model method, in modern spectrum estimation. Walker used Yule method to study the decaying sinusoidal time series and obtained the Yule–Walker equation.

In 1930, the famous control theory expert Wiener first defined the autocorrelation function and power spectral density (PSD) of a random process, and established the famous Wiener–Khintchine theorem.

In 1948, Bartlett first proposed a method to estimate the power spectrum using the autoregressive model coefficients. According to the structural characteristics of the Toeplitz matrix, Levinson proposed a fast algorithm for solving the Yule–Walker equation.

In 1949, according to the Wiener–Khintchine theorem, Tukey proposed an auto-correlation spectrum estimation method for finite data samples.

In 1958, Blackman and Tukey studied the autocorrelation method for the classical spectrum estimation, known as the BT method. The periodogram method and the BT method can be realized by FFT, and both methods are still the commonly used methods of spectrum estimation.

In 1965, Cooley and Tukey proposed the FFT algorithm, which effectively promoted the development of spectrum estimation.

Generally speaking, modern spectrum estimation began in 1967, which is mainly aimed at solving the problems of poor resolution and variance of classical spectrum estimation. Modern spectrum estimation can be divided into two classes, parameter model spectrum estimation and nonparametric model spectrum estimation. The former includes the AR model, MA model, ARMA model, PRONY index model, and so on. The latter includes the minimum variance method, multiple signal classification (MUSIC) method, and so on. In 1967, Berg proposed the maximum entropy spectral analysis method, inspired by the linear prediction method. In 1968, Parzen formally proposed the autoregressive (AR) spectral estimation method. In 1971, van den Bos proved the equivalence of the maximum entropy spectrum analysis and the AR spectrum estimation. At this point, other spectral estimation methods are further studied. In 1972, Prony proposed a spectral estimation method equivalent to the autoregression method. In 1973, Pisarenko proposed the harmonic decomposition method for estimating the frequency of the sinusoidal signal. In 1981, Schmidt proposed the MUSIC algorithm.

The contents of modern spectrum estimation are very wide ranging, whose application fields are very extensive. The classification of PSD estimation methods is summarized in Figure 10.3.

10.4.2 The Periodogram Method

The periodogram method is a kind of classical PSD estimation method. The main advantage of this method is that the FFT can be used to estimate the spectrum. This method can be applied in the case of long sequence. When the sequence length is long enough, we can get a better power spectrum estimation by using the improved method. The discrete-time Fourier transform of a finite-length real stationary random sequence $x(n)$ ($\Delta t = 1$ is the sampling interval) is as follows:

$$X(e^{j\omega}) = \sum_{n=0}^{N-1} x(n) e^{-j\omega n} \tag{10.45}$$

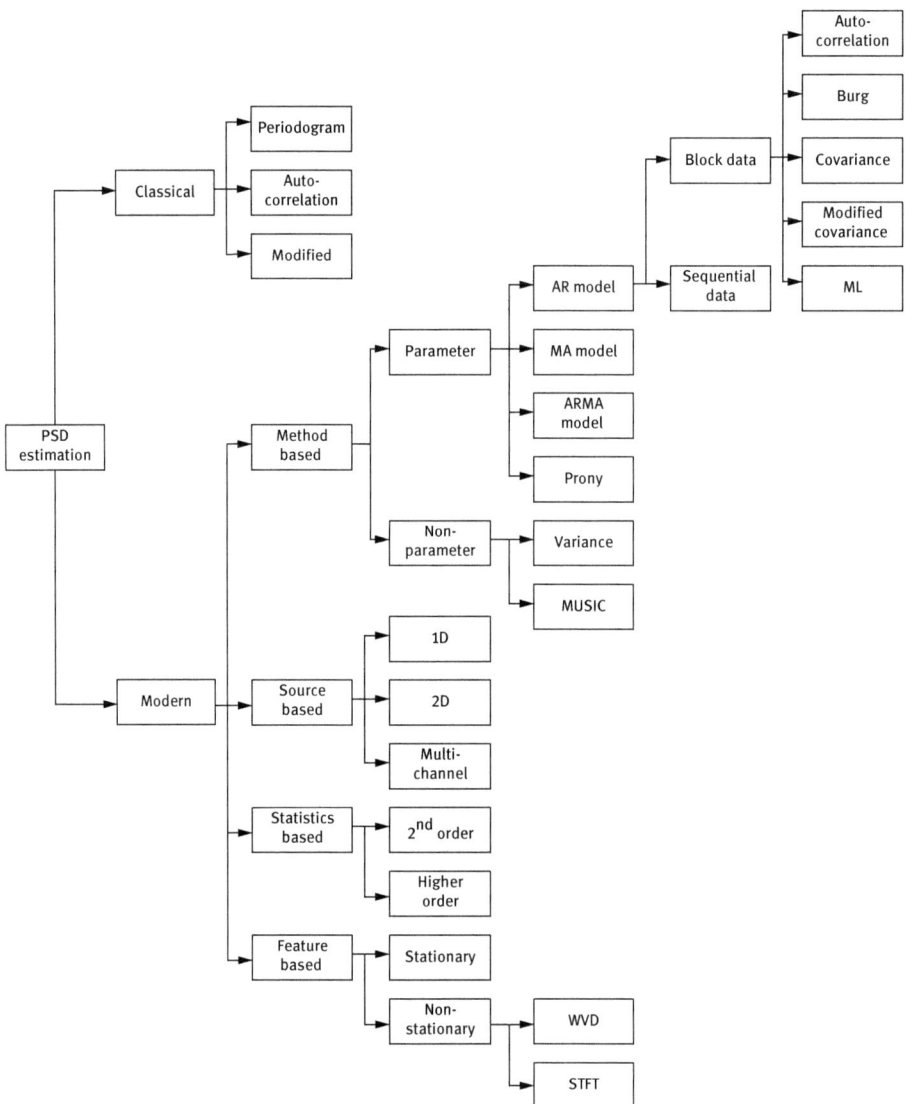

Figure 10.3: The classification of PSD estimation methods.

The periodogram $P_{\text{per}}(e^{j\omega})$ is given as

$$P_{\text{per}}(e^{j\omega}) = \frac{1}{N}X(e^{j\omega})X^*(e^{j\omega}) = \frac{1}{N}|X(e^{j\omega})|^2 \tag{10.46}$$

The above equation is called the direct form of classical spectrum estimation. The PSD can also be calculated by Fourier transform of the autocorrelation function, which is called the indirect form of classical spectrum estimation, denoted as the BT method. The biased estimate of the autocorrelation function is

$$\hat{R}_N^{(2)}(m) = \frac{1}{N} \sum_{n=0}^{N-1-|m|} x(n)x(n+m), \quad -(N-1) \le m \le N-1$$

Taking Fourier transform, we obtain

$$P_{\text{per}}(e^{j\omega}) = \sum_{m=-\infty}^{+\infty} \hat{R}_N^{(2)}(m)e^{-j\omega m} = \frac{1}{N}|X(e^{j\omega})|^2$$

Example 10.2: Suppose $x(n)$ is a discrete-time Gaussian white noise with zero mean and unit variance. Generate such noise sequence with a MATLAB program and calculate its autocorrelation function. Find its PSD using the periodogram algorithm. *Solution:* Three required curves are shown in Figure 10.4.

We see from Figure 10.4 that the PSD estimation result obtained from the periodogram has a large gap with the power spectrum of white noise in theory. There are two main reasons for this gap: first, the length of the original data sequence is too short (only 60 sample points), not enough to fully express the characteristics of white noise; second, the truncation of the original data with a rectangular window also has a certain impact on the result. Therefore, increasing the length of the sequence and

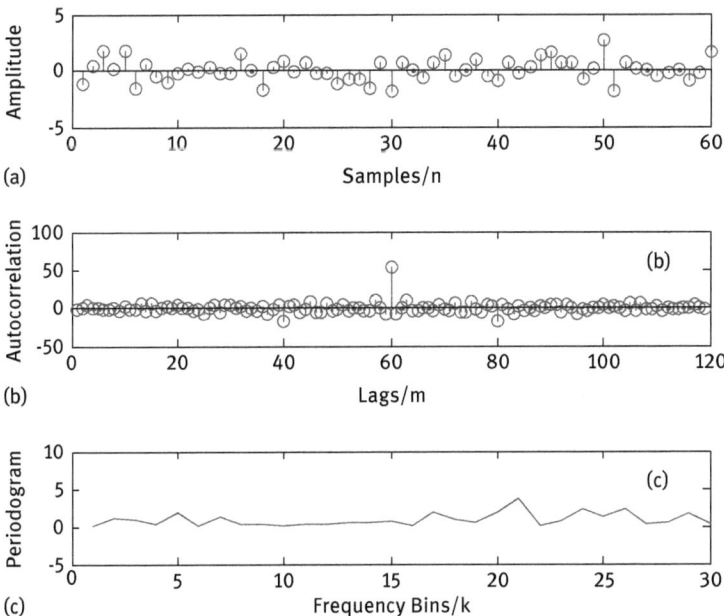

(a)

(b)

(c)

Figure 10.4: The generated white noise, its correlation function, and the periodogram: (a) a Gaussian white noise, (b) its autocorrelation function, and (c) the periodogram.

selecting the appropriate window function are two basic methods to improve the performance of the spectrum estimation.

10.4.3 The Performance of Periodogram for PSD Estimation

1. Mathematical expectation of Periodogram for PSD Estimation

The mathematical expectation of $P_{\text{Per}}(e^{j\omega})$ is

$$E[P_{\text{per}}(e^{j\omega})] = \sum_{m=-(N-1)}^{N-1} E[\hat{R}_N^{(2)}(m)]e^{-j\omega m} = \sum_{m=-(N-1)}^{N-1} \frac{N-|m|}{N} R_{xx}(m)e^{-j\omega m}$$

$$= \sum_{m=-\infty}^{+\infty} w_B(m)R_{xx}(m)e^{-j\omega m} \tag{10.47}$$

where

$$w_B(m) = \begin{cases} 1-\frac{|m|}{N}, & |m| \le N-1 \\ 0, & |m| \ge N \end{cases} \tag{10.48}$$

is a triangle window function, denoted as the Bartlett window, whose Fourier transform is given as

$$W_B(e^{j\omega}) = \frac{1}{N}\left[\frac{\sin(N\omega/2)}{N\sin(\omega/2)}\right]^2 \tag{10.49}$$

Equation (10.47) means that the mathematical expectation $E[P_{\text{per}}(e^{j\omega})]$ of the periodogram is the Fourier transform of the product between the window function $w_B(n)$ and the autocorrelation function $R_{xx}(m)$. Based on the convolution theorem, this product corresponds to the convolution of $W_B(e^{j\omega})$ and $P_{xx}(e^{j\omega})$ in frequency domain, which is shown as

$$E[P_{\text{per}}(e^{j\omega})] = \frac{1}{2\pi}P_{xx}(e^{j\omega}) \star W_B(e^{j\omega}) \tag{10.50}$$

Since $W_B(e^{j\omega})$ is not a unit impulse function, generally

$$E[P_{\text{per}}(e^{j\omega})] \ne P_{xx}(e^{j\omega}) \tag{10.51}$$

This means that $P_{\text{per}}(e^{j\omega})$ is a biased estimation of $P_{xx}(e^{j\omega})$. However, when $N \to \infty$, $W_B(e^{j\omega})$ converges to a unit impulse function. Thus,

$$\lim_{N\to\infty} E[P_{\text{per}}(e^{j\omega})] = P_{xx}(e^{j\omega}) \tag{10.52}$$

This means that the periodogram is an asymptotically unbiased estimation.

2. Variance of Periodogram for PSD Estimation

In general, the exact expression for the variance of periodogram requires the calculation of the fourth-order moment of the random signal, which is rather difficult. Therefore, the approximate expression is generally used. We assume that the random signal $x(n)$, $n = 0, 1, \ldots, N-1$, is a real Gaussian white noise sequence with the zero mean, whose periodogram is expressed as

$$P_{per}(e^{j\omega}) = \frac{1}{N}|X(e^{j\omega})|^2 = \frac{1}{N}\sum_{l=0}^{N-1}\sum_{m=0}^{N-1}x(l)x(m)e^{j\omega m}e^{-j\omega l} \tag{10.53}$$

The covariance of the periodogram at ω_1 and ω_2 is

$$\text{Cov}[P_{per}(e^{j\omega_1}), P_{per}(e^{j\omega_2})] = E[P_{per}(e^{j\omega_1})P_{per}(e^{j\omega_2})] - E[P_{per}(e^{j\omega_1})]E[P_{per}(e^{j\omega_2})] \tag{10.54}$$

where

$$E[P_{per}(e^{j\omega_1})P_{per}(e^{j\omega_2})] = \frac{1}{N^2}\sum_{k=0}^{N-1}\sum_{l=0}^{N-1}\sum_{m=0}^{N-1}\sum_{n=0}^{N-1}E[x(k)x(l)x(m)x(n)]e^{j[\omega_1(k-l)+\omega_2(m-n)]} \tag{10.55}$$

Based on the formula of higher order statistics for the multivariate random variables under the Gaussian white noise condition, we have

$$\begin{aligned}E[x(k)x(l)x(m)x(n)] = &\ E[x(k)x(l)]E[x(m)x(n)] \\ &+ E[x(k)x(m)]E[x(l)x(n)] + E[x(k)x(n)]E[x(l)x(m)]\end{aligned} \tag{10.56}$$

Since $x(n)$ is the white noise sequence,

$$E[x(k)x(l)x(m)x(n)] = \begin{cases} \sigma_x^4, & k=l, \ m=n, \text{ or } k=m, \ l=n, \text{ or } k=n, \ l=m \\ 0, & \text{others} \end{cases} \tag{10.57}$$

Substituting eq. (10.57) into eq. (10.55), we obtain after arrangement

$$E[P_{per}(e^{j\omega_1})P_{per}(e^{j\omega_2})] = \sigma_x^4\left\{1 + \left[\frac{\sin(\omega_1+\omega_2)N/2}{N\sin(\omega_1+\omega_2)/2}\right]^2 + \left[\frac{\sin(\omega_1-\omega_2)N/2}{N\sin(\omega_1-\omega_2)/2}\right]^2\right\} \tag{10.58}$$

Since

$$E[P_{per}(e^{j\omega_1})] = E[P_{per}(e^{j\omega_2})] = \sigma_x^2 \tag{10.59}$$

substituting eqs. (10.58) and (10.59) into eq. (10.54), we obtain

$$\text{Cov}[P_{\text{per}}(e^{j\omega_1}), P_{\text{per}}(e^{j\omega_2})] = \sigma_x^4 \left\{ \left[\frac{\sin(\omega_1 + \omega_2)N/2}{N\sin(\omega_1 + \omega_2)/2} \right]^2 + \left[\frac{\sin(\omega_1 - \omega_2)N/2}{N\sin(\omega_1 - \omega_2)/2} \right]^2 \right\}$$

(10.60)

If $\omega_1 = \omega_2 = \omega$, then the variance of the periodogram is

$$\text{Var}[P_{\text{per}}(e^{j\omega})] = \sigma_x^4 \left\{ 1 + \left[\frac{\sin \omega N}{N \sin \omega} \right]^2 \right\}$$

(10.61)

Obviously, if N tends to infinity, the variance of the periodogram does not tend to zero. Therefore, the periodogram is not the consistent estimation of the power spectrum density.

3. Leakage of the periodogram

In the periodogram estimation, the finite length of the random signal with data length N can be regarded as an infinite random signal truncated by a rectangular window, which is equivalent to the product between an infinite signal sequence and a window function. The Fourier transform of that product is in the form of $\frac{\sin N\pi f}{\sin \pi f}$. Therefore, the spectrum of the finite-length sequence equals the convolution of the spectrum of the ideal sequence and the spectrum of the window function. If the true power of the signal is concentrated in a narrowband, the convolution operation will extend the power of this narrowband to the adjacent frequency band. This phenomenon is called the "spectrum leakage." In addition to the distortion of the spectrum estimation, the leakage phenomenon has a harmful effect on the power spectrum estimation and the measurement of the sinusoidal component. The main lobe of a weak signal is easily leaking to the adjacent side lobe, which causes the vagueness and distortion of the spectrum estimation. In addition, the convolution operation causes the main lobe of the signal to be widened, and the increased width is determined by the main lobe of the window. For the rectangular window, the width of the main lobe of the Fourier transform is approximately equal to the reciprocal of the observation time. Therefore, the resolution of the power spectrum estimation is not high for the observation signal with short data length.

Example 10.3: A random signal is composed of a linear combination of a random phase sinusoidal signal and a white noise as $x(n) = A\sin(\omega_0 n + \varphi) + v(n)$, in which φ is a uniformly distributed random variable in the range of $[-\pi, \pi]$ and $v(n)$ is a white noise with variance $\sigma_v^2 = 1$. The parameters $A = 5$ and $\omega_0 = 0.4\pi$ are set. The data lengths are taken as $N = 64$ and $N = 256$, respectively. Please estimate the PSD of the signal with the periodogram using MATLAB programming. Operate the estimation for 50 times and plot the results.

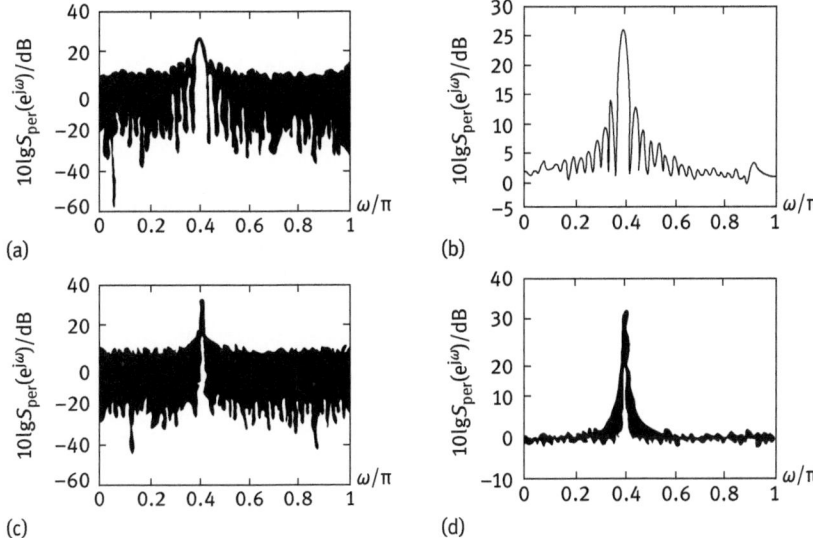

Figure 10.5: An example of PSD estimation with the periodogram: (a) $N = 64$ with 50 runs, (b) $N = 64$ and averaged, (c) $N = 256$ with 50 runs, and (d) $N = 256$ and averaged.

Solution: Based on the requirement of the problem, we obtain the PSD estimation as shown in Figure 10.5.

We see the spectrum leakage in Figure 10.5. We also see in the figure that when the data length is increased, the main lobe of the spectrum becomes narrowed.

10.4.4 Methods for Improving the Performance of Periodogram

There are many improved methods for the periodogram algorithm.

1. Averaged periodogram
Suppose $x_i(n)$, $i = 0, 1, \ldots, K - 1$, are K uncorrelated realizations of a random process $x(n)$, and the data length of each $x_i(n)$ is M, i.e., $n = 0, 1, \ldots, M - 1$. Thus, the periodogram of $x_i(n)$ is shown as

$$P_{\text{per}}^{(i)}(e^{j\omega}) = \frac{1}{M} \left| \sum_{n=0}^{M-1} x_i(n) e^{-j\omega n} \right|^2 \quad i = 1, 2, \ldots, K \tag{10.62}$$

Take the average to such independent periodogram as

$$P_{\text{per}}^{(\text{av})}(e^{j\omega}) = \frac{1}{K} \sum_{i=1}^{K} P_{\text{per}}^{(i)}(e^{j\omega}) \tag{10.63}$$

It can be proved that $P_{\text{per}}^{(\text{av})}(e^{j\omega})$ is the asymptotically unbiased estimation of $P_{xx}(e^{j\omega})$, whose variance is

$$\text{Var}[P_{\text{per}}^{(\text{av})}(e^{j\omega})] = \frac{1}{K}\text{Var}[P_{\text{per}}^{(i)}(e^{j\omega})] \approx \frac{1}{K}P_{xx}^2(e^{j\omega}) \tag{10.64}$$

Furthermore, Bartlett proposed that a random signal of length N can be divided into K segments, each of length M. Thus, each subsignal is denoted as $x_i(n) = x(n + iM)$, $n = 0, 1, \ldots, M - 1$; $i = 0, 1, \ldots, K - 1$. Taking the periodogram to every subsignal and then taking the average, we obtain

$$P_{\text{per}}^{(\text{BT})}(e^{j\omega}) = \frac{1}{M}\sum_{i=0}^{K-1}\left|\sum_{n=0}^{M-1}x(n+iM)e^{-j\omega n}\right|^2 \tag{10.65}$$

The above equation is referred to as the Bartlett periodogram.

2. Window function methods

The window functions are generally used to improve the performance of the PSD estimation. The periodogram can be written as follows:

$$P_{\text{per}}(e^{j\omega}) = \frac{1}{N}\sum_{n=-\infty}^{+\infty}|x(n)w_R(n)e^{-j\omega n}|^2 \tag{10.66}$$

where the rectangular window function $w_R(n)$ is defined as

$$w_R(n) = \begin{cases} 1, & 0 \le n \le N-1 \\ 0, & \text{others} \end{cases} \tag{10.67}$$

Taking the mathematical expectation to eq. (10.66), we obtain

$$E[P_{\text{per}}(e^{j\omega})] = \sum_{m=-\infty}^{+\infty}R_{xx}(m)w_B(m)e^{-j\omega m} \tag{10.68}$$

where $w_B(m)$ is the same as the triangular window defined in eq. (10.48). In fact, $w_B(n)$ is the convolution of $w_R(n)$ and $w_R(-n)$:

$$w_B(n) = \frac{1}{N}w_R(n) \star w_R(-n) \tag{10.69}$$

The smoothness and the main lobe leakage are determined by the type of window function. The commonly used windows include the rectangular window, triangular window, Hanning window, Hamming window, and Blackman window. Table 6.2 and Figure 6.37 demonstrate the frequency characteristics of the windows.

3. Modified periodogram: Welch algorithm

The Welch algorithm modifies the periodogram in two aspects. First, it improves the segmentation method, which allows each segment and its adjacent segment to have a certain overlap, such as 50% overlap. Thus, the number of segments changes as

$$K = \frac{N - M/2}{M/2} \qquad (10.69)$$

where M is the data length of each segment and N is the total data length. Second, use different window functions to improve the spectrum distortion.

The PSD for each segment estimation by Welch algorithm is expressed as

$$P_W^{(i)}(e^{j\omega}) = \frac{1}{MU} \left| \sum_{n=0}^{M-1} x_i(n) w(n) e^{-j\omega n} \right|^2 \qquad (10.70)$$

where $w(n)$ is the window function and $x_i(n)$ denotes the data sequence in the ith segment. The parameter U is defined as

$$U = \frac{1}{M} \sum_{n=0}^{M-1} w^2(n) \qquad (10.71)$$

Thus, the PSD estimation by Welch algorithm is expressed as

$$P_W(e^{j\omega}) = \frac{1}{K} \sum_{i=0}^{K-1} P_W^{(i)}(e^{j\omega}) = \frac{1}{KMU} \sum_{i=0}^{K-1} \left| \sum_{n=0}^{M-1} x_i(n) w(n) e^{-j\omega n} \right|^2 \qquad (10.72)$$

The mathematical expectation of the Welch algorithm is

$$E[P_W(e^{j\omega})] = \frac{1}{K} \sum_{i=0}^{K-1} E[P_W^{(i)}(e^{j\omega})] = E[P_W^{(i)}(e^{j\omega})] \qquad (10.73)$$

When the overlap of the data is 50%, its estimation variance is about

$$\text{Var}[P_W(e^{j\omega})] \approx \frac{9}{8K} P_{xx}^2(e^{j\omega}) \qquad (10.74)$$

We see from eqs. (10.73) and (10.74) that the average of the Welch algorithm does not change the mean of the estimation, while the variance can be improved.

Example 10.4: The combination of two sinusoidal signals and a Gaussian white noise is given. Try to use MATLAB programming to implement PSD estimation by using the periodogram, the windowed averaging periodogram, and Welch algorithm, respectively. Plot the estimated PSD curves.

Solution: The MATLAB program is shown as follows:

```
clear all; clc;
fs=1000; N=1024; Nfft=256; N1=256;
n=0:N-1; t=n/fs; w=hanning(256)'; noverlap=128; dflag='none';
ss=zeros(1,N); ss=sin(2*pi*50*t)+2*sin(2*pi*120*t);
randn('state',0); nn=randn(1,N);
% Periodogram
xn1=ss+nn; Pxx1=10*log10(abs(fft(xn1,Nfft).^2)/N);
% Averged Periodogram
xn2=xn1; Pxx21=abs(fft(xn2(1:256),N1).^2)/N; Pxx22=abs(fft(xn2
  (257:512),N1).^2)/N;
Pxx23=abs(fft(xn2(513:768),N1).^2)/N; Pxx24=abs(fft(xn2(769:1024),
  N1).^2)/N;
Pxx2=10*log10((Pxx21+Pxx22+Pxx23+Pxx24))/4;
% Windowed averaging Periodogram
xn3=xn1; Pxx31=abs(fft(w.*xn3(1:256),N1).^2)/norm(w)^2; Pxx32=abs(fft
  (w.*xn3(257:512),N1).^2)/norm(w)^2;
Pxx33=abs(fft(w.*xn3(513:768),N1).^2)/norm(w)^2; Pxx34=abs(fft(w.
  *xn3(769:1024),N1).^2)/norm(w)^2;
Pxx3=10*log10((Pxx31+Pxx32+Pxx33+Pxx34))/4;
% Welch Algorithm
xn4=xn1; xn4=sin(2*pi*50*t)+2*sin(2*pi*120*t)+randn(1,N); Pxx4=pwelch
  (xn4,window);
% Normalized
Pxx1=Pxx1/max(Pxx1); Pxx2=Pxx2/max(Pxx2); Pxx3=Pxx3/max(Pxx3);
  Pxx4=Pxx4/max(Pxx4);
% Plot
figure(1)
subplot(221); plot(Pxx1(1:N1/2)); axis([1 N1/2 -0.8 1.2]);
xlabel('Frequency/Hz'); ylabel('Normalized PSD estimation');
title('Periodogram'); grid on;
subplot(222); plot(Pxx2(1:N1/2)); axis([1 N1/2 -0.4 1.2]);
xlabel('Frequency/Hz'); ylabel('Normalized PSD estimation');
title('Averaged Periodogram'); grid on;
subplot(223); plot(Pxx3(1:N1/2)); axis([1 N1/2 -0.4 1.2]);
xlabel('Frequency/Hz'); ylabel('Normalized PSD estimation');
title('Windowed averaging Periodogram'); grid on;
subplot(224); plot(Pxx4(1:N1/2)); axis([1 N1/2 -0.4 1.2]);
xlabel('Frequency/Hz'); ylabel('Normalized PSD estimation');
title('Welch Algorithm'); grid on;
```

Figure 10.6 demonstrates the PSD estimation curves of different algorithms.

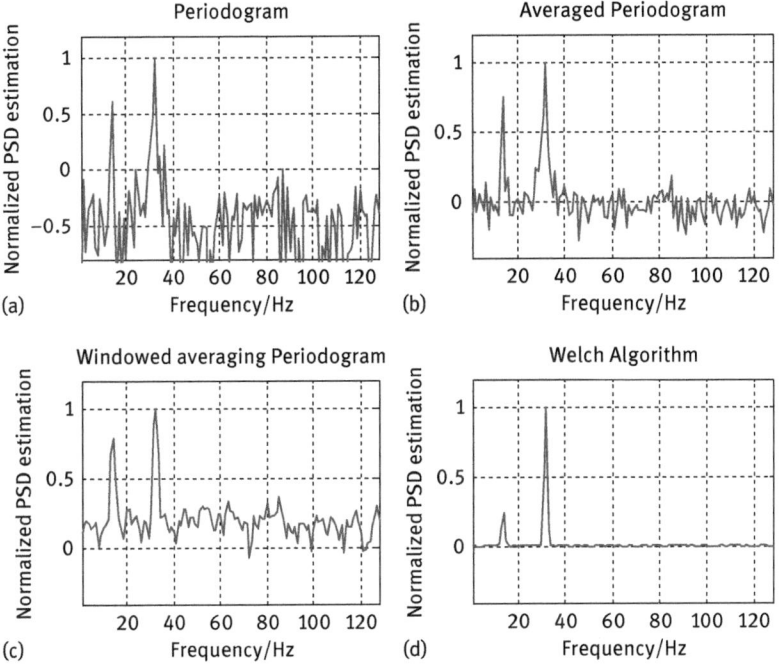

Figure 10.6: The PSD estimation results of different methods: (a) periodogram, (b) averaged periodogram, (c) windowed averaging periodogram, and (d) Welch algorithm.

10.5 Modern Methods for PSD Estimation

10.5.1 Drawbacks of Classical PSD Estimation Methods

From the previous discussion we know that although the classical power spectrum estimation and its improvement methods have the advantage of higher computational efficiency, these methods still have some disadvantages: first, the variance performance of these methods is poor, mainly due to the fact that the power spectrum estimation cannot realize the ideal expectation and limitation operation, leading to an estimate of error; second, the resolution of these methods is low, due to the assumption that the data outside the window are all supposed as zero; and third, these methods have a certain degree of spectrum leakage problem, caused by the windows which are not easy to be eliminated. Since the 1960s, the parameter model- and nonparameter model-based methods for PSD estimation have been referred to as the modern spectrum estimation.

The modern spectrum estimation is very rich in contents, mainly including the parameter model methods, such as AR, MA, ARMA, and PRONY exponential model methods, and the non-parameter model methods, such as the multiple signal classification (MUSIC) algorithm, variance algorithm, and so on. In recent years, many new theories, such as higher order spectrum estimation, multidimensional spectrum estimation, and multichannel spectrum estimation, have been proposed. Among them, the bispectrum and the trispectrum estimations have been widely used.

10.5.2 The PSD Estimation with AR Model

1. AR spectrum estimation

The AR spectrum estimation method is a modern spectral estimation method based on AR parameter estimation. Its basic ideas are as follows:

(1) Select a suitable model for the estimation according to the analysis and under-standing of the signal. The AR model is usually employed.
(2) Estimate the model parameters based on the known observation data. For the AR model, the parameters a_k, $k = 1, 2, \ldots, p$, and σ_w^2 should be estimated.
(3) Estimate the PSD of the random signal based on the estimated AR model parameters.

The basic concept and theory of the AR model and its estimation have been intro-duced in Section 9.7 of Chapter 9. Here, a brief review is given as follows.

Suppose $x(n)$ is a wide-sense stationary random signal, whose AR model can be expressed as

$$x(n) = - \sum_{k=1}^{p} a_k x(n-k) + w(n) \tag{10.75}$$

where a_k, $k = 1, 2, \ldots, p$ denote the AR parameters, p is the order of the model, and $w(n)$ is a white noise. Based on the autocorrelation of $x(n)$, we obtain the Yule–Walker equation as

$$\begin{bmatrix} R_{xx}(0) & R_{xx}(-1) & \cdots & R_{xx}(-p) \\ R_{xx}(1) & R_{xx}(0) & \cdots & R_{xx}(1-p) \\ \vdots & \vdots & \ddots & \vdots \\ R_{xx}(p) & R_{xx}(p-1) & \cdots & R_{xx}(0) \end{bmatrix} \begin{bmatrix} 1 \\ a_1 \\ \vdots \\ a_p \end{bmatrix} = \begin{bmatrix} \sigma_w^2 \\ 0 \\ \vdots \\ 0 \end{bmatrix} \tag{10.76}$$

where $R_{xx}(\cdot)$ represents the autocorrelation function of $x(n)$. Solving the above Yule–Walker equation by the Levinson–Durbin recursive algorithm or other methods, we obtain the estimations to the model parameters $\hat{a}_k$, $k = 1, 2, \ldots, p$, and the variance $\hat{\sigma}_w^2$.

According to eq. (10.75), we state the white noise $w(n)$ as the input of a linear system, and $x(n)$ as its output. Thus, we can obtain the relation between the input power and the output power via Fourier transform as

$$P_{xx}(z) = \sigma_w^2 |H(z)|^2 = \frac{\sigma_w^2}{|A(z)|^2} \tag{10.77}$$

or

$$P_{xx}(e^{j\omega}) = \frac{\sigma_w^2}{|A(e^{j\omega})|^2} \tag{10.78}$$

where $A(e^{j\omega}) = 1 + \sum\limits_{k=1}^{p} a_k e^{-j\omega k}$ is the Fourier transform of a_k. Thus, the PSD can be expressed as

$$P_{AR}(e^{j\omega}) = \frac{\sigma_w^2}{\left|1 + \sum\limits_{k=1}^{p} a_k e^{-j\omega k}\right|^2} \tag{10.79}$$

Substituting the estimated parameters $\hat{a}_k$, $k = 1, 2, \ldots, p$, and $\hat{\sigma}_w^2$ into the above equation, we obtain

$$\hat{P}_{AR}(e^{j\omega}) = \frac{\hat{\sigma}_w^2}{\left|1 + \sum\limits_{k=1}^{p} \hat{a}_k e^{-j\omega k}\right|^2} \tag{10.80}$$

Taking into account the periodicity of the power spectrum of the discrete-time signal, if we sample N frequency bins in the range of $-\pi < \omega \leq \pi$, eq. (10.80) can be written as

$$\hat{P}_{AR}(e^{j2\pi l/N}) = \frac{\hat{\sigma}_w^2}{\left|1 + \sum\limits_{k=1}^{p} \hat{a}_k e^{-j2k\pi l/N}\right|^2} \tag{10.81}$$

Example 10.5: Suppose AR(4) model is expressed as

$$x(n) = 2.7607x(n-1) - 3.8106x(n-2) + 2.6535x(n-3) - 0.9237x(n-4) + w(n)$$

where $w(n)$ is a Gaussian white noise with zero mean and unit variance. One hundred observation samples are obtained from the model. Please estimate the PSD function of the signal by using the periodogram and AR methods.

Solution: The PSD functions can be obtained based on eqs. (10.46) and (10.81), respectively. The order of the AR model is $p = 7$. Figure 10.7 demonstrates the estimation results of both methods.

2. The performance of AR spectral estimation

(1) Stability
The sufficient and necessary condition for the stability of the AR model is that all poles of its system function $H(z)$ are inside the unit circle. If the Yule–Walker

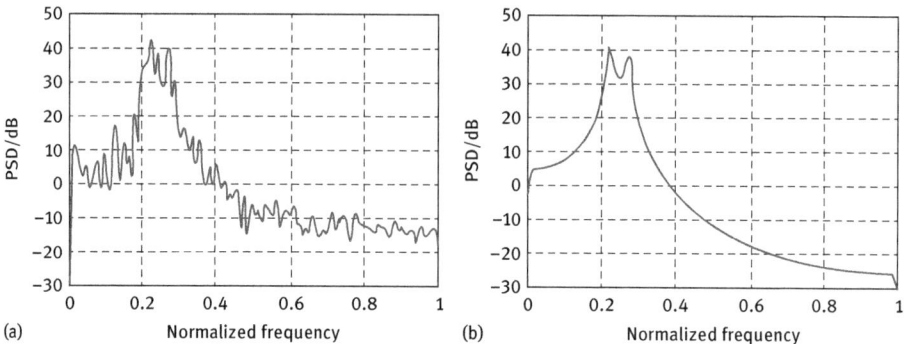

Figure 10.7: The PSD estimation results by periodogram and the AR model: (a) result of periodogram and (b) result of the AR model.

equation is positive definite, then all the roots of $A(z)$ constructed by solution a_k, $k = 1, 2, \ldots, p$, are inside the unit circle.

(2) Smoothness

Since the AR model is a rational fraction, the estimated spectrum is rather smooth. Figure 10.8 demonstrates a comparison between the AR spectrum and the periodogram estimation. Obviously, the AR spectrum is much smoothed.

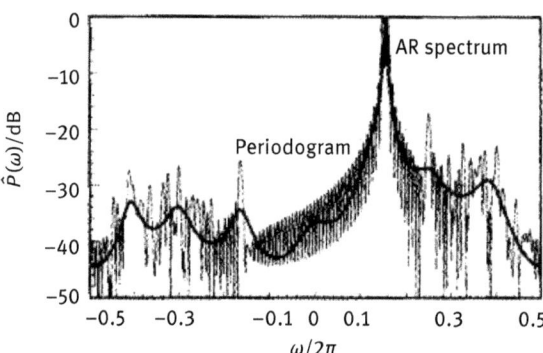

Figure 10.8: A comparison between the AR spectrum and the periodogram estimation. (The thin line represents the result of periodogram, while the thick line represents the result of AR estimation.)

(3) Resolution

The concept of the time-bandwidth product of the signal shows that the frequency resolution of Fourier transform for the signal of data length N is roughly equal to F_s/N, where F_s is the sampling frequency. The resolution of classical PSD estimation is proportional to $2\pi k/N$, which is the width of the main lobe of the window function. In other words, the resolution of classical PSD estimation is inversely proportional to the length of the signal.

Unlike classical spectrum estimation, the modern spectrum estimation implies extrapolation of the data and the autocorrelation function, which may be beyond the given length. For example, the AR model fits the given data in the minimal mean square sense as $\hat{x}(n) = -\sum\limits_{k=1}^{p} a_k x(n-k)$, resulting in a longer length of $\hat{x}(n)$.

Furthermore, $\hat{x}(n)$ can be extrapolated further if $x(n)$ is substituted by $\hat{x}(n)$, which may improve the resolution of AR spectrum estimation further.

On the other hand, the AR spectrum corresponds to an infinite autocorrelation sequence, denoted by $R_a(m)$, which is shown as

$$P_{AR}(e^{j\omega}) = \frac{\sigma_w^2}{\left|1 + \sum\limits_{k=1}^{p} a_k e^{-j\omega k}\right|^2} = \sum\limits_{m=-\infty}^{\infty} R_a(m)e^{-j\omega m}$$

It can be proved that the relation between $R_a(m)$ and $R_{xx}(m)$ is shown as

$$R_a(m) = \begin{cases} R_{xx}(m), & |m| \le p \\ -\sum\limits_{k=1}^{p} a_k R_a(m-k), & |m| > p \end{cases} \qquad (10.82)$$

It is obvious if $|m| > p$ is satisfied, $R_a(m)$ can be extrapolated with eq. (10.82). In the classical spectrum estimation, the value of the autocorrelation when $|m| > p$ is regarded as zero, causing the resolution limit by the window function.

3. Problems in AR spectrum estimation

There are still some problems in the AR spectrum estimation method, mainly including the following:

(1) Spectral line splitting problem: in the spectrum estimation, one spectrum peak may split into two peaks.
(2) Spectral peak frequency shift: the spectrum peak departs from its true position.
(3) Noise problem: If the observed data contain noise, the resolution of AR spectral estimation will decrease because the additional noise makes the estimated spectral peaks widen, smooth, and deviate from the true positions.

Many solutions for solving above problems have been proposed, which can be found in literature works.

10.5.3 The PSD Estimation with Maximum Entropy Method

1. The concepts of information content and entropy

The so-called information content is a measure of information. The information content $I(x_i)$ is usually expressed by the probability $P(x_i)$ of the event x_i, $i = 1, 2, \ldots, N$:

$$I(x_i) = \log_a \frac{1}{P(x_i)} = -\log_a P(x_i) \qquad (10.83)$$

If $a = 2$, the unit is "bit"; if $a = e$, the unit is "nat"; and if $a = 10$, the unit is "Hartley."

Entropy is a measure of the system uncertainty, which is widely used in many areas. In information theory, the entropy is the averaged amount of information, defined as

$$H = \sum_{i=1}^{N} P(x_i)I(x_i) = - \sum_{i=1}^{N} P(x_i)\ln P(x_i) \tag{10.84}$$

The bigger the entropy, the greater the averaged amount of information, too, and, as a result, the greater the uncertainty or randomness of the random signal.

2. Maximum entropy spectrum estimation

As a typical modern spectrum estimation method, the maximum entropy spectral estimation extrapolates the autocorrelation sequence to get the maximal entropy and then estimates the spectrum based on the extrapolated autocorrelation sequence. As a result, the resolution of the spectral estimation becomes better.

Suppose a zero-mean Gaussian stationary random sequence $x(n)$ is given. Its $M + 1$ values of the autocorrelation function are $R_{xx}(0)$, $R_{xx}(1)$, $\ldots$, $R_{xx}(M)$. We can obtain its autocorrelation matrix:

$$\mathbf{R}_M = \begin{bmatrix} R_{xx}(0) & R_{xx}(1) & \cdots & R_{xx}(M) \\ R_{xx}(1) & R_{xx}(0) & \cdots & R_{xx}(M-1) \\ \vdots & \vdots & \ddots & \vdots \\ R_{xx}(M) & R_{xx}(M-1) & \cdots & R_{xx}(0) \end{bmatrix} \tag{10.85}$$

The entropy of an $(M + 2)$-dimensional zero-mean Gaussian random vector is shown as

$$H = \ln (2\pi e)^{\frac{M+2}{2}} [\det[\mathbf{R}_M]]^{1/2} \tag{10.86}$$

where $\det[\cdot]$ denotes the determinant of the matrix. The extrapolated autocorrelation matrix is shown as

$$\hat{\mathbf{R}}_{M+1} = \begin{bmatrix} R_{xx}(0) & R_{xx}(1) & \cdots & R_{xx}(M) & \hat{R}_{xx}(M+1) \\ R_{xx}(1) & R_{xx}(0) & \cdots & R_{xx}(M-1) & R_{xx}(M) \\ \vdots & \vdots & \ddots & \vdots & \vdots \\ R_{xx}(M) & R_{xx}(M-1) & \cdots & R_{xx}(0) & R_{xx}(1) \\ \hat{R}_{xx}(M+1) & R_{xx}(M) & \cdots & R_{xx}(1) & R_{xx}(0) \end{bmatrix}$$

$$= \begin{bmatrix} & & & & \hat{R}_{xx}(M+1) \\ & \mathbf{R}_M & & & R_{xx}(M) \\ & & & & \vdots \\ & & & & R_{xx}(1) \\ \hat{R}_{xx}(M+1) & R_{xx}(M) & \cdots & R_{xx}(1) & R_{xx}(0) \end{bmatrix} \tag{10.87}$$

In order to obtain the maximum entropy in eq. (10.87), we need to have $\det[\hat{\boldsymbol{R}}_{M+1}]$ gone up to its maximum. Define a new vector as

$$\boldsymbol{C} = \left[\, \hat{R}_{xx}(M+1) \quad R_{xx}(M) \quad \cdots \quad R_{xx}(1) \quad R_{xx}(0) \,\right]^{\mathrm{T}} \tag{10.88}$$

By using the matrix identity, $\det[\hat{\boldsymbol{R}}_{M+1}]$ can be rewritten as

$$\det[\hat{\boldsymbol{R}}_{M+1}] = \det[\boldsymbol{R}_M][R_{xx}(0) - \boldsymbol{C}^{\mathrm{T}}\boldsymbol{R}_M^{-1}\boldsymbol{C}] \tag{10.89}$$

Taking the derivative of the above equation and letting it as zero, we obtain

$$[1 \quad 0 \quad \cdots \quad 0 \quad 0]\boldsymbol{R}_M^{-1}\boldsymbol{C} = 0 \tag{10.90}$$

Solving the above equation, we obtain the value of $\hat{R}_{xx}(M+1)$. And then by continuing using the same method, we can obtain $\hat{R}_{xx}(M+2)$ and other estimates of the autocorrelation function. This extrapolation of the autocorrelation function in the maximum entropy principle expands the information of the autocorrelation function, so it can improve the resolution of the spectrum estimation. This method is called the maximum entropy spectrum estimation method.

3. Equivalence with AR spectrum estimation
The Yule–Walker equation of the AR(M) signal is shown as

$$\begin{bmatrix} R_{xx}(0) & R_{xx}(1) & \cdots & R_{xx}(M) \\ R_{xx}(1) & R_{xx}(0) & \cdots & R_{xx}(M-1) \\ \vdots & \vdots & \ddots & \vdots \\ R_{xx}(M) & R_{xx}(M-1) & \cdots & R_{xx}(0) \end{bmatrix} \begin{bmatrix} 1 \\ a_1 \\ \vdots \\ a_M \end{bmatrix} = \begin{bmatrix} \sigma_w^2 \\ 0 \\ \vdots \\ 0 \end{bmatrix} \tag{10.91}$$

where σ_w^2 is the variance of zero-mean white noise $w(n)$ and $a_1, \ldots, a_M$ are AR parameters. The above equation can be rewritten as

$$[1 \quad 0 \quad \cdots \quad 0]\boldsymbol{R}_M^{-1} = \frac{1}{\sigma_w^2}[1 \quad a_1 \quad \cdots \quad a_M] \tag{10.92}$$

where $\boldsymbol{R}_M$ is the $M \times M$ autocorrelation matrix. Multiplying $x(n-M-1)$ on both sides of $x(n) = -\sum_{k=1}^{M} a_k x(n-k) + w(n)$ and taking the expectation, we obtain

$$R_{xx}(M+1) + \sum_{k=1}^{M} a_k R_{xx}(M+1-k) = 0 \tag{10.93}$$

If the autocorrelation values $R_{xx}(0), R_{xx}(1), \ldots, R_{xx}(M)$ are known, then the autocorrelation function estimation value $\hat{R}_{xx}(M+1)$ can be obtained with the above equation as

$$\hat{R}_{xx}(M+1) = -\sum_{k=1}^{M} a_k R_{xx}(M+1-k) \qquad (10.94)$$

If we substitute eq. (10.92) into eq. (10.90), we can also obtain eq. (10.94). Since the estimated values of the extrapolated autocorrelation function are equivalent for both methods, it is shown that the maximum entropy spectrum estimation method and the AR parameter model spectrum estimation method are equivalent.

In this way, the maximum entropy power spectrum theoretical expression and the maximum entropy spectrum estimation expression, respectively, are shown as follows:

$$P_{MEM}(e^{j\omega}) = \frac{\sigma_w^2}{\left|1 + \sum\limits_{k=1}^{p} a_k e^{-j\omega k}\right|^2} \qquad (10.95)$$

$$\hat{P}_{MEM}(e^{j\omega}) = \frac{\hat{\sigma}_w^2}{\left|1 + \sum\limits_{k=1}^{p} \hat{a}_k e^{-j\omega k}\right|^2} \qquad (10.96)$$

Obviously, eqs. (10.95) and (10.96) are the same as eqs. (10.79) and (10.80).

Example 10.6: Try to realize the maximum entropy spectrum estimation (AR spectrum estimation) with MATLAB program.
Solution: The MATLAB program is as follows:

```
clear all;
Fs=1000; N=1024; Nfft=256; n=0:N-1; t=n/Fs;
noverlap=0; dflag='none'; window=hanning(256);
randn('state',0);
x=sin(2*pi*50*t)+2*sin(2*pi*120*t); xn=x+randn(1,N);
[Pxx1,f]=pmem(xn,14,Nfft,Fs); %Maximal entropy or AR PSD estimation
figure(1)
subplot(211); plot(f,10*log10(Pxx1)); xlabel('Frequency/Hz'); ylabel
    ('PSD/dB');
title('Maximal entropy or AR PSD estimation'); grid on;
subplot(212); psd(xn,Nfft,Fs,window,noverlap,dflag); xlabel('Frequency/
    Hz'); ylabel('PSD/dB');
title('PSD estimation by Welch method'); grid on
```

Figure 10.9 demonstrates the result of the maximal entropy (or AR) spectral estimation, as well as the result of the Welch method. It is obvious that the former is much smoother.

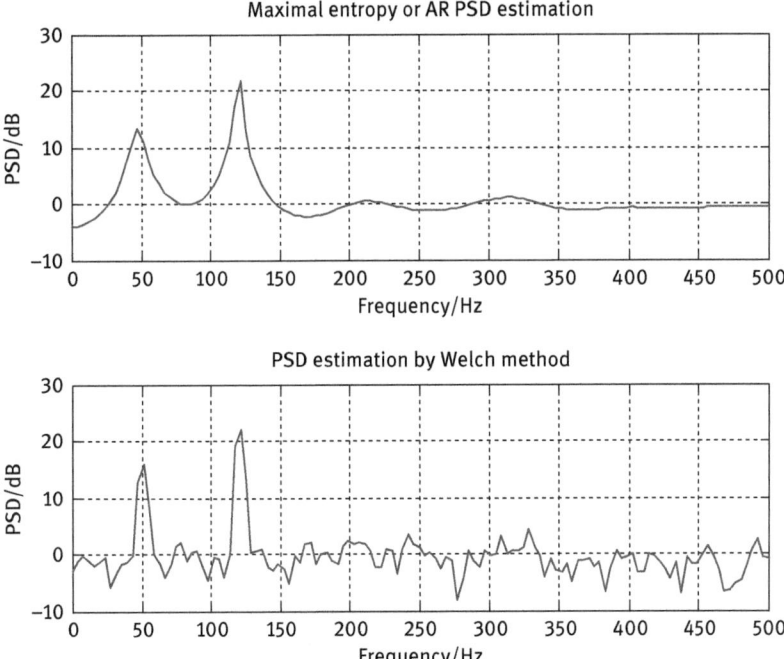

Figure 10.9: The maximal entropy (or AR) spectral estimation (the upper figure) and the Welch spectral estimation (the lower figure).

10.5.4 The PSD Estimation with MA and ARMA Models

1. MA spectrum estimation
As an all-zero model, the MA model is shown as follows:

$$x(n) = \sum_{k=0}^{q} b_k w(n-k) \tag{10.97}$$

Since the MA model does not have any pole, the narrowband PSD estimation will not get a high resolution. However, for the stochastic processes represented by the MA model, the MA model spectrum estimation may get a better result, which is mainly caused by the MA random process itself.

The MA model PSD estimation needs to get the autocorrelation estimation $\hat{R}_{xx}(m)$, $|m| \le q$, based on the observation data first, and then to get the PSD estimation as follows:

$$\hat{P}_{MA}(e^{j\omega}) = \sum_{m=-q}^{q} \hat{R}_{xx}(m) e^{-j\omega m}, \quad |m| \le q \tag{10.98}$$

It can be seen that the MA spectral estimation is in fact the periodogram estimation.

2. ARMA spectrum estimation

If the noise is heave, the performance of the AR spectrum estimation may deteriorate. In this case, the ARMA spectrum estimation may get a better result.

The AR model of a wide-sense stationary random signal $x(n)$ is given as

$$x(n) = \sum_{k=0}^{q} b_k w(n-k) - \sum_{k=1}^{p} a_k x(n-k) \tag{10.99}$$

The system function of the ARMA model can be obtained as

$$H(z) = \frac{B(z)}{A(z)} = \frac{\sum\limits_{k=0}^{q} b_k z^{-k}}{\sum\limits_{k=0}^{p} a_k z^{-k}} \tag{10.100}$$

where $a_0 = 1$, $b_0 = 1$. The PSD estimation of the ARMA model can be further obtained as

$$\hat{P}_{\mathrm{ARMA}}(e^{j\omega}) = \sigma_w^2 \left| \frac{\hat{B}(e^{j\omega})}{\hat{A}(e^{j\omega})} \right|^2 \tag{10.101}$$

where

$$\hat{A}(e^{j\omega}) = \sum_{k=0}^{p} \hat{a}_k e^{-j\omega k}, \quad \hat{B}(e^{j\omega}) = \sum_{k=0}^{q} \hat{b}_k e^{-j\omega k} \tag{10.102}$$

AR parameters $\hat{a}_k$, $k = 1, 2, \ldots, p$, can be estimated by using the method of solving Yule–Walker equation or the overdetermined linear equations. The estimation of MA parameters $\hat{b}_k$, $k = 1, 2, \ldots, q$, can be obtained by using the spectral decomposition method. Substituting $\hat{a}_k$ and $\hat{b}_k$ into eq. (10.101), we can obtain the ARMA PSD estimation.

10.5.5 The PSD Estimation with Minimum Variance Method

The minimum variance spectrum estimation (MVSE) was proposed in 1968 by Capon.

Suppose the output signal $y(n)$ of an FIR filter $A(z) = \sum\limits_{k=0}^{p} a_k z^{-k}$ for the input $x(n)$ is

$$y(n) = \sum_{k=0}^{p} a_k x(n-k) = X^T a \qquad (10.103)$$

where X and a are the signal vector and the coefficient vector of the filter, respectively. The power of the output signal is

$$\rho = E[|y(n)|^2] = E[a^H X^* X^T a] = a^H E[X^* X^T] a = a^H R_p a \qquad (10.104)$$

where the symbols "H" and "*" denote the conjugate transpose and conjugate of the matrix, respectively. R_p is the autocorrelation matrix of $y(n)$. If the mean value of $y(n)$ is zero, then it also represents the variance of $y(n)$.

In order to get the filter coefficients, we need

$$\sum_{k=0}^{p} a_k e^{-j\omega_i k} = e^H(\omega_i) a = 1 \qquad (10.105)$$

which means that the signal can pass through the filter at the frequency of the filter passband ω_i. In the above equation,

$$e^H(\omega_i) = \begin{bmatrix} 1 & e^{j\omega_i} & \cdots & e^{j\omega_i p} \end{bmatrix}^T$$

At the same time, the power ρ is also required to achieve its minimum at the frequencies close to ω_i under the condition of eq. (10.105) holding, known as the minimum variance. If the above conditions are guaranteed, it can be proved that the filter coefficients are as follows:

$$a_{MV} = \frac{R_p^{-1} e(\omega_i)}{e^H(\omega_i) R_p^{-1} e(\omega_i)} \qquad (10.106)$$

The minimum variance is shown as

$$\rho_{MV} = \frac{1}{e^H(\omega_i) R_p^{-1} e(\omega_i)} \qquad (10.107)$$

Thus, the minimum variance spectrum estimation is given as

$$P_{MV}(e^{j\omega}) = \frac{1}{e^H(\omega) R_p^{-1} e(\omega)} \qquad (10.108)$$

Based on the relation between the minimum variance spectrum estimation and the AR spectrum estimation, as long as the coefficients of the AR model are obtained, the minimum variance spectrum estimation can be calculated by the following equation:

$$\hat{P}_{MV}(e^{j\omega}) = \frac{1}{\sum\limits_{m=-p}^{p} \hat{R}_{MV}(m)e^{-j\omega m}} \tag{10.109}$$

where $\hat{R}_{MV}(m)$ is the estimated value of the autocorrelation sequence, obtained by using the AR model coefficients, which is shown as

$$\hat{R}_{MV}(m) = \frac{1}{\sigma_m^2} \sum_{i=0}^{p-m} (p+1-m-2i)a_p(m+i)a_p^*(i), \quad m = 0, 1, \ldots, p \tag{10.110}$$

10.5.6 Pisarenko Harmonic Decomposition

For the mixed sinusoidal spectrum estimation problem in white noise, the Pisarenko decomposition method is used to deal with it as a special ARMA process, and the eigendecomposition technique is used to solve the problem.

Suppose the signal $s(n)$ is a combination of M real-valued sinusoidal random signals shown as

$$s(n) = \sum_{i=1}^{M} q_i \sin(n\omega_i + \varphi_i) \tag{10.111}$$

where the initial phase φ_i is a normal distributed random variable in $(-\pi, \pi)$. The frequency ω_i (or f_i) for each sine wave is determined by $1 - (2\cos\omega_i)z^{-1} + z^{-2} = 0$. Then for M frequencies, we have

$$\prod_{i=1}^{M} [1 - (2\cos\omega_i)z^{-1} + z^{-2}] = 0 \tag{10.112}$$

The above equation is the $2M$ th-order polynomial expressed as

$$\sum_{k=1}^{2M} a_k z^{-k} = 0 \tag{10.113}$$

in which $a_0 = 1$. The above equation can be written as

$$\sum_{k=1}^{2M} a_k z^{2M-k} = \prod_{i=1}^{M} (z - z_i)(z - z_i^*) \tag{10.114}$$

where $|z_i| = 1$ and $a_k = a_{2M-k}$, $k = 0, 1, \ldots, M$. Thus, the process composed of M real sine signals can be expressed as

$$s(n) = -\sum_{i=1}^{2M} a_i x(n-i) \tag{10.115}$$

If the combined sinusoidal process in white noise is

$$x(n) = s(n) + w(n) = \sum_{i=1}^{M} q_i \sin(n\omega_i + \varphi_i) + w(n) \tag{10.116}$$

where $w(n)$ is the white noise, satisfying $E[w(n)] = 0$, $E[w(n)w(m)] = \sigma_w^2 \delta_{nm}$, $E[s(n)w(m)] = 0$, then $x(n)$ can be expressed as

$$x(n) = -\sum_{i=1}^{2M} a_i s(n-i) + w(n) \tag{10.117}$$

Substituting $s(n-i) = x(n-i) - w(n-i)$ into the above equation, we have

$$x(n) + \sum_{i=1}^{2M} a_i x(n-i) = w(n) + \sum_{i=1}^{2M} a_i w(n-i) \tag{10.118}$$

We see the above equation as a kind of ARMA$(2M, 2M)$.

From eq. (10.118) and by using the eigendecomposition technique, we can employ the roots $z_i = e^{\pm j2\pi f_i \Delta t}$ of the eigenpolynomial

$$1 + a_1 z^{-1} + \ldots + a_{2p} z^{-2M} = 0 \tag{10.119}$$

to obtain the frequencies of M sinusoidal signals f_i, $i = 1, 2, \ldots, M$. The amplitudes (or powers) and the variance of the white noise can be obtained by the correlation analysis.

10.5.7 The PSD Estimation Based on the Matrix Eigenvector Decomposition

The PSD estimation based on the eigendecomposition of the autocorrelation matrix plays an important role in modern spectral estimation, which improves the stability of spectrum estimation (i.e., the ability to resist noise interference). Similar to the Pisarenko method, this method is suitable for the spectral estimation of sinusoidal signals in white noise. The basic idea of the method can be described as follows. The sinusoidal signal and white noise have different contributions to the decomposition of the autocorrelation matrix. By using the eigendecomposition of the autocorrelation matrix, the noisy signal is divided into a signal subspace and a noise subspace. The performance of PSD estimation can be improved by eliminating the influence of noise.

1. Eigendecomposition of autocorrelation matrix

Suppose a random signal $x(n)$ is composed of M sine signals and a white noise:

$$x(n) = s(n) + w(n) \tag{10.120}$$

where $s(n) = \sum_{i=1}^{M} q_i e^{j\omega_i n}$. The autocorrelation function of $x(n)$ is

$$R_{xx}(m) = \sum_{i=1}^{M} A_i e^{j\omega_i n} + \sigma_w^2 \delta_{nm} \tag{10.121}$$

where $A_i = q_i^2$ and ω_i denote the power and frequency of the ith complex sinusoidal signal, respectively. σ_w^2 denotes the power of the white noise. The $(p+1) \times (p+1)$ autocorrelation matrix of $x(n)$ is shown as

$$\boldsymbol{R}_p = \begin{bmatrix} R_{xx}(0) & R_{xx}^*(1) & \cdots & R_{xx}^*(p) \\ R_{xx}(1) & R_{xx}(0) & \cdots & R_{xx}^*(p-1) \\ \vdots & \vdots & \ddots & \vdots \\ R_{xx}(p) & R_{xx}(p-1) & \cdots & R_{xx}(0) \end{bmatrix} \tag{10.122}$$

Define the signal vector as

$$\boldsymbol{e}_i = \begin{bmatrix} 1 & e^{j\omega_i} & \cdots & e^{j\omega_i p} \end{bmatrix}^{\mathrm{T}}, \quad i = 1, 2, \ldots, M \tag{10.123}$$

Thus,

$$\boldsymbol{R}_p = \sum_{i=1}^{M} A_i \boldsymbol{e}_i \boldsymbol{e}_i^{\mathrm{H}} + \sigma_w^2 \boldsymbol{I} \stackrel{\Delta}{=} \boldsymbol{S}_p + \boldsymbol{W}_p \tag{10.124}$$

where $\boldsymbol{I}$ is a $(p+1) \times (p+1)$ unit matrix. $\boldsymbol{S}_p$ and $\boldsymbol{W}_p$ are the signal matrix and the noise matrix, respectively. Since the maximal rank of the signal matrix is M and if $p > M$, then $\boldsymbol{S}_p$ is singular. Because of the existence of noises, the rank of $\boldsymbol{R}_p$ is still $p + 1$. Taking the eigendecomposition to $\boldsymbol{R}_p$, we obtain

$$\boldsymbol{R}_p = \sum_{i=1}^{M} \lambda_i \boldsymbol{V}_i \boldsymbol{V}_i^{\mathrm{H}} + \sigma_w^2 \sum_{i=1}^{p+1} \boldsymbol{V}_i \boldsymbol{V}_i^{\mathrm{H}} = \sum_{i=1}^{M} (\lambda_i + \sigma_w^2) \boldsymbol{V}_i \boldsymbol{V}_i^{\mathrm{H}} + \sigma_w^2 \sum_{i=M+1}^{p+1} \boldsymbol{V}_i \boldsymbol{V}_i^{\mathrm{H}} \tag{10.125}$$

where λ_i, $i = 1, 2, \ldots, M$, are eigenvalues of R_p, which are also the eigenvalues of S_p, satisfying $\lambda_1 \geq \lambda_2 \geq \cdots \geq \lambda_M > \sigma_w^2$. V_i are the eigenvectors corresponding to λ_i, $i = 1, 2, \ldots, M$, and the eigenvectors are orthogonal to each other.

We see from eq. (10.125) that all the eigenvectors of R_p comprise a $(p+1)$-dimensional space. This vector space can be divided into two subspaces: one is the signal subspace spanned by eigenvectors V_1, V_2, $\ldots$, V_M of S_p, whose eigenvalues are $(\lambda_1 + \sigma_w^2)$, $(\lambda_2 + \sigma_w^2)$, $\ldots$, $(\lambda_M + \sigma_w^2)$; the other subspace is the noise subspace spanned by eigenvectors V_{M+1}, V_{M+2}, $\ldots$, V_{p+1} of W_p, whose eigenvalues are all σ_w^2. Based on the concept of signal and noise subspaces, we can make the PSD estimation in the signal subspace and the noise subspace separately, so as to eliminate or weaken the influence of noise and obtain better results.

2. Power spectrum estimation based on the signal subspace

If we discard the eigenvectors corresponding to the noise subspace and retain the eigenvectors corresponding to the signal subspace, then we can replace R_p with $\hat{R}_p$, whose rank is M, which is shown as

$$\hat{R}_p = \sum_{i=1}^{M} (\lambda_i + \sigma_w^2) V_i V_i^H \tag{10.126}$$

Such process improves the signal-to-noise ratio of $x(n)$, resulting in a better PSD estimation result.

3. Multiple signal classification (MUSIC) spectrum estimation

Based on the fact that the signal vectors e_i, $i = 1, 2, \ldots, M$, are orthogonal to the noise vector, the multiple signal classification (MUSIC) algorithm extends it to the orthogonality between e_i and the linear combination of the eigenvectors of each noise as

$$e_i \left(\sum_{k=M+1}^{p+1} a_k V_k \right) = 0, \quad i = 1, 2, \ldots, M \tag{10.127}$$

Let $e(\omega) = \begin{bmatrix} 1 & e^{j\omega} & \cdots & e^{j\omega M} \end{bmatrix}^T$; then $e(\omega) = e_i$ holds. From eq. (10.127), we have

$$e^H(\omega) \left[\sum_{k=M+1}^{p+1} a_k V_k V_k^H \right] e(\omega) = \sum_{k=M+1}^{p+1} a_k |e^H(\omega) V_k|^2 \tag{10.128}$$

Since the above equation equals zero at $\omega = \omega_i$,

$$\hat{P}(\omega) = \frac{1}{\displaystyle\sum_{k=M+1}^{p+1} a_k |e^H(\omega) V_k|^2} \tag{10.129}$$

should be infinity at $\omega = \omega_i$. However, since V_k is derived from the decomposition of the autocorrelation matrix and the autocorrelation matrix is generally estimated by the observed data, there will be an error. Therefore, $\hat{P}(\omega)$ will present sharp peaks but not infinity at $\omega = \omega_i$. The frequencies corresponding to the peaks are the frequencies of the sinusoidal signals. The resolution of this method is better than that of the AR spectral estimation.

If $\omega_k = 1$, $k = M + 1$, $M + 2$, $\ldots$, $p + 1$, are taken, then the estimation formula is the famous MUSIC spectrum estimation shown as

$$\hat{P}_{\text{MUSIC}}(\omega) = \frac{1}{e^H(\omega)\left(\sum_{k=M+1}^{p+1}|V_k V_k^H|\right)e(\omega)} \tag{10.130}$$

If $\omega_k = 1/\lambda_k$, $k = M + 1$, $M + 2$, $\ldots$, $p + 1$, are taken, then the estimation formula is the eigenvector (EV) spectrum estimation, which is shown as

$$\hat{P}_{\text{EV}}(\omega) = \frac{1}{e^H(\omega)\left(\sum_{k=M+1}^{p+1}\left|\frac{1}{\lambda_k}V_k V_k^H\right|\right)e(\omega)} \tag{10.131}$$

We see from eqs. (10.130) and (10.131) that the eigenvectors for both MUSIC and EV spectrum estimations are all V_{M+1}, V_{M+2}, $\ldots$, V_{p+1} in the noise subspace, so such methods are referred to as the noise subspace-based spectrum estimation methods.

Example 10.7: Suppose that the signal $s(n)$ is composed of two sinusoidal signals, whose frequencies are 50 Hz and 150 Hz, respectively. The signal is contaminated by a Gaussian white noise. Try to estimate the PSD of the signal by using the MUSIC and EV algorithms with a MATLAB program.

Solution: The MATLAB program is shown as follows:

```
clear all
Fs=1000; %sampling frequency
n=0:1/Fs:1;
xn=sin(2*pi*200*n)+sin(2*pi*220*n)+0.1*randn(size(n));
order=30; nfft=1024; p=25;
figure(1); pmusic(xn,p,nfft,Fs); title('PSD estimation with MUSIC
algorithm')
figure(2); peig(xn,p,nfft,Fs); title('PSD estimation with EV
algorithm');
```

Figure 10.10 demonstrates the PSD estimations with both algorithms.

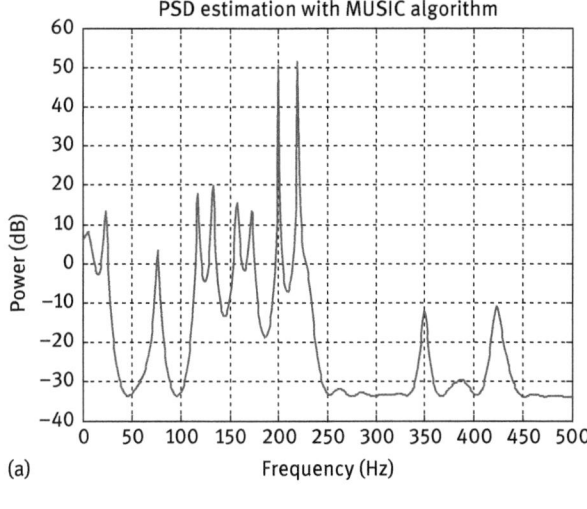

(a)

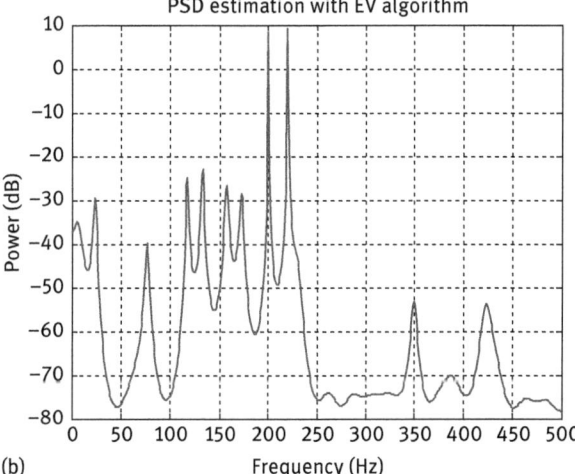

(b)

Figure 10.10: PSD estimations with MUSIC and EV algorithms: (a) MUSIC and (b) EV.

10.5.8 A Comparison of Various PSD Estimation Methods

Figure 10.11 demonstrates a comparison among various PSD estimation methods.

From Figure 10.11, we see the following:

(1) The AR PSD estimations with autocorrelation method (see Figure 10.11(b) and (c)) do not have good spectrum detection abilities when the order is low, while their resolution and spectrum detection ability are improved when the order increases.

(2) The resolution in Figure 10.11(d) is very good.

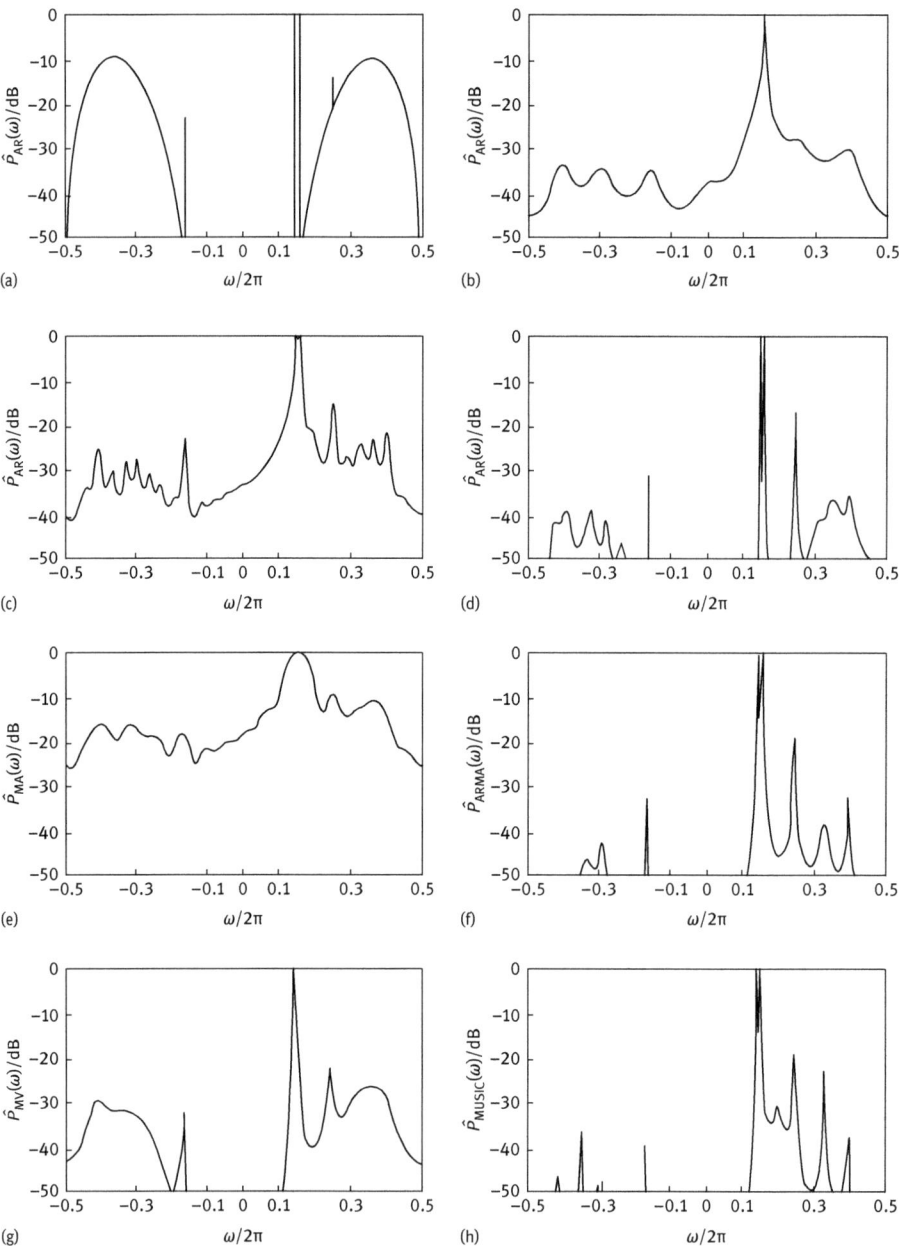

Figure 10.11: A comparison among various PSD estimation methods: (a) the truth PSD curve; (b) AR PSD estimation, $p = 10$; (c) AR PSD estimation, $p = 30$; (d) AR PSD estimation with Burg algorithm $p = 13$; (e) MA PSD estimation, $q = 16$; (f) ARMA PSD estimation, $p = 10$, $q = 13$; (g) minimal variance PSD estimation; and (h) eigendecomposition PSD estimation.

(3)   The resolution of MA PSD estimation is bad.

(4)   The resolution of ARMA PSD estimation is better than that of MA estimation, but not as good as AR estimation. However, its noise is smoothed.

(5)   The resolution of the minimal variance PSD estimation is not as good as that of AR estimation.

10.6 The Cepstrum Analysis of Signals

10.6.1 The Concept of the Cepstrum

The so-called cepstrum is in fact to conduct the spectrum analysis of the spectrum again. Cepstrum analysis is very effective for the detection of periodic components in a complex spectrum, and it has the ability of deconvolution and source signal separation. Therefore, it is widely used in many engineering areas, such as vibration analysis, noise source identification, machine fault diagnosis and prediction, speech analysis, and echo cancellation.

1. The concept of real cepstrum
Suppose that the unilateral PSD of $x(t)$ is $G_{xx}(\Omega)$, whose real cepstrum is defined as

$$C_p(q) = |F^{-1}\{\ln G_{xx}(\Omega)\}|^2 \tag{10.132}$$

where $F^{-1}\{\cdot\}$ denotes the inverse Fourier transform. The independent variable q is named as "quefrency," with a unit of s or ms. A bigger q value denotes a faster fluctuation in the cepstrum. The real cepstrum has some other definitions as

$$C_a(q) = |F^{-1}\{\ln G_{xx}(\Omega)\}| \tag{10.133}$$

$$C(q) = F^{-1}\{\ln G_{xx}(\Omega)\} \tag{10.134}$$

2. The concept of complex cepstrum
The Fourier transform of the continuous-time signal $x(t)$ is expressed as

$$X(j\Omega) = X_R(j\Omega) + jX_I(j\Omega) \tag{10.135}$$

where $X_R(j\Omega)$ and $X_I(j\Omega)$ are the real and imaginary parts of $X(j\Omega)$, respectively. Then, the complex cepstrum is defined as

$$C_c(q) = F^{-1}\{\ln X(j\Omega)\} \tag{10.136}$$

We can find from eq. (10.136):

$$C_c(q) = F^{-1}\{\ln X(j\Omega)\} = F^{-1}\{\ln[A(j\Omega)e^{j\Phi(j\Omega)}]\}$$
$$= F^{-1}\{\ln A(j\Omega)\} + jF^{-1}\{\Phi(j\Omega)\}$$

(10.137)

where $A(j\Omega)$ and $\Phi(j\Omega)$ are the magnitude and phase of $X(j\Omega)$, respectively.

We see from the above definitions that the cepstrum is the inverse Fourier transform of the logarithmic spectrum, which transforms the spectrum from the frequency domain to a new domain, referred to as the quefrency domain. The purposes of logarithmic operation are to extend the dynamic range of the spectrum, to make for the deconvolution, and to help the separation of signals and noises.

10.6.2 Homomorphic Filtering and Applications of the Cepstrum

1. The homomorphic filtering

The so-called homomorphic filtering is a kind of generalized linear filtering technique, which is mainly applied in the case of multiplicative noise in the signal. Suppose that the discrete-time signal is shown as follows:

$$x(n) = s(n)v(n)$$

(10.138)

It is not easy to eliminate the multiplicative noise $v(n)$ in the signal by using the commonly used linear filter. The basic idea of the homomorphic filtering is to transform the multiplicative relation in eq. (10.138) into a linear relation by taking a logarithmic operation as

$$\ln[x(n)] = \ln[s(n)] + \ln[v(n)]$$

(10.139)

And let

$$\tilde{x}(n) = \ln[x(n)], \quad \tilde{s}(n) = \ln[s(n)], \quad \tilde{v}(n) = \ln[v(n)]$$

Then,

$$\tilde{x}(n) = \tilde{s}(n) + \tilde{v}(n)$$

(10.140)

Thus, the multiplicative noise is transformed into the additive form. Then, the linear filtering can be used to eliminate $\tilde{v}(n)$ in $\tilde{x}(n)$, resulting in a rather pure $\tilde{s}(n)$. Then, the exponential operation is used to obtain

$$s(n) = \exp[\tilde{s}(n)] \qquad (10.142)$$

2. Deconvolution by the cepstrum

In the fields of radar, sonar, ultrasonic imaging, and speech processing, the signal $s(n)$ and noise $v(n)$ are often given with a convolution relation as

$$x(n) = s(n) \star v(n) \qquad (10.143)$$

The separation of the signal and noise is a deconvolution problem. Taking Fourier transform on both sides of the above equation, we obtain

$$X(e^{j\omega}) = S(e^{j\omega})V(e^{j\omega}) \qquad (10.144)$$

And then we rewrite it as the PSD form:

$$|X(e^{j\omega})|^2 = |S(e^{j\omega})|^2|V(e^{j\omega})|^2 \qquad (10.145)$$

Taking the inverse Fourier transform on both sides of the above equation, we obtain

$$\begin{aligned} C_x(q) &= F^{-1}\{\ln|X(e^{j\omega})|^2\} = F^{-1}\{\ln|S(e^{j\omega})|^2\} + F^{-1}\{\ln|V(e^{j\omega})|^2\} \\ &= C_s(q) + C_v(q) \end{aligned} \qquad (10.146)$$

This is the cepstrum form of $x(n)$. Obviously, after Fourier transform, the convolution relation between the original signal and the noise becomes the multiplicative relation. The logarithmic operation transforms the multiplicative relation into the additive relation, which is much easier for the separation of signal and noise.

3. Applications of the cepstrum

Example 10.8: An example of cepstrum analysis-based rolling bearing fault diagnosis is given as follows. Figure 10.12 shows the rolling bearing test and the analysis

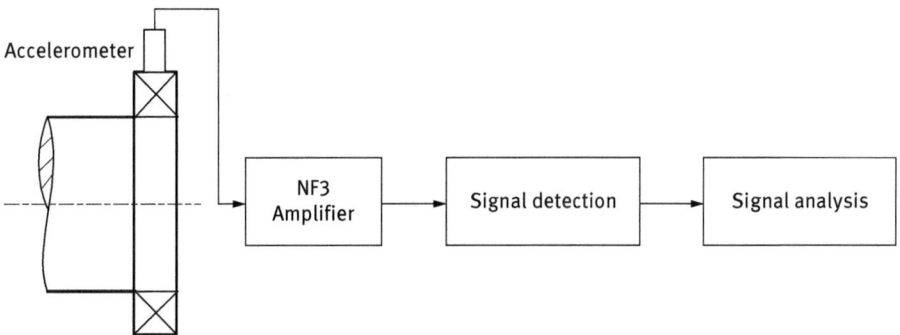

Figure 10.12: The diagram of test and analysis system for rolling bearing.

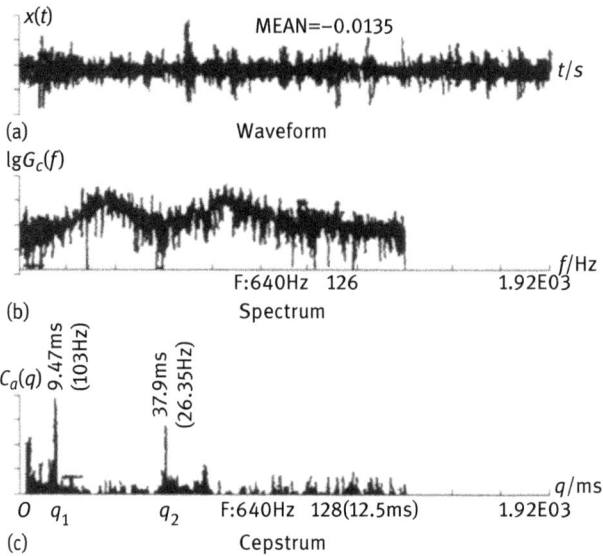

Figure 10.13: The vibration analysis of the rolling bearing signal: (a) waveform of the signal, (b) spectrum, and (c) cepstrum.

system. Try to use the cepstrum to determine the inner fault characteristic frequency f_1 and the ball bearing fault characteristic frequency f_2.

Solution: Taking FFT to detect the vibration signal, we obtain its spectral information. Then we calculate its cepstrum. Figure 10.13 shows the waveform, spectrum, and cepstrum of the vibration signal.

Obviously, it is not easy to find the fault characteristics from the waveform in Figure 10.13(a). However, we can find two peaks in the cepstrum in Figure 10.13(c), marked as $q_1 = 9.47$ms and $q_2 = 30.90$ms. We can further obtain the spectral intervals as $\Delta f_1 = 1/q_1 = 1,000/9.47 = 105.60$Hz and $\Delta f_2 = 1/q_2 = 1,000/37.90 = 26.38$Hz, respectively, which are close to the theoretic values of $f_1 = 106.35$Hz and $f_2 = 26.35$Hz. Therefore, we believe that both peaks reflect the faults of bearing inner ring and ball, respectively.

Example 10.9: An acoustic attenuation signal $s(t) = e^{-at} \sin(\Omega t + \phi)$ with exponential decaying is shown in Figure 10.14(a). In practice, the received signal $x(t)$ is usually a superposition of multiple arrival signals caused by multipath effect. Such signal is expressed as $x(t) = s(t) \star [1 + b_1\delta(t - \tau_1) + b_2\delta(t - \tau_2)]$, which is shown in Figure 10.14 (b). Try to design a cepstrum filter to resume the original signal $s(t)$ by removing the multipath effect in $x(t)$ with a MATLAB program.

Solution: The MATLAB program contains three parts: (1) generating the original signal $s(n)$ and its superposition of multiple arrivals, (2) filtering the signal with the cepstrum filter, and (3) resuming the original signal. The MATLAB program is shown as follows:

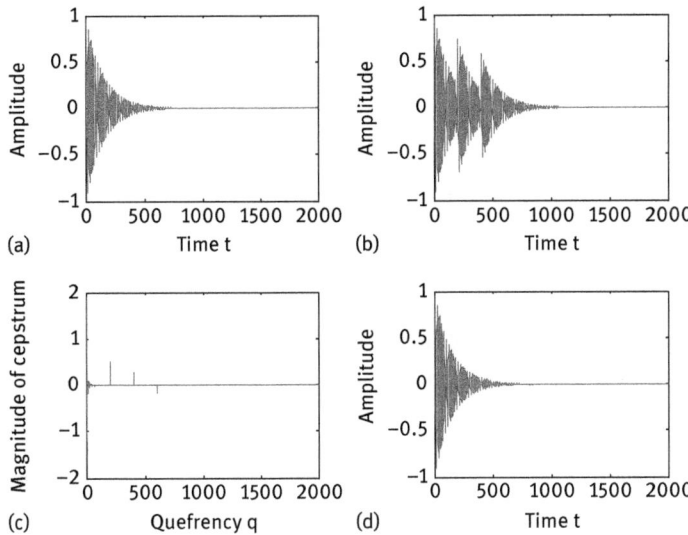

Figure 10.14: Removing the multipath effect by the cepstrum filtering: (a) original signal, (b) multiple arrived signal, (c) cepstrum, and (d) resumed signal.

```
clear all;
% Generating multiple arrived signal
n=2000; nn=1:1:n; f1=50; thr=0.01; xs=zeros(n,1); ys=zeros(n,1); xs=sin
    (2*pi*f1/1000.*nn'); ys=exp(-nn/150);
xe=ys'.*xs; xe1=zeros(n,1); xe2=zeros(n,1);
xe1(201:n)=0.5*xe(1:n-200); xe2(401:n)=0.4*xe(1:n-400); x=xe+xe1+xe2;
% Calculating the cepstrum and the cepstrum filtering
[xhat,nd,xhat1]=cceps(x);
for i=100:n
        if abs(xhat(i))>thr; xhat(i)=0; end
end
% Resuming the original signal
xer=icceps(xhat,nd);
% Output results
figure(1)
subplot(221); plot(xe); xlabel('Time t'); ylabel('Amplitude'); text
    (1700,0.7,'(a)');
subplot(222); plot(x); xlabel('Time t'); ylabel('Amplitude'); text
    (1700,0.7,'(b)');
subplot(223); plot(xhat1); xlabel('Quefrency q'); ylabel('Mugnitude of
    cepstrum'); text(1700,1.4,'(c)');
subplot(224); plot(xer); xlabel('Time t'); ylabel('Amplitude'); text
    (1700,0.7,'(d)');
```

Figure 10.14 demonstrates the curves of the original signal, the multiple arrived signal, the complex cepstrum, and the resumed signal. It is obvious that the original signal is well resumed.

10.7 Applications of PSD Estimation and Analysis

10.7.1 Examples of PSD Estimation in Signal Analysis

1. Determination of natural frequency of large structures

For determining the natural frequency of large engineering structures, we generally use random signals of the earth pulsation as the excitation to measure the response and carry out the spectral analysis. The frequency corresponding to the peak value of the spectrum is the natural frequency f_0 of the engineering structure, as shown in Figure 10.15.

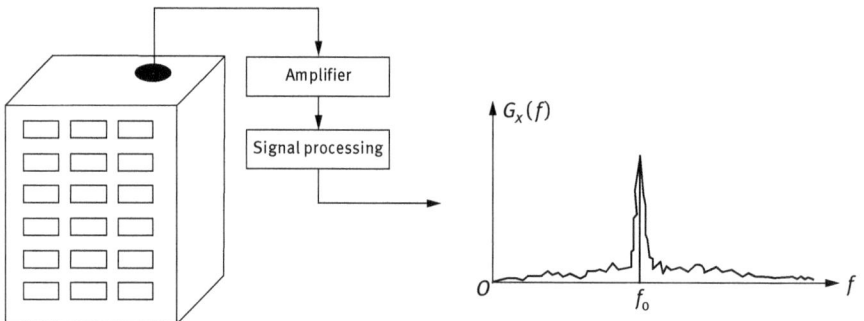

Figure 10.15: Schematic diagram of natural frequency measurement of a large engineering structure.

Such method is simple, is economical and practical, and is suitable for the study and analysis of the dynamic characteristics of nuclear power plants, dams, and large buildings.

2. Motor noise source identification

Generally, noises of large induction motors increase with time. The conventional sound level meter can measure the sound pressure, but it cannot determine the cause of the noise. If the noise PSD is analyzed, the corresponding technical measures can be taken to reduce the noise. Figure 10.16 shows the schematic diagram of the motor noise analysis system.

As shown in Figure 10.16, the motor noise and the vibration signal are obtained by the sound level meter and the accelerator sensor (NP-300), respectively. The signals are analyzed by the FFT analyzer. Figure 10.17 demonstrates the PSDs of noisy and processed vibration signals.

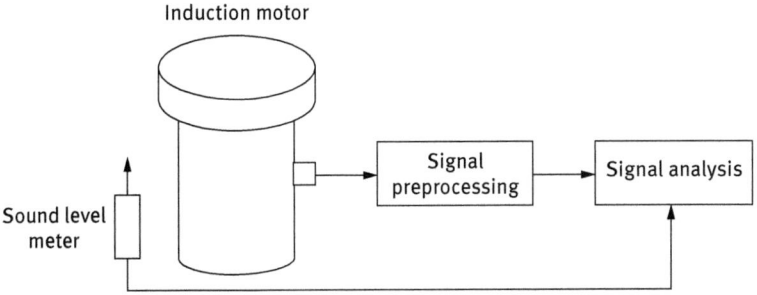

Figure 10.16: Schematic diagram of the motor noise analysis system.

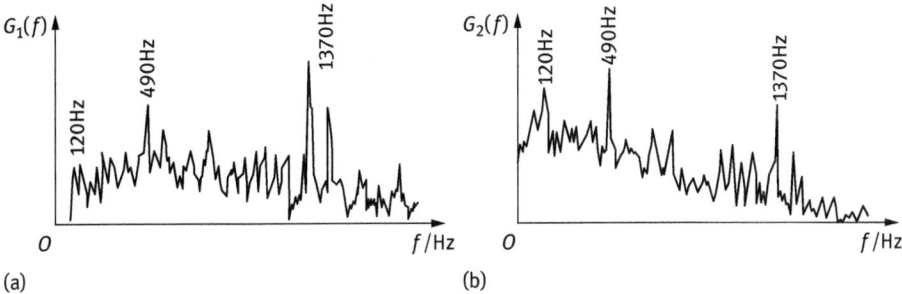

Figure 10.17: The PSDs of noisy and processed vibration signals: (a) PSD of the noisy signal and (b) PSD of the processed vibration signal.

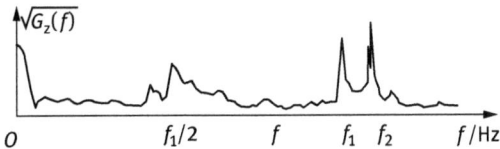

Figure 10.18: The root PSD of the engine vibration signal.

As shown in Figure 10.17, the frequency at 120 Hz is the frequency twice of the power frequency. The frequency at 490 Hz is the characteristic frequency of the motor bearing. The frequency at 1,370 Hz is the high-frequency noise caused by the internal clearance of the motor. In order to reduce the noise level of the motor, we have to reduce the values of the spectrum at 120 Hz, 490 Hz, and 1,370 Hz, in which the frequency at 490 Hz can be reduced by replacing the bearing and two other components of noises can be reduced by noise insulation materials.

3. Engine fault diagnosis

The power spectrum analysis of an engine vibration signal is of great importance to the aviation flight safety. Figure 10.18 shows the root PSD of the engine vibration signal.

In the figure, f_1 and f_2 are low-pressure and high-pressure rotor unbalance components of the engine, respectively. Both strengths are within normal range. We can also see from the figure that there is a spectral component group nearby $f_1/2$, which may be the main factor causing the engine trouble. Based on the above analysis, the proper measure can be taken.

10.7.2 Examples of PSD Estimation in Medical Auxiliary Diagnosis

1. The PSD analysis of the pulse wave

The use of modern signal processing methods to analyze the pulse condition in traditional Chinese medicine is very worthy of study. When taking the pulse, the sensor is pressed on the "cun" and "guan" positions of the left and right wrist parts, as shown in Figure 10.19. The signals obtained are then analyzed with the spectrum analysis.

Before the spectrum analysis with Welch method, a band-pass filter is used to remove the components of higher than 50 Hz and lower than 1 Hz. The spectrum in the range of 1–50 Hz is divided into five segments, each segment containing 10 Hz. The mean of each segment is calculated and denoted as E_1, E_2, E_3, E_4, and E_5. The energy ratio is defined as

$$ER = \frac{E_1}{E_2 + E_3 + E_4 + E_5}$$

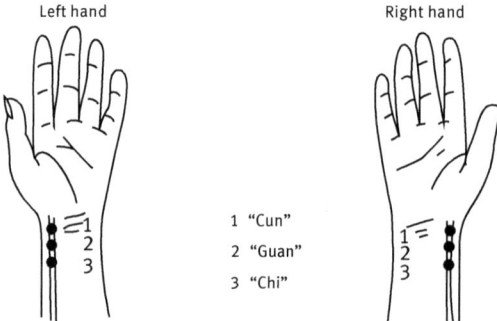

Left hand	Right hand

1 "Cun"
2 "Guan"
3 "Chi"

Figure 10.19: Positions of the sensor when taking the pulses.

Experiments show that the ER values for normal persons and patients with acute hepatitis, heart disease, and gastrointestinal disorders are significantly different. The results are listed as follows:

(1)   The ER values at "cun" of the left wrist for the patients with heart disease are less than 100.
(2)   The ER values at "guan" of the left wrist for the patients with acute hepatitis are less than 100.
(3)   The ER values at "guan" of the right wrist for the patients with gastrointestinal disorders are less than 100.

The above results show that the ER value may be used to express the health status of the human body, and the taken position of the pulse may reflect the situation of different organs.

2. Identification of dyslexia based on EEG spectrum analysis
The rhythm of EEG signal can be analyzed by PSD estimation technique. The spectrum estimation plays an important role in the classification of sleep and anesthetic depth, the relationship between intellectual activities and EEG signals, the reflection of brain injury in an electroencephalogram, and the influence of the environment (such as noise) to the human body.

Some children, with "word reading difficulties" (dyslexia), have a normal abstract thinking ability, but often confuse in mirroring and opposite letters, such as b and d, and m and w. It is not easy to determine the abnormal phenomenon. However, the power spectrum of the EEG signal can reflect such kind of anomaly. Figure 10.20 shows the average PSD of the bipolar EEG spectrum for a group of normal children and

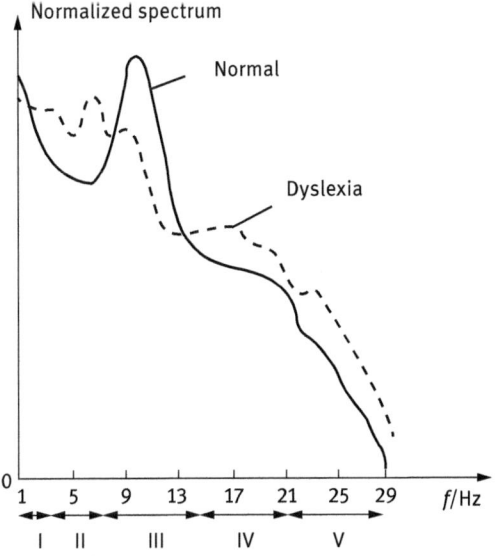

Figure 10.20: The average PSD of the bipolar EEG spectra of normal children and children with reading difficulties in parietal and occipital lobes at resting state.

children with reading difficulties in parietal and occipital lobes at resting state. If the PSD is divided into five bands, it can be seen that the spectrum of the children with dyslexia is significantly higher than that of normal children in II, IV, and V bands, while the value of the spectrum of the normal children in band III is much significant. If we define a characteristic vector based on the normalized spectrum, the classification accuracy of the two kinds of children can reach over 90%.

ⓘ Exercises

10.1 Describe the concepts of the correlation function and power spectrum density of random signals.

10.2 Describe the method and performance of the unbiased estimation for the autocorrelation sequence.

10.3 Describe the method and performance of the biased estimation for the auto-correlation sequence.

10.4 Describe the fast estimation method for the autocorrelation sequence.

10.5 Describe the periodogram PSD estimation method.

10.6 Point out the problems of the periodogram PSD estimation method and give the improved methods.

10.7 Point out the problems of the classical PSD estimation methods.

10.8 Explain the method and performance of the AR spectrum estimation.

10.9 Explain the method and performance of the maximum entropy spectrum estimation.

10.10 Explain the equivalence for the maximum entropy spectrum estimation and the AR spectrum estimation.

10.11 Explain the methods of the MA spectrum estimation and the ARMA spectrum estimation.

10.12 Explain the method of the minimal variance spectrum estimation.

10.13 Explain the Pisarenko spectrum analysis method.

10.14 Describe the multiple signal classification (MUSIC) spectrum estimation method.

10.15 Compare performances of various spectrum estimation methods.

10.16 Explain the concept of the cepstrum.

10.17 Describe the principle of the cepstrum analysis.

10.18 $x(n)$ is given as an ergodic random signal. Now $\hat{r}(m) = \frac{1}{N-|m|}$ $\sum\limits_{n=0}^{N-1-|m|} x_N(n)x_N(n+m)$ is used to estimate its autocorrelation function. Find the mean value and variance of the estimation.

10.19 The autocorrelation function of a random process is $r(m) = 0.8^{|m|}, m = 0, \pm 1, \ldots$ Now we try to estimate $\hat{r}(m)$ with 100 points of data samples. Determine the bias between $\hat{r}(m)$ and $r(m)$ under the following conditions:(1) $m = 0$, (2) $m = 10$, (3) $m = 50$, and (4) $m = 80$.

10.20 The signal data contain N samples, whose sampling frequency is $f_s = 1,000$Hz. The average method is used to improve the spectrum estimation of the periodogram when the signal is divided into nonoverlapped K segments, each with the length of $M = N/K$. It is assumed that the interval between two spectral peaks in the spectrum is 0.04π. What should be the value of M in order to distinguish them?

10.21 A discrete-time signal contains N data samples. Try to use the averaged periodogram to estimate its spectrum. Suppose the signal is divided into nonoverlapped K segments, each with the length of $M = N/K$.

(1)   Suppose we know that there is a peak in its truth PSD shown in Figure E10.1. If N is big, what should be the value of M in order to make the main lobe of the spectral window narrower than the peak?

(2)   If there are two peaks in the power spectrum which are separated by 2 rad/s, what is the minimum number of M that can resolve the peaks?

(3)   In the above two cases, what are the disadvantages if M is too big?

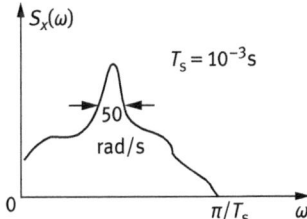

Figure E10.1

10.22 It is known that the upper limit of the ECG is about 50 Hz, so the sampling frequency is taken as $f_s = 200$Hz. If the required frequency resolution is $\Delta f = 2$Hz, determine the number of data points for each segment when doing the spectral estimation.

10.23 Try to prove that if the output $x(n)$ of a p th-order AR model under a white noise excitation is a stationary random process, then the poles of the AR model must be located inside the unit circle.

10.24 An AR(2) process is shown as follows:

$$x(n) = -a_1x(n-1) - a_2x(n-2) + u(n)$$

Try to determine its stability condition.

10.25 Suppose that the first four values of the autocorrelation functions of a stationary random signal are

$$r_x(0) = 1, \quad r_x(1) = -0.5, \quad r_x(2) = 0.625, \quad r_x(3) = -0.6875$$

and $r_x(m) = r_x(-m)$. Try to establish the first-order, second-order, and third-order AR models by using these autocorrelation values and to determine the coefficients of the model and corresponding mean square errors (note: by solving the Yule–Walker equation).

10.26 Suppose that five observation values for an autoregressive process are known as $\{1, 1, 1, 1, 1\}$.
   (1)   Find the first- and second-order reflection coefficients.
   (2)   Estimate the PSD of the autoregressive process.

10.27 The reason why the resolution of the AR spectrum estimation is higher than that of the classical spectrum estimation is that the former implies the extrapolation of the autocorrelation function. When we conduct the AR(p) estimation, suppose the first $p+1$ values of the autocorrelation function are $\{R_0, R_1, \ldots, R_p\}$, try to determine the extrapolated autocorrelation function.

10.28 Given the transfer function of an ARMA$(1, 1)$ process as $H(z) = \frac{1+b(1)z^{-1}}{1+a(1)z^{-1}}$. If $H(z)$ is approximated by the AR(∞) model, whose transfer function is $H_{AR}(z) =$

$\frac{1}{1+c(1)z^{-1}+c(2)z^{-2}+\ldots}$, try to prove $c(k) = \begin{cases} 1, k=0 \\ [a(1) - b(1)][-b(1)]^{k-1}, k \geq 1 \end{cases}$.

10.29 Generate a white noise sequence with 256 samples and estimate the spectrum by using the Welch method. Suppose the length of each segment is 64 samples, and the overlapping is 32 samples. Try to find the averaged PSD estimation and the PSD estimation with the total 256 samples once.

10.30 A method for generating a set of test data on a computer is as follows. First, generate a set of zero-mean white noise data $w(n)$, and let its power be σ^2. Send $w(n)$ into a third-order FIR system whose transfer function is $H(z) = 1 - 0.1z^{-1} + 0.09z^{-2} + 0.648z^{-3}$; the PSD of the output $y(n)$ is $P_y(e^{j\omega}) = \sigma^2 |H(e^{j\omega})|^2$. Add three real sinusoidal signals whose normalized frequencies are $f_1' = 0.1$, $f_2' = 0.25$, and $f_3' = 0.26$. Adjust σ^2 and the amplitudes of the sinusoidal signals to make the signal-to-noise ratio at f_1', f_2', f_3' are 10 dB, 50 dB and 50 dB, respectively. Thus, the PSD of the given signal $x(n)$ can be obtained.
   (1)   Let the length of the test data be $N = 256$; plot its waveform.
   (2)   Plot the truth PSD $P_x(e^{j\omega})$ of the test signal.

10.31 Using the data in Problem 10.29, perform the following calculation:
   (1)   Estimate the PSD with the AR model parameters obtained by the autocorrelation method. Suppose the orders of the model are $p = 8$, $p = 11$, and $p = 14$, respectively.
   (2)   Perform the same calculation with the Burg algorithm.

11 The Optimal Filtering for Random Signals

11.1 Introduction

11.1.1 The Classical Filter and the Statistical Optimal Filter

In the 6th chapter, the concepts of filtering and filter, especially the theory and design methods of digital filters, were introduced. There are a wide variety of filters and their applications are very extensive.

Although the types of filters vary, but on the basis of the filter design and implementation, filters can be divided into two categories: the classical filter and the statistical optimal filter, the latter is often known as the modern filter. The so-called classical filter usually assumes that the desired component and the unwanted component in the input signal $x(n)$ occupy different frequency bands, where the latter can be removed and the former can be retained by the filter. However, if the spectra of unwanted and desired components overlap, then the classical filter does not yield desired results.

Different from classical filters, the statistical optimal filter does not depend on the frequency difference between the signal and the noise for noise suppression and signal extraction. The basic idea of the statistical optimal filter is to estimate the parameters of the useful signal from the noisy observed signal based on some statistical optimal criteria. Once the useful signal is estimated, it will have a higher signal-to-noise ratio than the original signal. In general, such filtering technique that is designed and that runs according to the statistical optimal criterion is called statistical optimal filtering. The theoretical research of the statistical optimal filter originated from the work of Wiener in the 1940s. The Wiener filter is a typical example of the statistical filter. Other important statistical optimal filters include the Kalman filter and the linear predictor. The adaptive filter described in the next chapter also belongs to the category of the statistical optimal filter. The singular value decomposition algorithm and the eigenvalue decomposition algorithm also come under this kind of filter.

This chapter introduces the basic theories and methods of the Wiener filter and Kalman filter, and gives some applications of the statistical optimal filter, including the basic principle of the causal Wiener filter, the derivation of Wiener–Hopf equation, the basic principle of Wiener prediction, and the basic concept, principle, and applications of Kalman filter.

11.1.2 Two Important Statistical Optimal Filters

1. The concept of a Wiener filter

A Wiener filter is a common name for a class of linear optimal filter, named after Norbert Wiener, a mathematician and the founder of cybernetics in the United States.

https://doi.org/10.1515/9783110465082-011

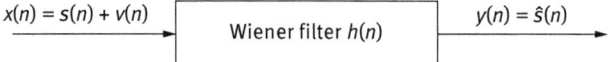

Figure 11.1: The system block diagram of the Wiener filter.

The basic task of Wiener filter is to extract the waveform of useful signal or the parameter of estimation from the noise-contaminated observation signal. The Wiener filter theory assumes that the input of the linear filter is the sum of the desired signal and the noise, where both are wide-sense stationary random processes, and their second-order statistics (such as mean, variance, correlation function, etc.) are known. The Wiener filter calculates the optimal coefficients for the linear filter based on the minimum mean square error (MSE) criterion of the input and output signals. On the basis of Wiener's study, a number of optimal criteria, including the maximum output signal-to-noise ratio and the statistical detection, are proposed, and some equivalent optimal linear filters are obtained. The Wiener filter is one of the most important theoretical achievements of linear filtering theory in the twentieth century. Figure 11.1 shows the system block diagram of Wiener filter.

2. The concept of Kalman filter

A Kalman filter is a recursive filter for linear time-varying systems, named after Rudolph E. Kalman. This filter incorporates past measurement error estimates into new measurement errors to estimate future errors that can be described by a differential equation (or difference equation) containing orthogonal state variables. When the input signal is a random signal generated by a white noise, the Kalman filter can minimize the root–mean–square error between the desired output and the actual output.

The first implementation of the Kalman filter was done by Schmidt. At NASA research center, Kalman found that the filter was of interest in solving the orbital prediction of the Apollo program and was actually used in Apollo's navigational computers. The earliest research papers on a Kalman filter were published by Swerling and Kalman and by Kalman and Bucy in 1960 and 1961, respectively. Over the years, the Kalman filter has attracted attention in the field of signal processing, and there are also some new developments in recent years, including Schmidt expansion filter, innovation filter and square root filter, and so on. The widely used phase-locked loop technology in electronic engineering and computer technology is actually a Kalman filter.

A typical application of Kalman filter is to estimate the signal parameters from the limited noisy observations. For example, in radar applications, it is of interest to be able to track parameters such as position, velocity, and acceleration of the target. However, because of the impact of observation noise, the resulting measurements are often not accurate enough. If Kalman filter is used to process the observed data, we can get a better estimation of the target parameters.

11.2 The Principle of Wiener Filter

11.2.1 Causal Wiener Filter

Let the unit impulse response of the linear discrete system be $h(n)$, whose input signal is $x(n)$, which is the combination of desired signal $s(n)$ and additive noise $v(n)$, given as

$$x(n) = s(n) + v(n) \tag{11.1}$$

The output $y(n)$ of the system is a convolution between the input signal and the system, given as

$$y(n) = x(n) \star h(n) = \sum_{m=-\infty}^{+\infty} h(m)x(n-m) \tag{11.2}$$

The task of the Wiener filter is to make the output signal $y(n)$ as close as possible to the desired signal $s(n)$. In general, $y(n)$ is referred to as the estimation of $s(n)$, denoted by $y(n) = \hat{s}(n)$. In this way, the Wiener filter is also called the estimator of the desired signal $s(n)$. For example, if $h(n)$ is causal, then the output $y(n) = \hat{s}(n)$ can be viewed as a linear combination of observations at the current time $x(n)$ and in the past $x(n-1)$, $x(n-2)$, ... Generally speaking, if the current and past observations are used to estimate the current output signal, it is known as the filter problem. If the past observations are used to estimate the current or future output signal, it is known as the prediction problem. If the past observation data are used to estimate the past output signal, it is known as the smoothing problem.

Referring to the block diagram of the Wiener filter shown in Figure 11.1, it is generally not possible for the system to get the estimated output signal $y(n) = \ddot{s}(n)$ the same as the desired signal $s(n)$. The error function $e(n)$ is usually used to express the error between $\hat{s}(n)$ and $s(n)$:

$$e(n) = s(n) - \hat{s}(n) \tag{11.3}$$

Obviously, the error function is random. The Wiener filter uses the minimum MSE criterion to estimate the desired signal. The MSE criterion is expressed as

$$E[e^2(n)] = E[(s(n) - \hat{s}(n))^2] \tag{11.4}$$

For a causal system, eq. (11.2) changes as

$$y(n) = x(n) \star h(n) = \hat{s}(n) = \sum_{m=0}^{+\infty} h(m)x(n-m) \tag{11.5}$$

Thus, the MSE expression is written as

$$E[e^2(n)] = E[(s(n) - \sum_{m=0}^{+\infty} h(m)x(n-m))^2] \tag{11.6}$$

To minimize the MSE, we take the partial derivative to eq. (11.6) with respect to $h(m)$, $m = 0, 1, \cdots$, and let the derivative be 0,

$$2E[(s(n) - \sum_{m=0}^{+\infty} h(m)x(n-m))x(n-l)] = 0, \quad l = 0, 1, \ldots \tag{11.7}$$

which is

$$E[(s(n)x(n-l)] = \sum_{m=0}^{+\infty} h(m)E[x(n-m)x(n-l)], \quad l = 0, 1, \ldots \tag{11.8}$$

If the correlation function is used to express the aforementioned equation, it yields

$$R_{xs}(l) = \sum_{m=0}^{+\infty} h(m)R_{xx}(l-m), \quad l = 0, 1, \ldots \tag{11.9}$$

Equation (11.9) is referred to as Wiener–Hopf equation. The Wiener–Hopf equation is a commonly used method to study various mathematical and physical problems. Its basic idea is to convert the original equation into a functional equation by using the integral transformation, and then solve it by the factorization method.

If the autocorrelation function $R_{xx}(\cdot)$ and the cross-correlation function $R_{xs}(\cdot)$ between the desired signal and the observed signal are available, the system impulse response $h(n)$ can be solved by the Wiener–Hopf equation as $h_{opt}(n)$, which is the optimum in the sense of the least MSE. The minimum MSE is

$$
\begin{aligned}
E[e^2(n)]_{\min} &= E[(s(n) - \sum_{m=0}^{+\infty} h_{opt}(m)x(n-m))^2] \\
&= E[s^2(n) - 2s(n)\sum_{m=0}^{+\infty} h_{opt}(m)x(n-m) \\
&\quad + \sum_{m=0}^{+\infty}\sum_{r=0}^{+\infty} h_{opt}(m)x(n-m)h_{opt}(r)x(n-r)] \\
&= R_{ss}(0) - 2\sum_{m=0}^{+\infty} h_{opt}(m)R_{xs}(m) + \sum_{m=0}^{+\infty} h_{opt}(m)\left[\sum_{r=0}^{+\infty} h_{opt}(r)R_{xx}(m-r)\right]
\end{aligned}
$$

$$\tag{11.10}$$

Simplifying eq. (11.10) further, we get

$$E[e^2(n)]_{min} = R_{ss}(0) - \sum_{m=0}^{+\infty} h_{opt}(m)R_{xs}(m) \qquad (11.11)$$

11.2.2 The Solution of the Wiener–Hopf Equation

1. Finite impulse response method

Equation (11.5) implies that the unit impulse response of the system is an infinite sequence. Since in practice we cannot obtain such a sequence, we have to use a finite length sequence to approximate it. Thus, eq. (11.5) can be rewritten as

$$y(n) = \hat{s}(n) = \sum_{m=0}^{N-1} h(m)x(n-m) \qquad (11.12)$$

where the length of $h(n)$ is assumed to be N. Similarly, eq. (11.6) can also be rewritten as

$$E[e^2(n)] = E[(s(n) - \sum_{m=0}^{N-1} h(m)x(n-m))^2] \qquad (11.13)$$

Applying the derivative to eq. (11.13) with respect to $h(m)$, and letting it to be equal to 0, we obtain

$$2E[(s(n) - \sum_{m=0}^{N-1} h(m)x(n-m))x(n-l)] = 0, \quad l = 0, 1, \dots, N-1 \qquad (11.14)$$

and

$$E[(s(n)x(n-l)] = \sum_{m=0}^{N-1} h(m)E[x(n-m)x(n-l)], \quad l = 0, 1, \dots, N-1 \qquad (11.15)$$

Thus, we have

$$R_{xs}(l) = \sum_{m=0}^{N-1} h(m)R_{xx}(l-m), \quad l = 0, 1, \dots, N-1 \qquad (11.16)$$

From eq. (11.16), we can get N linear equations as

$$\begin{cases} R_{xs}(0) = h(0)R_{xx}(0) + h(1)R_{xx}(1) + \cdots + h(N-1)R_{xx}(N-1) \\ R_{xs}(1) = h(0)R_{xx}(1) + h(1)R_{xx}(0) + \cdots + h(N-1)R_{xx}(N-2) \\ \vdots \qquad\qquad\qquad \vdots \\ R_{xs}(N-1) = h(0)R_{xx}(N-1) + h(1)R_{xx}(N-2) + \cdots + h(N-1)R_{xx}(0) \end{cases} \tag{11.17}$$

Rewriting it into a matrix form, we obtain

$$\begin{bmatrix} R_{xx}(0) & R_{xx}(1) & \cdots & R_{xx}(N-1) \\ R_{xx}(1) & R_{xx}(0) & \cdots & R_{xx}(N-2) \\ \vdots & \vdots & \ddots & \vdots \\ R_{xx}(N-1) & R_{xx}(N-2) & \cdots & R_{xx}(0) \end{bmatrix} \begin{bmatrix} h(0) \\ h(1) \\ \vdots \\ h(N-1) \end{bmatrix} = \begin{bmatrix} R_{xs}(0) \\ R_{xs}(1) \\ \vdots \\ R_{xs}(N-1) \end{bmatrix} \tag{11.18}$$

or

$$\boldsymbol{R}_{xx}\boldsymbol{H} = \boldsymbol{R}_{xs} \tag{11.19}$$

where

$$\boldsymbol{H} = \begin{bmatrix} h(0) & h(1) & \cdots & h(N-1) \end{bmatrix}^{\mathrm{T}} \tag{11.20}$$

$$\boldsymbol{R}_{xx} = \begin{bmatrix} R_{xx}(0) & R_{xx}(1) & \cdots & R_{xx}(N-1) \\ R_{xx}(1) & R_{xx}(0) & \cdots & R_{xx}(N-2) \\ \vdots & \vdots & \ddots & \vdots \\ R_{xx}(N-1) & R_{xx}(N-2) & \cdots & R_{xx}(0) \end{bmatrix} \tag{11.21}$$

$$\boldsymbol{R}_{xs} = \begin{bmatrix} R_{xs}(0) & R_{xs}(1) & \cdots & R_{xs}(N-1) \end{bmatrix}^{\mathrm{T}} \tag{11.22}$$

where $\boldsymbol{H}$ is the unit impulse response of the Wiener filter to be obtained, $\boldsymbol{R}_{xx}$ is the autocorrelation matrix of the signal $x(n)$, and $\boldsymbol{R}_{xs}$ is the cross-correlation vector between $x(n)$ and $s(n)$. If the autocorrelation matrix is nonsingular, the unit impulse response can be obtained by the matrix inversion operation:

$$\boldsymbol{H} = \boldsymbol{R}_{xx}^{-1}\boldsymbol{R}_{xs} \tag{11.23}$$

Furthermore, we can obtain the minimum MSE as

$$E[e^2(n)]_{\min} = R_{ss}(0) - \sum_{m=0}^{N-1} h_{\mathrm{opt}}(m)R_{xs}(m) \tag{11.24}$$

Obviously, if the autocorrelation function $R_{xx}(m)$ and the cross-correlation function $R_{xs}(m)$ are known, the unit impulse response of the optimal system can be obtained by solving the Wiener–Hopf equation as shown in eq. (11.16) or (11.19), realizing the optimal Wiener filtering. On the other hand, if the order N of the Wiener–Hopf equation is large, the computational complexity will be very large, and eq. (11.19) will involve a matrix inversion operation. However, the Wiener–Hopf equation can be solved by a method similar to the AR model parameter estimation (e.g., Levinson–Durbin algorithm).

If the signal and noise are uncorrelated, namely,

$$R_{sv}(m) = R_{vs}(m) = 0 \tag{11.25}$$

then

$$R_{xs}(m) = E[x(n)s(n+m)] = E[s(n)s(n+m) + v(n)s(n+m)] = R_{ss}(m)$$
$$R_{xx}(m) = E[x(n)x(n+m)] = E[(s(n) + v(n))(s(n+m) + v(n+m))] = R_{ss}(m) + R_{vv}(m) \tag{11.26}$$

Thus, eqs. (11.16) and (11.24) get converted to

$$R_{xs}(l) = \sum_{m=0}^{N-1} h(m)[R_{ss}(l-m) + R_{vv}(l-m)], \quad l=0, 1, \ldots, N-1 \tag{11.27}$$

$$E[e^2(n)]_{\min} = R_{ss}(0) - \sum_{m=0}^{N-1} h_{\text{opt}}(m)R_{ss}(m) \tag{11.28}$$

Example 11.1: Suppose that the wide-sense stationary random signal $x(n)$ is composed of desired signal $s(n)$ and noise $v(n)$, denoted as $x(n) = s(n) + v(n)$, where $s(n)$ and $v(n)$ are independent of each other. The autocorrelation function is given as $R_{ss}(m) = 0.6^{|m|}$, and $v(n)$ is the white noise with a zero mean and unit variance.
(1) Design a Wiener filter with $N = 2$ to estimate the desired signal $s(n)$.
(2) Find the minimum MSE of the Wiener filter.

Solution: (1) On the basis of the given conditions, we have $R_{ss}(m) = 0.6^{|m|}$ and $R_{vv}(m) = \delta(m)$. Substituting the given conditions into the Wiener–Hopf equation, we obtain

$$\begin{cases} 2h(0) + 0.6h(1) = 1 \\ 0.6h(0) + 2h(1) = 0.6 \end{cases}$$

Then the solution can be obtained as $h(0) = 0.451, \quad h(1) = 0.165$.
(2) Substituting values of $h(0)$ and $h(1)$ into the minimum MSE expression (11.28), we get

$$E[e^2(n)]_{\min} = R_{ss}(0) - \sum_{m=0}^{N-1} h_{opt}(m)R_{ss}(m) = 1 - h(0) - 0.6h(1) = 0.45$$

If the value of $x(n)$ is known, then the estimate of $s(n)$ is available. If the order of the Wiener filter is increased, then the estimation accuracy of the system can be improved and the MSE can be reduced.

2. Prewhitening method

The prewhitening method was proposed by Bode and Shannon. The key to this method is to convert the input signal $x(n)$ into white noise $w(n)$ by using a prewhitening filter and further solve the Wiener–Hopf equation.

In general, the input signal (i.e., the observed signal) $x(n)$ is often not white. The input signal first passes through a whitening filter $H_W(e^{j\omega})$, and then passes through the optimal filter $G(e^{j\omega})$ under the white noise condition as shown in Figure 11.2:

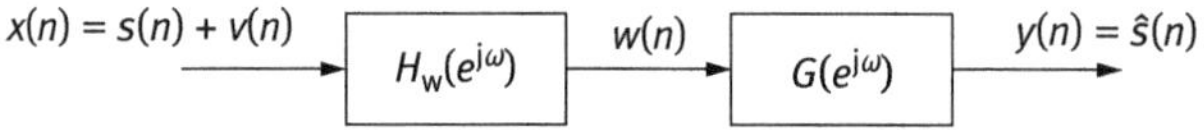

Figure 11.2: The block diagram of the prewhitening and Wiener filter.

In the figure, $w(n)$ denotes the white noise obtained by prewhitening the input $x(n)$. Thus, as long as a whitening filter $H_W(e^{j\omega})$ is obtained, the prewhitening can be achieved and the optimal estimation of the input signal can be further determined.

From the stochastic signal parametric modeling approach, we know that the random signal $x(n)$ can generally be viewed as the response of a linear system (such as autoregressive, moving average, and autoregressive moving average models) to white noise excitation. Suppose that the system function in the z-domain of the linear system is $B(z)$, and the z-transform of $x(n)$ and $w(n)$ are $X(z)$ and $W(z)$, respectively. According to the principle of random signals passing through a linear system, we have

$$R_{xx}(z) = \sigma_w^2 B(z)B(z^{-1}) \tag{11.29}$$

where $R_{xx}(z)$ denotes the PSD of $x(n)$ in the z-domain, and $B(z)$ and $B(z^{-1})$ correspond to the parts of $R_{xx}(z)$ whose poles and zeros are inside or outside the unit circle, respectively.

Since all the zeros and poles are inside the unit circle, $B(z)$ is a minimum phase system, and $1/B(z)$ is also a minimum phase system. In this manner, eq. (11.30) can be used for prewhitening:

$$W(z) = \frac{1}{B(z)}X(z) \tag{11.30}$$

Thus, $x(n)$ and $w(n)$ are taken as the system input and output, respectively, achieving the prewhitening processing. The system function $1/B(z)$ in eq. (11.30) is actually the whitening filter $H_W(e^{j\omega})$ as shown in Figure 11.2, which can be written as

$$H_W(z) = \frac{1}{B(z)} \tag{11.31}$$

In contrast to Figure 11.1, the Wiener filtering problem is essentially the problem of solving the optimal filter $h_{opt}(n)$ under the minimum MSE criterion. Now, according to the prewhitening process, Figure 11.2 is further refined as Figure 11.3:

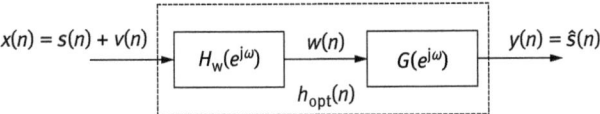

Figure 11.3: Pre-whitening Wiener filtering.

In Figure 11.3, the portion in the dashed box is the same as the Wiener filter in Figure 11.1, denoted as $h_{opt}(n)$, which is the cascade of the prewhitening filter $H_W(e^{j\omega})$ and the optimal linear filter $G(e^{j\omega})$. Applying the z-transform to $h_{opt}(n)$, we obtain

$$H_{opt}(z) = H_W(z)G(z) = \frac{G(z)}{B(z)} \tag{11.32}$$

As can be seen from the above equation, as long as the whitening filter $H_W(z) = 1/B(z)$ and the optimal linear filter $G(z)$ are known, we can get the solution of the Wiener filter, where $B(z)$ can be obtained from the autocorrelation function of the observation signal $x(n)$. The excitation signal of $G(z)$ is a white noise, which can be solved in accordance with the following method.

From Figure 11.3, we have

$$y(n) = \hat{s}(n) = \sum_{m=0}^{+\infty} g(m)w(n-m) \tag{11.33}$$

The MSE is

$$E[e^2(n)] = E\left[\left(s(n) - \sum_{m=0}^{+\infty} g(m)w(n-m)\right)^2\right]$$

$$= E\left[s^2(n) - 2s(n)\sum_{m=0}^{+\infty} g(m)w(n-m) + \sum_{m=0}^{+\infty}\sum_{r=0}^{+\infty} g(m)w(n-m)g(r)w(n-r)\right]$$

$$= R_{ss}(0) - 2\sum_{m=0}^{+\infty} g(m)R_{ws}(m) + \sum_{m=0}^{+\infty} g(m)\left[\sum_{r=0}^{+\infty} g(r)R_{ww}(m-r)\right]$$

$$\tag{11.34}$$

where $g(n)$ is the unit impulse response of $G(z)$. Substituting $R_{ww}(m) = \sigma_w^2 \delta(m)$ into above equation and rearranging, we obtain

$$
\begin{aligned}
E[e^2(n)] &= R_{ss}(0) - 2 \sum_{m=0}^{+\infty} g(m)R_{ws}(m) + \sigma_w^2 \sum_{m=0}^{+\infty} g^2(m) \\
&= R_{ss}(0) + \sum_{m=0}^{+\infty} \left[\sigma_w g(m) - \frac{R_{ws}(m)}{\sigma_w} \right]^2 - \frac{1}{\sigma_w^2} \sum_{m=0}^{+\infty} R_{ws}^2(m)
\end{aligned}
\tag{11.35}
$$

To minimize the MSE $E[e^2(n)]$ is equivalent to minimizing the middle term in eq. (11.35). Let $\sigma_w g(m) - \frac{R_{ws}(m)}{\sigma_w} = 0$, we obtain

$$
g_{opt}(m) - \frac{R_{ws}(m)}{\sigma_w^2}, \quad m \geq 0
\tag{11.36}
$$

We noticed that $g_{opt}(m)$ is causal. Applying the unilateral z-transform to the above equation, we get

$$
G_{opt}(z) = \frac{[R_{ws}(z)]_+}{\sigma_w^2}
\tag{11.37}
$$

where $[R_{ws}(z)]_+$ denotes the unilateral z-transform of $R_{ws}(m)$. The system function of Wiener –Hopf equation can be expressed as

$$
H_{opt}(z) = \frac{G_{opt}(z)}{B(z)} = \frac{[R_{ws}(z)]_+}{\sigma_w^2 B(z)}
\tag{11.38}
$$

Since the cross-correlation function between the observed signal $x(n)$ and the desired signal $s(n)$ can be expressed as

$$
\begin{aligned}
R_{xs}(m) &= E[x(n)s(n+m)] = E\left[\sum_{k=-\infty}^{+\infty} b(k)w(n-k)s(n+m) \right] \\
&= \sum_{k=-\infty}^{+\infty} b(k)R_{ws}(m+k) = R_{ws}(m) \star b(-m)
\end{aligned}
\tag{11.39}
$$

where $b(n)$ is the impulse response of $B(z)$. Applying the z-transform to the above equation, we get

$$
R_{xs}(z) = R_{ws}(z)B(z^{-1})
\tag{11.40}
$$

Thus, the Wiener filter in eq. (11.38) can be rewritten as

$$
H_{opt}(z) = \frac{G_{opt}(z)}{B(z)} = \frac{[R_{ws}(z)]_+}{\sigma_w^2 B(z)} = \frac{[R_{xs}(z)/B(z^{-1})]_+}{\sigma_w^2 B(z)}
\tag{11.41}
$$

The minimal MSE of the causal Wiener filter is given as

$$E[e^2(n)]_{min} = R_{ss}(0) - \frac{1}{\sigma_w^2}\sum_{m=0}^{+\infty} R_{ws}^2(m) \tag{11.42}$$

By using Parseval's theorem, we can get the minimal MSE in the z-domain as

$$E[e^2(n)]_{min} = \frac{1}{2\pi j}\oint_c [R_{ss}(z) - H_{opt}(z)R_{xs}(z^{-1})]\frac{dz}{z} \tag{11.43}$$

where the unit circle can be used in the contour integral.

The steps for solving the Wiener–Hopf equation by the prewhitening method are listed as follows:

Step 1: Find $R_{xx}(z)$ by applying the z-transform to the autocorrelation $R_{xx}(m)$ of the observation signal $x(n)$.

Step 2: Find the minimum phase system $B(z)$ by using $R_{xx}(z) = \sigma_w^2 B(z)B(z^{-1})$ in eq. (11.29).

Step 3: Find the causal $G_{opt}(z)$ based on the minimal MSE criterion, and get $H_{opt}(z) = \frac{G_{opt}(z)}{B(z)} = \frac{[R_{ws}(z)]_+}{\sigma_w^2 B(z)}$.

Step 4: Based on $H_{opt}(z) = \frac{G_{opt}(z)}{B(z)} = \frac{[R_{ws}(z)]_+}{\sigma_w^2 B(z)} = \frac{[R_{xs}(z)/B(z^{-1})]_+}{\sigma_w^2 B(z)}$ in eq. (11.41), find the

system function $H_{opt}(z)$ of the Wiener–Hopf equation. The optimal unit impulse response $h_{opt}(n)$ of Wiener filter can be obtained by the inverse z-transform.

Example 11.2: The input signal of a linear time-invariant system is known as $x(n) = s(n) + v(n)$, where the autocorrelation function of $s(n)$ is $R_{ss}(m) = 0.8^{|m|}$, and $v(n)$ is a white noise with zero mean and unit variance, independent of $s(n)$. Design a causal Wiener filter to estimate $s(n)$, and find the minimal MSE.

Solution: From the given conditions we know that $R_{ss}(m) = 0.8^{|m|}$, $R_{vv}(m) = \delta(m)$, $R_{sv}(m) = 0$, and $R_{xs}(m) = R_{ss}(m)$.

Step 1: Find $R_{xx}(z)$ based on $R_{xx}(m)$. Applying the z-transform to $R_{xx}(m) = R_{ss}(m) + R_{vv}(m)$, we obtain

$$R_{xx}(z) = R_{ss}(z) + R_{vv}(z) = \frac{0.36}{(1-0.8z^{-1})(1-0.8z)} + 1$$

$$= 1.6 \times \frac{(1-0.5z^{-1})(1-0.5z)}{(1-0.8z^{-1})(1-0.8z)}, \quad 0.8 < |z| < 1.25$$

Step 2: Find the minimum phase system $B(z)$ and the variance of the white noise. Since $R_{xx}(z) = \sigma_w^2 B(z)B(z^{-1})$, based on the distribution of zeros and poles, we can get $B(z)$ and the variance of the white noise as

$$B(z) = \frac{1-0.5z^{-1}}{1-0.8z^{-1}}, \quad |z|>0.8; \quad B(z^{-1}) = \frac{1-0.5z}{1-0.8z}, \quad |z|<1.25; \quad \sigma_w^2 = 1.6$$

Step 3: Find the causal system $G_{opt}(z)$ and then $H_{opt}(z)$, based on the minimal MSE criterion. From eq. (11.41), we have

$$H_{opt}(z) = \frac{[R_{xs(z)}/B(z^{-1})]_+}{\sigma_w^2 B(z)} = \frac{1-0.8z^{-1}}{1.6(1-0.5z^{-1})}\left[\frac{0.36}{(1-0.8z^{-1})(1-0.5z)}\right]_+$$

where the convergence region in the square brackets is $0.8<|z|<2$. Applying the inverse z-transform to it, we obtain

$$Z^{-1}\left[\frac{0.36}{(1-0.8z^{-1})(1-0.5z)}\right] = 0.6 \times (0.8)^n u(n) + 0.6 \times 2^n u(-n-1)$$

where $u(n)$ is the unit step signal. Taking the causal part (the first term) of the above equation, the corresponding unilateral z-transformation is given as

$$\left[\frac{0.36}{(1-0.8z^{-1})(1-0.5z)}\right]_+ = 0.6 \times \frac{1}{1-0.8z^{-1}}$$

Thus,

$$H_{opt}(z) = \frac{[R_{xs}(z)/B(z^{-1})]_+}{\sigma_w^2 B(z)} = \frac{1-0.8z^{-1}}{1.6 \times (1-0.5z^{-1})}\left[\frac{0.6}{1-0.8z^{-1}}\right] = \frac{3/8}{1-0.5z^{-1}}$$

$$h_{opt}(n) = 0.375 \times (0.5)^n, \quad n \geq 0$$

Step 4: Find the minimal MSE based on eq. (11.42). The minimal MSE can be obtained by $E[e^2(n)]_{min} = R_{ss}(0) - \frac{1}{\sigma_w^2}\sum_{m=0}^{+\infty} R_{ws}^2(m)$, given as

$$E[e^2(n)]_{min} = 0.375$$

Example 11.3: Let the pure sinusoidal signal be $s(n)$, and the additive noise $v(n)$ is obtained by an MA model exited by a Gaussian white noise $v(n)$. The input signal $x(n)$ is the combination of both $s(n)$ and $v(n)$ as $x(n) = s(n) + v(n)$. Set $d(n) = s(n)$. Process $x(n)$ with the Wiener filter by the MATLAB programming.
Solution: The MATLAB program is given as follows:

```
clear all;
n = (1:1000)'; N=1000; s = sin(0.1*pi*n); d=s; v = 0.8*randn(1000,1);
```

```
ma = [1, -0.8, 0.4, -0.3]; x = filter(ma,1,v)+s; M = 13;
b = firwiener(M-1,x,d); y = filter(b,1,x); e = d - y;
figure(1);
subplot(221); plot(n(900:end),x(900:end));
axis([900,1000,-4,4]); xlabel('Sample numbers'); ylabel('Amplitude');
    title('Original input signal');
subplot(222),plot(n(900:end),[y(900:end),d(900:end)]);
axis([900,1000,-2,2]); xlabel('Sample numbers'); ylabel('amplitude');
    title('Filtered and pure signals');
subplot(223),plot(n(900:end),e(900:end));
axis([900,1000,-4,4]); xlabel('Sample numbers'); ylabel('Amplitude');
    title('Error');
subplot(224),stem(b);
axis([0,14,-0.2,0.2]); xlabel('Sample numbers'); ylabel('Amplitude');
    title('Impulse response of Wiener filter');
```

Figure 11.4 shows the comparison between the filtered and original signals, and also shows the unit impulse response of the Wiener filter.

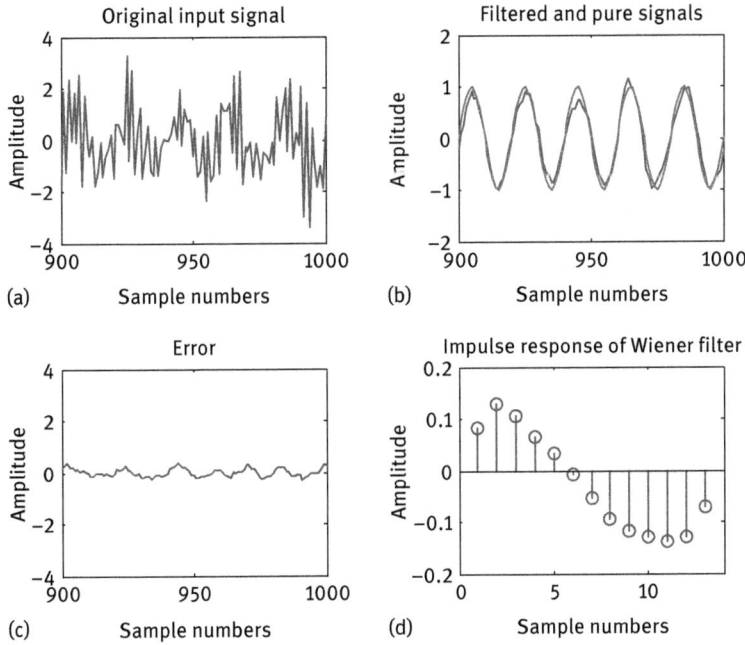

Figure 11.4: Wiener filtering: (a) original input signal; (b) filtered and pure signals; (c) error signal; and (d) impulse response of the Wiener filter.

11.3 Wiener Predictor

Similar to the Wiener filter, the Wiener predictor is a minimal MSE estimator that uses past observations to estimate the current and future values of the signal.

11.3.1 Causal Wiener Predictor

Figure 11.5 shows the model of the Wiener predictor, where, $y_d(n)$ and $y(n)$ denote the expected output and the actual output of the system, respectively.

If the model shown in Figure 11.5 is assumed to be a causal system, then

$$y(n) = \hat{s}(n + N) = \sum_{m=0}^{+\infty} h(m)x(n - m) \tag{11.44}$$

The MSE is

$$E[e^2(n)] = E[(\hat{s}(n + N) - \sum_{m=0}^{+\infty} h(m)x(n - m))^2] \tag{11.45}$$

Applying the derivative to the above equation respect to $h(m)$, and letting it to be equal to 0, we have

$$R_{xx}(M + l) = \sum_{m=0}^{+\infty} h(m)R_{xx}(l - m), \quad l \geq 0 \tag{11.46}$$

Since $y_d(n) = s(n + N)$, then

$$R_{xy_d}(m) = E[x(n)s(n + m + N)] = R_{xs}(m + N)$$

Applying the z-transform to the above equation, we have

$$R_{xy_d}(z) = z^N R_{xs}(z), \quad R_{xy_d}(z^{-1}) = z^{-N} R_{xs}(z^{-1}) \tag{11.47}$$

Since the desired output $y_d(n) = s(n)$ of the Wiener filter corresponds to the desired output $y_d(n) = s(n + N)$ of the Wiener predictor, the system function of the Wiener predictor can be obtained by referring the result of the Wiener filter as

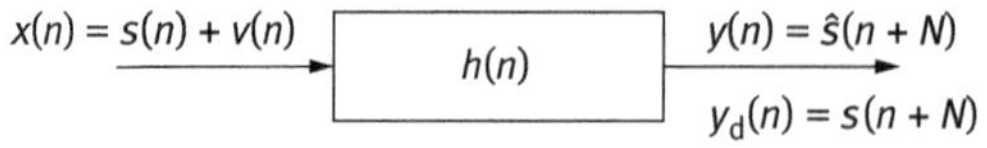

$x(n) = s(n) + v(n)$ → | $h(n)$ | → $y(n) = \hat{s}(n + N)$
$y_d(n) = s(n + N)$

Figure 11.5: The model of the Wiener predictor.

$$H_{\text{opt}}(z) = \frac{[R_{xy_d}(z)/B(z^{-1})]_+}{\sigma_w^2 B(z)} = \frac{[z^N R_{xs}(z)/B(z^{-1})]_+}{\sigma_w^2 B(z)} \qquad (11.48)$$

The minimal MSE is

$$E[e^2(n+N)]_{\min} = R_{ss}(0) - \frac{1}{\sigma_w^2} \sum_{m=0}^{+\infty} R_{wy_d}^2(m) \qquad (11.49)$$

By using Parseval's theorem, it can be expressed as

$$E[e^2(n+N)]_{\min} = \frac{1}{2\pi j} \oint_c [R_{ss}(z) - H_{\text{opt}}(z)z^{-N} R_{xs}(z^{-1})] \frac{dz}{z} \qquad (11.50)$$

Example 11.4: The observed signal is given as $x(n) = s(n) + v(n)$, where the autocorrelation sequence of $s(n)$ is $R_{ss}(m) = 0.8^{|m|}$, and $v(n)$ is a white noise with zero mean and unit variance. Both $s(n)$ and $v(n)$ are independent of each other. Design a causal Wiener predictor to estimate $s(n+N)$ and find the minimum MSE.

Solution: According to the given conditions $R_{ss}(m) = 0.8^{|m|}$, $R_{vv}(m) = \delta(m)$, $R_{sv}(m) = 0$, and $R_{xs}(m) = R_{ss}(m)$, we have $R_{xx}(m) = R_{ss}(m) + R_{vv}(m)$. Applying the z-transform to it, we get

$$R_{xx}(z) = R_{ss}(z) + R_{vv}(z)$$
$$= \frac{0.36}{(1-0.8z^{-1})(1-0.8z)} + 1 = 1.6 \times \frac{(1-0.5z^{-1})(1-0.5z)}{(1-0.8z^{-1})(1-0.8z)}, \quad 0.8 < |z| < 1.25$$

Since $R_{xx}(z) = \sigma_w^2 B(z)B(z^{-1})$, based on the distribution of zeros and poles of the system, we can get the minimum phase system $B(z)$ and the variance of the white noise as

$$B(z) = \frac{1-0.5z^{-1}}{1-0.8z^{-1}}, \quad |z| > 0.8; \quad B(z^{-1}) = \frac{1-0.5z}{1-0.8z}, \quad |z| < 1.25; \quad \sigma_w^2 = 1.6$$

From eq. (11.48), when $N = 1$, we have

$$H_{\text{opt}}(z) = \frac{[zR_{xs(z)}/B(z^{-1})]_+}{\sigma_w^2 B(z)} = \frac{1-0.8z^{-1}}{1.6(1-0.5z^{-1})} \left[\frac{0.36z}{(1-0.8z^{-1})(1-0.5z)} \right]_+$$

where the convergence region in the square brackets is $0.8 < |z| < 2$. Applying the z-transform to this part of expression, we obtain

$$Z^{-1} \left[\frac{0.36z}{(1-0.8z^{-1})(1-0.5z)} \right] = 0.48 \times (0.8)^n u(n) + 1.2 \times 2^n u(-n-1)$$

where $u(n)$ is the unit step signal. Taking the causal part (or the first term in the right-hand side) of the above equation, the z-transform is

$$\left[\frac{0.36z}{(1-0.8z^{-1})(1-0.5z)}\right]_{+} = 0.48 \times \frac{1}{1-0.8z^{-1}}$$

Thus,

$$H_{\text{opt}}(z) = \frac{[zR_{xs}(z)/B(z^{-1})]_{+}}{\sigma_w^2 B(z)} = \frac{1-0.8z^{-1}}{1.6 \times (1-0.5z^{-1})}\left[\frac{0.48}{1-0.8z^{-1}}\right] = \frac{0.3}{1-0.5z^{-1}}$$

$$h_{\text{opt}}(n) = 0.3 \times (0.5)^n, \quad n \geq 0$$

The above equation can be rewritten in the form of a difference equation as

$$s(n+1) = 0.3x(n) + 0.5s(n)$$

The minimal MSE is

$$E[e^2(n+1)]_{\min} = \frac{1}{2\pi j}\oint_c [R_{ss}(z) - H_{\text{opt}}(z)z^{-1}R_{xs}(z^{-1})]\frac{dz}{z}$$

$$= \frac{1}{2\pi j}\oint_c \left[\frac{0.36}{(z-0.5)(1-0.8z)}\right]dz = 0.6$$

11.3.2 *N*-Step Pure Predictor

The so-called N-step pure predictor refers to the prediction to $s(n+N)$ when the observation signal does not contain noise. We consider the prewhitening method to solve the Wiener system. Since $R_{xx}(z) = \sigma_w^2 B(z)B(z^{-1})$, we have

$$R_{xx}(z) = R_{ss}(z) = R_{xs}(z) = \sigma_w^2 B(z)B(z^{-1}) \tag{11.51}$$

Substituting eq. (11.51) into eqs. (11.49) and (11.50), we have

$$H_{\text{opt}}(z) = \frac{[z^N R_{xs}(z)/B(z^{-1})]_{+}}{\sigma_w^2 B(z)} = \frac{[z^N \sigma_w^2 B(z)B(z^{-1})/B(z^{-1})]_{+}}{\sigma_w^2 B(z)} = \frac{[z^N B(z)]_{+}}{B(z)} \tag{11.52}$$

$$E[e^2(n+N)]_{\min} = \frac{1}{2\pi j}\oint_c [R_{ss}(z) - H_{\text{opt}}(z)z^{-N}R_{xs}(z^{-1})]\frac{dz}{z}$$

$$= \frac{\sigma_w^2}{2\pi j}\oint_c [B(z)B(z^{-1}) - [z^N B(z)]_{+}z^{-N}B(z^{-1})]\frac{dz}{z} \tag{11.53}$$

If the unit impulse response $b(n)$ corresponding to $B(z)$ is a real-valued sequence, then we can get the simplified MSE expression as

$$E[e^2(n+N)]_{\min} = \sigma_w^2 \left[\sum_{n=-\infty}^{+\infty} b^2(n) - \sum_{n=-\infty}^{+\infty} b(n+N)u(n)b(n+N) \right]$$

Because $B(z)$ is a minimum phase system, and it must be causal, then the above equation can be further simplified as follows:

$$E[e^2(n+N)]_{\min} = \sigma_w^2 \left[\sum_{n=0}^{+\infty} b^2(n) - \sum_{n=0}^{+\infty} b^2(n+N) \right] = \sigma_w^2 \sum_{n=0}^{N-1} b^2(n) \qquad (11.54)$$

The above equation shows that the minimum MSE increases with the number of predicted steps N, that is, the farther the prediction, the greater will be the prediction error.

11.3.3 One-Step Linear Wiener Predictor

In general, the prediction problem needs to predict the current value or the future value according to p past observation values. For the one-step prediction, we can get the linear prediction equation as

$$\hat{x}(n) = \sum_{m=1}^{p} h(m)x(n-m) \qquad (11.55)$$

It can be written in the form of AR model:

$$\hat{x}(n) = - \sum_{m=1}^{p} a_{m,p}x(n-m) \qquad (11.56)$$

where $a_{m,p} = -h(m)$ denotes the m th coefficient of the p th-order AR model. By using this method for solving the AR model parameters, we can get a set of Yule–Walker equations:

$$R_{xx}(l) + \sum_{m=1}^{p} a_{m,p}R_{xx}(l-m) = 0, \quad l = 1, 2, \ldots, p \qquad (11.57)$$

Furthermore, we can get the minimal prediction error as

$$E[e^2(n)]_{\min} = R_{xx}(0) + \sum_{m=1}^{p} a_{m,p}R_{xx}(m) \qquad (11.58)$$

Example 11.5: Given $x(n) = s(n)$ and $R_{ss}(m) = 0.8^{|m|}$, design a one-step linear predictor and find the minimum MSE.

Solution: From the given conditions, we have $R_{xx}(m) = R_{ss}(m) = 0.8^{|m|}$, $R_{xx}(0) = 0$, $R_{xx}(\pm 1) = 0.8$, and $R_{xx}(\pm 2) = 0.64$. From eq. (11.57), we obtain

$$
\begin{cases}
0.8 + a_{1,p} + 0.8 a_{2,p} = 0 \\
0.64 + 0.8 a_{1,p} + a_{2,p} = 0
\end{cases}
$$

The solution is $a_{1,p} = -0.8$, $a_{2,p} = 0$. Then

$$\hat{x}(n) = 0.8x(n-1)$$

The minimum MSE is

$$E[e^2(n)]_{\min} = R_{xx}(0) + \sum_{m=1}^{p} a_{m,p} R_{xx}(m) = 1 - 0.64 = 0.36$$

11.4 A Brief Introduction to Kalman Filter

11.4.1 The Principle of Kalman Filter

The Wiener filter solves the problem of optimal linear filtering based on the minimum MSE criterion, but it still has a few limitations. For example, this filter requires the input signal to be stationary. Moreover, it is essentially a frequency domain method since this optimal filter is determined by a number of correlation functions, or PSD functions. Furthermore, if the observed data are in a vector form, the spectral decomposition problem is difficult to deal with, and sometimes cannot even be solved.

A Kalman filter can be considered as generalization of a Wiener filter. It can be applied not only to stationary processes but also to nonstationary processes. It can be used not only for linear filtering problem but also for nonlinear control problems, and can even be used for the multiinput–multioutput system. The basic characteristics of the Kalman filter include the analysis in both time domain and state space domain. The Kalman filter is widely used in the areas of target locating and tracking, noise suppression, and signal estimation, since it has many excellent performances.

A discrete-time linear system with vector input (or excitation vector) $w(k)$ and output $z(k)$ can be described in terms of two equations, called the state equation and the output equation, as given in eqs. (11.59) and (11.60), respectively,

$$x(k+1) = \boldsymbol{\Phi}(k+1, k)x(k) + \boldsymbol{\Gamma}(k+1, k)w(k), \quad k = 0, 1, \ldots \tag{11.59}$$

$$z(k+1) = H(k+1)x(k+1) + v(k+1), \quad k = 0, 1, \ldots \tag{11.60}$$

where the excitation vector $w(k)$ and the measurement error vector $v(k)$ are all white noises. For both equations, the following definitions can be given:

$x(k)$ is an $n \times 1$ vector called the state vector. $w(k)$ is a $p \times 1$ vector called the excitation vector. $\Phi(k+1, k)$ is an $n \times n$ matrix called the state transition matrix from time k to time $k+1$. $\Gamma(k+1, k)$ is an $n \times p$ matrix called the excitation transition matrix from time k to time $k+1$. $z(k+1)$ is an $m \times 1$ vector called the measurement vector at time $k+1$. $H(k+1)$ is an $m \times n$ matrix called the measurement matrix. $v(k+1)$ is an $m \times 1$ vector called the measurement error. $x(0)$ is an $n \times 1$ vector called the initial condition or the initial state vector at time $k = 0$.

In the given time $1, 2, \ldots, j$ and observation conditions $z(1), z(2), \ldots, z(j)$, we consider that the state vector $x(k)$ is estimated from these observations. Suppose that $\hat{x}(k \mid j)$ is used to represent the estimation of $x(k)$ when $z(1), z(2), \ldots, z(j)$ are given, then we have

$$\hat{x}(k \mid j) = g[z(1), z(2), \ldots, z(j)] \tag{11.61}$$

According to the relationship between k and j, we can divide the estimation problem into three main types, namely, if $k > j$, it is the prediction problem; if $k = j$, it is the filtering problem; and if $k < j$, it is the smoothing problem.

Figure 11.6 shows the structure of the state equation and the output equation of a linear discrete-time system. From eq. (11.59) we can get the state $x(k+1)$ at time $k+1$ from the state $x(k)$ at time k. In fact, after the recursive operation, the state $x(k)$ of the system at arbitrary time k can be written in a form composed of the initial state of the system, the state transition matrix, the excitation vector, and the excitation transition matrix, as given in eq. (11.62):

$$x(k) = \Phi(k, 0)x(0) + \sum_{i=1}^{k} \Phi(k, i)\Gamma(i, i-1)w(i-1), k = 1, 2, \ldots \tag{11.62}$$

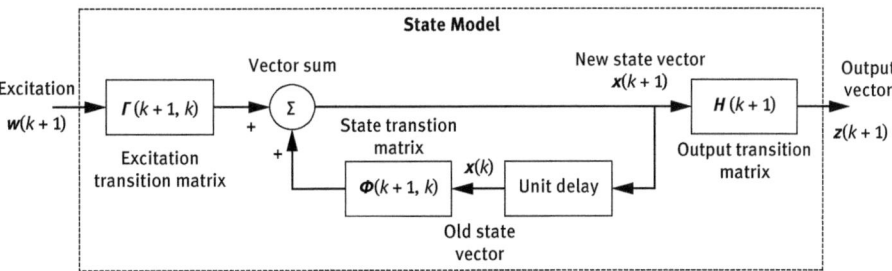

Figure 11.6: The block diagram showing the next state and output equation of a linear system.

Under the assumption that both $\boldsymbol{\Phi}(k+1, k)$ and $\boldsymbol{\Gamma}(k+1, k)$ are deterministic, by using eqs. (11.59) and (11.62), we can get the covariance matrix of the state vector $\boldsymbol{x}(k+1)$ as

$$\boldsymbol{P}(k+1|k) = \boldsymbol{\Phi}(k+1, k)\boldsymbol{P}(k|k)\boldsymbol{\Phi}^{\mathrm{T}}(k+1, k) + \boldsymbol{\Gamma}(k+1, k)\boldsymbol{Q}(k)\boldsymbol{\Gamma}^{\mathrm{T}}(k+1, k), \quad k = 0, 1, \ldots \tag{11.63}$$

where $\boldsymbol{Q}(k)$ is the known covariance matrix of the excitation vector. The initial condition $\boldsymbol{P}(0|0) = \boldsymbol{P}(0)$ is also assumed.

Combining the prediction result of the current state of the system with the measured state value, we can get the optimal estimate of the system state as

$$\hat{\boldsymbol{x}}(k+1|k+1) = \boldsymbol{\Phi}(k+1, k)\hat{\boldsymbol{x}}(k|k) + \boldsymbol{K}(k+1)[\boldsymbol{z}(k+1) - \boldsymbol{H}(k+1)\boldsymbol{\Phi}(k+1, k)\hat{\boldsymbol{x}}(k|k)] \tag{11.64}$$

The initial condition is $\hat{\boldsymbol{x}}(0|0) = 0, \quad k \geq 0$, where the Kalman gain matrix is expressed as

$$\begin{aligned} \boldsymbol{K}(k+1) = \ & \boldsymbol{P}(k+1|k)\boldsymbol{H}^{\mathrm{T}}(k+1)[\boldsymbol{H}(k+1)\boldsymbol{P}(k+1|k)\boldsymbol{H}^{\mathrm{T}}(k+1) \\ & + \boldsymbol{R}(k+1)]^{-1}, \quad k = 0, 1, \ldots \end{aligned} \tag{11.65}$$

where $\boldsymbol{R}(k+1)$ is the covariance matrix of the measurement error, which is an $m \times m$ matrix.

In fact, eq. (11.64) gives the optimal estimate of the system state at time k. However, to keep the Kalman filter running until the end of the process, we need to continue to update its covariance matrix, namely,

$$\boldsymbol{P}(k+1|k+1) = [\boldsymbol{I} - \boldsymbol{K}(k+1)\boldsymbol{H}(k+1)]\boldsymbol{P}(k+1|k), k = 0, 1, \ldots \tag{11.66}$$

where $\boldsymbol{I}$ is the unit matrix.

In the Kalman filter, the state vector $\boldsymbol{x}(k)$ is a function of the observed data $\boldsymbol{z}(k)$. After obtaining the observed value $z(1), z(2), \ldots, z(k)$, the state estimate at time k is represented by $\hat{\boldsymbol{x}}(k|k)$. As a function of the observed value, the Kalman filter continuously estimates the state, as shown in Figure 11.7.

Figure 11.8 shows the block diagram of the Kalman filter estimation determined by eq. (11.64).

If we merge the $\hat{\boldsymbol{x}}(k|k)$ terms in eq. (11.64), we can get another form of the Kalman filter, expressed as

$$\begin{aligned} \hat{\boldsymbol{x}}(k+1|k+1) = \ & [\boldsymbol{\Phi}(k+1, k) - \boldsymbol{K}(k+1)\boldsymbol{H}(k+1)\boldsymbol{\Phi}(k+1, k)]\hat{\boldsymbol{x}}(k|k) \\ & + \boldsymbol{K}(k+1)\boldsymbol{z}(k+1), k \geq 0 \end{aligned} \tag{11.67}$$

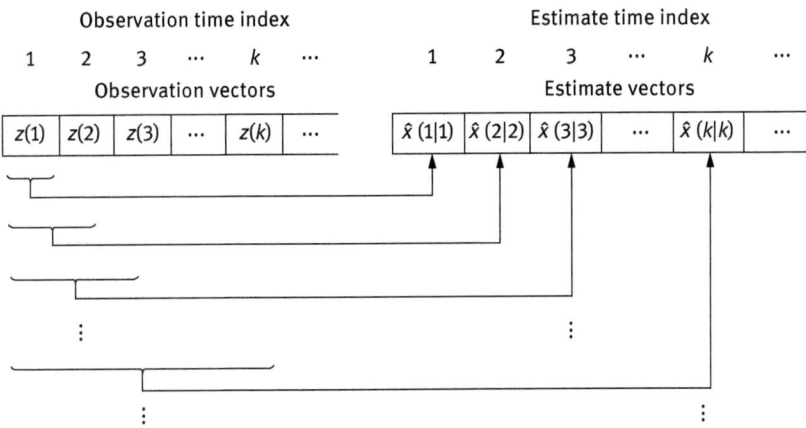

Figure 11.7: Timing diagram for the discrete Kalman filter.

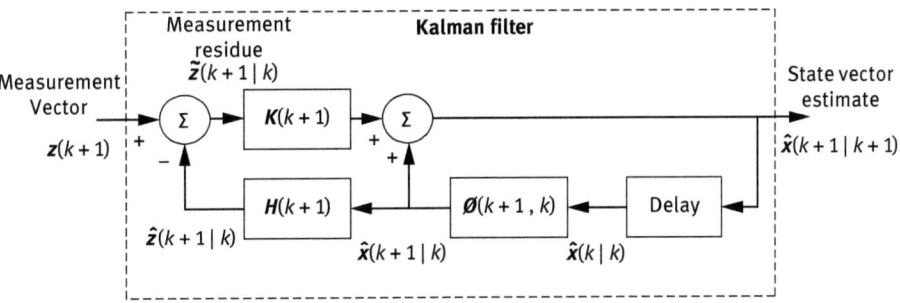

Figure 11.8: A vector block diagram for the Kalman filter in terms of measurement residue.

Similar to eq. (11.64), the initial condition of the above equation is $\hat{x}(0|0) = \mathbf{0}$. If we define

$$B(k|K) = \boldsymbol{\Phi}(k+1, k) - K(k+1)H(k+1)\boldsymbol{\Phi}(k+1, k)$$

The structure of this Kalman filter is shown in Figure 11.9.

11.4.2 The Analysis of Kalman Filter

We see from eq. (11.64) that the estimate at time $k+1$ obtained by $k+1$ observations is the function of the optimal estimation at time k from k observation values, as shown in Figure 11.10. The observation error $\tilde{z}(k+1|k)$ in the figure is defined as

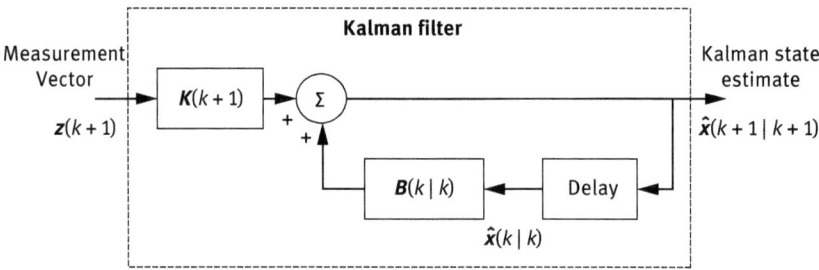

Figure 11.9: An alternative vector form of the Kalman filter.

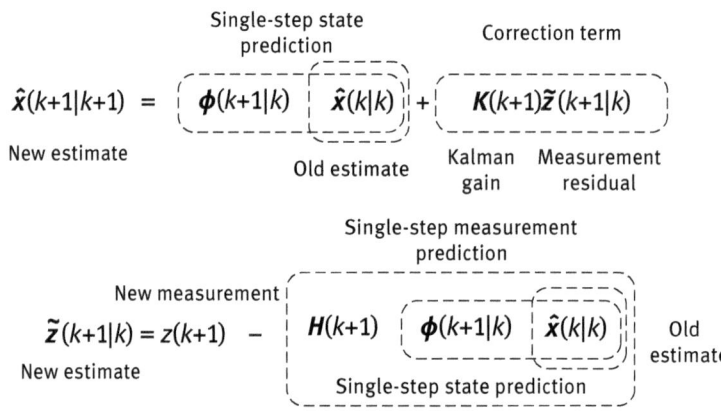

Figure 11.10: A predictor–corrector structure of the Kalman filtering equation.

$$\tilde{z}(k+1 \mid k) = z(k+1) - \hat{z}(k+1 \mid k) \tag{11.68}$$

The filtering equations in Figure 11.10 show how the Kalman filter works as a predictor–corrector. First, a new state is predicted (with single-step prediction), and then it is corrected by adding a product of the Kalman gain and the measurement error. The measurement error is the difference between the new measurement and the one-step prediction. If both of these are the same, the residue is zero and no correction is required.

11.4.3 The Calculation of Kalman Filter

It can be seen from eq. (11.64) that the new estimate is a function of the state transition matrix $\boldsymbol{\Phi}(k+1 \mid k)$, the measurement matrix $\boldsymbol{H}(k+1)$, the old estimate $\hat{x}(k \mid k)$, the new measurement $z(k+1)$, and the Kalman gain $\boldsymbol{K}(k+1)$, where we need to calculate only the Kalman gain. The Kalman gain matrix (which may be

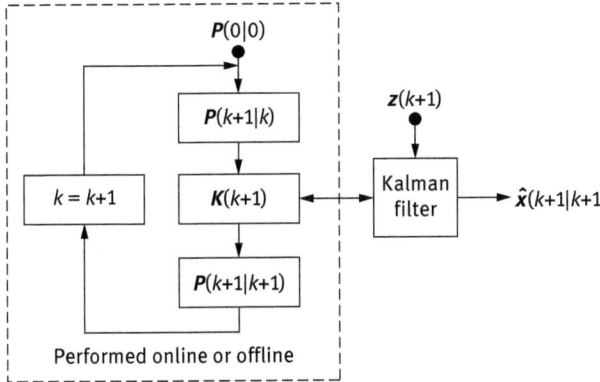

Figure 11.11: Recursive calculation procedure for the prediction error covariance, the Kalman gain, and the filtered covariance matrix.

different at different time) is determined by eqs. (11.63), (11.65), and (11.66). It can be obtained by the recursive calculation to $P(k+1 \mid k), K(k+1)$, and $P(k+1 \mid k+1)$, as shown in Figure 11.11. Note that the calculation for $P(k+1 \mid k), K(k+1)$, and $P(k+1 \mid k+1)$ must be conducted in accordance with the order in the figure.

We see from eqs. (11.63), (11.65), and (11.66) that $P(k+1|k), K(k+1)$, and $P(k+1|k+1)$ are independent of the observations and estimates. These matrices can, therefore, be calculated "offline" before they are needed, or in "real-time" as the observed data arrive. If they are calculated offline in advance, there must be enough space to store the values of $K(k+1)$ at all times. If the real-time calculation is performed when the observed values are obtained, it is necessary to reduce the sampling rate as much as possible to reduce the amount of calculation.

A one-dimensional (or scalar) Kalman filter is given as an example, as follows.

Example 11.6: The state and measurement models of the known Kalman filter are expressed as

$$x(k+1) = -0.7x(k) + w(k), \quad k = 0, 1, \ldots$$
$$z(k+1) = x(k+1) + v(k+1), \quad k = 0, 1, \ldots$$

The covariance of the excitation signal and the measurement error covariance are given as

$$E[w(j)w(k)] = Q\delta(j-k) = \delta(j-k)$$
$$E[v(j)v(k)] = R\delta(j-k) = \tfrac{1}{2}\delta(j-k)$$

Estimate the value $x(k+1)$ at moment $k+1$.

Solution: Since the state vector is in one dimension, the Kalman filter in eq. (11.64) is in the scalar form, whose initial condition is $\hat{x}(0 \mid 0) = 0$.

$$\hat{x}(k+1 \mid k+1) = -0.7\hat{x}(k \mid k) + K(k+1)[z(k+1) + 0.7\hat{x}(k \mid k)], k = 0, 1, \ldots$$

where the values of $K(k+1), P(k+1 \mid k)$, and $P(k+1 \mid k+1)$ at times $k = 0, 1, 2, \ldots$ are determined by eqs. (11.63), (11.65), and (11.66).

$$P(k+1 \mid k) = -0.7P(k \mid k)(-0.7) + 0.5, P(0 \mid 0) = P(0)$$
$$K(k+1) = P(k+1 \mid k)[P(k+1 \mid k) + 0.5]^{-1}$$
$$P(k+1 \mid k+1) = [1 - K(k+1)]P(k+1 \mid k)$$

On the basis of the above results and $P(0 \mid 0) = 10$, we can obtain $P(k+1 \mid k), K(k+1)$, and $P(k+1 \mid k+1)$, as shown in Figure 11.12.

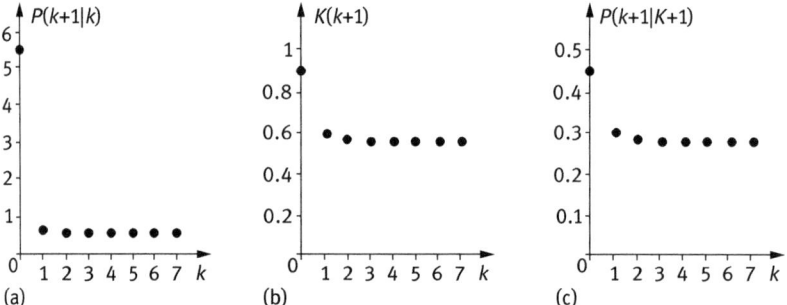

Figure 11.12: The convergence of (a)$P(k+1 \mid k)$, (b)$K(k+1)$, and (c)$P(k+1 \mid k+1)$.

From the figure we found that the Kalman gain fluctuates in the initial phase, but quickly converges to a steady-state value of 0.5603575. The steady-state value of single-step prediction error is 0.6372876. The initial value of $P(k+1 \mid k+1)$ is relatively large, but quickly converges to a steady-state value of 0.2801788, which represents the MSE of the filter estimation at steady state.

we see that the form of the Kalman filter does not relate to the observed data, and so if necessary, the offline calculation can be done before the observation.

Example 11.7: Assume that the state equation and the measurement equation of the Kalman filter are as follows:

$$x(k+1) = 0.8x(k) + w(k), \quad k = 0, 1, \ldots$$
$$z(k+1) = x(k+1) + v(k+1), \quad k = 0, 1, \ldots$$

where $w(n)$ and $v(n)$ are independent white noise sequences, satisfying $Q = \sigma_w^2 = 0.36$ and $R = \sigma_v^2 = 1$. Find the recursive expression of $x(k)$ by the MATLAB programming.

Solution: Some functions related to the Kalman filter are set in MATLAB, such as dlqe, dlqr, dlqew, lqe, lqed, lqew, kalman, kalmd, and so on. Using dlqe function, the key parameters of recursive Kalman filter can be obtained. MATLAB program is as follows:

```
clear all;
deltam=0.8; gammam=1; % The coefficients of the state equation;
hm=1; % The coefficient of the measurement equation;
qk=0.36; % The scalar form of the covariance matrix of the excitement
   vector;
rk=1; % The scalar form of the covariance matrix of the measurement error
   vector;
[k_gain, pk1,pk]=dlqe(deltam,gammam,hm,qk,rk); % The design of the
   discrete-time Kalman filter;
disp([k_gain, pk1,pk]) % Display the result
```

From the MATLAB program, we get $P(k+1 \mid k) = 0.600$, $P(k+1) = 0.375$, and $K(k+1) = 0.375$. Thus, the above parameters can be substituted in eq. (11.64), and the optimal estimate of the recursive Kalman filter is

$$\hat{x}(k+1 \mid k+1) = 0.8\hat{x}(k \mid k) + 0.375[z(k+1) - 1 \times 0.8\hat{x}(k \mid k)]$$
$$= 0.5\hat{x}(k \mid k) + 0.375z(k+1)$$

For the above recursive estimation formula, given the observation signal, one can recursively find the optimal recursive estimation.

Exercises

11.1 Describe the basic concept and principle of a Wiener filter.

11.2 Describe the basic concept and principle of a Kalman filter.

11.3 Derive the causal Wiener filter.

11.4 Derive the noncausal Wiener filter based on the relevant literature.

11.5 Describe the methods and steps involved in solving the Wiener–Hopf equation.

11.6 Describe the concept of a Wiener predictor.

11.7 Describe the mean of various variables and parameters in the Kalman filter.

11.8 Explain the state equation and measurement equation of the Kalman filter.

11.9 Describe the recursive solving method for the Kalman filter.

11.10 Assume that the PSD of the signal $s(t)$ is $P_{ss}(s) = \frac{1}{1-s^2}$, and the PSD of noise $v(t)$ is $P_{vv}(s) = \frac{1}{4-s^2}$. The signal and noise are uncorrelated. Find the transfer function of the continuous-time causal Wiener filter.

11.11 Perform the linear prediction for $s(t + \tau)$ in accordance with the minimum MSE criterion using the following form of linear estimation

$$\hat{d}(t) = as(t) + b\frac{ds}{dt}$$

Determine the coefficients a and b by using the orthogonal principle and the fractional extremum method. Prove the consistency of both methods, and find the MSE.

11.12 Suppose that the input signal of a linear filter is $x(t) = s(t) + v(t)$, satisfying $E[s(t)] = E[v(t)] = 0$. And $R_{ss}(\tau) = e^{-|\tau|}$, $R_{vv}(\tau) = e^{-2|\tau|}$, and $R_{sv}(\tau) = 0$ are given. Find the transfer function of the continuous-time causal Wiener filter.

11.13 Interpolation problem. Suppose that $x(0)$ and $x(T)$ are given. We need to estimate $x(t)$ in the range of $0 < t < T$ by using the linear estimation as $x(t) = ax(0) + bx(T)$. Determine parameters a and b.(Note : $x(0)$ and $x(T)$ are the values of $x(t)$ at time $t = 0$ and $t = T$, which are random variables.)

11.14 Suppose $x(t)$ is a band-limited random process. It satisfies $S_x(\omega) = 1$ in the range of $|\omega| < \omega_c = \frac{\pi}{T_s}$, shown as in Figure E11.1.

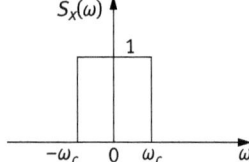

Figure E11.1:

Now we need to estimate $\hat{x}(t)$ by using the sampling sequence $x(kT_s)$ at $t = kT_s$, that is,

$$\hat{x}(t) = \sum_{k=-\infty}^{+\infty} a_k(t)x(kT_s)$$

Prove $a_k = \frac{\sin \omega_c(t - kT_s)}{\omega_c(t - kT_s)}$ based on the minimum MSE criterion.

11.15 The PSDs of the signal $s(t)$ and noise $v(t)$ are given as $P_{ss}(\omega) = \frac{2a}{\omega^2 + a^2}$ and $P_{vv}(\omega) = 1$, respectively. Both signal and noise are independent of each other. Design a noncausal Wiener filter to estimate $s(t)$ and $s(t + \tau)$, $\tau > 0$.

11.16 Given a received signal as $x(t) = s(t) + v(t)$, where $s(t)$ is the pure signal and $v(t)$ is noise, satisfying $E[s(t)] = E[v(t)] = 0$. Both signal and noise are independent of each other.

(1) Estimate $s'(t) = \frac{ds(t)}{dt}$ based on the orthogonal principle, and prove that the optimal $s'(t)$ can be obtained from the following equation: $R_{xs^*} = \int_0^{+\infty} R_x(\tau - \lambda)\hat{h}(\lambda)d\lambda, \quad \tau = 0 \sim +\infty.$

(2) If the causality is not required, prove $\hat{H}(j\omega) = \frac{j\omega S_{xs}(\omega)}{S_x(\omega)}$.

(3) If $R_{ss}(\tau) = e^{-|\tau|}$ and $R_n(\tau) = 2\delta(\tau)$ are given, find the noncausal optimal estimation of $s'(t)$.

11.17 Suppose the system model is $x(n+1) = 0.6x(n) + w(n)$, the measurement equation is $z(n) = x(n) + v(n)$, where $w(n)$ is a white noise with the variance of $\sigma_w^2 = 0.82$, and $v(n)$ is also a white noise with the variance of $\sigma_v^2 = 1$. Both $v(n)$ and $w(n)$ are uncorrelated. Find the discrete-time Wiener filter.

11.18 The autocorrelation function of signal $s(n)$ is $R_{ss}(m) = 0.7^{|m|}$, $m = 0, \pm 1, \pm 2, \ldots$. The signal is polluted by white noise $v(n)$ whose variance is $\sigma_v^2 = 0.4$. Both $s(n)$ and $v(n)$ are independent.

(1) Design a finite impulse response (FIR) filter with data length 3, whose output $y(n)$ has $E[(y(n) - s(n))^2]$ reaching its minimal.

(2) Design a noncausal optimal filter.

11.19 Suppose that the constant system model is given as

$$x(k+1) = 2x(k) + w(k)$$
$$z(k+1) = x(k+1) + v(k+1)$$

which satisfies $E[w(k)] = 0$, $E[w^2(k)] = 4$, $E[v(k)] = 0$, $E[v^2(k)] = 8$. Give the changing rule of $P(k)$ in the Kalman filter.

11.20 A time-invariant system is given as

$$x(k+1) = \tfrac{1}{2}x(k) + w(k)$$
$$z(k+1) = x(k+1) + v(k+1)$$

which satisfies the following conditions:

$$E[w(k)] = 0, \quad Cov[w(k), w(j)] = E[w(k)w(j)] = Q_k\delta_{kj}$$
$$E[v(k)] = 0, \quad Cov[v(k), v(j)] = E[v(k)v(j)] = R_k\delta_{kj}$$
$$E[x(0)] = \bar{x}(0), \quad Var[x(0)] = P_0, \quad Cov[x(0), w(k)] = 0$$
$$Cov[w(k), v(j)] = 0, \quad Cov[x(0), v(k)] = 0$$

and $\bar{x}(0) = 0$, $P_0 = 1$, $Q_k = 1$, $R_k = 2$ are given. The first two measurement values are $z(1) = 2$, $z(2) = 1$. Find the optimal values $\hat{x}(1)$ and $\hat{x}(2)$.

12 Adaptive Filtering

12.1 Introduction

Adaptive filtering or adaptive filter is a very important branch of signal processing. Since Widrow put forward the concept of the adaptation in 1959, adaptive filtering theory has been paid more attention and has been developed and perfected. Especially in recent years, with the rapid development of very large-scale integration (VLSI) and computer technologies, there have been many high-performance signal processing chips and high-performance general-purpose computers, providing an important material basis for the developments and applications of signal processing, especially for adaptive filters. On the other hand, the development of signal processing theory and application also provides the necessary theoretical foundation for the further development of adaptive filtering theory. It can be said that the adaptive filtering theory will continue its applications in various areas such as communications, radar, sonar, automatic control, image and speech processing, pattern recognition, biomedical engineering, and seismic exploration, and will promote the progress of these areas.

"Adaptive" has the meaning of initiative to adapt to the external environment. As the name suggests, an adaptive filter is a kind of digital filter that can automatically adjust its performance according to the input signal. Its most essential characteristic is the ability of self-learning and self-adjustment, the so-called self-adaptation. In general, adaptive filters can automatically adjust their parameters and/or structures to adapt to changing circumstances based on some optimal criterion during an iterative process to achieve the optimal filtering.

12.2 Structure of Transversal Adaptive Filter and Random Gradient Method

The block diagram of the adaptive filter is shown in Figure 12.1.

In Figure 12.1, $x(n)$, $y(n)$, and $d(n)$ represent the input, output, and reference (or desired) signals, respectively, at time n. $e(n)$ is the error signal. The system parameters of the adaptive filter are automatically controlled and adjusted by the error signal $e(n)$ to adapt to $x(n+1)$ at next time $n+1$, so that the output signal $y(n+1)$ is closer to the desired signal and the error signal $e(n+1)$ is further reduced.

https://doi.org/10.1515/9783110465082-012

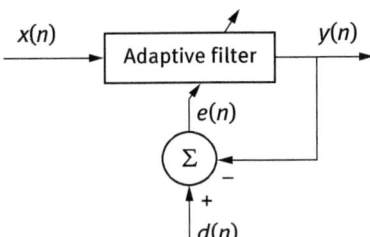

Figure 12.1: The block diagram of the adaptive filter.

12.2.1 The Structure and Performance Function of Transversal Adaptive Filter

1. Transversal adaptive filter

The transversal adaptive filter is a kind of basic adaptive filter, generally divided into two structures, the single-input and multi-input structures, as shown in Figures 12.2 and 12.3, respectively.

In both Figures 12.2 and 12.3, the adaptive weight vector is given as

$$w(n) = [w_0(n) \ w_1(n) \ \cdots \ w_M(n)]^{\mathrm{T}} \tag{12.1}$$

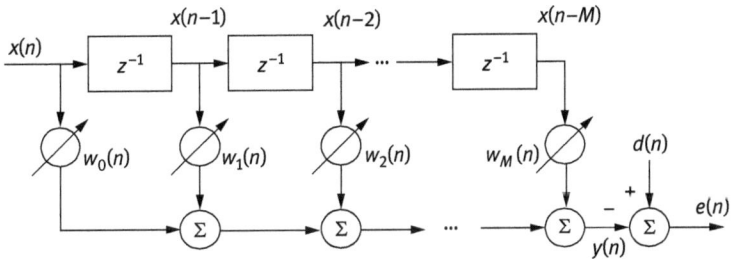

Figure 12.2: The single-input transversal adaptive filter.

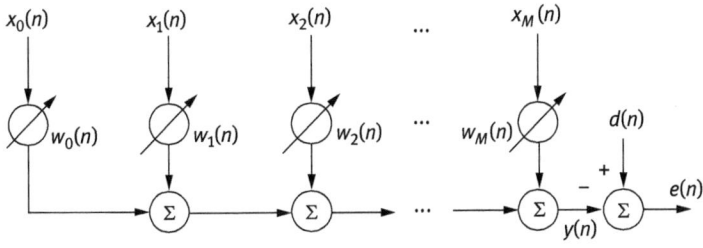

Figure 12.3: The multi-input transversal adaptive filter.

where the superscript T denotes the transpose operation. For the single-input structure, the input vector $\boldsymbol{x}(n)$ comes from a single source, denoted as

$$\boldsymbol{x}(n) = [x(n)\ x(n-1)\ \ldots\ x(n-M)]^{\mathrm{T}} \tag{12.2}$$

While the input vector $\boldsymbol{x}(n)$ for the multi-input structure is from $M+1$ different signal sources, denoted as

$$\boldsymbol{x}(n) = [x_0(n)\ x_1(n)\ \ldots\ x_M(n)]^{\mathrm{T}} \tag{12.3}$$

In the above figures, z^{-1} denotes the unit delay, $d(n)$ is the desired response of the filter, $y(n)$ is the output signal, and $e(n)$ is the difference between $d(n)$ and $y(n)$, which is known as the error signal.

If the input signal vector $\boldsymbol{x}(n)$ is multiplied by the weight vector $\boldsymbol{w}(n)$ of the filter, the output signal $y(n)$ at time n is obtained, which is expressed as

$$y(n) = \boldsymbol{x}^{\mathrm{T}}(n)\boldsymbol{w}(n) = \boldsymbol{w}^{\mathrm{T}}(n)\boldsymbol{x}(n) \tag{12.4}$$

The error signal of the adaptive system is expressed as

$$\begin{aligned} e(n) &= d(n) - y(n) \\ &= d(n) - \boldsymbol{w}^{\mathrm{T}}(n)\boldsymbol{x}(n) = d(n) - \boldsymbol{x}^{\mathrm{T}}(n)\boldsymbol{w}(n) \end{aligned} \tag{12.5}$$

The error signal is then fed back as a control signal for the adjustment of the weighting coefficients of the adaptive filter. In fact, the so-called adaptive ability of the adaptive filter is achieved relying on this error control or structural adjustment. If the input signal of the adaptive filter is a stationary random sequence, we get the following relation by taking the square and mathematical expectation operations to both sides of eq. (12.5):

$$E[e^2(n)] = E[d^2(n)] + \boldsymbol{w}^{\mathrm{T}}(n)E[\boldsymbol{x}(n)\boldsymbol{x}^{\mathrm{T}}(n)]\boldsymbol{w}(n) - 2E[d(n)\boldsymbol{x}^{\mathrm{T}}(n)\boldsymbol{w}(n)] \tag{12.6}$$

The input autocorrelation matrix $\boldsymbol{R}$ is defined as

$$\boldsymbol{R} = E[\boldsymbol{x}(n)\boldsymbol{x}^{\mathrm{T}}(n)] = E\begin{bmatrix} x^2(n) & x(n)x(n-1) & \cdots & x(n)x(n-M) \\ x(n-1)x(n) & x^2(n-1) & \cdots & x(n-1)x(n-M) \\ \vdots & \vdots & \cdots & \vdots \\ x(n-M)x(n) & x(n-M)x(n-1) & \cdots & x^2(n-M) \end{bmatrix}$$

$$\tag{12.7}$$

or

$$R = E[x(n)x^T(n)] = E \begin{bmatrix} x_0^2(n) & x_0(n)x_1(n) & \cdots & x_0(n)x_M(n) \\ x_1(n)x_0(n) & x_1^2(n) & \cdots & x_1(n)x_M(n) \\ \vdots & \vdots & \cdots & \vdots \\ x_M(n)x_0(n) & x_M(n)x_1(n) & \cdots & x_M^2(n) \end{bmatrix} \qquad (12.8)$$

where eq. (12.7) corresponds to the case of single input, and eq. (12.8) corresponds to the case of multi-input. The cross-correlation vector between the input signal and the expected response is defined as

$$p = E[d(n)x(n)] = E[d(n)x(n) \; d(n)x(n-1) \; \cdots \; d(n)x(n-M)]^T \qquad (12.9)$$

or

$$p = E[d(n)x(n)] = E[d(n)x_0(n) \; d(n)x_1(n) \; \cdots \; d(n)x_M(n)]^T \qquad (12.10)$$

for either single-input case or multi-input case. Thus eq. (12.6) changes as

$$E[e^2(n)] = E[d^2(n)] + w^T(n)Rw(n) - 2p^T(n)w(n) \qquad (12.11)$$

2. The performance of adaptive filters

It is customary to call the mean square error (MSE) $E[e^2(n)]$ as the performance function or the performance surface of the adaptive filter, denoted as ξ, J or MSE, that is,

$$MSE = \xi = J = E[e^2(n)] \qquad (12.12)$$

From eq. (12.11), when the input signal $x(n)$ and the expected response $d(n)$ are stationary random processes, the performance function ξ is exactly the quadratic function of the weight vector $w(n)$. The two-dimensional MSE function is a bowl-shaped paraboloid. If the dimension of the weight vector is greater than 2, the performance function is a super-paraboloid. Since the autocorrelation matrix is positive definite, the super-paraboloid faces upward, indicating that the MSE function has a unique minimum. The weight vector corresponding to the minimum value is the optimal weight vector w_{opt} of the adaptive filter, which is equal to the weight vector h_{opt} of the Wiener filter.

Figure 12.4 shows a typical 2D MSE function. If the number of weights of the adaptive filter is greater than 2, the super-paraboloid of the performance surface still has a unique global optimal point.

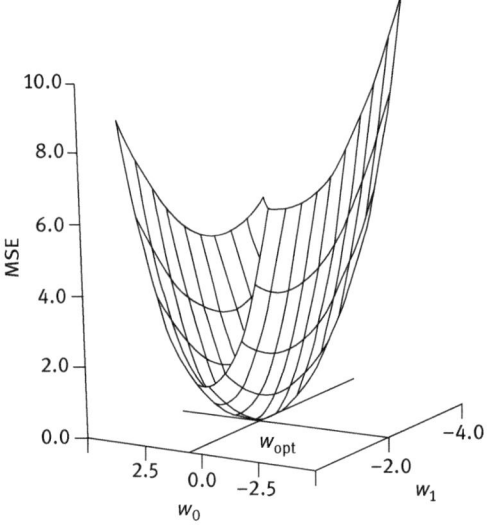

MSE

10.0

8.0

6.0

4.0

2.0

0.0

2.5   0.0 -2.5

w_0

w_{opt}

-4.0

-2.0

w_1

Figure 12.4: Two-dimensional mean square error function.

12.2.2 The Search on the Quadratic Performance Surface

The purpose of searching on the performance surface is to find the minimum value of the performance function and get the optimal weight vector corresponding to the minimum value. Thus, the problem to search for the minimum value in the quadratic performance surface is in fact the use of derivatives to obtain the extreme value in mathematics. For the performance function, we need to find the gradient and the minimum value corresponding to the zero of the gradient.

Finding the gradient for the MSE function in eq. (12.11), we get

$$\nabla = \frac{\partial \xi}{\partial w} = \frac{\partial E[e^2(n)]}{\partial w} = \left[\frac{\partial \xi}{\partial w_0} \; \frac{\partial \xi}{\partial w_1} \; \cdots \; \frac{\partial \xi}{\partial w_M} \right]^{\mathrm{T}} = 2Rw - 2p \tag{12.13}$$

Let the above equation be equal to 0, the optimal weight coefficient vector can be obtained as

$$w_{\text{opt}} = R^{-1}p \tag{12.14}$$

Equation (12.14) is called Wiener–Hopf equation. Obviously, eq. (12.14) has the same form and meaning as the Wiener filter shown in eq. (11.23) in the previous chapter. Substituting eq. (12.14) into eq. (12.11), we can get the minimum MSE (MMSE) of the adaptive filter as

$$\xi_{\min} = E[e^2(n)]_{\min} = E[d^2(n)] + w_{\mathrm{opt}}^{\mathrm{T}}Rw_{\mathrm{opt}} - 2p^{\mathrm{T}}w_{\mathrm{opt}}$$
$$= E[d^2(n)] + [R^{-1}p]^{\mathrm{T}}RR^{-1}p - 2p^{\mathrm{T}}R^{-1}p \tag{12.15}$$

Using the matrix operation rules, we can simplify the above equation as

$$\xi_{\min} = E[d^2(n)] + p^{\mathrm{T}}R^{-1}p = E[d^2(n)] - p^{\mathrm{T}}w_{\mathrm{opt}} \tag{12.16}$$

From eq. (12.14), we know that if the autocorrelation matrix R of the input signal and the cross-correlation vector p between the desired response and input signal can be obtained, we can get the optimal weight vector w_{opt} directly. However, in practice, w_{opt} in eq. (12.14) is often difficult to solve. On the one hand, it is often difficult to obtain a priori knowledge about the signal and noise. On the other hand, when the order of R is high, it is difficult to calculate the inverse matrix of R directly. Therefore, the iteration method is usually used for the realization of the optimal weight vector, achieving the minimum mean square value and the optimal weight vector ultimately.

Commonly used performance surface search methods include the gradient descending iterative algorithms, such as Newton method, conjugate gradient method, and steepest descent method. This section briefly introduces the Newton method and the steepest descent method.

1. Newton method

Newton method is a mathematical method for finding the zero-crossing point of function $f(x)$ by iteration method, that is, finding the solution of $f(x) = 0$. Assuming $f(x)$ as a function of variable x, the solving process of Newton method is as follows: starting from the initial estimate x_0 and using the first derivative $f'(x_0)$ at x_0 to calculate the new value x_1 as follows:

$$x_1 = x_0 - \frac{f(x_0)}{f'(x_0)} \tag{12.17}$$

Then, we can calculate next value x_2 by the derivative $f'(x_1)$ at x_1 and $f(x_1)$. The general iteration equation is

$$x_{k+1} = x_k - \frac{f(x_k)}{f'(x_k)}, \quad k = 0, 1, \dots \tag{12.18}$$

The convergence of the Newton method is related to the initial estimate x_0 and the function $f(x)$ itself. It is known that the convergence of the Newton method for a large class of functions is quite fast. On the other hand, it is often necessary to estimate the derivative using the following equation:

$$f'(x_k) = \frac{f(x_k) - f(x_{k-1})}{x_k - x_{k-1}} \tag{12.19}$$

Thus the Newton method can be expressed as

$$x_{k+1} = x_k - \frac{x_k - x_{k-1}}{f(x_k) - f(x_{k-1})} f(x_k), \quad k = 0, 1, \cdots \tag{12.20}$$

Searching on the performance surface by Newton's method is actually a search for the minimum of the performance function, that is, a point whose first derivative (or gradient) is zero. Define $\xi'(w_m)$, $m = 0, 1, \ldots, M$, as the first derivative of the m th weight coefficient. The iteration equation is shown as

$$w_m(n+1) = w_m(n) - \frac{\xi'(w_m(n))}{\xi''(w_m(n))}, \quad m = 0, 1, \ldots, M \tag{12.21}$$

where $\xi'(w_m(n))$ and $\xi''(w_m(n))$ are the first-order and the second-order derivatives for the m th weight coefficient of the MSE function. The gradient of the performance function can be written in the vector form as

$$\nabla(n) = \xi'(\mathbf{w}(n)) = \left[\frac{\partial \xi}{\partial w_0} \frac{\partial \xi}{\partial w_1} \cdots \frac{\partial \xi}{\partial w_M} \right]^{\mathrm{T}} \tag{12.22}$$

The second-order derivative of its performance function is

$$\nabla'(n) = \xi''(\mathbf{w}(n)) = \left[\frac{\partial^2 \xi}{\partial w_0^2} \frac{\partial^2 \xi}{\partial w_1^2} \cdots \frac{\partial^2 \xi}{\partial w_M^2} \right]^{\mathrm{T}} \tag{12.23}$$

On the other hand, the gradient of the MSE performance function is expressed as

$$\nabla(n) = 2\mathbf{R}\mathbf{w}(n) - 2\mathbf{p} \tag{12.24}$$

Taking the left multiplication to both sides of the above equation with $\frac{1}{2}\mathbf{R}^{-1}$, and according to $\mathbf{w}_{\mathrm{opt}} = \mathbf{R}^{-1}\mathbf{p}$, we get

$$\mathbf{w}_{\mathrm{opt}} = \mathbf{w}(n) - \frac{1}{2}\mathbf{R}^{-1}\nabla(n) \tag{12.25}$$

The above equation can also be written as an adaptive form as

$$\mathbf{w}(n+1) = \mathbf{w}(n) - \frac{1}{2}\mathbf{R}^{-1}\nabla(n) \tag{12.26}$$

The above equations show that when the performance function is a quadratic function, the Newton method can reach the optimal solution with one-step iteration.

In practice, the calculation of Newton method is much more complex. This is because we must estimate the matrix R and vector p, and the performance function may also be nonquadratic. These factors directly affect the performance of the Newton method. In general, a convergence factor is needed to introduce to adjust the speed of adaptive iteration. So eq. (12.26) becomes

$$w(n+1) = w(n) - \mu R^{-1}\nabla(n), \quad 0 < \mu < 1 \tag{12.27}$$

2. Steepest descent method

The steepest descent method is a classical and very useful method of finding the extremum by iteration. Geometrically, the iterative adjustment of the weight vector results in a reduction for the MSE of the system in the opposite direction of its gradient and finally reaches the MMSE ξ_{min}. When the MMSE is achieved, the weight vector becomes the optimal weight vector w_{opt}. In the search process to the performance surface of the adaptive filter, the gradient vector can be expressed as

$$\nabla(n) = \frac{\partial \xi(w(n))}{\partial w(n)} \tag{12.28}$$

Thus, the steepest descent method can be expressed as

$$w(n+1) = w(n) - \mu\nabla(n) \tag{12.29}$$

where μ is a positive constant, called the convergence factor, which is used to adjust the adaptive iteration step length. It is also known as the adaptive step size.

In order to prove that the steepest descent method satisfies $\xi(w(n+1)) < \xi(w(n))$, the first-order Taylor expansion of the performance function is performed and eq. (12.29) is used to get

$$\xi(w(n+1)) \approx \xi(w(n)) - \mu||\nabla(n)||^2 \tag{12.30}$$

Since the convergence factor μ is a positive constant, the performance function $\xi(w(n))$ tends to decrease as the increase of n. When $n \to \infty$, the performance function tends to the minimum ξ_{min}.

The adaptive iterative equation of the steepest descent method can be obtained by substituting eq. (12.24) into eq. (12.29):

$$w(n+1) = w(n) + 2\mu[p - Rw(n)] \tag{12.31}$$

The stability of the steepest descent method depends on two factors: the convergence factor and the autocorrelation matrix. Define the weight error vector $c(n)$ as

$$c(n) = w(n) - w_{\mathrm{opt}} \qquad (12.32)$$

Using the above equation and $w_{\mathrm{opt}} = R^{-1}p$, we eliminate the cross-correlation vector p in eq. (12.31) and get

$$c(n+1) = (I - 2\mu R)c(n) \qquad (12.33)$$

where I is the unit matrix. Equation (12.33) emphasizes that the stability of the steepest descent method is controlled by μ and R. Using the orthogonal similarity transformation, the autocorrelation matrix R can be expressed as

$$R = Q\Lambda Q^{-1} \qquad (12.34)$$

where Q is the orthogonal matrix, satisfying

$$Q^{-1} = Q^{\mathrm{T}} \qquad (12.35a)$$

and

$$QQ^{-1} = I \qquad (12.35b)$$

Each column vector of the matrix Q is an eigenvector corresponding to the eigenvalues of the autocorrelation matrix R. Λ is a diagonal matrix whose diagonal elements are eigenvalues of matrix R. These eigenvalues are often expressed as $\lambda_0, \lambda_1, \ldots, \lambda_M$, which are all positive real values. Each eigenvalue corresponds to a column eigenvector in the matrix Q. Substituting eq. (12.34) into eq. (12.33), we have

$$c(n+1) = [I - 2\mu\, Q\, \Lambda\, Q^{-1}]c(n) \qquad (12.36)$$

Multiplying Q^{-1} on the left side to the above equation, and using the properties of the orthogonal matrix, we get

$$Q^{-1}c(n+1) = [I - 2\mu\, \Lambda]Q^{-1}c(n) \qquad (12.37)$$

Define

$$c'(n) = Q^{-1}c(n) = Q^{-1}[w(n) - w_{\mathrm{opt}}] \qquad (12.38)$$

We have

$$\boldsymbol{c}'(n+1) = [\boldsymbol{I} - 2\mu\,\boldsymbol{\Lambda}]\boldsymbol{c}'(n) \tag{12.39}$$

Set the initial value of $\boldsymbol{c}'(n)$ as

$$\boldsymbol{c}'(0) = \boldsymbol{Q}^{-1}[\boldsymbol{w}(0) - \boldsymbol{w}_{\text{opt}}] \tag{12.40}$$

and suppose the initial value of the adaptive weight vector is $\boldsymbol{w}(0) = 0$, the above equation is simplified as

$$\boldsymbol{c}'(0) = -\boldsymbol{Q}^{-1}\boldsymbol{w}_{\text{opt}} \tag{12.41}$$

Considering the m th mode of vector $\boldsymbol{c}'(n)$, the iteration formula of the steepest descent method shown in eq. (12.39) becomes

$$c'_m(n+1) = [1 - 2\mu\lambda_m]c'_m(n), \quad m = 0, 1, \ldots, M \tag{12.42}$$

where λ_m is the m th eigenvalue of $\boldsymbol{R}$ and $c'_m(n)$ is the m th element of $\boldsymbol{c}'(n)$. The above equation is the first-order homogeneous equation. If the initial value of $c'_m(n)$ is set as $c'_m(0)$, then the solution of the difference equation is

$$c'_m(n) = [1 - 2\mu\lambda_m]^n c'_m(0), \quad m = 0, 1, \ldots, M \tag{12.43}$$

Since $\boldsymbol{R}$ is a positive definite matrix whose eigenvalues are all positive real values, $c'_m(n)$, $n = 0, 1, \ldots$ constructs a geometric series with a common ratio of $1 - 2\mu\lambda_m$. In order to ensure a stable convergence of the steepest descent method, there must be

$$-1 < 1 - 2\mu\lambda_m < 1, \quad m = 0, 1, \ldots, M \tag{12.44}$$

When the number of iterations $n \to \infty$, each mode of the steepest descent tends to 0, regardless of the initial state. This means when $n \to \infty$, the weight vector of the adaptive filter tends to be the optimal weight vector $\boldsymbol{w}_{\text{opt}}$. Rewriting eq. (12.42) in a vector form, we have

$$\boldsymbol{c}'(n) = [\boldsymbol{I} - 2\mu\,\boldsymbol{\Lambda}]^n \boldsymbol{c}'(0) \tag{12.45}$$

From eq. (12.44), the convergence factor μ of the steepest descent method is obtained as

$$0 < \mu < \frac{1}{\lambda_{\text{max}}} \tag{12.46}$$

where λ_{max} is the largest eigenvalue of the autocorrelation matrix $\boldsymbol{R}$.

The main advantage of the steepest descent method is its simplicity. However, this method needs a lot of iterations to converge to the point which is close to the optimal solution. This is because the steepest descent method is based on a first-order approximation of the performance surface around the current point. In practical applications, if the simplicity of the calculation is relatively important, it is appropriate to select the steepest descent method. However, if the rate of convergence is more important, then Newton method and its improved method should be used.

12.3 The Least Mean Square Algorithm of Adaptive Filter

12.3.1 Least Mean Square Algorithm

In the steepest descent method, if we can get the exact value of the gradient $\nabla(n)$ at each step of the iterative process and the convergence factor μ is chosen appropriately, this method definitely converges to the optimal Wiener solution. However, it is difficult to measure the gradient vector accurately at each step of the iteration, since this requires a priori knowledge about the autocorrelation matrix R and the cross-correlation vector p. In practical applications, the gradient vector needs to be estimated from the data at each step of the iteration. The least mean square (LMS) algorithm proposed by Widrow et al. is an MMSE algorithm between the expected response and the output signal of the filter. The gradient vector is estimated according to the input signal during the iterative process to achieve optimal coefficients. As the steepest descent method, the LMS algorithm does not need to calculate the corresponding correlation matrix, and does not need to perform matrix operations.

As a linear adaptive filtering algorithm, LMS algorithm basically includes two processes: one is the filtering process, and the other is the adaptive process. During the first process, the adaptive filter calculates its response to the input signal and results in an estimated error signal, by comparison with the expected response. During the second process, the error signal automatically adjusts the parameters of the filter itself. Both processes form a feedback loop as shown in Figure 12.5.

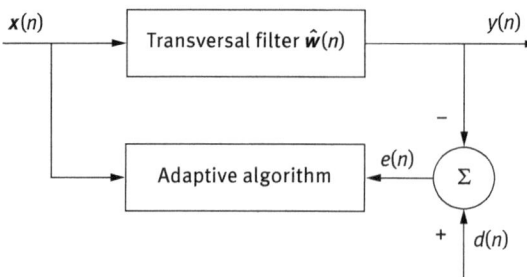

Figure 12.5: The block diagram of the adaptive transverse filter.

From Figure 12.5, we get the error signal of the adaptive filter as

$$e(n) = d(n) - y(n) \tag{12.47}$$

where $y(n)$ is the output signal of the adaptive filter,

$$y(n) = x^T(n)w(n) = w^T(n)x(n) \tag{12.48}$$

where $x(n)$ is the input vector, defined as $x(n) = [x(n)\ x(n-1)\ L\ x(n-M)]^T$ for the single- input structure, or $x(n) = [x_0(n)\ x_1(n)\ K\ x_M(n)]^T$ for the multi-input structure.

We need to estimate the gradient $\nabla(n)$ of the MSE $\xi = E[e^2(n)]$ and replace the truth-value $\nabla(n)$ with the estimated value $\hat{\nabla}(n)$. LMS algorithm replaces its mean squared value with the instantaneous squared value of each iteration of the error signal and estimates the gradient as

$$\hat{\nabla}(n) = \left[\frac{\partial e^2(n)}{\partial w_0(n)}\ \frac{\partial e^2(n)}{\partial w_1(n)}\ \cdots\ \frac{\partial e^2(n)}{\partial w_M(n)} \right]^T \tag{12.49a}$$

It can also be rewritten into a vector form as

$$\hat{\nabla}(n) = \frac{\partial e^2(n)}{\partial w(n)} \tag{12.49b}$$

Substituting eqs. (12.47) and (12.48) into eq. (12.49b), we get

$$\hat{\nabla}(n) = 2e(n) \frac{\partial e(n)}{\partial w(n)} = -2e(n)x(n) \tag{12.50}$$

Replacing the truth gradient value $\nabla(n)$ with its estimation $\hat{\nabla}(n)$, we get

$$w(n+1) = w(n) + \mu(-\hat{\nabla}(n)) = w(n) + 2\mu e(n)x(n) \tag{12.51}$$

where μ is the convergence factor of the adaptive filter. Equation (12.51) is the iteration formula of the famous LMS algorithm. It can be seen that the weight vector of the adaptive iteration at next moment can be obtained from the weight vector at current moment plus the input vector with the error signal as the scaling factor. Figure 12.6 shows the flowchart for implementing the LMS algorithm.

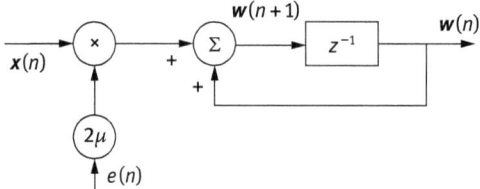

Figure 12.6: The flowchart for implementing the LMS algorithm.

12.3.2 Performance Analysis of LMS Algorithm

1. Convergence of LMS algorithm

Convergence is a very important index of the adaptive filter. In order to test the convergence of the LMS algorithm, we first prove that the gradient estimate shown in eq. (12.50) is unbiased.

Taking the mathematical expectation to eq. (12.50), and using eqs. (12.47) and (12.48), we have

$$E[\hat{\nabla}(n)] = -2E[e(n)\boldsymbol{x}(n)] = -2E[(d(n)-y(n))\boldsymbol{x}(n)]$$
$$= -2E[d(n)\boldsymbol{x}(n)-\boldsymbol{x}(n)\boldsymbol{x}^{\mathrm{T}}(n)\boldsymbol{w}(n)] = 2[\boldsymbol{R}\boldsymbol{w}(n)-\boldsymbol{p}] \qquad (12.52)$$
$$= \nabla(n)$$

From the above result, we see that the LMS algorithm is unbiased in estimating the gradient.

For the sake of convenience, it is assumed that the two successive iterations of the LMS algorithm are long enough to ensure that the input signals $\boldsymbol{x}(n)$ and $\boldsymbol{x}(n+1)$ are not correlated with each other. From eq. (12.51), the weight vector $\boldsymbol{w}(n)$ is only a function of the input vector $\boldsymbol{x}(n-1)$, $\boldsymbol{x}(n-2)$, K, $\boldsymbol{x}(0)$. Since these input vectors are uncorrelated, $\boldsymbol{w}(n)$ and $\boldsymbol{x}(n)$ are also uncorrelated. Thus, taking mathematical expectation to eq. (12.51), we have

$$E[\boldsymbol{w}(n+1)] = E[\boldsymbol{w}(n)] + 2\mu E[e(n)\boldsymbol{x}(n)]$$
$$= E[\boldsymbol{w}(n)] + 2\mu\{E[d(n)\boldsymbol{x}(n)] - E[\boldsymbol{x}(n)\boldsymbol{x}^{\mathrm{T}}(n)\boldsymbol{w}(n)]\} \qquad (12.53)$$

Furthermore, we get

$$E[\boldsymbol{w}(n+1)] = E[\boldsymbol{w}(n)] + 2\mu\{\boldsymbol{p} - \boldsymbol{R}E[\boldsymbol{w}(n)]\}$$
$$= (\boldsymbol{I} - 2\mu\boldsymbol{R})E[\boldsymbol{w}(n)] + 2\mu\boldsymbol{p} \qquad (12.54)$$

where $\boldsymbol{I}$ is the identity matrix with the same dimension as $\boldsymbol{R}$. Let the initial value of the weight vector be $\boldsymbol{w}(0)$, then through $n+1$ iterations we get

$$E[w(n+1)] = (I - 2\mu R)^{n+1}w(0) + 2\mu \sum_{j=0}^{n} (I - 2\mu R)^{j}p \qquad (12.55)$$

Using the orthogonal similarity transformation of the matrix and referring to eq. (12.39), we have

$$E[c'(n+1)] = [I - 2\mu \Lambda]^{n}c'(0) \qquad (12.56)$$

where $c'(n)$ is the principal coordinate form of the weight vector $w(n)$. Λ is the diagonal matrix of the autocorrelation matrix R whose diagonal elements are eigenvalues of R, that is,

$$\Lambda = \begin{bmatrix} \lambda_0 & & & 0 \\ & \lambda_1 & & \\ & & \ddots & \\ 0 & & & \lambda_M \end{bmatrix} \qquad (12.57)$$

Since R is positive definite, all eigenvalues are positive real values. For the diagonal matrix $(I - 2\mu \Lambda)$, as long as its every diagonal element has a value less than 1, we get

$$\lim_{n \to \infty} [I - 2\mu \Lambda]^{n} = 0 \qquad (12.58)$$

Thus, $c'(n)$ reaches its optimum weight vector as

$$c'_{opt} = E[c'(n)] = 0 \qquad (12.59)$$

The convergence factor should satisfy the following convergence condition:

$$0 < \mu < \frac{1}{\lambda_{max}} \qquad (12.60)$$

where λ_{max} is the largest eigenvalue of R, it is also the largest diagonal element in Λ. Since

$$\lambda_{max} \le tr[\Lambda] = tr[R] \qquad (12.61)$$

The constraint condition of the convergence factor shown in eq. (12.60) can be rewritten as follows:

$$0 < \mu < \frac{1}{tr[R]} \qquad (12.62)$$

or

$$0 < \mu < \frac{1}{(M+1)P_{\text{in}}}$$

(12.63)

where tr[·] is the trace of the matrix, P_{in} is the power of the input signal. In general, eq. (12.63) is more convenient than eq. (12.62). This is because the power of the input signal is easier to estimate than the eigenvalues of its autocorrelation matrix.

From eq. (12.59), if the convergence condition $0 < \mu < \frac{1}{\lambda_{\text{max}}}$ is satisfied, the weight vector $c'(n)$ of the adaptive filter associated with the principal axis coordinate will converge to 0 vector finally. If $c'(n)$ is transformed to its original coordinate, we know that $c'_{\text{opt}} = 0$, which corresponds to

$$w_{\text{opt}} = R^{-1}p$$

(12.64)

This indicates that the LMS algorithm eventually converges to the Wiener filter.

In the above discussion, the uncorrelated assumption between two input samples is very harsh. In fact, the implementation of this kind of adaptive filter shows that even if there is a large correlation between the input samples, the mathematical expectation of the weight coefficient vector can converge to the Wiener solution, but the MSE will become significant.

2. Adaptive time constant and learning curve

Another expression of the MSE function can be obtained from eqs. (12.11) and (12.16) as

$$\xi = \xi_{\text{min}} + (w(n) - w_{\text{opt}})^{\text{T}} R(w(n) - w_{\text{opt}})$$

(12.65)

According to the definition of eq. (12.32), the above equation is rewritten as

$$\xi = \xi_{\text{min}} + c^{\text{T}}(n)Rc(n)$$

(12.66)

And then through the orthogonal similarity transform, the coordinate axis is rotated to the main coordinate system, we get

$$\xi = \xi_{\text{min}} + c'^{\text{T}}(n)\Lambda c'(n)$$

(12.67)

Substituting eq. (12.45) into the above equation, we have

$$\begin{aligned}
\xi &= \xi_{\text{min}} + [(I - 2\mu\Lambda)^n c'(0)]^{\text{T}} \Lambda [(I - 2\mu\Lambda)^n c'(0)] \\
&= \xi_{\text{min}} + c'^{\text{T}}(0)[(I - 2\mu\Lambda)^n]^{\text{T}} \Lambda (I - 2\mu\Lambda)^n c'(0)
\end{aligned}$$

(12.68)

Since both matrix $(I - 2\mu\Lambda)$ and matrix Λ are the diagonal matrices, we get

$$\xi = \xi_{\min} + c'^{\mathrm{T}}(0)(I - 2\mu\Lambda)^{2n}\Lambda c'(0) \tag{12.69}$$

It can also be written in a scalar form as

$$\xi = \xi_{\min} + \sum_{m=0}^{M} c_m'^2(0)\lambda_m(1 - 2\mu\lambda_m)^{2n} \tag{12.70}$$

where $c'_m(0)$ is the m th component of vector $c'(0)$, and λ_m is the m th diagonal element of the diagonal matrix Λ.

Equation (12.70) is the adaptive learning curve of the LMS algorithm, showing that the MSE function ξ is an exponential function of the iteration number n. As long as the convergence condition eq. (12.62) is satisfied, the MSE decreases exponentially with the iteration and eventually converges to the MMSE $\xi_{\min}$. Figure 12.7 shows the learning curve of the MSE function.

Define

$$r_m = 1 - 2\mu\lambda_m, \quad m = 0, 1, \cdots, M \tag{12.71}$$

In fact, r_m is the ratio of the geometric progression shown in eq. (12.43). If the exponential envelope curve is used to fit the geometric progression, the time constant can be obtained as follows:

$$r_m = \exp\left(-\frac{1}{\tau_m}\right), \quad m = 0, 1, \cdots, M \tag{12.72}$$

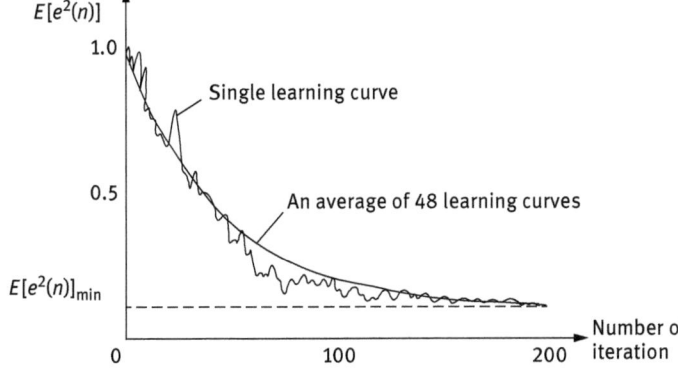

Figure 12.7: The learning curve of the LMS algorithm.

$$r_m = 1 - \frac{1}{\tau_m} + \frac{1}{2!\tau_m^2} - \frac{1}{3!\tau_m^3} + \cdots, \quad m = 0, 1, \cdots, M \tag{12.73}$$

If we take the first two terms of eq. (12.73), we have

$$r_m \approx 1 - \frac{1}{\tau_m}, \quad m = 0, 1, \cdots, M \tag{12.74}$$

Taking a comparison between eqs. (12.71) and (12.74), we have

$$\tau_m = \frac{1}{2\mu\lambda_m}, \quad m = 0, 1, \cdots, M \tag{12.75}$$

Equation (12.75) is the time constant of the m th weight coefficient of the LMS algorithm. By the definition of common ratio r_m and the adaptive learning curve in eq. (12.70), we get the relation between the MSE time constant τ_{mse} and the weight coefficient time constant τ as follows:

$$(r_{\mathrm{mse}})_m = r_m^2, \quad m = 0, 1, \cdots, M \tag{12.76a}$$

$$(\tau_{\mathrm{mse}})_m = \frac{\tau_m}{2}, \quad m = 0, 1, \cdots, M \tag{12.76b}$$

Thus, the MSE time constant of the m th mode is

$$(\tau_{\mathrm{mse}})_m \approx \frac{1}{4\mu\lambda_m}, \quad m = 0, 1, \cdots, M \tag{12.77}$$

In the above equations, τ_m and $(\tau_{\mathrm{mse}})_m$ denote the time constant and MSE time constant of the m th weight coefficient and the m th mode, respectively. $(r_{\mathrm{mse}})_m$ is the common ratio of the learning curve shown in eq. (12.70), defined as

$$(r_{\mathrm{mse}})_m = (1 - 2\mu\lambda_m)^2, \quad m = 0, 1, \cdots, M \tag{12.78}$$

In the LMS algorithm each gradient estimate is based on an input data sample, such that the time constant $(T_{\mathrm{mse}})_m$ for the input data is equal to the MSE time constant $(\tau_{\mathrm{mse}})_m$ of the algorithm.

$$(T_{\mathrm{mse}})_m = (\tau_{\mathrm{mse}})_m \approx \frac{1}{4\mu\lambda_m}, \quad m = 0, 1, \cdots, M \tag{12.79}$$

The size of the time constant determines the length of the adaptive learning process or the speed of convergence. In general, each λ_m, $m = 0, 1, K, M$ may not be necessarily equal. In this way, the convergence rate of each weight coefficient or each mode may not be equal. Only when each weight coefficient converges, the whole adaptive filter converges. If all λ_m, $m = 0, 1, \cdots, M$ are equal, then the time constants of these weighting coefficients will also be equal, which is equivalent to that all the input signals are uncorrelated and have equal powers. The learning curve at this point is a true exponential function curve.

3. Weight misadjustment in LMS adaptive algorithm

When the gradient truth-value $\nabla(n)$ is replaced by its estimate $\hat{\nabla}(n)$, the gradient estimation noise is generated, as shown in eq. (12.80):

$$\mathbf{v}(n) = \hat{\nabla}(n) - \nabla(n) \tag{12.80}$$

The existence of the gradient estimation noise can cause the weight misadjustment, which affects the performance of the adaptive filter, so that the weight vector of the filter cannot reach the Wiener vector and cannot achieve the minimum MSE. The misadjustment coefficient is defined as

$$M_{\mathrm{d}} = \frac{\xi_{\mathrm{ss}} - \xi_{\min}}{\xi_{\min}} \tag{12.81}$$

where

$$\xi_{\mathrm{ss}} = \lim_{n \to \infty} E[e^2(n)] \tag{12.82}$$

is the steady-state MSE.

   (1) Gradient estimation noise

   In the LMS algorithm, the gradient estimate is

$$\hat{\nabla}(n) = -2e(n)\mathbf{x}(n) = \nabla(n) + \mathbf{v}(n) \tag{12.83}$$

From eq. (12.52), the gradient estimation noise has zero mean. In addition, when n is sufficiently large, the variance matrix of the gradient estimation noise is

$$\mathrm{Var}[\mathbf{v}(n)] = E[\mathbf{v}(n)\mathbf{v}^{\mathrm{T}}(n)] \approx 4E[e^2(n)\mathbf{x}(n)\mathbf{x}^{\mathrm{T}}(n)] = 4E[e^2(n)]E[\mathbf{x}(n)\mathbf{x}^{\mathrm{T}}(n)]$$
$$\approx 4\xi_{\mathrm{ss}}\mathbf{R} \tag{12.84}$$

Taking the orthogonal coordinate transform to the gradient estimation noise $\mathbf{v}(n)$, we have

$$\mathbf{v}'(n) = \mathbf{Q}\mathbf{v}(n) \tag{12.85}$$

The variance matrix of $\mathbf{v}'(n)$ is

$$\mathrm{Var}[\mathbf{v}'(n)] = E[\mathbf{Q}\mathbf{v}(n)\mathbf{v}^{\mathrm{T}}(n)\mathbf{Q}^{\mathrm{T}}] = \mathbf{Q}\{\mathrm{Var}[\mathbf{v}(n)]\}\mathbf{Q}^{-1} \tag{12.86}$$

From the above study, we get the following three conclusions:

First, when n is sufficiently large, the gradient estimation noise $\mathbf{v}(n)$ can be approximated as a generalized stationary random sequence. Because of its average $E[\mathbf{v}(n)] = 0$ is independent of time, its second moment $\mathrm{Var}[\mathbf{v}(n)] \approx 4\xi_{\mathrm{ss}}\mathbf{R}$ also has nothing to do with time.

Second, when n is sufficiently large, the gradient estimation noise is uncorrelated at each iteration time, i.e.

$$E[\mathbf{v}(n)\mathbf{v}(n+k)] = 0, \quad k \neq 0 \tag{12.87}$$

Third, when n is sufficiently large, the variance of the gradient estimation noise after the orthogonal transform is a diagonal matrix.

(2) Weight noise vector

When the gradient estimation noise is considered, the LMS algorithm can be expressed as

$$\mathbf{w}(n+1) = \mathbf{w}(n) - \mu\hat{\nabla}(n) = \mathbf{w}(n) - \mu[\nabla(n) + \mathbf{v}(n)] \tag{12.88}$$

Let the difference $\mathbf{c}(n)$ between the weight vector $\mathbf{w}(n)$ and the optimal weight vector $\mathbf{w}_{\mathrm{opt}}$ for the n iteration be called the weight error vector or weight noise vector. From the gradient expression in eq. (12.13),

$$\begin{aligned}\nabla(n) &= -2\mathbf{p} + 2\mathbf{R}\mathbf{w}(n) = -\mathbf{p} + 2\mathbf{R}[\mathbf{w}_{\mathrm{opt}} - \mathbf{c}(n)] \\ &= 2\mathbf{R}\mathbf{c}(n)\end{aligned} \tag{12.89}$$

Subtracting $\mathbf{w}_{\mathrm{opt}}$ from both sides in eq. (12.88), and substituting eq. (12.89) into eq. (12.88), we get

$$\mathbf{c}(n+1) = \mathbf{c}(n) - 2\mu\mathbf{R}\mathbf{c}(n) - \mu\mathbf{v}(n) = (\mathbf{I} - 2\mu\mathbf{R})\mathbf{c}(n) - \mu\mathbf{v}(n) \tag{12.90}$$

Equation (12.90) is a first-order vector difference equation with respect to $\mathbf{c}(n)$ with an excitation function $[-\mu\mathbf{v}(n)]$. Since the excitation function is stochastic, the response is also stochastic. This shows that the randomness of the weight vector comes from the randomness of gradient estimation. In addition, when n is sufficiently large, it can be deduced that $\mathbf{c}(n)$ is stationary since $\mathbf{v}(n)$ is stationary. Taking an orthogonal coordinate transform to eq. (12.90), we get

$$\begin{aligned}\mathbf{c}'(n+1) &= \mathbf{Q}\mathbf{c}(n+1) = \mathbf{Q}[\mathbf{Q}^{-1}\mathbf{Q} - 2\mu\mathbf{Q}^{-1}\mathbf{\Lambda}\mathbf{Q}]\mathbf{c}(n) - \mu\mathbf{Q}\mathbf{v}(n) \\ &= (\mathbf{I} - 2\mu\mathbf{\Lambda})\mathbf{c}'(n) - \mu\mathbf{v}'(n)\end{aligned} \tag{12.91}$$

The above equation can be written as

$$c'_m(n+1) = (1 - 2\mu\lambda_m)c'_m(n) - \mu v'_m(n), \quad m = 0, 1, \cdots, M \tag{12.92}$$

When n is large enough, different $v'_m(n)$ is uncorrelated with each other, so $c'_m(n)$ is also uncorrelated with each other. The variance matrix $\mathrm{Var}[c'(n)]$ is a diagonal matrix. Since $c'(n)$ is stationary when n is large enough, so

$$\mathrm{Var}[c'(n)] = E[c'(n)c'^{\mathrm{T}}(n)] = E[c'(n+1)c'^{\mathrm{T}}(n+1)] \tag{12.93}$$

Thus, from eq. (12.91), we have

$$\mathrm{Var}[c'(n)] = (I - 2\mu\Lambda)\mathrm{Var}[c'(n)](I - 2\mu\Lambda) + \mu^2\mathrm{Var}[v'(n)] \tag{12.94}$$

Since $\mathrm{Var}[c'(n)]$ is a diagonal matrix, it leads to

$$\mathrm{Var}[c'(n)] = \mu^2[4\mu\Lambda - \mu^2\Lambda^2]^{-1}\mathrm{Var}[v'(n)] \tag{12.95}$$

In practice, the LMS algorithm always takes the smallest μ value, so that $\mu\Lambda < < I$. The square term of $\mu\Lambda$ in eq. (12.95) is negligible:

$$\mathrm{Var}[c'(n)] = \frac{\mu}{4}\Lambda^{-1}\mathrm{Var}[v'(n)] \tag{12.96}$$

Substituting eq. (12.86) into eq. (12.96), we get

$$\mathrm{Var}[c'(n)] = \frac{\mu}{4}\Lambda^{-1}(4\xi_{ss}\Lambda) = \mu\xi_{ss}I \tag{12.97}$$

Since $c'(n) = Qc(n)$ and $Q^{\mathrm{T}}Q = I$, the error vector of the above equation on the original coordinate system also holds as

$$\mathrm{Var}[c(n)] = \mu\xi_{ss}I \tag{12.98}$$

Its scalar form is

$$E[c_m^2(n)] = \mu\xi_{ss}, \quad m = 0, 1, \cdots, M \tag{12.99}$$

It can be seen that the noise components of the weight vector have equal variance and are uncorrelated.

(3) Misadjustment caused by the gradient estimation noise

If the weight vector does not have any noise and converges to the Wiener solution, then the MSE is minimized as ξ_{min}. When the weight vector appears random noise, the steady-state solution of the weight vector will deviate its

Wiener solution and cause the excessive MSE, which makes the steady-state MSE ξ_{ss} be greater than the minimum MSE ξ_{min}. From eq. (12.67), we get the excessive MSE as

$$\xi_{ss} - \xi_{min} = \lim_{n \to \infty} E[c'^T(n)\Lambda c'(n)] = \lim_{n \to \infty} \sum_{m=0}^{M} \lambda_m E[c'^2_m(n)]$$

$$= \mu \xi_{ss} \text{tr}[R] \tag{12.100}$$

From eq. (12.100) we get

$$\xi_{ss} = \frac{\xi_{min}}{1 - \mu \text{tr}[R]} \tag{12.101}$$

Substituting the above equation into eq. (12.81), we have

$$M_d = \frac{\xi_{ss} - \xi_{min}}{\xi_{min}} = \frac{\mu \text{tr}[R]}{1 - \mu \text{tr}[R]} \tag{12.102}$$

As can be seen from the above equation, in order to maintain a small value of the misadjustment coefficient M_d, $\mu \text{tr}[R] \ll 1$ should be selected. Based on the requirement of the weight vector, $0 < \mu < \frac{1}{\lambda_{max}}$ should be selected. Thus

$$0 < \mu < \frac{1}{\text{tr}[R]} \tag{12.103}$$

At such a small value of μ, the misadjustment coefficient M_d can be expressed as

$$M_d = \mu \text{tr}[R] = \mu \sum_{m=0}^{M} \lambda_m \tag{12.104}$$

where the upper bound of the summation represents the dimension of the weight vector. It is useful to express the misadjustment coefficient into a function of the adaptive speed and weight coefficients. From eq. (12.77), we have

$$\mu \lambda_m = \frac{1}{4(\tau_{mse})_m}, \quad m = 0, 1, \cdots, M \tag{12.105}$$

Substituting the above equation into eq. (12.104), we get

$$M_d = \sum_{m=0}^{M} \frac{1}{4(\tau_{mse})_m} = \frac{M+1}{4} \left[\frac{1}{(\tau_{mse})_m} \right]_{av} \tag{12.106}$$

where

$$\left[\frac{1}{(\tau_{mse})_m}\right]_{av} = \frac{1}{M+1}\sum_{m=0}^{M}\frac{1}{(\tau_{mse})_m} \tag{12.107}$$

When all the eigenvalues of $\boldsymbol{R}$ are equal, the time constants are also equal, denoted as τ_{mse}, and the misadjustment coefficient becomes

$$M_d = \frac{M+1}{4\tau_{mse}} \tag{12.108}$$

which means the misadjustment coefficient is directly proportional to the number of weight coefficients. The above equation shows that if a large time constant is chosen, the misadjustment coefficient will be reduced. Whereas for a given time constant, the misadjustment coefficient increases proportionally to the weighted coefficient numbers.

The main advantage of the LMS adaptive filter is that its convergence performance is stable and the algorithm is relatively simple. However, as a kind of gradient algorithm, LMS algorithm has its inherent shortcomings. First of all, this method generally cannot reach the extreme points through the shortest path from any initial point. Second, when the eigenvalues of the autocorrelation matrix $\boldsymbol{R}$ are dispersed, the performance of this method tends to deteriorate, resulting in a slow convergence or even divergence. In fact, there have been many improvements to adaptive filter algorithm. However, LMS adaptive filter still has very important theoretical significance and application value.

Example 12.1: Suppose the adaptive filter structure is shown in Figure 12.8. (1) Write its expression of the performance function; (2) determine its convergence range; and (3) write its iteration expression of the LMS algorithm.
Solution: (1) From $e(n) = d(n) - y(n)$ and $y(n) = \boldsymbol{w}^T(n)\boldsymbol{x}(n) = w_0(n)x(n) + w_1(n)x(n-1)$, we have

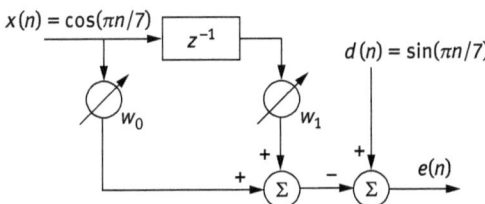

$$\xi = E[e^2(n)] = E[d^2(n)] - 2w_0 E[d(n)x(n)] - 2w_1 E[d(n)x(n-1)] + w_0^2 E[x^2(n)]$$
$$+ 2w_0 w_1 E[x(n)x(n-1)] + w_1^2 E[x^2(n-1)]$$

where $E[d^2(n)] = E[\sin^2(\pi n/7)] = 0.5$, $E[d(n)x(n-1)] = E[\sin(\pi n/7)\cos(\pi n/7)] = 0$, $E[d(n)x(n-1)] = E[\sin(\pi n/7)\cos(\pi(n-1)/7)] = 0.2169$, $E[x^2(n)] = E[\cos^2(\pi n/7)] = 0.5$, $E[x(n)x(n-1)] = E[\cos(\pi n/7)\cos(\pi(n-1)/7)] = 0.4505$, $E[x^2(n-1)] = E[\cos^2(\pi(n-1)/7)]$, $= 0.5$. Then $\xi = E[e^2(n)] = 0.5(w_0^2 + w_1^2) + 0.901 w_0 w_1 - 0.4338 w_1 + 0.5$.

(2) Since $\mathbf{R} = E\begin{bmatrix} x^2(n) & x(n)x(n-1) \\ x(n-1)x(n) & x^2(n-1) \end{bmatrix} = \begin{bmatrix} 0.5 & 0.4505 \\ 0.4505 & 0.5 \end{bmatrix}$, we have

$\text{tr}[\mathbf{R}] = 0.5 + 0.5 = 1$. Thus $0 < \mu < 1/\text{tr}[\mathbf{R}] = 1$.

(3) The iteration expression of the LMS algorithm is $\mathbf{w}(n+1) = \mathbf{w}(n) + 2\mu e(n)\mathbf{X}(n)$, where $e(n) = d(n) - y(n) = d(n) - [w_0 x(n) + w_1 x(n-1)]$.

Example 12.2: Try to realize the LMS adaptive filter by MATLAB programming, and give the convergence curve of the system MSE.
Solution: MATLAB program is as follows:

```
clear all;
N1=1000; N=1000; M=15; mu1=0.001; mu2=0.01;
for k=1:N
x=randn(1,N1); noise=0.1*randn(1,N1); d=filter(6,1,x)+noise;
ha1 = adaptfilt.lms(M,mu1); [y1,e1] = filter(ha1,x,d); e21(k,:)=e1;
ha2 = adaptfilt.lms(M,mu2); [y2,e2] = filter(ha2,x,d); e22(k,:)=e2;
end
E_mu1=sum(e21.^2,1)./N; E_mu2=sum(e22.^2,1)./N;
Figure; plot(E_mu1); grid; hold on; plot(E_mu2); xlabel('Iteration
  numbers'); ylabel('MSE')
```

Figure 12.9 shows the convergence curves of the adaptive filter under two different convergence conditions. Obviously, when the convergence factor is larger, the adaptive system converges faster.

12.3.3 Improved LMS Algorithms

1. Normalized LMS (NLMS) adaptive algorithm
It is known that the stability, convergence, and steady-state performance of the LMS algorithm are directly related to the number of coefficients of the adaptive filter and the power of the input signal. In order to ensure the stable convergence of the adaptive filter, an NLMS algorithm is presented to normalize the

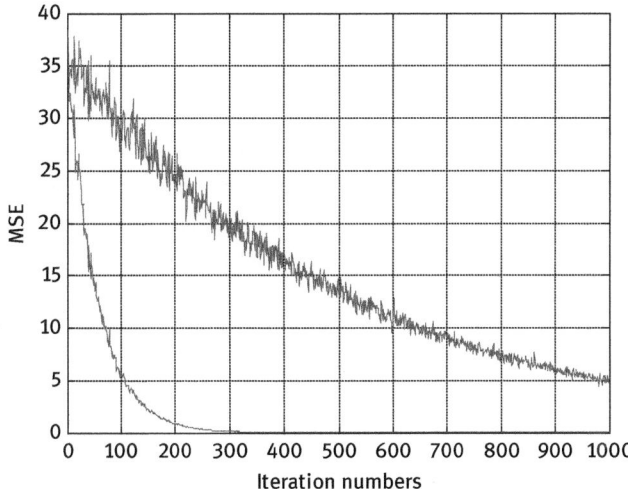

Figure 12.9: The convergence curves of the adaptive filter under two different convergence conditions.

convergence factor. The normalized convergence factor of this algorithm is denoted as

$$\mu' = \frac{\mu}{\sigma_x^2} \tag{12.109}$$

where σ_x^2 is the variance of the input signal $x(n)$. In general, the following time average is used instead of the statistical variance:

$$\hat{\sigma}_x^2(n) = \sum_{i=0}^{M} x^2(n-i) = x^{\mathrm{T}}(n)x(n) \tag{12.110}$$

where $\hat{\sigma}_x^2(n)$ is the estimated variance at time n. For the stationary random input signal $x(n)$, $\hat{\sigma}_x^2(n)$ is the unbiased and consistent estimation. Substituting the normalized convergence factor into the LMS algorithm, we get

$$w(n+1) = w(n) + 2\frac{\mu}{x^{\mathrm{T}}(n)x(n)} e(n)x(n) \tag{12.111}$$

In order to avoid a 0 value in the denominator of the above equation, we usually add a small positive constant c in the denominator:

$$w(n+1) = w(n) + 2\frac{\mu}{c + x^{\mathrm{T}}(n)x(n)} e(n)x(n) \tag{12.112}$$

Thus as long as the convergence condition

$$0 < \mu < 1 \tag{12.113}$$

is satisfied, the adaptive filter will be convergent if n is large enough. Since the normalized convergence factor $\mu' = \frac{\mu}{c + x^T(n)x(n)}$ in eq. (12.112) changes with time in the iterative process, it is actually a normalized variable step size algorithm. In this kind of algorithm, we can use different methods to estimate the variance of the input signal, thus forming different normalized variable step size algorithms. For example, the first-order AR model can be used in the form of adaptive iteration process for the recursive estimation as follows:

$$\hat{\sigma}_x^2(n) = \beta \hat{\sigma}_x^2(n-1) + (1-\beta)x^2(n) \tag{12.114}$$

where β is the smooth parameter, a positive value close to 1.

2. Leakage LMS algorithm

The iteration equation of the leakage LMS algorithm is shown as follows:

$$w(n+1) = \gamma w(n) + 2\mu e(n)x(n) \tag{12.115}$$

where γ is a positive constant, and satisfies

$$0 < \gamma < 1 \tag{12.116}$$

Usually, the value of γ is taken approximately as 1. If $\gamma = 1$, then the leakage LMS algorithm becomes LMS algorithm. For a conventional LMS algorithm, when μ suddenly changes to 0, the weight vector coefficients will not change again. For the leakage LMS algorithm, when μ becomes 0, the weight vector changes gradually and eventually becomes a $\mathbf{0}$ vector. This process is called leakage. The leakage LMS algorithm is used in adaptive differential pulse code modulation to reduce channel errors. It is also used to eliminate side-lobe effects in adaptive arrays. In fact, the performance of the leakage LMS algorithm is not as good as that of the conventional LMS algorithm under noiseless conditions:

$$\begin{aligned} w(n+1) &= \gamma w(n) + 2\mu x(n)(d(n) - x^T(n)(n)) \\ &= [\gamma I - 2\mu x(n)x^T(n)]w(n) + 2\mu d(n)x(n) \end{aligned} \tag{12.117}$$

If the input signal and the weight vector are independent of each other, then

$$E[w(n+1)] = (\gamma I - 2\mu R)E[w(n)] + 2\mu p \tag{12.118}$$

or

$$E[w(n+1)] = \left[I - 2\mu\left(R + \frac{1-\gamma}{2\mu}\right)I\right]E[w(n)] + 2\mu p \tag{12.119}$$

To ensure the stability of the algorithm, we need to have

$$\lim_{n\to\infty}\{E[w(n)]\} = \left[R + \frac{1-\gamma}{2}I\right]^{-1}p \tag{12.120}$$

Obviously, the above equation deviates from the optimal weight vector $w_{\text{opt}} = R^{-1}p$. Therefore, the leakage LMS algorithm is a biased LMS algorithm. If γ is close to 1, the deviation will be small. It can be proved that the stability condition of the leakage LMS algorithm is

$$1 < \mu < \frac{1}{\lambda_{\min} + \frac{1-\gamma}{2\mu}} \tag{12.121}$$

Since the matrix $\left[R + \frac{1-\gamma}{2}I\right]$ is strictly positive definite, there is no eigenvalue of zero. In addition, the time constant of the i th weight coefficient of the leakage LMS algorithm is

$$\tau_i^{(L)} = \frac{1}{2\mu\lambda_i + (1-\gamma)} < \tau_i \tag{12.122}$$

where $\tau_i^{(L)}$ denotes the time constant of the i th weight coefficient of the leakage LMS algorithm. Obviously, the leakage LMS algorithm is possible to converge faster than the LMS algorithm with a smaller time constant $\tau_i^{(L)}$.

3. Polarity LMS algorithm
In some applications, especially in the field of high-speed communications, the computational cost of algorithms are required strictly. Thus, a class of adaptive algorithms called polarity (or symbolic) LMS algorithms are generated. This kind of algorithm can significantly reduce the amount of computation of the adaptive filter, and thereby effectively simplifying the corresponding hardware circuit and program calculation. The iterative equation of the three different implementations of this algorithm is shown as follows:

$$\begin{aligned}
w(n+1) &= w(n) + 2\mu\text{sgn}[e(n)]x(n) \\
w(n+1) &= w(n) + 2\mu e(n)\text{sgn}[x(n)] \\
w(n+1) &= w(n) + 2\mu\text{sgn}[e(n)]\text{sgn}[x(n)]
\end{aligned} \tag{12.123}$$

where the symbolic function is defined as

$$\text{sgn}(t) = \begin{cases} 1, & t > 0 \\ 0, & t = 0 \\ -1, & t < 0 \end{cases} \tag{12.124}$$

The main advantage of the polar LMS algorithm is its small computational complexity. Obviously, this algorithm reduces the N bits operation of a data sample to a one-bit operation, that is, the operation of sign or polarity. On the other hand, compared with the basic LMS algorithm, this algorithm degrades the performance of the gradient estimation, leading to the decrease of the convergence rate and the increase of the steady-state error.

4. Smoothing of gradient estimation for LMS algorithm

It may be a significant drawback of the LMS algorithm to replace the gradient truth-value with a noisy gradient estimate in the iteration equation. It is possible to improve the performance of the LMS algorithm if the instantaneous values of gradient estimation are replaced by a smooth value of successive gradient estimates. Many methods can be used to smooth a time series, which can be divided into two types, linear smoothing and nonlinear smoothing.

The adaptive-iterative algorithm for the gradient estimation of the smooth LMS algorithm is

$$\boldsymbol{w}(n+1) = \boldsymbol{w}(n) + 2\mu \boldsymbol{b}(n) \tag{12.125}$$

where

$$\boldsymbol{b}(n) = [b_0(n) \ b_1(n) \ \cdots \ b_M(n)]^{\mathrm{T}} \tag{12.126}$$

For linear smoothing, an efficient smooth method is the neighborhood averaging method, that is

$$b(n) = \frac{1}{N} \sum_{j=n-N+1}^{n} e(j)\boldsymbol{x}(j) \tag{12.127}$$

where N is the number of sample points that participate in the smooth gradient estimate. Another effective smoothing method is the low-pass filtering method, which uses a low-pass filter for linear smoothing:

$$b_i(n) = \text{LPF}\{e(n)x(n-i) \ e(n)x(n-i+1) \ K \ e(n)x(n-i+N)\} \tag{12.128}$$

where LPF$\{\cdot\}$ denotes the low-pass filter.

For nonlinear smoothing, the median filtering is often used. Suppose the i th element $b_i(n)$ in the vector $\boldsymbol{b}(n)$ is shown as

$$b_i(n) = \text{Med}[e(n)x(n-i)]_N \tag{12.129}$$

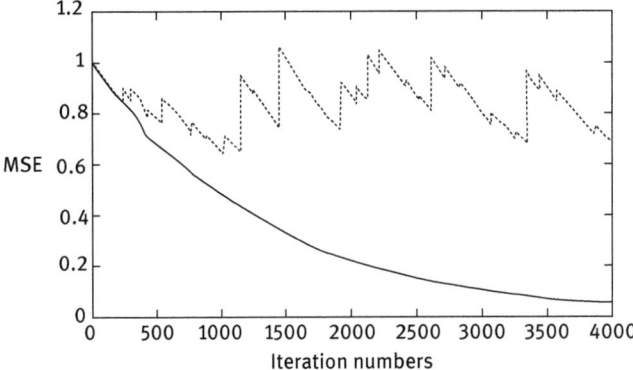

Figure 12.10: A comparison between the smoothing median LMS algorithm and the basic LMS algorithm.(— smooth median LMS algorithm; · · · · · · the basic LMS algorithm)

or

$$b_i(n) = \text{Med}[e(n)x(n-i) \ \cdots \ e(n-N+1)x(n-i-N+1)] \tag{12.130}$$

where Med represents the median operation. Median smoothing can be used to eliminate the noise of gradient estimation as well as the linear smoothing. It has little effect on the "edge" component of the signal. Figure 12.10 shows the results of the median smoothing LMS algorithm. Obviously, the basic LMS algorithm cannot converge well due to the effect of pulse-like noise, while the smooth median LMS algorithm is very good.

5. Decorrelation LMS algorithm

One of the main drawbacks of LMS algorithm is its slow convergence rate, which is mainly due to the correlation of each element of the input signal vector of the algorithm. It has been shown that the decorrelation of the input vector can effectively accelerate the convergence rate of the LMS algorithm.

Define the correlation coefficient between $x(n)$ and $x(n-1)$ at time n as

$$c(n) = \frac{x^T(n)x(n-1)}{x^T(n-1)x(n-1)} \tag{12.131}$$

By definition, if $c(n) = 1$, then $x(n)$ is called the coherent signal of $x(n-1)$. If $c(n) = 0$, then $x(n)$ and $x(n-1)$ are uncorrelated. If $0 < c(n) < 1$, then $x(n)$ and $x(n-1)$ are correlated. The bigger the value of $c(n)$ is, the stronger the correlation between $x(n)$ and $x(n-1)$ will be.

In fact, $c(n)x(n-1)$ represents the correlated part between $x(n)$ and $x(n-1)$. If this part is subtracted from $x(n)$, it is equivalent to a decorrelation operation. The decorrelation direction vector is defined as

$$v(n) = x(n) - c(n)x(n-1) \tag{12.132}$$

On the other hand, suppose the convergence factor satisfies the following minimization problem:

$$\mu(n) = \arg\min_{\mu} J[w(n-1) + \mu v(n)] \tag{12.133}$$

Thus the time-varying convergence factor is obtained as

$$\mu(n) = \frac{e(n)}{x^T(n)v(n)} \tag{12.134}$$

Thus, the iterative equation of the decorrelation LMS adaptive algorithm is

$$w(n+1) = w(n) + \mu(n)v(n) \tag{12.135}$$

12.3.4 Practical Considerations

In applications of the LMS adaptive filter, we need to pay attention to the following problems.

1. Finite word length

The implementation of an adaptive digital filter involves a finite word-length operation, resulting in a performance difference between the actual adaptive filter and the ideal adaptive filter of infinite accuracy. The main factors causing this performance difference include the quantization of the input and desired signals, the quantization of the filter coefficients, and the rounding errors in the filter operation.

It is assumed that w_{ip} and w_{fp} are used to represent the weight vectors of the infinite precision and the finite precision of the adaptive filter, respectively. The adaptive filter is considered to be numerically stable if the difference vector $w_{ip} - w_{fp}$ of the two vectors is always bounded, that is, the rounding error propagation system is always stable. The numerical stability is an inherent characteristic of adaptive algorithms and cannot be changed by increasing the digital accuracy. Increasing word length or reorganizing operations may delay the divergence of the adaptive filter. Only by improving the performance of the rounding error propagation system, the algorithm can make a substantial change, so as to achieve the stability of the adaptive filter.

The numerical accuracy of the adaptive digital filter can be used to measure the deviation of the actual value from the theoretical value due to the rounding error at steady state. If the numerical accuracy does not meet the requirements, the system output error will increase. By increasing the word length, such error can be reduced. However, if the numerical stability is not good, and if the accumulation of rounding errors cannot be effectively suppressed, it may have disastrous consequences, that is, the algorithm divergence or collapse.

2. Robustness of LMS adaptive filter
The robustness of the adaptive filter can be expressed as the sensitivity of the adaptive system to the initial condition $w(0)$, the system optimal residuals and other error $e_0(n)$. The energy of the interference $E_d(n)$ and the energy of the estimation error $E_e(n)$ are defined as

$$E_d(n) = \frac{1}{2\mu} \|w(0)\|^2 + \sum_{j=0}^{n} |e_0(n)|^2 \tag{12.136a}$$

$$E_e(n) = \frac{1}{2\mu} \|w(n)\|^2 + \sum_{j=0}^{n} |y(n)|^2 \tag{12.136b}$$

Assume that the convergence factor satisfies $0 < 2\mu \le \|x(n)\|^2$, it can be proved that the coefficient vector determined by the LMS algorithm satisfies the following condition:

$$E_e(n) \le E_d(n) \tag{12.137}$$

The above equation indicates that the energy of the interference is the upper bound of the energy of the system residual, thus showing the robustness of the LMS algorithm. On the other hand, for all finite energy interferences, the LMS algorithm minimizes the maximum possible difference between the two energies.

3. Convergence factor and system error
Convergence factor is an important parameter of the LMS adaptive filter, which controls the balance between the convergence rate and the steady-state misadjustment. In general, small convergence factors result in a slow convergence rate and small steady-state misadjustment. In a digital adaptive system, however, when the iterative increment is smaller than half of the LSB, that is,

$$|2\mu e(n)x(n-i)| \le \frac{\text{LSB}}{2} \tag{12.138}$$

the adaptive iteration of the LMS algorithm will stop. Therefore, the reduction of μ will result in a reduction in system performance. From eq. (12.139) we can see the relationship between the convergence factor and the system error:

$$|e(n)| \le \frac{\text{LSB}}{4\mu X_{\text{rms}}} \triangleq \text{DRE} \tag{12.139}$$

where X_{rms} is the root mean square value of the input signal, and DRE is the digitized residue. For a given word length, the DRE will increase significantly if the convergence factor is reduced. Therefore, in practice, the convergence factor of the LMS algorithm cannot be reduced indefinitely, and its lower bound is determined by the degree of the quantization and finite precision on the system.

5. Block LMS algorithm

In some applications, practical problems require that the adaptive filter be arranged in a high order. For example, in a teleconferencing system, the order of the adaptive filter for echo cancellation can be as high as 8,000. In this case, it is quite difficult to implement the adaptive echo cancellation in real time. If the block LMS algorithm is adopted, the computational complexity of the above-mentioned application can be effectively reduced and the performance can be improved.

The so-called block adaptive filter is to process a data block at each time. The block diagram of the block LMS filter is shown in Figure 12.11. This filter has the following characteristics: high numerical accuracy, easy to achieve parallel computing, and data transmission. One can use FFT to calculate the convolution and correlation matrix, thereby reducing the computational complexity.

6. LMS algorithm in the transform domain

The main factor that affects the convergence speed of the LMS adaptive filter is the ratio of the largest eigenvalue to the smallest eigenvalue $\lambda_{max}/\lambda_{min}$ of the input signal autocorrelation matrix. One of the most effective ways to improve the convergence speed is to whiten the input signal. The transform domain LMS adaptive filter is proposed based on this concept.

Figure 12.12 shows the block diagram of the transform domain LMS adaptive filter, where the N-dimensional input vector $x(n)$ is transformed into another N-dimensional input vector $z(n) = [z_0(n)\ z_1(n)\ \cdots\ z_{N-1}(n)]^T$ via the $N \times N$ orthogonal normalized transformation matrix T

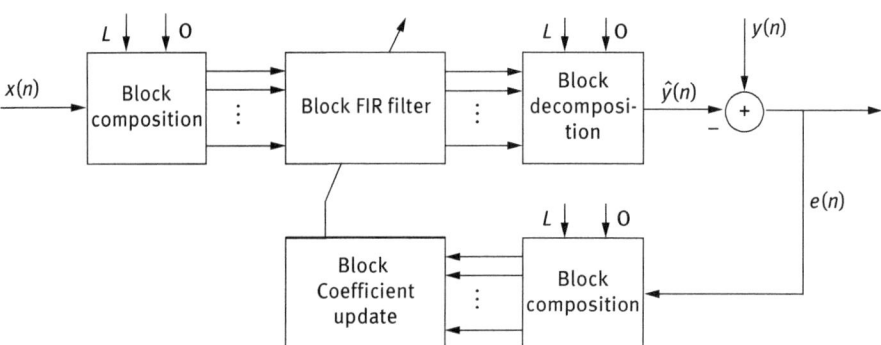

Figure 12.11: The structure of the block adaptive filter.

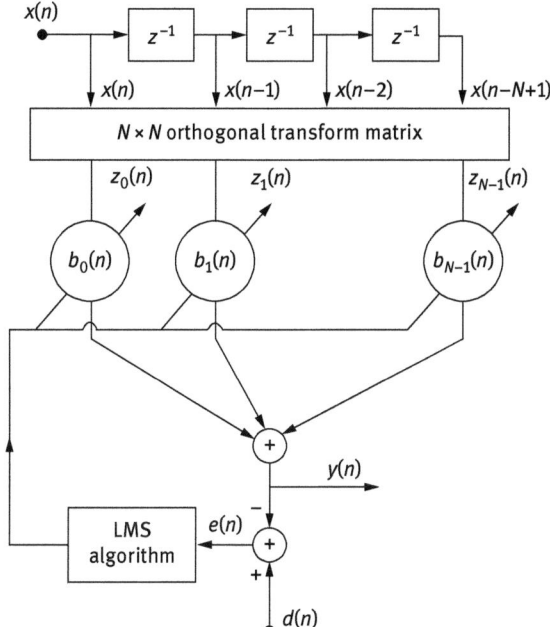

Figure 12.12: The schematic of the transform domain LMS adaptive filter.

$$\mathbf{z}(n) = \mathbf{T}\mathbf{x}(n) \tag{12.140}$$

where

$$\mathbf{T}\mathbf{T}^{\mathrm{T}} = \mathbf{I} \tag{12.141}$$

Let the transform domain weight vector $\mathbf{b}(n)$ be

$$\mathbf{b}(n) = [b_0(n) \ b_1(n) \ \cdots \ b_{N-1}(n)]^{\mathrm{T}} \tag{12.142}$$

The output of the adaptive filter is

$$y(n) = \mathbf{z}^{\mathrm{T}}(n)\mathbf{b}(n) = \mathbf{b}^{\mathrm{T}}(n)\mathbf{z}(n) \tag{12.143}$$

If $d(n)$ is the desired response, then the output error is

$$e(n) = d(n) - y(n) = d(n) - \mathbf{b}^{\mathrm{T}}(n)\mathbf{z}(n) \tag{12.144}$$

The transform domain LMS adaptive algorithm is shown as

$$b_i(n+1) = b_i(n) + 2\mu_i e(n)z_i(n), \quad i = 0, 1, 2, \cdots, N-1 \tag{12.145}$$

where the convergence factor is

$$\mu_i = \frac{\mu}{E[z_i^2(n)]} \tag{12.146}$$

The iteration equation of the transform domain LMS algorithm can also be written as a vector form

$$\boldsymbol{b}(n+1) = \boldsymbol{b}(n) + 2\mu\boldsymbol{\Lambda}^{-2}e(n)\boldsymbol{z}(n) \tag{12.147}$$

where

$$\boldsymbol{\Lambda}^2 = \text{diag}[E[z_0^2(n)], E[z_1^2(n)], \ldots, E[z_{N-1}^2(n)]] \tag{12.148}$$

In practical applications, the transform domain adaptive filter usually uses Fourier transform or discrete cosine transform as the orthogonal transform.

The Wiener solution $\boldsymbol{b}_{\text{opt}}$ and the minimum MSE of the transform domain LMS adaptive filter are

$$\boldsymbol{b}_{\text{opt}} = \boldsymbol{R}_{zz}^{-1}\boldsymbol{p}_{zd} \tag{12.149}$$

$$e_{\min}^{\text{TR}}(n) = E[d^2(n)] - \boldsymbol{p}_{zd}^{\text{T}}\boldsymbol{b}_{\text{opt}} \tag{12.150}$$

where

$$\boldsymbol{R}_{zz} = E[\boldsymbol{z}(n)\boldsymbol{z}^{\text{T}}(n)] = \boldsymbol{T}\boldsymbol{R}_{xx}\boldsymbol{T}^{\text{T}} \tag{12.151a}$$

$$\boldsymbol{p}_{zd} = E[\boldsymbol{z}(n)d(n)] = \boldsymbol{T}\boldsymbol{p}_{xd} \tag{12.151b}$$

are the autocorrelation matrix of the signal $\boldsymbol{z}(n)$ and the cross-correlation vector between $\boldsymbol{z}(n)$ and $d(n)$, where the matrix $\boldsymbol{T}$ is the orthogonal transformation matrix. Substituting eq. (12.150) into eqs. (12.148) and (12.149), respectively, we get

$$\boldsymbol{h}_{\text{opt}} = (\boldsymbol{T}\boldsymbol{R}_{xx}\boldsymbol{T}^{\text{T}})^{-1}\boldsymbol{T}\boldsymbol{p}_{xd} = \boldsymbol{T}\boldsymbol{R}_{xx}^{-1}\boldsymbol{p}_{xd} = \boldsymbol{T}\boldsymbol{w}_{\text{opt}} \tag{12.152}$$

and

$$\begin{aligned} e_{\min}^{\text{TR}}(n) &= E[d^2(n)] - \boldsymbol{p}_{xd}^{\text{T}}\boldsymbol{T}^{\text{T}}\boldsymbol{T}\boldsymbol{w}_{\text{opt}} \\ &= E[d^2(n)] - \boldsymbol{p}_{xd}^{\text{T}}\boldsymbol{w}_{\text{opt}} = e_{\min}(n) \end{aligned} \tag{12.153}$$

where $\boldsymbol{w}_{\text{opt}} = \boldsymbol{R}_{xx}^{-1}\boldsymbol{p}_{xd}$ and $e_{\min}(n)$ are the Wiener vector and the corresponding minimum MSE of the conventional LMS adaptive filter, respectively. Equations

(12.152) and (12.153) show that the relationship between the Wiener vectors in both conventional LMS algorithm and the transform domain LMS algorithm is determined by the orthogonal transform used. Regardless of the orthogonal transform, the minimum MSE of the transform domain LMS algorithm is always equal to the minimum MSE of the conventional LMS algorithm, as long as the number of weighting coefficients is the same.

As to the convergence properties of the transform domain LMS algorithm, it can be shown that the transform domain LMS adaptive algorithm has better convergence performance than the conventional LMS adaptive algorithm, as long as the orthogonal transform matrix is properly selected.

12.4 The Recursive Least Square Algorithm of Adaptive Filters

The least square (LS) method is a typical data processing method to infer unknown parameters from observed data. Its basic idea is to minimize the sum of the squares of the errors between the observed and calculated values. Since its introduction by famous mathematician Gauss in 1795, the LS method has been widely used in many fields and has become one of the basic algorithms in the areas of system identification, parameter estimation, and adaptive signal processing.

In the LMS adaptive algorithm, the gradient of the instantaneous square of the output error $e^2(n)$ is used to approximate the gradient of the MSE $E[e^2(n)]$ in order to facilitate its implementation. In fact, we can directly inspect the average power of the error signal over a period of time for the adaptive system when the input is a stationary signal. For example, the average power is minimized as a criterion for measuring the performance of the adaptive system. This is the recursive least square (RLS) algorithm introduced in this section.

12.4.1 The Principle of the Linear LS Algorithm

The structure of the linear combiner is shown in Figure 12.13.
The problem now is to estimate the expected response $y(n)$ of the linear combiner:

$$\hat{y}(n) = \sum_{k=1}^{M} w_k(n)x_k(n) = \boldsymbol{w}^{\mathrm{T}}(n)\boldsymbol{x}(n) \tag{12.154}$$

The estimation error of the above equation is defined as

$$e(n) = y(n) - \hat{y}(n) = y(n) - \boldsymbol{w}^{\mathrm{T}}(n)\boldsymbol{x}(n) \tag{12.155}$$

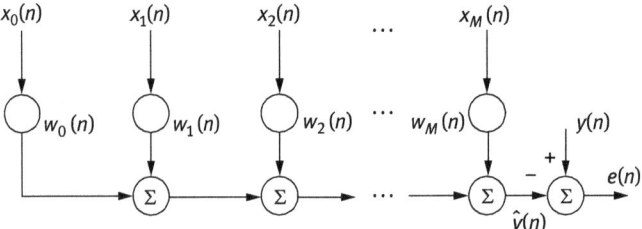

Figure 12.13: The structure of the linear combiner.

The sum of the square of the error $e(n)$ is

$$E = \sum_{n=0}^{N-1} |e(n)|^2 \tag{12.156}$$

Assume that the coefficient vector remains constant, and the coefficient vector obtained when the square error is minimum is optimal to estimate the expected response under the LS criterion. In the above equation, $w(n)$ is known as the regression vector and $e(n)$ is the residual. The regression equation shown in eq. (12.155) can be written as the vector form:

$$e = y - Xw \tag{12.157}$$

where $e = [e(0)\ e(1)\ \cdots\ e(N-1)]^{\mathrm{T}}$ and $y = [y(0)\ y(1)\ \cdots\ y(N-1)]^{\mathrm{T}}$ are the $N \times 1$ error vector and the expected response, respectively, $X = [x(0)\ x(1)\ \cdots\ x(N-1)]^{\mathrm{T}}$ is the $N \times M$ input matrix, and $w = [w_1\ w_2\ \ldots\ w_M]^{\mathrm{T}}$ is the parameter vector of the linear combiner.

By using the vector form of the regression equation, the error energy can be written as

$$\begin{aligned} E = e^{\mathrm{T}}e &= (y^{\mathrm{T}} - w^{\mathrm{T}}X^{\mathrm{T}})(y - Xw) \\ &= y^{\mathrm{T}}y - w^{\mathrm{T}}X^{\mathrm{T}}y - y^{\mathrm{T}}Xw + w^{\mathrm{T}}X^{\mathrm{T}}Xw \\ &= E_y - w^{\mathrm{T}}\hat{p} - \hat{p}^{\mathrm{T}}w + w^{\mathrm{T}}\hat{R}w \end{aligned} \tag{12.158}$$

where

$$E_y = y^{\mathrm{T}}y = \sum_{n=0}^{N-1} |y(n)|^2 \tag{12.159}$$

$$\hat{R} = X^T X = \sum_{n=0}^{N-1} x(n) x^T(n) \tag{12.160}$$

$$\hat{p} = X^T y = \sum_{n=0}^{N-1} x(n) y(n) \tag{12.161}$$

Obviously, both the LS method and the MMSE method are based on the quadratic cost function. If the time averaging operator $\sum_{n=0}^{N-1}$ is used to replace the expectation operator $E[\cdot]$, the derived formulas of the two methods are identical. If the time-averaged correlation matrix $\hat{R}$ is positive definite, the LS estimation w_{LS} can be obtained by solving the following canonical equations:

$$\hat{R} w_{LS} = \hat{p} \tag{12.162}$$

The minimum square error is

$$E_{LS} = E_y - p^T \hat{R}^{-1} \hat{p} = E_y - \hat{p}^T w_{LS} \tag{12.163}$$

There are various methods to solve the canonical equation. For example, Cholesky decomposition method and singular value decomposition method can be used. Please refer to relative references.

12.4.2 RLS Adaptive Filter

1. LS adaptive filter
The RLS adaptive filter is a filter based on the LS criterion, in which the square error of the coefficients is minimized at every moment of its operation. The iterative cost function is

$$E(n) = \sum_{j=0}^{n} \lambda^{n-j} |e(j)|^2 = \sum_{j=0}^{n} \lambda^{n-j} |y(j) - w^T(j) x(j)|^2 \tag{12.164}$$

where $y(j)$ and $\hat{y}(j) = w^T(j) x(j)$ represent the desired response and the output signal of the adaptive filter, $e(j)$ is the instantaneous error signal, and $\lambda < 1$ is a positive constant, referred to as the forgetting factor. The effect of the forgetting factor is to ensure that the filter can keep only the "nearest" data and forget the "past" data so that the algorithm is suitable for nonstationary environment. Figure 12.14 shows the effect of the forgetting factor.

From eq. (12.164) we can get the following three conclusions: First, the cost function $E(n)$ is a function of n that changes at each step of the iteration to reflect

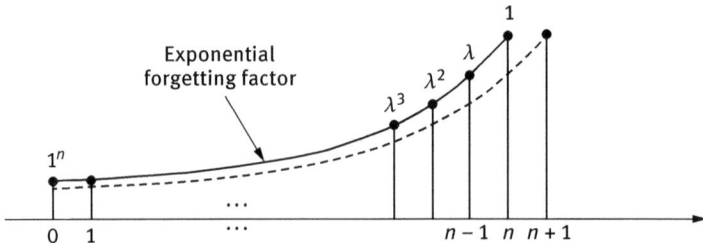

Figure 12.14: The effect of the forgetting factor.

the impact of the new data sample. Second, the optimal criterion having $E(n)$ to its minimum is the weighted LS method. Third, if $\lambda = 1$, the criterion is the ordinary LS method.

Taking the derivative to the cost function in eq. (12.164) with respect to the weight vector of the adaptive filter and letting it be 0 yields the canonical equation as

$$\hat{R}(n)w(n) = \hat{p}(n) \tag{12.165}$$

where

$$\hat{R}(n) = \sum_{j=0}^{n} \lambda^{n-j} x(j) x^{\mathrm{T}}(j)$$
$$\hat{p}(n) = \sum_{j=0}^{n} \lambda^{n-j} x(j) y(j) \tag{12.166}$$

If $\hat{R}(n)$ is of full rank, the canonical equation shown in eq. (12.165) can be solved by matrix inversion, and then the weight vector $w(n)$ of the adaptive system is obtained. However, this method is very time consuming. With the continuous input of new observations, the calculation is very troublesome. In fact, it is possible to recursively calculate $\hat{R}(n)$ iteratively, that is,

$$\hat{R}(n) = \lambda \hat{R}(n-1) + x(n) x^{\mathrm{T}}(n) \tag{12.167}$$

The above equation shows that the new correlation matrix is obtained by weighting the correlation matrix obtained from last iteration with the forgetting factor λ and then adding the new information. Similarly, we can also get the iterative expression of the cross-correlation vector

$$\hat{p}(n) = \lambda \hat{p}(n-1) + x(n) y(n) \tag{12.168}$$

In this way, we can use the two recursive equations and new observation data to recursively find the weight vector of the adaptive filter without finding the solution of the canonical equation in eq. (12.165). Substituting eqs. (12.167) and (12.168) into eq. (12.165), we have

$$[\hat{R}(n) - x(n)x^T(n)]w(n-1) = \hat{p}(n) - x(n)y(n) \tag{12.169}$$

After a simple arrangement, we have

$$\hat{R}(n)w(n-1) + x(n)e(n) = \hat{p}(n) \tag{12.170}$$

where the estimation error is

$$e(n) = y(n) - w^T(n-1)x(n) \tag{12.171}$$

If the correlation matrix $\hat{R}(n)$ is invertible, and $\hat{R}^{-1}(n)$ is used to left-multiply both sides of eq. (12.170), we get

$$w(n+1) + \hat{R}^{-1}(n)x(n)e(n) = \hat{R}^{-1}(n)\hat{p}(n) = w(n) \tag{12.172}$$

Define the adaptive gain vector $g(n)$ as

$$\hat{R}(n)g(n) \overset{\Delta}{=} x(n) \tag{12.173}$$

Thus, eq. (12.172) can be written as

$$w(n) = w(n-1) + g(n)e(n) \tag{12.174}$$

2. RLS algorithm

The adaptive gain vector $g(n)$ in eq. (12.174) is generally calculated by a recursive way. The basic idea is to find a recursive formula, and to correct recursively the inverse of the autocorrelation matrix $P(n) = \hat{R}^{-1}(n)$; then you can effectively simplify the calculation of $g(n)$. Based on the inverse matrix lemma:

$$(A + uv^T)^{-1} = A^{-1} - \frac{A^{-1}uv^T A^{-1}}{1 + v^T A^{-1}u} \tag{12.175}$$

where A is an $N \times N$ invertible matrix, u and v are $N \times 1$ vectors. The recursive formula of $P(n)$ can be obtained as shown in eq. (12.176):

$$P(n) = \lambda^{-1}P(n-1) - g(n)\bar{g}^T(n) \tag{12.176}$$

In fact, according to the matrix at the previous moment $P(n-1)$ and the new observation $x(n)$ and $y(n)$, the new matrix $P(n)$ can be calculated by the following steps:

$$\bar{g}(n) = \lambda^{-1}P(n-1)x(n)$$
$$\alpha(n) = 1 + \bar{g}^T(n)x(n)$$
$$g(n) = \frac{\bar{g}(n)}{\alpha(n)} \tag{12.177}$$
$$P(n) = \lambda^{-1}P(n-1) - g(n)\bar{g}^T(n)$$

The algorithm shown in eq. (12.177) is often called the RLS algorithm. Combining eqs. (12.171), (12.177), and (12.174), the complete adaptive gain calculation and the weight coefficients update of the adaptive RLS algorithm are constructed.

12.4.3 Practical Considerations

1. Computational complexity
The computational complexity of the RLS algorithm is mainly determined by the amount of computation required by one time of correction. Since $P(n)$ is a Hermitian matrix, the amount of computation required by each iteration is $2M^2 + 4M$. The computation complexity of $\bar{g}(n)$ and the update of $P(n)$ is $O(M^2)$. And the computational complexity of the scalar product and the vector product is $O(M)$.

2. The initialization of the algorithm
In practical applications, the initial values of $P(n)$ and $w(n)$ are usually determined as $P(-1) = \delta^{-1}I$ and $w(-1) = 0$, where δ is a very small positive number.

3. The effect of the limited word length
In practical applications, the theoretical accuracy of the RLS algorithm is affected by the limited word-length effect, which may lead to the numerical instability. For example, the key part of the RLS algorithm represented by eqs. (12.171), (12.174), and (12.177) is to update $P(n)$ with eq. (12.176). If the symmetry or positive definiteness of $P(n)$ is lost due to the limited word-length effect, the instability of the algorithm may be caused. In fact, we can simply calculate the upper triangular (or the lower triangular) part of eq. (12.176), and then supplement the rest part based on its symmetry, so as to preserve the symmetry of the Hermitian matrix. Another method is to replace $P(n)$ with $[P(n) + P^T(n)]/2$.

12.5 The Main Structures of Adaptive Filters

12.5.1 Adaptive Noise Cancellation and Its Applications

1. The fundamentals of adaptive noise cancellation
Adaptive noise cancellation (ANC) is an adaptive method for extracting useful signals from noise by the correlation of noises. Figure 12.15 is a block diagram of a typical ANC system.

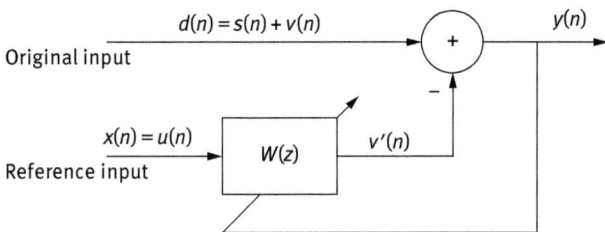

Figure 12.15: The block diagram of a typical adaptive noise cancellation system.

In the above figure, the original input signal $d(n)$ is the sum of the desired signal $s(n)$ and the noise $v(n)$, and the reference input signal $x(n)$ is the noise $u(n)$ correlated with $v(n)$. Assume that $s(n)$, $v(n)$, and $u(n)$ are zero mean stationary random processes, and $s(n)$ is uncorrelated with $v(n)$ and $u(n)$. As can be seen from Figure 12.15, the output $v'(n)$ of the adaptive filter is a filtered noise $u(n)$. The output of the entire ANC system $y(n)$ is

$$y(n) = s(n) + v(n) - v'(n) \tag{12.178}$$

We get from the above equation:

$$y^2(n) = s^2(n) + (v(n) - v'(n))^2 + 2s(n)(v(n) - v'(n)) \tag{12.179}$$

Taking the mathematical expectation to eq. (12.179), and concerning the uncorrelation between $s(n)$ and $v(n)$ or $u(n)$, it yields

$$E[y^2(n)] = E[s^2(n)] + E[(v(n) - v'(n))^2] \tag{12.180}$$

Since the signal power $E[s^2(n)]$ is independent of the adaptation of the adaptive filter, the minimized $E[y^2(n)]$ is equivalent to minimizing $E[(v(n) - v'(n))^2]$. Thus by eq. (12.178), we get

$$v(n) - v'(n) = y(n) - s(n) \tag{12.181}$$

It can be seen when $E[(v(n) - v'(n))^2]$ reaches its minimum, $E[(y(n) - s(n))^2]$ also reaches its minimum, then the MSE between $y(n)$ and $s(n)$ achieves the minimum.

In the ideal case, if $v(n) = v'(n)$, then $y(n) = s(n)$. Thus, the adaptive filter automatically adjusts its weighting coefficients to process $u(n)$ into $v(n)$, so that the noise in the output signal $y(n)$ is completely canceled, leaving only the desired signal $s(n)$.

The necessary condition for an adaptive filter to perform the above tasks is that the reference input signal $x(n) = u(n)$ must be correlated with the noise $v(n)$ to be canceled.

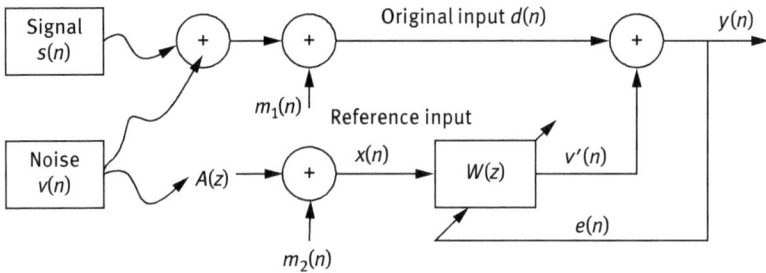

Figure 12.16: An adaptive noise cancellation system.

To further illustrate the principle of the ANC system, Figure 12.16 illustrates its working principle.

In Figure 12.16, the original input $d(n)$ consists of the sum of the desired signal $s(n)$ and two noises $v(n)$ and $m_1(n)$, the reference input $x(n)$ consists of other two noises $u(n) = v(n) \star a(n)$ and $m_2(n)$, where $a(n)$ is the unit impulse response of the transmission channel, the corresponding transfer function is $A(z)$. Both $v(n)$ and $u(n)$ are correlated since they are from the common source. On the other hand, $v(n)$ and $s(n)$ are uncorrelated. Noises $m_1(n)$ and $m_2(n)$ are also uncorrelated, and they are uncorrelated with $s(n)$, $v(n)$, and $u(n)$. $d(n) = s(n) + v(n) + m_1(n)$ is the desired response of the filter, and $e(n) = y(n)$ is the error signal which is also the output of the ANC system. If the adaptive process is convergent and has a minimal MSE solution, then the adaptive filter is equivalent to the Wiener filter. In this case, the input power spectrum of the adaptive filter is

$$P_{xx}(z) = P_{m_2 m_2}(z) + P_{vv}(z)|A(z)|^2 \tag{12.182}$$

The cross-power spectrum $P_{xd}(z)$ between the filter input and the desired response is dependent only on the correlation component of its original input and the reference input:

$$P_{xd}(z) = P_{vv}(z)A(z^{-1}) \tag{12.183}$$

Thus we have

$$W_{\text{opt}}(z) = \frac{P_{vv}(z)A(z^{-1})}{P_{m_2 m_2}(z) + P_{vv}(z)|A(z)|^2} \tag{12.184}$$

It can be seen from eq. (12.184) that $W_{\text{opt}}(z)$ is independent of the power spectrum $P_{ss}(z)$ of the desired signal and the power spectrum $P_{m_1 m_1}(z)$ of the uncorrelated noise. If the additive noise $m_2(n)$ in the reference input is zero, then $P_{m_2 m_2}(z)$ is zero, and the optimal transfer function of the filter becomes

$$W_{\text{opt}}(z) = \frac{1}{A(z)} \qquad (12.185)$$

The above equation shows that the optimal transfer function $W_{\text{opt}}(z)$ of the adaptive filter is equal to the inverse of the transfer function $A(z)$ of the reference input transmission channel. At this time, the adaptive filter can make the noise $v(n)$ in the ANC system output as zero, but the original uncorrelated noise $m_1(n)$ cannot be canceled.

2. Application of ANC

Example 12.3: Try to use the ANC system to recover the noise-polluted sinusoidal signal by MATLAB programming.

Solution: MATLAB program is listed as follows:

```
clear; N=1000; M=15; s=sin(2*pi*0.05*[0:N-1]'); noise=0.5*randn(1,N);
g=fir1(M-1,0.4); fnoise=filter(g,1,noise); d=s.'+fnoise; mu = 0.015;
ha = adaptfilt.lms(M+1,mu); [y,e] = filter(ha,noise,d);
figure(1)
subplot(221);plot(800:999,s(801:1000)); axis([800,1000,-2,2]); ylabel
  ('Amplitude'); xlabel('Time');
title('Pure signal');
subplot(222);plot(800:999,d(801:1000)); axis([800,1000,-2,2]); ylabel
  ('Amplitude'); xlabel('Time');
title('Noisy signal');
subplot(223);plot(800:999,e(801:1000)); axis([800,1000,-2,2]); ylabel
  ('Amplitude'); xlabel('Time');
title('Signal after ANC processing');
subplot(224);plot(800:999,s(801:1000),800:999,e(801:1000)); axis([800,
  1000,-2,2]);
ylabel('Amplitude'); xlabel('Time'); title('Comparison between
  processed and pure signals');
```

Figure 12.17 shows the results of the adaptive noise cancellation.

Example 12.4: Extraction of fetal electrocardiogram (ECG) on the mother abdomen electrode.

ECG monitoring of the fetus during pregnancy is one of the important technical means to ensure the safety of mother and child. With fetal ECG observation, clinicians can understand whether the fetal position and heart rate are normal or not, so that doctors can predict the physiological conditions of the fetus in the womb. Here we

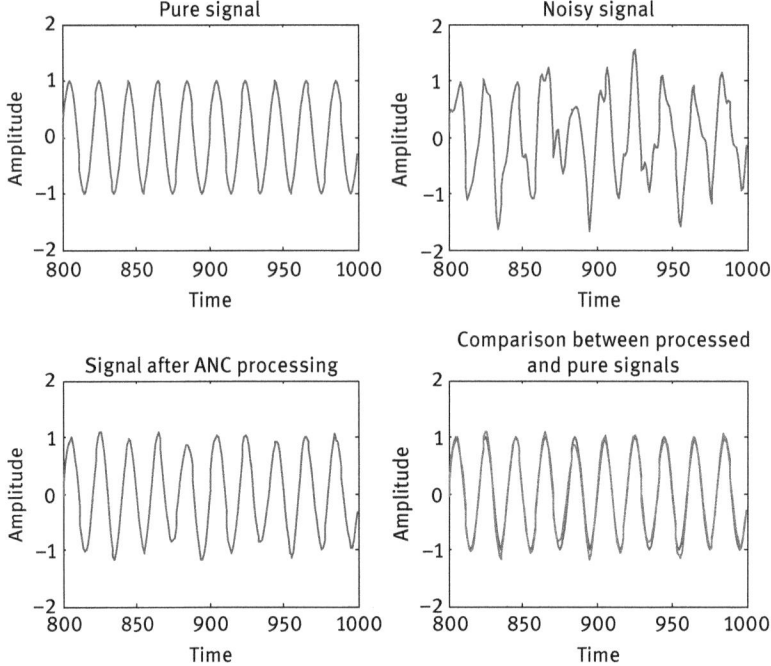

Figure 12.17: The results of the adaptive noise cancellation: (a) pure sinusoidal signal; (b) noise-contaminated signal; (c) signal after adaptive cancellation; and (d) comparison.

take the fetal ECG signal processing as an example to illustrate the application of ANC system.

Fetal ECG is measured in the maternal abdominal wall, known as the abdominal wall fetal ECG. The signal measured from the maternal abdominal wall can be expressed as

$$x(t) = s(t) + m(t) + v(t) \tag{12.186}$$

where $s(t)$ is the ECG signal of the fetus, $m(t)$ is the ECG signal of the mother, $v(t)$ is the noise. Figure 12.18 shows a schematic representation of fetal ECG measurements.

As the fetal ECG signal $s(t)$ is weak, coupled with the maternal ECG and noise interference, it is difficult to visually identify the fetal ECG signal. Figure 12.19 shows a typical fetal ECG obtained from the maternal abdominal wall. Obviously, the fetal ECG signal is basically submerged by the maternal ECG signal and noise interference.

If the ANC system is adopted, the maternal ECG signal obtained from the mother chest is used as the reference signal and the maternal abdominal wall signal as the original input signal, the maternal ECG signal on the fetal ECG

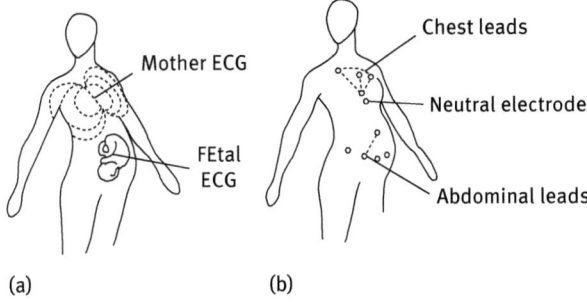

(a)   (b)

Figure 12.18: A schematic representation of fetal electrocardiogram measurements.

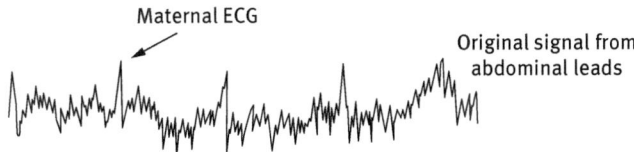

Figure 12.19: Fetal ECG measured from the maternal abdominal wall.

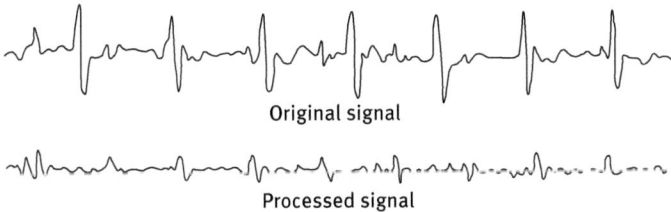

Figure 12.20: The result obtained from the ANC method to extract the fetal ECG signal. Original signal (upper figure) and the processed signal (lower figure).

signal can be effectively eliminated. Figure 12.20 shows a result obtained by using the ANC method to extract the fetal ECG signal.

12.5.2 Adaptive Line Enhancement and Its Applications

1. Fundamentals of adaptive line enhancement

The adaptive line enhancement (ALE) is an adaptive method for detecting weak sinusoidal signals or narrowband signals in wideband noise. Figure 12.21 shows the block diagram of the ALE.

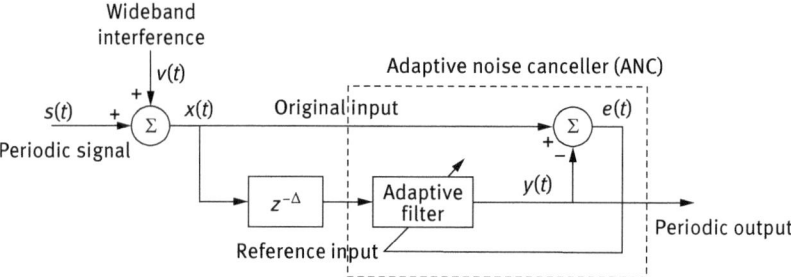

Figure 12.21: The block diagram of the ALE.

As can be seen, the ALE is actually an ANC system without the external reference input signal. Its role is to suppress broadband noise, and to enhance or highlight the narrowband or sinusoidal signal as far as possible for spectral analysis and other subsequent processing. Let the original input signal $x(t) = s(t) + v(t)$ of the ANC system in the right half of the diagram be a narrowband or periodic signal, $v(t)$ denotes a broadband noise. $x(t)$ is delayed Δ sampling intervals and then is sent into the ANC system. The adaptive coefficients are adjusted in accordance with the LMS criterion. In order to minimize the MSE $E[e^2(t)]$, the output $y(t)$ should cancel out the $s(t)$ component in $x(t)$ as much as possible, leaving only the wideband noise component $v(t)$ in the error signal $e(t)$. In this way, when the adaptive algorithm converges, $y(t)$ is the optimal approximation of $s(t)$ and the narrowband or sinusoidal signal is extracted.

The key to guaranteeing the ALE system works properly is to ensure that $v(t - \Delta)$ in the delayed signals $x(t - \Delta)$ is not correlated with $v(t)$, and at the same time to ensure that $s(t - \Delta)$ and $s(t)$ are still correlated. Since the sinusoidal or periodic signal has a periodic correlation, the correlation remains after delay Δ intervals, and the wideband noise loses its correlation due to the delay. Therefore, the ALE can effectively enhance the sinusoidal signal or the narrowband signal, and suppress the impact of broadband noise. It should be noted that the reasonable choice of delay Δ is of great significance for improving the spectral line enhancement effect. In general, a better choice is to make the equivalent phase shift of a sine wave exactly equal to 360°.

2. Application of ALE

Example 12.5: An ALE is used to enhance the spectral lines of narrowband signals.

Figure 12.22 shows the results of enhancing narrowband signal by the ALE. As can be seen from the figure, if the input signal is a single sine wave plus white noise (a); or half of the noise power is white and the other half is colored ((b) and (c)). In these three cases, the sine wave obtained by the ALE fluctuates around the zero value of the amplitude, while the spectrum obtained by the conventional Fourier transform has a stronger background noise spectrum. Obviously, the ALE effectively suppresses the spectra of white and colored noise, and achieves the purpose of enhancing signal lines.

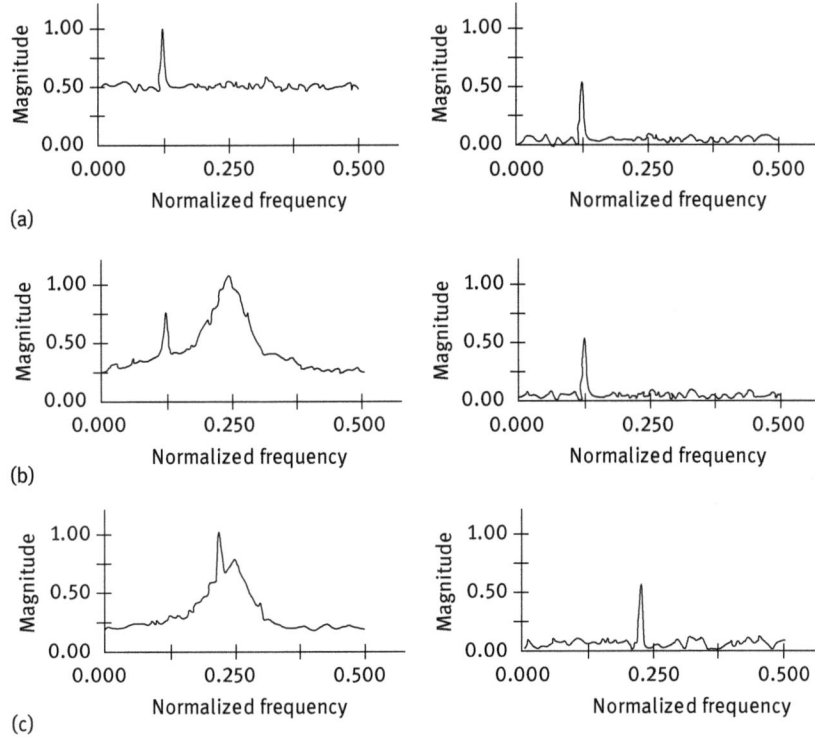

Figure 12.22: Comparison between classical Fourier transform (left) and ALE (right) for power spectrum estimation: (a) white noise as the background; (b) white and colored noises as the background, each with 50% power; (c) same as (b) with different signal spectrum.

The ALE can be combined with AR spectral estimation for dynamic spectral estimation of Doppler echoes, that is, estimate dynamically the power spectrum for non-stationary random signals. This method can better reflect the time-varying characteristics of the power spectrum than the conventional segmented Fourier transform method.

Example 12.6: A linear combination of two sinusoidal signals is given. The combined signal is contaminated by Gaussian white noise. Try to write a MATLAB program to implement the ALE algorithm to eliminate the white noise in the combined signal.

Solution: The MATLAB program is shown as follows:

```
clear; delay=1; N= 5000;
s=0.5*sin(2*pi*0.05*[0:N+delay-1])+sin(2*pi*0.1*[0:N+delay-1]);
noise=2*randn(1,N+delay); x = s(1:N);
d=s(1+delay:N+delay)+noise(1+delay:N+delay);
mu=0.001; ha=adaptfilt.lms(32,mu); [y,e]=filter(ha,x,d);
```

```
[pdd,w]=pwelch(d(N-1000:N)); [pyy,w]=pwelch(y(N-1000:N)); [pss,w]=pwel
  ch(s(N-1000:N));
subplot(321); plot(N-100:N,s(N-100:N)); axis([4900,5000,-3,3]); title
  ('Pure signal'); ylabel('Amplitude');
subplot(323); plot(N-100:N,d(N-100:N)); axis([4900,5000,-8,8]); title
  ('Noisy signal'); ylabel('Amplitude');
subplot(325); plot(N-100:N,y(N-100:N));
axis([4900,5000,-3,3]); title('Enhanced signal'); ylabel('Amplitude');
  xlabel('Samples');
subplot(322); plot(w/pi/2,10*log10(pss)); axis([0,0.5,-50,50]);
  ylabel('Magnitude'); title('PSD of pure signal');
subplot(324); plot(w/pi/2,10*log10(pdd)); axis([0,0.5,-50,50]);
  ylabel('Magnitude'); title('PSD of noisy signal');
subplot(326); plot(w/pi/2,10*log10(pyy));
axis([0,0.5,-50,50]); ylabel('Magnitude'); xlabel('Frequency'); title
  ('PSD of enhanced signal');
```

Figure 12.23 shows the results of the ALE algorithm.

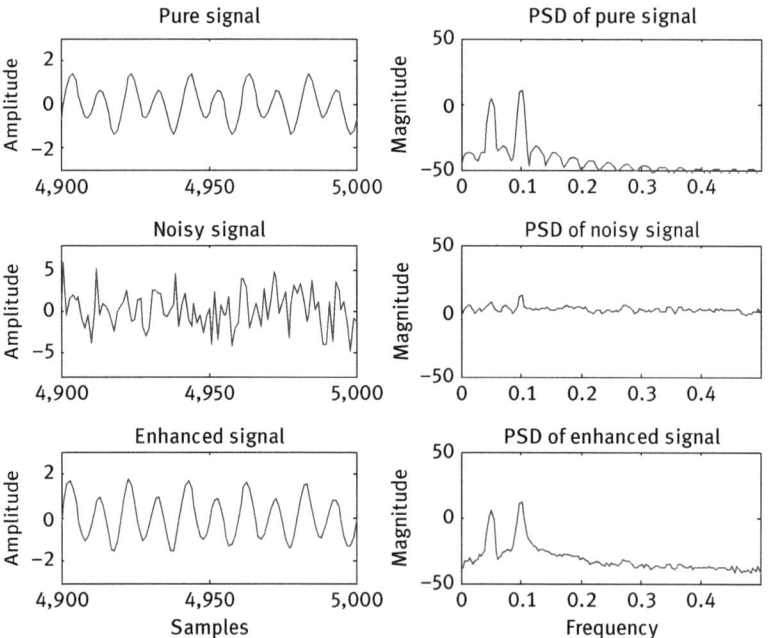

Figure 12.23: Results of the ALE algorithm. CG/TS: Insert comma as thousand separator for all 4 and 5 digit numbers in the artwork of Fig. 12.23.Waveforms are shown on the left, and spectra are shown on the right.

12.5.3 Adaptive System Identification and Its Applications

1. The principle of adaptive system identification (ASI)

In some cases, we may not necessarily understand the internal structure of a real physical system. We just need to understand its input and output characteristics, or its transmission characteristics. The so-called system identification is to determine the behavior of the system based on the input and output characteristics. The ASI is the use of the adaptive filter to achieve the identification of the system. ASI is also called adaptive simulation. The block diagram of ASI is shown in Figure 12.24.

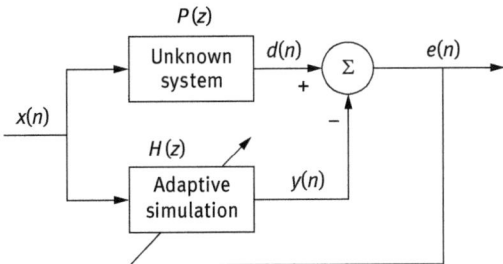

Figure 12.24: The block diagram of adaptive system identification.

In the figure, it is assumed that the system function of the unknown system is expressed by $P(z)$, the system function of the adaptive filter is represented as $H(z)$ when it reaches the steady state. In the adaptive iterative process, the error signal $e(n) \rightarrow$ min indicates that the error between $d(n)$ and $y(n)$ reaches the minimum, that is, $y(n) \approx d(n)$. Since the unknown system $P(z)$ has the same input signal $x(n)$ and approximately the same output signal as the adaptive system $H(z)$ does, $H(z)$ can be regarded as a good approximation of $P(z)$, called the simulation or identification.

2. The application of ASI

Example 12.7: Adaptive control of blood pressure.

The adaptive adjustment or control of blood pressure has a great importance in clinical applications, especially for postoperative monitoring of patients or long-term shock patients. The adaptive control of the blood pressure can be realized by using the ASI technology. Figure 12.25 shows a blood pressure-adaptive control schematic.

As shown in the figure, the blood pressure is continuously monitored by the blood pressure monitor, and the average blood pressure is taken as the output of the system. The injection volume of the vasoconstrictor is used as the input of the system, by controlling the valve shown in the figure to achieve the adjustment of injection volume. The function of the ASI is to approach the model of the physiological system by monitoring the physiological system and adjusting the adaptive system. When the simulated system is basically consistent with the real physiological system, the

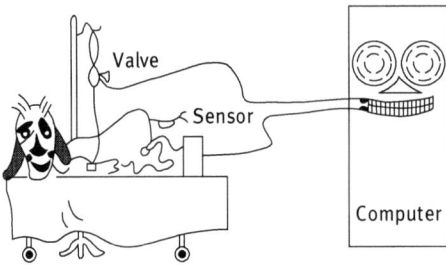

Computer   **Figure 12.25:** The blood pressure-adaptive control schematic.

measured value of blood pressure can be roughly equal to the expected value, so that the blood pressure is automatically maintained at the expected value.

🅰 Exercises

12.1 Describe the concept of the performance function of the adaptive filter.
12.2 Describe the method of searching on the performance surface.
12.3 Explain the Newton method and steepest descent method.
12.4 Describe the basic principle of the LMS algorithm for the adaptive filter.
12.5 Summarize the performances of the LMS algorithm.
12.6 Summarize the improved LMS algorithms and their respective characteristics.
12.7 Describe the RLS algorithm of the adaptive filter.
12.8 Describe the basic principle and applications of ANC system.
12.9 Describe the basic principle and applications of ALE.
12.10 Describe the basic principle and applications of ASI.
12.11 An adaptive linear combiner is given in Figure E12.1. Let $N = 10$.

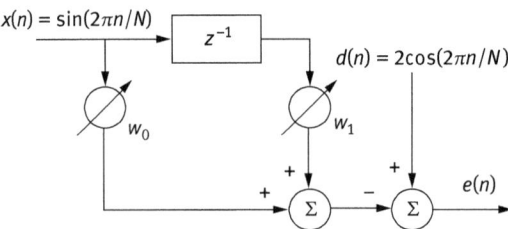

Figure E12.1

(1)   Find its optimal weight vector.
(2)   Derive the expression of $y(n)$ by using the result of (1).
(3)   Prove $y(n) = d(n)$ by using the result of (2).
12.12 Prove that all eigenvalues of the autocorrelation matrix for a white noise are equal.

12.13 On the performance surface in Exercise 12.1, if $w_1 = 0$, $N = 8$, and the mean square value of $e(n)$ is 0.2, find the gradient vector of the system.

12.14 Consider the single-weight adaptive linear combiner in Figure E12.2. The following relations are satisfied: $E[x^2(n)] = 1$, $E[x(n)x(n-1)] = 0.5$, $E[d^2(n)] = 4$, $E[d(n)x(n)] = -1$, $E[d(n)x(n-1)] = 1$. Derive the expression of the performance function of the system, and plot its curve.

12.15 If switch S in Figure E12.2 is closed, derive the expression and plot the performance function again based on the above conditions and requirements.

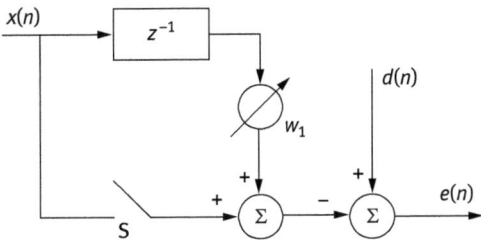

Figure E12.2

12.16 Assume that the convergence factor of an adaptive system is $\mu = 0.1$, the initial value of the weight vector is $w(0) = [5\ 2]^T$, and the performance function is as follows:

$$\xi = 2w_0^2 + 2w_1^2 + 2w_0w_1 - 14w_0 - 16w_1 + 42$$

Try to use the modified Newton method $w(n+1) = w(n) - \mu R^{-1}\nabla(n)$ to find the first five weight vector values, and find $w(20)$.

12.17 The condition for the adaptive system is the same as that in Exercise 12.16.
(1) Try to use the steepest descent method to find the first five weight vector values, and find $w(20)$.
(2) Try to give the learning curve of the steepest descent method.

12.18 When the components involved in an adaptive processing are orthogonal to each other, try to prove the adjustment of the coefficients can be performed independently. In this case, the p th-order filter can be decomposed into p first-order filters.

12.19 Consider the adaptive system shown in Figure E12.3:
(1) If $R_{xx}(m) = E[x(n)x(n+m)]$, write the performance function expression of the system.
(2) If $x(n) = \sin\frac{\pi n}{5}$, write the expression of the performance surface.
(3) If $x(n) = \sin\frac{\pi n}{5}$ and μ takes its maximal value 1/5, express the LMS algorithm.

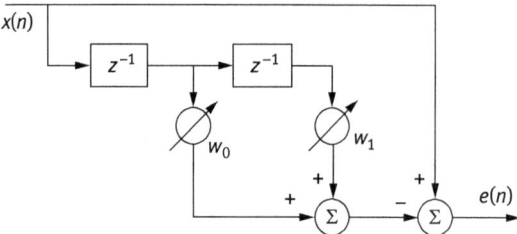

Figure E12.3

12.20 For the second-order recursive adaptive filter, try to prove that the weight coefficients w_0 and w_1 must be inside the triangle $\triangle ABC$ shown in Figure E12.4 to ensure the stability of the filter, that is, the triangle is equivalent to the unit circle on the z-plane.

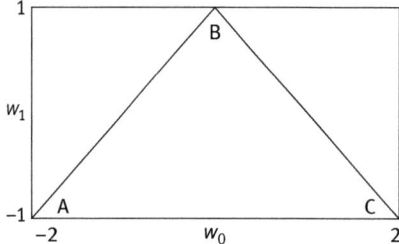

Figure E12.4

12.21 The adaptive system is shown in Figure E12.5. With known $x(n) = \sin\frac{2\pi}{15}$ and $L = 1$, find the performance function of the system.

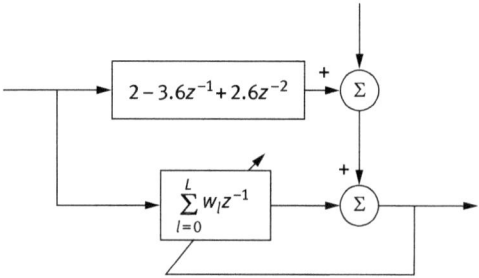

Figure E12.5

13 Higher-Order and Fractional Lower-Order Statistics

13.1 Higher-Order Statistics

13.1.1 Overview of Higher-Order Statistics

Many of signal-processing tools rely upon the so-called second-order statistics. The two most commonly used concepts involve the correlation function and the power spectral density function. The estimation of the power spectrum of discrete-time deterministic or stochastic signals is one of the most fundamental and useful tools in digital signal processing. The use of power spectrum spreads across radar, sonar, communication, speech, biomedical, geophysical, and other data-processing systems. In power spectral estimation, the signal under consideration is processed in such a way that the phase relationship among components is lost. It is necessary to utilize the higher-order statistics (HOS) to yield additional advantages, such as discrimination against additive Gaussian noise and preservation of phase information. HOS techniques were first applied to real signal processing problems in the 1970s, and since then they have continued to expand into different fields such as economics, speech, seismic data processing, plasma physics, optics, and biomedicine. Some of the motivations for using HOS in signal processing are as follows: (1) The HOS of non-Gaussian linear processes contains both amplitude and phase information. They have been used for time-series modeling and identification of nonminimum phase and noncausal systems. These applications include signal reconstruction from speckle images, seismic deconvolution, and channel equalization. (2) The HOS of Gaussian signals is statistically zero. Thus, the HOS can be used to measure non-Gaussianity and to separate additive mixtures of independent non-Gaussian signals and Gaussian noise. This feature can be exploited to detect and classify non-Gaussian signals and provide high noise immunity in applications where the signal source is corrupted with Gaussian noise. (3) The HOS is able to detect and characterize the nonlinear properties of mechanisms that generate time series via phase relations of their harmonic components. (4) The HOS is translation invariant because linear phase terms are canceled in the products of Fourier coefficients that define them. Functions that can serve as features for pattern recognition can be defined from higher-order spectra that satisfy other desirable invariance properties such as scaling, amplification, and rotation invariance.

13.1.2 The Characteristic Function

The characteristic function method is one of the main analytical tools of probability theory and mathematical statistics. It is easy to define HOS and derive their properties.

https://doi.org/10.1515/9783110465082-013

Definition 13.1: Let x be a real, discrete time, and nth-order stationary random process. If the distribution function of random variable x is $F(x)$,

$$\Phi(\omega) = E(e^{j\omega x}) = \int_{-\infty}^{+\infty} e^{j\omega x} dF(x) = \int_{-\infty}^{+\infty} e^{j\omega x} f(x) dx \qquad (13.1)$$

is called the first characteristic function of x. Here, $f(x)$ is the probability density function (pdf), which is the derivative of $F(x)$. ω is the frequency parameter of the characteristic function. The characteristic function $\Phi(\omega)$ is the Fourier transform of the probability density $f(x)$. Because $f(x) \geq 0$, $\Phi(\omega)$ has the maximum value at the origin, that is, $|\Phi(\omega)| \leq |\Phi(0)| = 1$.

The second characteristic function $\Psi(\omega)$ is defined as

$$\Psi(\omega) = \ln \Phi(\omega) \qquad (13.2)$$

Definition 13.2: If the n-dimensional random vector of $x = [x_1, x_2, \ldots, x_n]^T$ has a joint probability distribution function $F(x_1, x_2, \ldots, x_n)$, then the first joint characteristic function is

$$\begin{aligned}\Phi(\omega_1, \omega_2, \ldots, \omega_n) &= E[e^{j(\omega_1 x_1 + \omega_2 x_2 + \cdots + \omega_n x_n)}] \\ &= \int_{-\infty}^{\infty} \cdots \int_{-\infty}^{\infty} e^{j(\omega_1 x_1 + \omega_2 x_2 + \cdots + \omega_n x_n)} dF(x_1, x_2, \ldots, x_n)\end{aligned} \qquad (13.3)$$

It can be written as a vector form:

$$\Phi(\omega) = \int e^{j\omega^T x} f(x) dx \qquad (13.4)$$

where $f(x) = f(x_1, x_2, \ldots, x_n)$ is the joint pdf.

The second joint characteristic function of x is defined as

$$\Psi(\omega_1, \omega_2, \ldots, \omega_n) = \ln[\Phi(\omega_1, \omega_2, \ldots, \omega_n)] \qquad (13.5)$$

13.1.3 Statistical Definitions of Moments and Cumulants

1. For single random variable

(1) Higher order moment

The kth -order moment of the random variable x is defined as

$$m_k = E[x^k] = \int_{-\infty}^{\infty} x^k f(x) dx \qquad (13.6)$$

where $E[\ ]$ is the statistical expectation operator and the subscript k denotes the order of the moment. Obviously, $m_0 = 1$, $m_1 = \eta = E[x]$. The kth -order central moment of the random variable x is defined as

$$\mu_k = E[(x - \eta)^k] = \int_{-\infty}^{\infty} (x - \eta)^k f(x) dx \tag{13.7}$$

Then $\mu_0 = 1$, $\mu_1 = 0$, $\mu_2 = \sigma^2$.

If $m_k (k = 1, 2, \ldots, n)$ exists, the characteristic function $\Phi(\omega)$ of x can be expanded according to Taylor series, that is,

$$\Phi(\omega) = 1 + \sum_{k=1}^{n} \frac{m_k}{k!} (j\omega)^k + O(\omega^n) \tag{13.8}$$

And the relationship between m_k and the kth -order derivative of $\Phi(\omega)$ is

$$m_k = (-j)^k \frac{d^k \Phi(\omega)}{d\omega^k} \Big|_{\omega = 0} = (-j)^k \Phi^k(0), k \le n \tag{13.9}$$

(2) Higher-order cumulant

The second characteristic function $\Psi(\omega)$ of x can be expanded according to Taylor series:

$$\Psi(\omega) = \ln \Phi(\omega) = \sum_{k=1}^{n} \frac{c_k}{k!} (j\omega)^k + O(\omega^n) \tag{13.10}$$

The relationship between c_k and the kth -order derivative of $\Psi(\omega)$ is

$$c_k = \frac{1}{j^k} \left[\frac{d^k}{d\omega^k} \ln \Phi(\omega) \right] \Big|_{\omega = 0} = \frac{1}{j^k} \left[\frac{d^k \Psi(\omega)}{d\omega^k} \right] \Big|_{\omega = 0} = (-j)^k \Psi^k(0), k \le n \tag{13.11}$$

c_k is called the kth -order cumulant of the random variable x.

(3) The relationship between higher-order moment and cumulant

Let $n \to \infty$, $\Phi(\omega)$ can be written as

$$\Phi(\omega) = 1 + \sum_{k=1}^{\infty} \frac{m_k}{k!} (j\omega)^k = \exp\left[\sum_{k=1}^{\infty} \frac{c_k}{k!} (j\omega)^k \right]$$

$$= 1 + \sum_{k=1}^{\infty} \frac{c_k}{k!} (j\omega)^k + \frac{1}{2!} \left[\sum_{k=1}^{\infty} \frac{c_k}{k!} (j\omega)^k \right]^2 + \cdots + \frac{1}{n!} \left[\sum_{k=1}^{\infty} \frac{c_k}{k!} (j\omega)^k \right]^n + \cdots \tag{13.12}$$

The relationship between c_k and m_k can be obtained by comparing the coefficients with the same power of $(j\omega)^k (k = 1, 2, \ldots)$. Therefore, the kth-order cumulant may be expressed in terms of a combination of the kth moment and other lower-order moments. For example,

$$c_1 = m_1 = E[x] = \eta;$$
$$c_2 = m_2 - m_1^2 = E[x^2] - (E[x])^2 = E[(x - E[x])^2] = \mu_2;$$
$$c_3 = m_3 - 3m_1 m_2 + 2m_1^3 = E[x^3] - 3E[x]E[(x^2)] + 2(E[x])^3 = E[(x - E[x])^3] = \mu_3;$$
$$c_4 = m_4 - 3m_2^2 - 4m_1 m_3 + 12m_1^2 m_2 - 6m_1^4 \neq E[(x - E[x])^4] = \mu_4.$$

If the signal x is with zero mean ($m_1 = 0$), then the second- and third-order cumulants are identical to the second- and third-order moments, respectively: $c_1 = m_1 = 0$; $c_2 = m_2 = E[x^2]$; $c_3 = m_3 = E[x^3]$. If the process has nonzero mean, the mean may be subtracted from the process first and such operation is often conducted in practice with estimation from finite records. However, to generate the fourth-order cumulant, we need to have the knowledge of the fourth- and second-order moments, that is,
$$c_4 = m_4 - 3m_2^2 = E[x^4] - 3(E[x^2])^2.$$

Because of the unique linear property of the second characteristic function, using cumulants and cumulant spectra instead of moments is more common and preferable in the practical case of stochastic signals. However, it is noteworthy that estimates of cumulants are obtained in practice after computing estimates of moments from time-domain samples using their relationship.

2. For n -dimensional random variable

(1)   Higher-order moment
Given n -dimensional random variable $(x_1, x_2, \ldots, x_n)$, its first joint characteristic function is

$$\Phi(\omega_1, \omega_2, \ldots, \omega_n) = E\{\exp[j(\omega_1 x_1 + \omega_2 x_2 + \ldots + \omega_n x_n)]\} \tag{13.13}$$

Its second joint characteristic function is

$$\Psi(\omega_1, \omega_2, \ldots, \omega_n) = \ln \Phi(\omega_1, \omega_2, \ldots, \omega_n) \tag{13.14}$$

Expanding eqs. (13.13) and (13.14) in terms of Taylor series, respectively, the joint moment of order $r = k_1 + k_2 + \ldots + k_n$ can be defined by the first joint characteristic function $\Phi(\omega_1, \omega_2, \ldots, \omega_n)$ as

$$m_{k_1 k_2 \cdots k_n} = E[x_1^{k_1} x_2^{k_2} \cdots x_n^{k_n}] = (-j)^r \left[\frac{\partial^r \Phi(\omega_1, \omega_2, \cdots, \omega_n)}{\partial \omega_1^{k_1} \partial \omega_2^{k_2} \cdots \partial \omega_n^{k_n}} \right]_{\omega_1 = \omega_2 = \cdots = \omega_n = 0} \tag{13.15}$$

(2) Higher-order cumulant

Similarly, the joint cumulant of order $r = k_1 + k_2 + \ldots + k_n$ can be defined by the second joint characteristic function $\Psi(\omega_1, \omega_2, \ldots, \omega_n)$ as

$$
\begin{aligned}
c_{k_1 k_2 \ldots k_n} &= (-j)^r \frac{\partial \Psi(\omega_1, \omega_2, \ldots, \omega_n)}{\partial \omega_1^{k_1} \partial \omega_2^{k_2} \ldots \partial \omega_n^{k_n}} \bigg|_{\omega_1 = \omega_2 = \cdots = \omega_n = 0} \\
&= (-j)^r \frac{\partial^r \ln \Phi(\omega_1, \omega_2, \ldots, \omega_n)}{\partial \omega_1^{k_1} \partial \omega_2^{k_2} \ldots \partial \omega_n^{k_n}} \bigg|_{\omega_1 = \omega_2 = \cdots = \omega_n = 0}
\end{aligned}
\tag{13.16}
$$

(3) The relationship between the higher-order moments and cumulants

The joint cumulant $c_{k_1 k_2 \ldots k_n}$ can be represented by a polynomial of the joint moment $m_{k_1 k_2 \ldots k_n}$, but its general expression is quite complex. Hence, we only give the relationship among the second-, third-, and fourth-order joint cumulants and their corresponding order joint moments here.

Let $x_1, x_2, \ldots, x_n$ be a random process composed of real random variables with zero mean. The second-, third-, and fourth-order cumulants are given, respectively:

$$
\text{cum}(x_1, x_2) = E[x_1 x_2]
\tag{13.17a}
$$

$$
\text{cum}(x_1, x_2, x_3) = E[x_1 x_2 x_3]
\tag{13.17b}
$$

$$
\begin{aligned}
\text{cum}(x_1, x_2, x_3, x_4) &= E[x_1 x_2 x_3 x_4] - E[x_1 x_2] E[x_3 x_4] \\
&\quad - E[x_1 x_3] E[x_2 x_4] - E[x_1 x_4] E[x_2 x_3]
\end{aligned}
\tag{13.17c}
$$

where $\text{cum}(\cdot)$ represents the joint cumulant. For random variables with nonzero mean, x_i in eq. (13.17) can be replaced by $x_i - E[x_i]$.

3. Higher-order cumulant of stationary random process

Suppose that $x(t)$ is a stationary random process with zero mean and $\tau_1, \tau_2, \ldots, \tau_{k-1}$ are lags in time, then the cumulants of order k can be denoted by $c_{kx}(\tau_1, \tau_2, \ldots, \tau_{k-1})$. Setting $\tau_1 = \tau_2 = \cdots = \tau_{k-1} = \tau$, the second-, third-, and fourth-order cumulants can be rewritten as

$$
C_{2x}(\tau) = E[x(t) x(t + \tau)]
\tag{13.18a}
$$

$$
C_{3x}(\tau) = E[x(t) x^2(t + \tau)]
\tag{13.18b}
$$

$$
C_{4x}(\tau) = E[x(t) x^3(t + \tau)] - 3 C_{2x}(\tau) C_{2x}(0)
\tag{13.18c}
$$

By putting $\tau = 0$ into the aforementioned equations, we obtain the well-known parameters, such as variance, skewness, and kurtosis, which can be used to describe pdfs.

$$\text{Variance } \gamma_{2x} = c_{2x}(0) = E[x^2(k)] \tag{13.19a}$$

$$\text{Skewness } \gamma_{3x} = c_{3x}(0, 0) = E[x^3(k)] \tag{13.19b}$$

$$\text{Kurtosis } \gamma_{4x} = c_{4x}(0, 0, 0) = E[x^4(k)] - 3[\gamma_{2x}]^2 \tag{13.19C}$$

They describe the profile of the distribution. Zero skewness means that the signal obeys the symmetrical distribution, while nonzero skewness means that the signal obeys the asymmetric distribution. So the skewness is actually used to measure the degree of skewness that the signal distribution deviates from symmetry. Kurtosis can be used to measure "peakedness" of the distribution of the process.

13.1.4 The Properties of Higher-Order Cumulant

Higher-order cumulant has some important properties, which are as follows.
Property 13.1: Assume that $\lambda_i (i = 1, 2, \ldots, k)$ is a constant and $x_i (i = 1, 2, \ldots, k)$ is a random variable, then

$$\text{cum}(\lambda_1 x_1, \ldots, \lambda_k x_k) = \left(\prod_{i=1}^{k} \lambda_i \right) \text{cum}(x_1, \ldots, x_k) \tag{13.20}$$

Property 13.2: The additive property:

$$\text{cum}(x_0 + y_0, z_1, \ldots, z_k) = \text{cum}(x_0, z_1, \ldots, z_k) + \text{cum}(y_0, z_1, \ldots, z_k) \tag{13.21}$$

Property 13.3: If the random variable $x_i (i = 1, 2, \ldots, k)$ and the random variable $y_i (i = 1, 2, \ldots, k)$ are independent of one another, then

$$\text{cum}(x_1 + y_1, \ldots, x_k + y_k) = \text{cum}(x_1, \ldots, x_k) + \text{cum}(y_1, \ldots, y_k) \tag{13.22}$$

If a non-Gaussian signal $\{x(n)\}$ is contaminated by the additive colored Gaussian noise $\{e(n)\}$ that is independent of $\{x(n)\}$, that is, $y(n) = x(n) + e(n)$, then the higher-order cumulant of $\{y(n)\}$ is

$$\begin{aligned} \text{cum}_{ky}(\tau_1, \tau_2, \ldots, \tau_{k-1}) &= \text{cum}_{kx}(\tau_1, \tau_2, \ldots, \tau_{k-1}) + \text{cum}_{ke}(\tau_1, \tau_2, \ldots, \tau_{k-1}) \\ &= \text{cum}_{kx}(\tau_1, \tau_2, \ldots, \tau_{k-1}) \end{aligned} \tag{13.23}$$

That is to say, the higher-order cumulant can theoretically suppress the influence of Gaussian colored noise completely, but this important conclusion is not suitable for higher-order moment.

Property 13.4: The high-order cumulant is independent of the order of the variables; if $(i_1, \ldots, i_k)$ is a reordered sequence of $(1, \ldots, k)$, then

$$\text{cum}(x_1, \ldots, x_k) = \text{cum}(x_{i_1}, x_{i_2}, \ldots, x_{i_k}) \tag{13.24}$$

According to this property, the kth -order cumulant has $k!$ kinds of symmetric forms. Taking the third-order cumulant as an example, there are $3!=6$ symmetric forms:

$$\begin{aligned} c_{3x}(m, n) &= c_{3x}(n, m) = c_{3x}(-n, m-n) = c_{3x}(m-n, -m) \\ &= c_{3x}(m-n, -n) = c_{3x}(-m, n-m) \end{aligned} \tag{13.25}$$

Property 13.5: If α is a constant, then

$$\text{cum}(\alpha + z_1, z_2, \ldots, z_k) = \text{cum}(z_1, z_2, \ldots, z_k) \tag{13.26}$$

Because of the complexity, cumulants of fourth order or lower than fourth order are usually used in real applications. For a random process with symmetric distribution, the third-order cumulant is equal to zero so that the cumulants of higher than fourth order have to be used.

13.1.5 Estimation of Higher-Order Cumulant

In practical applications, it is impossible to know the true value of every order cumulant of the ergodic stationary random process. Therefore, the time average is used to replace assemble average. Given an N -sample vector $x(n)$, $n = 0, \ldots, N-1$, the following expressions describe three estimates for the second-, third-, and fourth-order cumulants, respectively,

$$\hat{C}_{2x}(\tau) = \frac{1}{N} \sum_{n=0}^{N-1} x(n)x[\text{mod}(n+\tau, N)] \tag{13.27a}$$

$$\hat{C}_{3x}(\tau) = \frac{1}{N} \sum_{n=0}^{N-1} x(n)x^2[\text{mod}(n+\tau, N)] \tag{13.27b}$$

$$\begin{aligned} \hat{C}_{4x}(\tau) = \frac{1}{N} \sum_{n=0}^{N-1} x(n)x^3[\text{mod}(n+\tau, N)] \\ - 3\frac{1}{N^2} \sum_{n=0}^{N-1} x(n)x^3[\text{mod}(n+\tau, N)] \sum_{n=0}^{N-1} x^2(n) \end{aligned} \tag{13.27c}$$

where $x \in \mathbb{R}$, $\tau = 0, 1, \ldots, N - 1$, and mod is the modulo operator (i.e., the remainder after integer division).

Another method through segmentation results in an asymptotically unbiased estimation.

Step 1: Divide $\{x(1), x(2), \ldots, x(N)\}$ into K segments (each segment contains M samples, and overlapping between two adjacent data segment is allowed), then subtract each mean value from every data segment respectively.

If the data sample corresponds to a deterministic energy signal, then the data segmentation is not appropriate. Similarly, if the process is cyclic, then M should be equal to the signal cycle or an integer multiple of the period.

Step 2: Assume that $\{x^{(i)}(k), \quad k = 0, 1, \ldots, M - 1\}$ is the ith ($i = 1, \ldots, K$) segmentation, then the higher-order moment is estimated as

$$m_n^{(i)}(\tau_1, \cdots, \tau_{n-1}) = \frac{1}{M} \sum_{k=M_1}^{M_2} x^{(i)}(k) x^{(i)}(k + \tau_1) \cdots x^{(i)}(k + \tau_{n-1}) \tag{13.28}$$

where $n = 2, 3, \ldots$, $i = 1, 2, \ldots K$, $\tau_n = 0, \pm 1, \pm 2, \ldots$, $M_1 = \max(0, -\tau_1, \ldots, -\tau_{n-1})$, $M_2 = \min(M - 1, M - 1 - \tau_1, \ldots, M - 1 - \tau_{n-1})$, and $|\tau_k| \leq L_n$. L_n decides the support area of the estimated nth -order moment function.

Step 3: Average all segments, that is,

$$\hat{m}_n^x(\tau_1, \ldots, \tau_{n-1}) = \frac{1}{K} \sum_{i=1}^{K} m_n^{(i)}(\tau_1, \ldots, \tau_{n-1}), \qquad n = 2, 3, \ldots, |\tau_k| \leq L_n \tag{13.29}$$

$\hat{m}_n^x(\tau_1, \ldots, \tau_{n-1})$ is considered to be a consistent estimation of $m_n^x(\tau_1, \ldots, \tau_{n-1})$. If it is a deterministic signal, then $\hat{m}_n^x(\tau_1, \ldots, \tau_{n-1}) = m_n^1(\tau_1, \ldots, \tau_{n-1})$, that is, $K = 1$, therefore $N = M$.

Step 4: For a random signal, use the relationship between cumulant and moment to generate the cumulant $\hat{c}_n^x(\tau_1, \ldots, \tau_{n-1})$. After subtracting the mean of every segment, there is

$$\hat{c}_2^x(\tau_1) = \hat{m}_2^x(\tau_1) \tag{13.30a}$$

$$\hat{c}_3^x(\tau_1, \tau_2) = \hat{m}_3^x(\tau_1, \tau_2) \tag{13.30b}$$

$$\begin{aligned} \hat{c}_4^x(\tau_1, \tau_2, \tau_3) = \hat{m}_4^x(\tau_1, \tau_2, \tau_3) &- \hat{m}_2^x(\tau_1)\hat{m}_2^x(\tau_3 - \tau_2) \\ &- \hat{m}_2^x(\tau_2)\hat{m}_2^x(\tau_3 - \tau_1) - \hat{m}_2^x(\tau_3)\hat{m}_3^x(\tau_2 - \tau_1) \end{aligned} \tag{13.30c}$$

where $|\tau_k| \leq L_n$, $k = 1, 2, 3$.

In MATLAB, the estimation process of HOS in HOSA toolbox includes cum2est, cum3est, and cum4est.

13.1.6 Higher-Order Spectra and Higher-Order Spectra Estimation

1. The definition and properties in the frequency domain

HOS data in the frequency domain are arranged multidimensionally. This polyspectra analysis yields three important benefits: (1) the suppression of Gaussian noise processes of unknown spectral characteristics; the bispectrum also suppresses noise with symmetrical probability distribution; (2) the reconstruction of the magnitude and phase response of systems; and (3) the detection and characterization of non-redundant, and the targeting of nonlinear interactions.

Polyspectra, or multidimensional spectra, are defined to be the Fourier transforms of the higher-order cumulant sequences. The mathematical foundations are based on the assumption that the cumulant sequences satisfy the bounding condition:

$$\sum_{\tau_1 = -\infty}^{\infty} \cdots \sum_{\tau_{k-1} = -\infty}^{\infty} |c_{kx}(\tau_1, \tau_2, \ldots, \tau_{k-1})| < \infty \tag{13.31}$$

Under this hypothesis, the rth-order cumulant spectrum is defined as the $(k-1)$-dimensional Fourier transforms of the ith-order cumulant:

$$S_{kx}(\omega_1, \omega_2, \ldots, \omega_{k-1}) = \sum_{\tau_1 = -\infty}^{\infty} \cdots \sum_{\tau_{k-1} = -\infty}^{\infty} c_{kx}(\tau_1, \tau_2, \ldots, \tau_{k-1}) \exp\left[-j \sum_{i=1}^{k-1} \omega_i \tau_i\right] \tag{13.32}$$

When $k = 2, 3, 4$, eq. (13.32) is simplified to power spectrum, bispectrum and trispectral, respectively. When $k = 2$, we have power spectrum:

$$S_{2x}(\omega) = \sum_{\tau_1 = -\infty}^{\infty} c_{2x}(\tau) \exp[-j\omega\tau_1] \tag{13.33}$$

When $k = 3$, we have bispectrum (third-order spectrum):

$$B_x(\omega_1, \omega_2) = \sum_{\tau_1 = -\infty}^{\infty} \sum_{\tau_2 = -\infty}^{\infty} c_{3x}(\tau_1, \tau_2) \exp[-j(\omega_1\tau_1 + \omega_2\tau_2)] \tag{13.34}$$

for $|\omega_1| \le \pi, |\omega_2| \le \pi$, and $|\omega_1 + \omega_2| \le \pi$.

As stated in eq. (13.34), the bispectrum is the 2D Fourier transform of the cumulant function. The couple (ω_1, ω_2) is called a bifrequency.

For a deterministic signal $x(n)$, the power spectrum can be expressed in terms of the Fourier transform of the signal as

$$S_{2x}(\omega) = X^*(\omega)X(\omega) \tag{13.35}$$

For a deterministic zero DC (direct current) signal, the bispectrum may be expressed in terms of the Fourier transform of the signal as

$$B_x(\omega_1, \omega_2) = \sum_{\tau = -\infty}^{\infty} \sum_{\tau = -\infty}^{\infty} \sum_{\tau = -\infty}^{\infty} x(n)x(n+\tau_1)x(n+\tau_2) \ \exp[-j(\omega\tau_1 + \omega\tau_2)] \qquad (13.36)$$

After setting $n + \tau_1 = m$ and $n + \tau_2 = k$, and splitting the exponent, eq. (13.36) is

$$B_x(\omega_1, \omega_2) = \left\{ \sum_{m=-\infty}^{\infty} x(m)e^{-j\omega_1 m} \right\} \left\{ \sum_{k=-\infty}^{\infty} x(k)e^{-j\omega_1 k} \right\} \left\{ \sum_{n=-\infty}^{\infty} x(n)e^{-j(\omega_1 + \omega_1)n} \right\} \qquad (13.37)$$

$$= X(\omega_1)X(\omega_2)X^*(\omega_1 + \omega_2)$$

When $k = 4$, we have trispectrum (fourth-order spectrum):

$$T_x(\omega_1, \omega_2, \omega_3) = \sum_{\tau_1 = -\infty}^{\infty} \sum_{\tau_2 = -\infty}^{\infty} \sum_{\tau_3 = -\infty}^{\infty} c_{4x}(\tau_1, \tau_2, \tau_3) \ \exp[-j(\omega_1\tau_1 + \omega_2\tau_2 + \omega_3\tau_3)]$$

$$(13.38)$$

It is notable that, for a deterministic signal, the bispectrum expression is similar to the periodogram expression of the power spectrum so that it is referred to as the higher-order periodogram (biperiodogram). It can be shown that the bispectrum of a random process can be estimated as the expected value of this biperiodogram over an ensemble of realizations of the process. In practice, only a single realization of the random process is available. Therefore, if the process is indeed stationary, this single realization can be divided into segments and the biperiodogram of each segment can be summed and then averaged to obtain a reliable estimate of the bispectrum.

Actually, the bispectrum is the spectral decomposition of skewness of the histogram of the time series, and the trispectrum is the spectral decomposition of the kurtosis of the histogram of the time series. In other words, whenever the histogram of a time series is skewed, the bispectrum must be different from zero. But if the time series is generated by a Gaussian process, both bispectrum and trispectrum are equal to zero.

The inverse transform formula of higher-order spectrum is

$$c_{kx}(\tau_1, \ldots, \tau_{k-1}) = \left(\frac{1}{2\pi}\right)^{k-1} \int_{-\pi}^{\pi} \cdots \int_{-\pi}^{\pi} S_{kx}(\omega_1, \ldots, \omega_{k-1})e^{\ j\left(\sum_{i=1}^{k-1}\omega_i\tau_i\right)} d\omega_1 \ldots d\omega_{k-1}$$

$$(13.39)$$

2. Properties of higher-order spectra
Property 13.6: Higher-order spectra are generally complex and multidimensional functions, which means that they have amplitudes and phases and can be written as

$$S_{kx}(\omega_1, \ldots, \omega_{k-1}) = |S_{kx}(\omega_1, \ldots, \omega_{k-1})|e^{j\theta_{kx}(\omega_1, \ldots, \omega_{k-1})} \tag{13.40}$$

This is a significant difference from the power spectrum. The power spectrum $P_x(\omega)$ of the random process $\{x(n)\}$ is real and $P_x(\omega) \geq 0$, which means that the power spectrum does not contain phase information.

Property 13.7 (Symmetry): This kind of symmetries is caused by the symmetry of higher-order cumulant. Taking the bispectrum as an example, we have

$$
\begin{aligned}
B_x(\omega_1, \omega_2) &= B_x^*(-\omega_1, -\omega_2) = B_x(\omega_2, \omega_1) = B_x(-\omega_1 - \omega_2, \omega_2) = B_x(-\omega_1 - \omega_2, \omega_1) \\
&= B_x(\omega_2, -\omega_1 - \omega_2) = B_x(\omega_1, -\omega_1 - \omega_2)
\end{aligned}
\tag{13.41}
$$

As shown in Figure 13.1, the spectrum defined region is divided into 12 sectors. Thus, knowledge of the bispectrum in the triangular region $\omega_2 \geq 0, \omega_1 \geq \omega_2, \omega_1 + \omega_2 \leq \pi$ is sufficient to describe the rest. This region (labeled 1) is often termed the principal region or the nonredundant region of computation of the bispectrum.

It has been proved that there are 96 symmetric regions in trispectrum.

3. Estimation of Higher-Order Spectra

Higher-order spectra are usually estimated by a sample sequence of $\{x(n)\}$. In practice, even if the underlying process is random and continuous, digital computations require discrete or finite length-sampled data. Just like the power spectra, there are two main approaches that can be used to estimate higher-order spectra: the conventional nonparametric methods (or "Fourier type") and the parametric approach, that is, based on the autoregressive model (AR), the moving average (MA), and the autoregressive and moving average (ARMA) models.

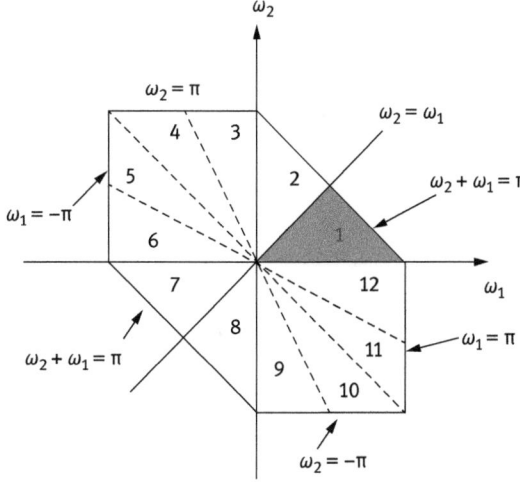

Figure 13.1: Symmetrical areas of bispectrum.

(1) Nonparametric methods

One direct method corresponds to the periodogram of the power spectrum. Its idea is similar to that of power spectrum estimation. The process includes segmenting observed data, calculating discrete Fourier transform (DFT) of each data segment by fast Fourier transform (FFT), estimating the moments of each data segment, and then obtaining the spectral estimation from the relationship between the cumulative spectrum and the moment spectrum. To reduce the variance of the estimation, the data must be windowed for smoothing. The concrete implementation is as follows:

Step 1: Divide $\{x(1), x(2), \ldots, x(N)\}$ into K segments. Each segment contains M (M is even) samples. Remove the mean for every segment. If the signal is deterministic, it does not need to be divided into segments. If we want to get the appropriate length of FFT, the data can be zero padded.

Step 2: Assuming that $\{x^{(i)}(k), k = 0, 1, \ldots, M - 1\}$ is the data of segment i, calculate the coefficient of DFT:

$$Y^{(i)}(\lambda) = \frac{1}{M} \sum_{k=0}^{M-1} x^{(i)}(k) e^{-j\frac{2\pi}{M}k\lambda} \tag{13.42}$$

$\lambda = 0, 1, \ldots, M/2 ; i = 1, 2, \ldots, K$.

Step 3: Set $M = M_1 N_0$, where M_1 is an odd integer, $M_1 = 2J_n + 1$, and then estimate the nth-order moment spectrum by the average in the frequency domain. Namely,

$$\hat{M}_n^{(i)}(\lambda_1, \ldots, \lambda_{n-1}) = \frac{1}{\Delta_n^{n-1}} \sum_{k_1 = -J_n}^{J_n} \cdots \sum_{k_{n-1} = -J_n}^{J_n} Y^{(i)}(\lambda_1 + k_1) \ldots Y^{(i)^*} \tag{13.43}$$

$$(\lambda_1 + \ldots + \lambda_{n-1} + k_1 + \ldots + k_{n-1})$$

where $i = 1, 2, \ldots, K$. Δ_n^{n-1} is the interval required for the frequency samples, $\Delta_n^{n-1} = f_s/N_n$, $\lambda_1 + \cdots + \lambda_{n-1} \le f_s/2$, and $0 \le \lambda_{n-1} \le \cdots \le \lambda_1$.

Step 4: Take the average of K segments data:

$$\hat{M}_n^{(i)}(\omega_1, \ldots, \omega_{n-1}) = \frac{1}{K} \sum_{i=1}^{K} \hat{M}_n^{(i)}(\omega_1, \ldots, \omega_{n-1}) \tag{13.44}$$

where $\omega_j \overset{\Delta}{=} (2\pi\Delta_n)\lambda_j$, $j = 1, 2, \ldots, n - 1$, $\lambda_j = 0, 1, \ldots, M - 1$.

The direct estimation of bispectrum can be achieved by bispecd.m and bispecdx. m in the HOSA toolbox of MATLAB.

Another efficient method is an indirect method – Blackman–Turkey method. It is obtained from the 2D Fourier transform of the third-order cumulant. After $\hat{c}_n^x(\tau_1, \ldots, \tau_{n-1})$ is obtained by the higher-order cumulant method described earlier, a windowed Fourier transform can be performed:

$$\hat{C}_n^x(\omega_1, \ldots, \omega_{n-1}) = \sum_{\tau_1 = -Ln}^{Ln} \cdots \sum_{\tau_{n-1} = -Ln}^{Ln} \hat{c}_n^x(\tau_1, \ldots, \tau_{n-1}) \tag{13.45}$$
$$\cdot w(\tau_1, \ldots, \tau_{n-1}) \cdot \exp\{-j(\omega_1 \tau_1 + \ldots + \omega_{n-1} \tau_{n-1})\}$$

bispeci.m function in HOSA toolbox of MATLAB is an example.

In eq. (13.45), $w(\tau_1, \ldots, \tau_{n-1})$ is a continuous window function of the bounded supporting domain. Similar to the conventional power spectrum estimation, to obtain a smooth estimation, an appropriate window must be chosen.

① In the bispectrum estimation of the real signal, the window function should satisfy

$$w(\tau_1, \tau_2) = w(\tau_2, \tau_1) = w(-\tau_1, \tau_2 - \tau_1) = w(\tau_1 - \tau_2, \tau_2). \tag{13.46}$$

② The window function needs to satisfy that the HOS estimation is zero when it is out of the supporting area. Namely,

$$w(\tau_1, \ldots, \tau_{n-1}) = 0, \quad |\tau_i| > Ln, \quad i = 1, 2, \ldots, n-1 \tag{13.47}$$

③ The window function is equal to 1 at the origin: $w(0, \ldots, 0) = 1$, which is termed as normalization condition.

④ The window function has real nonnegative Fourier transform, that is, $W(\omega_1, \ldots, \omega_{n-1}) \geq 0, |\omega_i| \leq \pi, i = 1, 2, \ldots, n-1$. And the window function should also have limited energy.

Taking the window function of bispectrum as an example, the window function $w(\tau_1, \tau_2)$ that satisfies the four constraints mentioned above can be constructed by one-dimensional lag window function $d(\tau_1)$:

$$w(\tau_1, \tau_2) = d(\tau_1)d(\tau_2)d(\tau_1 - \tau_2) \tag{13.48}$$

In eq. (13.48), the one-dimensional lag window $d(\tau_1)$ should satisfy: $d(\tau_1) = d(-\tau_1)$; $d(\tau_1) = 0, m > L, d(0) = 1$; the Fourier transform of $d(\tau_1)$ is $D(\omega) \geq 0, |\omega| \leq \pi$.

However, not all of the one-dimensional window functions satisfy the condition: for any ω, $D(\omega) \geq 0$. For example, the Hanning window has a negative side lobe in the frequency domain. It is easy to verify that the following three window functions satisfy the above constraints. The first one is an optimal window:

$$d_{opt}(\tau_1) = \begin{cases} \frac{1}{\pi} \left| \sin \frac{\pi m}{L} \right| \left[1 - \left| \frac{m}{L} \right| \right] \cos \frac{\pi m}{L}, & |m| \leq L \\ 0, & |m| > L \end{cases} \tag{13.49}$$

The second one is the Parzen window:

$$d_{\text{parzen}}(\tau_1) = \begin{cases} 1 - 6(|\frac{m}{L}|)^2 + 6(|\frac{m}{L}|)^3, & |m| \leq L/2 \\ 2(1 - |\frac{m}{L}|)^3, & L/2 < |m| \leq L \\ 0, & |m| > L \end{cases} \tag{13.50}$$

The third one is the uniform spectral domain window:

$$W_{\text{uniform}}(\omega_1, \omega_2) = \begin{cases} \frac{4\pi}{3P}, & |\tau_1| \leq P \\ 0, & |\tau_1| > P \end{cases} \tag{13.51}$$

where $|\omega| = \max(|\omega_1|, |\omega_2|, |\omega_1 + \omega_2|)$, $P = a_0/L$, and a_0 is a constant.

It is proved that the optimal window function is superior to the other two functions in terms of the supremum of deviation and variance.

(2)   Parametric methods (only for linear processes)

Similar to the power spectrum estimation, this method is based on the signal model of higher-order spectrum. It does not restrict the signal model to be a minimum phase system, but the input signal should be a white Gaussian noise process.

In Figure 13.2, the excitation is a non-Gaussian signal $e(n)$ and the additive noise $v(n)$ is a Gaussian-colored noise. Because $v(n)$ is statistically independent of $e(n)$, it is also statistically independent of the system output $x(n)$.

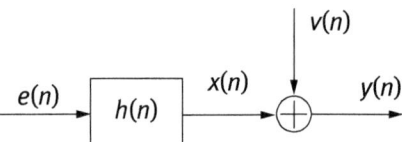

Figure 13.2: Single-input–single-output time-invariant system model for stationary non-Gaussian random process excitation.

First, let us consider the input signal be non-Gaussian i.i.d. Then the kth-order cumulant of $y(n)$ is

$$c_{ke}(\tau_1, \tau_2, \ldots, \tau_{k-1}) = \gamma_{ke}\delta(\tau_1)\delta(\tau_2)\ldots\delta(\tau_{k-1}) \tag{13.52}$$

Since the noise is Gaussian distributed, then

$$c_{ky}(\tau_1, \tau_2, \ldots, \tau_{k-1}) = \gamma_{ke}\sum_{n=0}^{\infty} h(n)h(n+\tau_1)\ldots h(n+\tau_{k-1}) \tag{13.53}$$

$$S_{ky}(\omega_1, \omega_2, \ldots, \omega_{k-1}) = r_{ke}H(\omega_1)H(\omega_2)\ldots H(\omega_{k-1})H\left(-\sum_{i=1}^{k-1}\omega_i\right) \tag{13.54a}$$

$$S_{ky}(z_1, z_2, \ldots, z_{k-1}) = \gamma_{ke}H(z_1)H(z_2)\ldots H(z_{k-1})H\left(\prod_{i=1}^{k-1}z_i^{-1}\right) \tag{13.54b}$$

Next, let us consider the input $e(n)$ is a non-Gaussian-colored signal and $v(k)$ is a Gaussian-colored noise. Then

$$c_{ky}(\tau_1, \tau_2, \ldots, \tau_{k-1}) = \sum_{m_0} \sum_{m_1} \cdots \sum_{m_{k-1}} c_{ke}(\tau_1 - m_1, \tau_2 - m_2, \ldots, \tau_{k-1} - m_{k-1})$$

$$\star h_k(m_1, m_2, \ldots, m_{k-1}) \tag{13.55}$$

Obviously, this is a multidimensional convolution, where $h_k(m_1, m_2, \ldots, m_{k-1}) = \sum_i h(i)h(i+m_1)\ldots h(i+m_{k-1})$ denotes a multidimensional correlation function. And also

$$H_k(z_1, z_2, \ldots, z_{k-1}) = H(z_1)H(z_2)\ldots H(z_{k-1})H\left(\prod_{i=1}^{k-1} z_i^{-1}\right) \tag{13.56}$$

$$S_{ky}(\omega_1, \ldots, \omega_{k-1}) = S_{ke}(\omega_1, \ldots, \omega_{k-1})H(\omega_1)H(\omega_2)\cdots H(\omega_{k-1})H\left(-\sum_{i=1}^{k-1}\omega_i\right) \tag{13.57a}$$

$$S_{ky}(z_1, \ldots, z_{k-1}) = S_{ke}(z_1, \ldots, z_{k-1})H(z_1)H(z_2)\ldots H(z_{k-1})H\left(\prod_{i=1}^{k-1} z_i^{-1}\right) \tag{13.57b}$$

Eqs. (13.55)–(13.57) are proposed by Bartlett, but he only considered the cases of $k = 2, 3, 4$. Brillinger and Rosenblatt extended these three equations to arbitrary order. Thus, these three equations are called Bartlett–Brillinger–Rosenblatt (BBR) formula. Particularly, when the input of the system $e(n)$ is an i.i.d. higher-order white noise, the BBR formula is

$$c_{ky}(\tau_1, \ldots, \tau_{k-1}) = \gamma_{ke} \sum_{i_1 = -\infty}^{\infty} \cdots \sum_{i_k = -\infty}^{\infty} h(i)h(i+\tau_1)\ldots h(i+\tau_k) \tag{13.58}$$

$$S_{kx}(\omega_1, \ldots, \omega_{k-1}) = \gamma_{ke}H(\omega_1)H(\omega_2)\ldots H(\omega_{k-1})H\left(-\sum_{i=1}^{k-1}\omega_i\right) \tag{13.59a}$$

$$S_{kx}(z_1, \ldots, z_{k-1}) = \gamma_{ke}H(z_1)H(z_2)\ldots H(z_{k-1})H\left(-\sum_{i=1}^{k-1} z_i\right) \tag{13.59b}$$

For the causal nonminimum phase system (the poles are inside the unit circle, but the zeros can be outside of the unit circle), the transfer function must converge outside of the unit circle, and the unit impulse response must be the causal sequence. Hence,

$$C_{kx}(\tau_1, \ldots, \tau_{k-1}) = \gamma_{ke} \sum_{i=0}^{\infty} h(i)h(i+\tau_1) \ldots h(i+\tau_{k-1}) \qquad (13.60)$$

Then the bispectrum and trispectrum estimations can be written as

$$C_{3x}(\tau_1, \tau_2) = \gamma_{3e} \sum_{i=0}^{\infty} h(i)h(i+\tau_1)h(i+\tau_2) \qquad (13.61a)$$

$$C_{4x}(\tau_1, \tau_2, \tau_3) = \gamma_{4e} \sum_{i=0}^{\infty} h(i)h(i+\tau_1)h(i+\tau_2)h(i+\tau_3) \qquad (13.61b)$$

$$B_x(\omega_1, \omega_2) = \gamma_{3e} H(\omega_1)H(\omega_2)H(-\omega_1-\omega_2) \qquad (13.62)$$

$$T_x(\omega_1, \omega_2, \omega_3) = \gamma_{4e} H(\omega_1)H(\omega_2)H(\omega_3)H(-\omega_1-\omega_2-\omega_3) \qquad (13.63)$$

In higher-order statistical analysis toolbox in MATLAB, cumtrue.m is the function of cumulants estimation, and bispect.m is the function of parametric estimation for bispectrum.

The process of higher-order spectral parameter estimation can be summarized as follows:

Step 1: Estimate the kth -order cumulant from a known sequence $\{x(n)\}$, typically $k \leq 4$.

Step 2: Establish an equation about the relation between the kth -order cumulant and the signal model, and then solve this equation to get the parameters of the model.

Step 3: Calculate the kth -order spectrum of the signal $\{x(n)\}$ by BBR formula.

In the aforementioned steps, the main problem is how to perform the second step for estimating model parameters. Different model corresponds to different method.

(1) MA model (equivalent to finite impulse response (FIR) system)

Nonminimum phase FIR systems can be identified by using HOS. The FIR system model is equivalent to a stationary non-Gaussian MA(q) process. Its model is

$$x(n) = \sum_{j=0}^{q} b(j)e(n-j) \qquad (13.64)$$

where q is known, $h(j) = b(j)$, $b(0) = 1$, $b(q) \neq 0$. $e(n)$ is an i.i.d. non-Gaussian white process with zero mean. Its kth -order cumulant is γ_{ke} and $E[e^2(n)] = \sigma_e^2 < \infty$. Then the system function is

$$H(z) = \sum_{j=0}^{q} b(i)z^{-j} \qquad (13.65)$$

It can be proved that the kth -order cumulant of the system output $\{x(n)\}$ is

$$C_{kx}(\tau_1, \ldots, \tau_{k-1}) = \gamma_{ke} \sum_{j=0}^{q} b(j)b(j+\tau_1) \ldots b(j+\tau_{k-1}) \tag{13.66}$$

Particularly,

$$C_{3x}(\tau_1, \tau_2) = \gamma_{3e} \sum_{j=0}^{q} b(j)b(j+\tau_1)b(j+\tau_2) \tag{13.67a}$$

$$C_{4x}(\tau_1, \tau_2, \tau_3) = \gamma_{4e} \sum_{j=0}^{q} b(j)b(j+\tau_1)b(j+\tau_2)b(j+\tau_3) \tag{13.67b}$$

Since $b(j) = 0, (j > q, j < 0)$, once any τ_m of $\tau_1, \cdots, \tau_{k-1}$ meets $|\tau_m| > q$, then $C_{kx}(\tau_1, \ldots, \tau_{k-1}) = 0$.

Figure 13.3 shows the regions of $c_{3x}(\tau_1, \tau_2) \neq 0$ of the q th -order MA system.

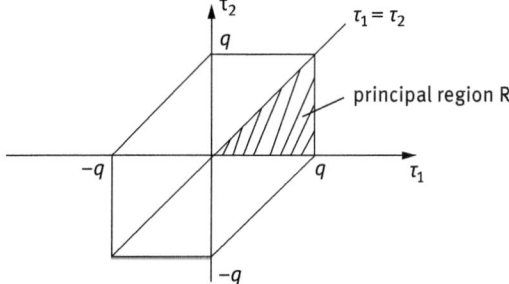

Figure 13.3: $C_{3x}(\tau_1, \tau_2) \neq 0$ region of the qth -order MA system.

The methods of identifying MA system with higher-order cumulants can be divided into three categories: closed-form solution, linear algebra, and nonlinear optimization solution. Here only linear algebraic solution will be introduced. The sensitivity to the sample cumulant estimation error is reduced via this method. The first popular method is Grouped Move (GM) algorithm, which is demonstrated by the following GM equations:

$$\sum_{i=0}^{q} b^2(i)R_x(m-i) = (\frac{\sigma_e^2}{\gamma_{3e}}) \sum_{i=0}^{q} b(i)c_{3x}(m-i, m-i), \quad -q \leq m \leq 2q \tag{13.68}$$

where $b^2(1), \ldots, b^2(q), \frac{\sigma_e^2}{\gamma_{3e}}, \frac{\sigma_e^2}{\gamma_{3e}}b(1), \ldots, \frac{\sigma_e^2}{\gamma_{3e}}b(q)$ are $2q + 1$ independent variables, and $R_x(m)$ is the correlation function. The linear least-squares method is used to solve the equations, and then the solution of $2q + 1$ variables are obtained. Consequently, the estimation of $b(1), \cdots, b(q)$ can be obtained. For the fourth-order cumulant, we can write

$$\sum_{i=0}^{q} b^3(i) R_x(m-i) = (\frac{\sigma_e^2}{\gamma_{4e}}) \sum_{i=0}^{q} b(i) c_{4x}(m-i, m-i, m-i), \ -q \leq m \leq 2q \qquad (13.69)$$

So we can use the linear least square method to obtain the estimation of $b(1), \cdots, b(q)$.

We need to be pay attention to some of the problems.

First, it is supposed that $b^2(i)$ is independent of $b(i)$ but it is not really true. So, the estimation of this method can only be suboptimal. Second, the system may not have a unique solution because the rank of the matrix in the matrix equations may not be equal to $2q+1$. Third, the algorithm is only applicable to the special case in which the additive noise is absent, that is, $R_y(\tau) = R_x(\tau)$. If the additive white noise exists, then $R_y(\tau) = R_x(\tau) + \sigma_e^2 \delta(m)$, which means that $2q$ equations corresponding to $-q \leq m \leq -1$ and $q+1 \leq m \leq 2q$ in GM equations are not affected by the noise (because these equations do not have $R_x(0)$). To eliminate the influence of noise, m cannot contain $0, 1, \ldots, q$. This leads to

$$\sum_{i=0}^{q} b^2(i) R_y(m-i) = (\frac{\sigma_e^2}{\gamma_{3e}}) \sum_{i=0}^{q} b(i) c_{3y}(m-i, m-i), \ -q \leq m \leq -1, q+1 \leq m \leq 2q$$

$$(13.70)$$

It can also be rewritten as

$$\sum_{i=1}^{q} b(i) c_{3y}(m-i, m-i) - \sum_{i=0}^{q} [\varepsilon b^2(i)] R_y(m-i) = -c_{3y}(m, m) \qquad (13.71)$$

where $-q \leq m \leq -1, q+1 \leq m \leq 2q, \varepsilon = \frac{\gamma_{3e}}{\sigma_e^2}$.

For the fourth-order cumulant, we can write

$$\sum_{i=1}^{q} b(i) c_{4y}(m-i, m-i, m-i) - \sum_{i=0}^{q} [\varepsilon b^3(i)] R_y(m-i) = -c_{4y}(m, m, m),$$
$$-q \leq m \leq -1, q+1 \leq m \leq 2q \qquad (13.72)$$

After removing the equation containing $R_x(0)$, only $2q$ equations are left, which are with $2q+1$ unknown parameters. In this case, the conventional methods are invalid.

Another outstanding method is Tugnait algorithm whose purpose is to apply the GM algorithm to the additive white noise case. On the basis of BBR formula, Tugnait proved the following equation:

$$\sum_{i=1}^{q} b(i)c_{3y}(i-m, q) - [\varepsilon b(q)]R_y(m) = -c_{3y}(-m, q), \quad 1 \le m \le q \tag{13.73}$$

$$\sum_{j=1}^{q} b(i)c_{4y}(j-m, q, 0) - [\varepsilon b(q)]R_y(m) = -c_{4y}(-m, q, 0), \quad 1 \le m \le q \tag{13.74}$$

Both eqs. (13.71) and (13.73) form a system of simultaneous equations, and similarly both eqs. (13.72) and (13.74) form another system of simultaneous equations. Thus, we can obtain $4q$ equations with $2q + 1$ unknown parameters so that the estimation of $b(1), \ldots, b(q), \varepsilon b(q)$ and $\varepsilon b^2(1), \ldots, \varepsilon b^2(q)$ and $\varepsilon b^3(1), \ldots b^3(q)$ is obtained.

$b(k)$ and $b^2(k)$ can be denoted as $b_1(k)$ and $b_2(k)$, respectively. If all the estimated $b_2(k)$ are nonnegative, then the final MA parameter is estimated as

$$\hat{b}(k) = sign[b_1(k)] \times \sqrt{0.5[b_1^2(k) + b_2(k)]} \tag{13.75}$$

otherwise $\hat{b}(k) = b_1(k)$.

Let the estimation of $b(k)$ and $b^3(k)$ be denoted by $b_1(k)$ and $b_3(k)$, respectively. If all the estimation of $b_3(k)$ and the corresponding $b_1(k)$ have the same negative and positive sign, the final parameter estimation of MA is

$$\hat{b}(k) = sign[b_1(k)] \times \frac{[|b_1(k)| + |b_3(k)|^{1/3}]}{2} \tag{13.76}$$

Otherwise, $\hat{b}(k) = b_1(k)$.

In higher order statistical analysis toolbox in MATLAB, the aforementioned method is corresponding to the function of maest.m.

(2)   ARMA System

Consider the model as

$$\sum_{i=0}^{p} a(i)x(n-i) = \sum_{i=0}^{q} b(i)e(n-i) \tag{13.77}$$

And the random process $\{x(n)\}$ of ARMA(p, q) is observed in the additive noise $v(n)$, that is, $y(n) = x(n) + v(n)$, and $a(0) = b(0) = h(0) = 1$, $b(q) \ne 0$.

Generally, some assumptions must be made for ARMA model:

C1: The system transfer function $H(z) = B(z)/A(z) = \sum_{i=0}^{\infty} h(i)z^{-i}$ does not have any zero and pole cancellation. This condition means that the system is causal and ARMA(p, q) cannot be further simplified.

C2: The input $e(n)$ is a non-Gaussian white noise i.i.d. processes with zero mean, and $E[e^2(n)] = \sigma_e^2 < \infty$, the kth -order cumulant is γ_{ke}.

C3: The observed noise $v(n)$ is Gaussian colored, and $v(n)$ and $e(n)$ are independent of each other.

Under the aforementioned three conditions, if $c_{kx}(\tau_1, \tau_2) = c_{kx}(\tau_1, \tau_2, 0, \ldots, 0)$, BBR formula changes into

$$\sum_{i=0}^{p} a(i)c_{kx}(m-i, n) = \gamma_{ke} \sum_{j=0}^{\infty} h^{k-2}(j)h(j+n) \sum_{i=0}^{p} b(i)h(j+m-i)$$

(13.78)

$$= \gamma_{ke} \sum_{j=0}^{\infty} h^{k-2}(j)h(j+n)b(j+m)$$

In eq. (13.78), $\sum_{i=0}^{p} a(i)h(n-i) = \sum_{j=0}^{q} b(i)\delta(n-i) = b(n)$ is applied.

Obviously, the AR system is a special case of ARMA system when $q=0$. Similar to ARMA system identification method based on autocorrelation function, the first step of ARMA identification method based on higher-order cumulant is to identify the parameter of AR, and then identify the parameter of MA part.

It can be proved that AR parameters of ARMA model that is depicted in eq. (13.79) can be uniquely identified by the least-squares solution under conditions C1–C3.

$$\sum_{i=0}^{p} a(i)c_{ky}(m-i, n) = 0, m = q+1, \ldots, q+p, n = q-p, \ldots, q$$

(13.79)

Owing to condition C3, $c_{ky}(\tau_1, \tau_2, \ldots, \tau_{k-1}) = c_{kx}(\tau_1, \tau_2, \ldots, \tau_{k-1})$, eq. (13.79) can be written as a matrix form:

$$Ca = c$$

(13.80)

Where $c = \begin{bmatrix} c_{kx}(q+1, q-p) \\ \vdots \\ c_{kx}(q+1, q) \\ \vdots \\ c_{kx}(q+p, q-p) \\ \vdots \\ c_{kx}(q+p, q) \end{bmatrix}$, $C = \begin{bmatrix} c_{kx}(q+1-p, q-p) & \cdots & c_{kx}(q, q-p) \\ \vdots & \vdots & \vdots \\ c_{kx}(q+1-p, q) & \cdots & c_{kx}(q, q) \\ \vdots & \vdots & \vdots \\ c_{kx}(q, q-p) & \cdots & c_{kx}(q+1-p, q-p) \\ \vdots & \vdots & \vdots \\ c_{kx}(q, q) & \cdots & c_{kx}(q+1-p, q) \end{bmatrix}$,

$$a = \begin{bmatrix} a(p) \\ a(p-1) \\ \vdots \\ a(1) \end{bmatrix}.$$

More directly, the cumulants of the AR process with noise satisfy:

$$\sum_{i=0}^{p} a(i)c_{2y}(m-i) = 0, \quad m > p \tag{13.81a}$$

$$\sum_{i=0}^{p} a(i)c_{3y}(m-i,\rho) = 0, \quad m > p \tag{13.81b}$$

$$\sum_{i=0}^{p} a(i)c_{3y}(m-i,\rho,\tau) = 0, \quad m > p \tag{13.81c}$$

The function of estimating the AR parameter in the higher-order statistical analysis toolbox is arrcest.m. The order of cumulant can be selected as 2, 3, or 4. AR parameter identified by this method has estimation bias. The bias is shown as an additive non-Gaussian noise. The residual time series $z(n)$ can be obtained by the following equation:

$$z(n) = \sum_{i=0}^{p} a(i)y(n-i) = \sum_{i=0}^{q} b(i)x(n-i) + \sum_{i=0}^{p} a(i)v(n-i) \tag{13.82}$$

The function of estimating ARMA parameters by using residual time series in the higher-order statistical analysis toolbox is armarts.m. It uses the normal equation to estimate AR parameter, and then the residual sequence of AR process is calculated and uses function maest.m to estimate MA parameter at last.

Another function to estimate ARMA parameters is armaqs.m. It is based on slices of q. First, AR parameters are estimated by using the normal equation, and then the impulse response can be estimated by

$$h(n) = \frac{\sum_{i=0}^{p} a(i)c_{3y}(q-i,n)}{\sum_{i=0}^{p} a(i)c_{3y}(q-i,0)}, \quad n = 1, \ldots, q \tag{13.83}$$

or

$$h(n) = \frac{\sum_{i=0}^{p} a(i)c_{4y}(q-i,n,0)}{\sum_{i=0}^{p} a(i)c_{4y}(q-i,0,0)}, \quad n = 1, \ldots, q \tag{13.84}$$

The final MA parameter can be estimated as

$$b(n) = \sum_{i=0}^{p} a(i)h(n-i) \tag{13.85}$$

For system identification, it is important to determine the order of the system. That is the problem of order determination. A number of methods have been proposed by using higher-order cumulant to solve this problem. The related functions in the higher-order statistical analysis toolbox in MATLAB are arorder.m and maorder.m.

(3) Examples
Estimating the model coefficients of the AR (2) process.

```
%Gaussian input
u=0.5*randn(1,1024);
%process of AR (2)
sig=filter([1],[1,-0.5,0.5],u); sig=sig-mean(sig);
%Setting parameters
norder=2; p=2; q=0;
%Estimated parameters
a=arrcest(sig,p,q,norder)
%End
```

13.2 Application of HOS for the Classification of Lung Sounds

Since antiquity, physicians have auscultated sounds inside the chest to identify signs of disease. Lung sound auscultation by stethoscope is a simple, quick, and noninvasive method to provide diagnostic information about a patient's lung. However, proper use of this method depends on the ability of the physician to recognize normal and abnormal sounds generated by the human body. In addition, lung sounds are nonstationary signals, which make them both difficult to analyze and hard to distinguish when using traditional auscultation methods. Thus, the use of the electronic stethoscope together with a pattern recognition system helps to overcome the limitations of traditional auscultation, providing an efficient method for clinical diagnosis. Lung sounds are classified as either normal (healthy individuals) or adventitious (abnormal). Adventitious sounds are divided into two categories: continuous sounds (wheezes and rhonchi) and discontinuous sounds (crackles).

Wheezing sounds are defined as high-pitched continuous sounds and rhonchi are low-pitched continuous sounds. It is specified that a wheeze contains a dominant frequency above 400 Hz, while rhonchi are characterized by a dominant frequency of about 200 Hz or less. The wheeze can be monophonic if it contains a single frequency, or polyphonic when several frequencies are simultaneously perceived. The diseases associated with wheezing sounds are asthma, pneumonia, and bronchitis. Crackles are discontinuous adventitious sounds caused by the sudden opening of small airways. Crackles are usually defined by their time-domain features such as initial deflection width and two-cycle duration. According to the ATS (American Thoracic Society), the average duration of initial deflection width and two-cycle duration of fine crackles are 0.7 and 5 ms, respectively. Those of coarse crackles are 1.5 and 10 ms, respectively. Fine crackles are heard in patients with pneumonia, pulmonary fibrosis, and congestive heart failure.

The application of HOS for lung sound classification used the parametric approach of bispectrum estimation (the double Fourier transform of the third-order moment sequence) to reveal information about lung sounds, for example, deviations from normality. Figure 13.4 shows the whole diagram for lung sound classification method based on HOS.

In Figure 13.4, the HOS is used as its feature extraction tool. Methods based on HOS are more suitable to deal with non-Gaussian processes and nonlinear systems. The major advantage of using HOS as a feature extraction method for pattern recognition is its noise immunity.

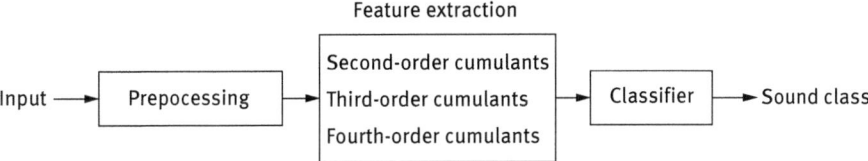

Figure 13.4: The diagram of lung sound classification method based on HOS.

Figure 13.5 illustrates the mean values (±2 standard deviations) of the second-, third-, and fourth-order cumulants extracted from the available vesicular and adventitious events. It can be inferred that each class has a different signature pattern according to their cumulants, showing that HOS is a promising feature extraction tool for lung sounds. It is clear from Figure 13.5 that although the extracted features of the lung sounds have different means, their large standard deviations can make classification difficult. The classification task is even harder when the objective is to distinguish between coarse and fine crackle or between monophonic and polyphonic wheeze.

13.3 Alpha-Stable Distribution and Fractional Lower-Order Statistics

13.3.1 Introduction

In virtually every engineering area, a significant component of the work that goes on is concerned with the measurement, interpretation, or processing of signals or noises. Because of the simplicity and closed form of pdf of the Gaussian process, it

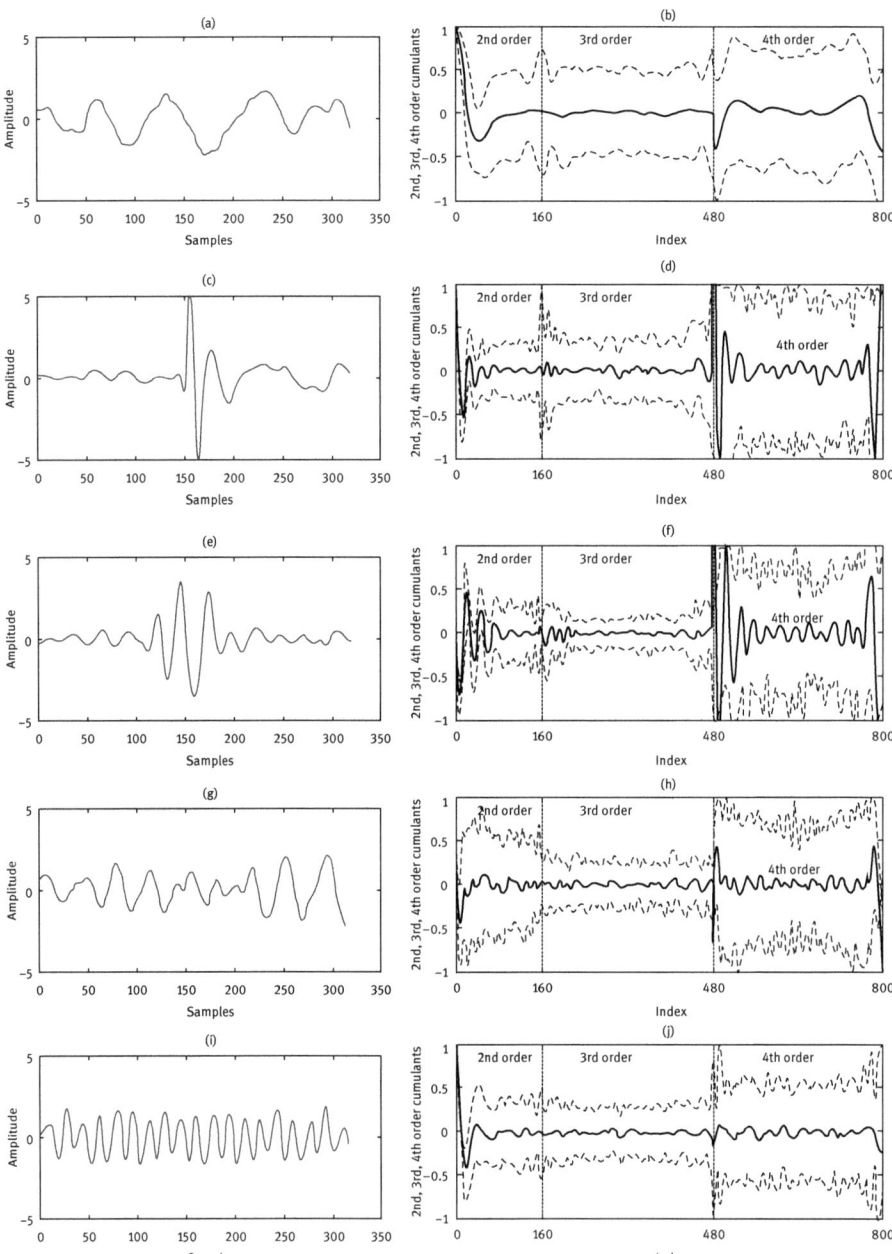

Figure 13.5: Lung sound events and their mean values (solid lines), third-, and fourth-order cumulants (dashed lines) (a) and (b) vesicular; (c) and (d) fine crackle; (e) and (f) coarse crackle; (g) and (h) monophonic wheeze; and (i) and (j) polyphonic wheeze.

is possible to design and implement optimal or nearly optimally statistical algorithms with low complexity. This makes the additive white Gaussian noise assumption very attractive in many fields, such as communication system, filtering, parameter estimation detection, system identification, and so on.

Nevertheless, measurements of the ambient noise in many physical applications show a non-Gaussian behavior. Experimental measurements in outdoor and indoor mobile radio channels, as well as digital subscriber loop systems, show the impulsive nature of the ambient noise due to man-made electromagnetic interference as well as natural noise. Car ignition and electromechanical switching are also shown to produce impulsive interference in a wide frequency band. Moreover, multiple access interference in some wireless and ultra-wideband networks is impulsive. Impulsive noise also appears in acoustic underwater channels, power line channels, radar/sea clutter, digital video broadcasting, electroencephalogram analysis, and financial market.

According to the theory of signal processing, to deeply analyze and discuss the performance of the algorithm, it is usually necessary to model the signal and noise based on a certain probability distribution theory. When the algorithm based on the ideal Gaussian model does not get the optimal solution in the non-Gaussian environment, it is necessary to carry out new theoretical development and analysis based on a more effective statistical model. The newly adopted model can not only describe the nature of signals and noises accurately but also make calculation and analysis simple. It has been shown that the non-Gaussian noise is also called impulsive noise, which is characterized by sharp, large-amplitude spikes that occur more frequently than in Gaussian noise. As will be explained later, the pdf of this noise decays in a tailless rapidly than the tail of the Gaussian pdf does. Therefore, the non-Gaussian noise or signal is best described by a heavy-tailed statistical distribution.

There are many heavy-tailed distributions that have been used to model non-Gaussian noise such as the Gaussian mixture, the generalized Gaussian distribution, the generalized Cauchy distribution, and the alpha-stable distribution. Multiple perspectives of the aforementioned four distributions are compared in Table 13.1. We can see the following: (1) The Gaussian mixture, the generalized Gaussian distribution, and the generalized Cauchy distribution have closed expressions of the pdf, which is relatively convenient and simple in the mathematical analysis of the algorithm. But the mixed Gaussian distribution and the generalized Gaussian distribution, as well as Gaussian distribution, are with exponential tail, rather than algebraic tail, which do not meet the actual situation in some cases, and the range of impulsive noise described is also limited. (2) By the controlling of the tail coefficient $p \in (0, 2]$, the generalized Cauchy distribution achieves the far wider description range of the impulsive noise, compared with the mixed Gaussian distribution and the Generalized Gaussian distribution. When $p = 2$, the generalized Cauchy distribution is transferred into the Cauchy distribution, which is also one of the special cases

Table 13.1: Comparison of several typical non-Gaussian distributions.

Distribution type	Closed expression of pdf	Characteristics	Tail
Gaussian mixture distribution	$f(x) = (1-\varepsilon)f_G(x;0;k\sigma^2) + \varepsilon f_G(x;0;\sigma^2)$ $f_G(x;0;\sigma^2)$ is the Gaussian pdf with zero mean and variance σ^2, $0 < \varepsilon < 1$, $0 < k < 1$	(1) mathematical description and analysis are simple;(2) the Gaussian distribution is its special case; (3) given the signal-to-noise ratio, the performance of the receiver may increase as ε increases	Exponential tail
Generalized Gaussian distribution	$f(x) = \frac{ka}{2\Gamma(1/k)} e^{-(a|x-\theta|)^k}$, $\Gamma(\bullet)$ is the gamma function, θ is the location parameter, a is the scale parameter (real; positive); k is the shape parameter (real; positive), controls the tail decay	(1) Mathematical description and analysis are simple; (2) the Gaussian distribution is its special case; (3) when $p < 1$, the pdf is not a bell shape but a sharp shape	Exponential tail
Generalized Cauchy distribution	$f(x) = a\sigma(\sigma^p + |x-\theta|^p)^{-2/p}$, $a = p\Gamma(2/p)/2(\Gamma(1/p))^2$, $\Gamma(\bullet)$ is the gamma function; p is the tail constant; θ is location parameter (real); σ is the scale parameter	(1) The Cauchy distribution and the Meridian distribution are its special cases; (2) the Cauchy distribution is not its special case	Algebraic tail
Alpha-stable distribution	None	(1) Satisfying the stability and generalized central limit theorem; (2) providing flexibility in impulsiveness modeling, with distributions ranging from light-tailed Gaussian to extremely impulsive; (3) the Gaussian distribution and the Cauchy distribution are its special cases	Algebraic tail

of alpha-stable distribution. It is worth noting that the Gaussian distribution is not the special case of the generalized Cauchy distribution. (3) The alpha-stable distribution has the advantage of being defined by a parameter $0 < \alpha \leq 2$ called the characteristic exponent, which controls the thickness of the pdf tails. When $\alpha = 2$, it is the Gaussian distribution. If $0 < \alpha < 2$, it is called the fractional lower order alpha-stable distribution. It is a typical heavy-tail distribution with algebraic tail, which makes it be very consistent with the actual data. And more importantly, it is the only non-Gaussian distribution model that satisfies the stability and generalized central limit theorem, which is more extensive and representative.

It has been verified that noise in the telephone line, atmospheric noise, and backscatter echoes in radar systems can be effectively described with the alpha-stable distribution, which can also be used to model economic time series.

13.3.2 General Description

For alpha-stable random variables, there is no explicit expression for the pdf in a general case. The most used characterization of stable variables, especially in practice, is based on the definition of the characteristic function that is nothing but the Fourier transform of its pdf.

A random variable X has an alpha-stable distribution if and only if there exist four unique parameters: $0 < \alpha \le 2$, $\gamma \ge 0$, $-1 \le \beta \le 1$, and a real a, such that the characteristic function of X has the following form:

$$\phi(t) = \exp(jat - \gamma|t|^{\alpha}[1 + j\beta sgn(t)w(t, \alpha)]) \tag{13.86}$$

$$\text{With } w(t, \alpha) = \begin{cases} \tan(\alpha\pi/2), & \alpha \neq 1 \\ (2/\pi)\log|t|, & \alpha = 1 \end{cases}, \quad \text{sign}(t) = \begin{cases} 1, & t > 0 \\ 0, & t = 0 \\ -1, & t < 0 \end{cases}$$

We denote an alpha-stable random variable as $X \sim S_\alpha(\gamma, a, \beta)$, where $\sim$ means "is distributed as." It is important to know the meaning of each parameter as well as its influence on the shape of the density or the distribution.

(1) $\alpha \in (0, 2]$ is called characteristic exponent and controls the tail thickness. For special case $\alpha = 1$, the distribution is Cauchy, and for $\alpha = 2$, the distribution is Gaussian, with dispersion equal to one-half of the variance. As α decreases from 2, the tails of the pdf decay algebraically rather than exponentially, and hence, there is a greater probability of the noise with significant amplitude. This implies that, when α is smaller, the realizations of an alpha-stable random variable become more impulsive and more variable.

(2) a is called position parameter and is analogous to the mean of the distribution. For $\alpha > 1$, a becomes the mean of the random variables. For $0 < \alpha < 1$, a represents the middle value.

(3) $\gamma > 0$, is called dispersion and is analogous to the variance. For Gaussian condition, it equals to half of the variance.

(4) $-1 < \beta < 1$ is called symmetry parameter and measures the amount of probability mass that is distributed to the left($\beta > 0$) or right($\beta < 0$) of the location parameter. $\beta = 0$ corresponds to a distribution that is symmetric around a, in which case the distribution is called symmetric α-stable (abbreviated as $S\alpha S$). An $S\alpha S$ is called standard distribution if $\gamma = 1$. The $S\alpha S$ pdfs present several similarities to the Gaussian pdf: they are smooth, single peak distribution, symmetry about middle value, bell shape, and satisfy the stability property. However, they also differ from the Gaussian pdf in several significant ways. For example, the $S\alpha S$ pdfs have sharper

maxima than the Gaussian pdf and algebraic (inverse power) tails in contrast to the exponential tails of the Gaussian pdf.

If $a = 0$, then the characteristic function in eq. (13.86) reduces to

$$\phi(t) = \exp(-\gamma|t|^{\alpha}) \tag{13.87}$$

The probability density function for the *SαS* case is given by the inversion formula of the characteristic function as

$$f_X(x) = \frac{1}{\pi} \int_0^{\infty} e^{\gamma t^{\alpha}} \cos(tx)dt \tag{13.88}$$

Close-form pdfs exist only for the following cases,

$$f_X(x) = \frac{\gamma}{\pi(x^2 + \gamma^2)} \quad for \ \alpha = 1 \ (Cauchy) \tag{13.89a}$$

$$f_X(x) = \frac{1}{2\sqrt{\pi\gamma}} e^{-\frac{x^2}{4\gamma}} \quad for \ \alpha = 2 \ (Gaussian) \tag{13.89b}$$

Figure 13.6 shows a plot of the pdfs for various values of *SαS* pdf with the different characteristic exponent α. As α decreases, the tails become heavier with respect to the Gaussian tail ($\alpha = 2$).

13.3.3 Fractional Lower- Order Statistics

Definition 13.3 (Fractional lower-order moment): The second-order moments exist only for Gaussian distribution since the variance is finite when $\alpha = 2$. However, the fractional lower-order moment p does exist for alpha-stable distribution with $\alpha < 2$ for the order of $p < \alpha$.

The fractional lower-order moment (FLOM) of *SαS* random variable $X \sim SαS(\gamma, a, 0)$ with zero location parameter ($\alpha = 2$) is given by

$$E[|X|^p] = C(p, \alpha)\gamma^{p/\alpha}, \quad 0 < p < \alpha \tag{13.90}$$

where $C(p, \alpha) = \frac{2^{p+1}\Gamma(\frac{p+1}{2})\Gamma(-p/\alpha)}{\alpha\sqrt{\pi}\Gamma(-p/2)}$, $\Gamma(\cdot)$ is the gamma function defined as $\Gamma(s) = \int_0^{\infty} x^{s-1}e^{-x}dx$.

Definition 13.4 (Minimum dispersion (MD) criterion): For second-order processes, the most commonly used criterion in estimation problems is the minimum mean square error criterion. For stable processes, the minimum mean square error criterion is no longer appropriate because there is no finite variance, and therefore criteria based on

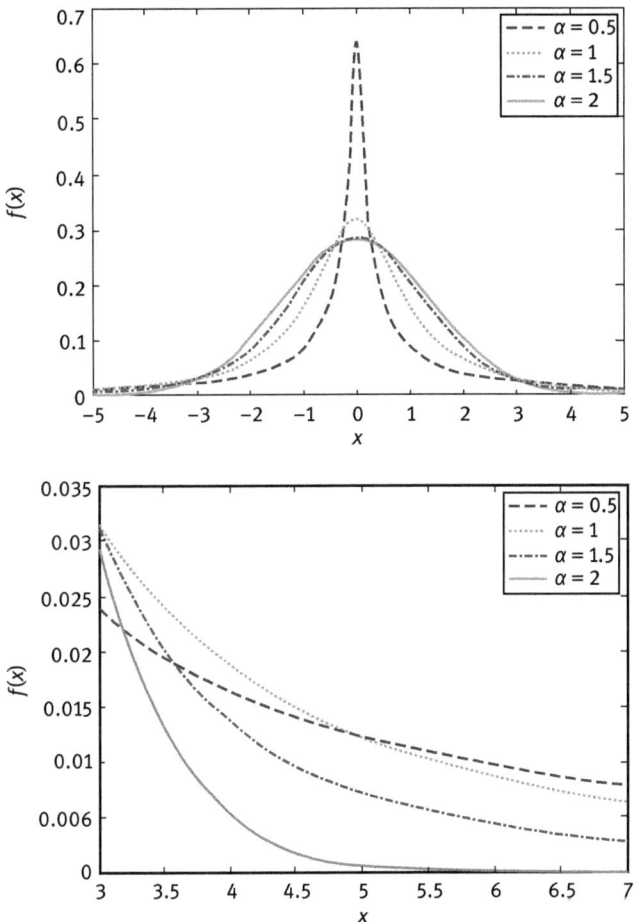

Figure 13.6: Plots of *SαS* pdf and its tails for various values of α (a) *SαS* pdf and (b) *SαS* pdf tails.

FLOM are employed. Alpha-stable signal processing methods based on FLOM introduce nonlinearity even into linear problems. This is because the linear space of
alpha-stable processes is a Banach space when $1 \leq \alpha < 2$ and a metric space when
$0 < \alpha < 1$. Estimation problems in Banach or metric spaces cannot be solved in the
same way as those in Hilbert spaces associated with Gaussian processes. The most
common criterion for solving estimation problems in symmetric stable processes is
the MD. Under the MD criterion, the linear estimation problem is formulated as
follows: assuming that the random variable X is *SαS*, find an element $\hat{X}$ in the linear
space $L(X(t), t \in T)$ such that

$$d_\alpha\left(X, \hat{X}\right) = E|X - \hat{X}|^\alpha = \inf_{Z \in L(X(t), t \in T)} E|X - Z|^\alpha \qquad (13.91)$$

$d_\alpha\left(X,\hat{X}\right)$ describes the distance between X and $\hat{X}$. According to the definition of FLOM, $d_\alpha\left(X,\hat{X}\right)$ is

$$d_\alpha(X, Y) = \begin{cases} \left[\frac{E(|X - Y|^p)}{C(p,\alpha)}\right]^{1/p}, & 0 < p < \alpha \quad 1 \leq \alpha \leq 2 \\ \left[\frac{E(|X - Y|^p)}{C(p,\alpha)}\right]^{\alpha/p}, & 0 < p < \alpha \quad 0 < \alpha < 1 \end{cases} \qquad (13.92)$$

Equation (13.92) shows that different p th-order moments are equivalent. For $1 < \alpha < 2$, the objective function in eq. (13.91) is convex and we can use Newton's method to obtain the best predictor. For $0 < \alpha < 1$, the objective function is concave, and we resort to global optimization algorithms, for example, simulated annealing. Although both methods are quite straightforward, they may present computational difficulties. Therefore, rather than using analytical means to find the optimum, employing an adaptive solution to eq. (13.91) is easier to implement. Limiting our interest to the case $1 < \alpha < 2$, in which the MD criterion is a convex function of the predictor parameters. The stochastic gradient method can be used to find the solution by minimizing the sample dispersion.

Definition 13.5 (Covariation): As described above, a non-Gaussian *SaS* random variable X, its p th-order moment $E[|X|^p]$ is finite only when $p < \alpha$. Covariation plays a role analogous to covariance. For two jointly *SaS* random variable X and Y with $1 < \alpha \leq 2$, the covariation is defined as

$$[X, Y]_\alpha = \int_S xy^{\langle \alpha - 1 \rangle} \mu(ds) \qquad (13.93)$$

where s is the unit circle, and $\mu(\cdot)$ represents the spectral measure of *SaS* random vector, (X, Y). The convention is $z^{\langle a \rangle} = \frac{|z|^{p+1}}{z}$. The commonly used estimator of the convention is the FLOM estimator:

$$[X, Y]_\alpha = \frac{E(XY^{\langle p - 1 \rangle})}{E(|Y|^p)} \gamma_y, \qquad 1 \leq p < \alpha \qquad (13.94)$$

where γ_y is the dispersion of *SaS* random variable Y. Specially,

$$[Y, Y]_\alpha = \|Y\|_\alpha^\alpha = \gamma_y \qquad (13.95)$$

In particular, if X and Y are jointly *SaS* with $\alpha = 2$ (Gaussian distribution), then $[X, Y]_\alpha = E(X, Y)$.

The difference between the covariation and the covariance is that there is no symmetry according to all α value except $\alpha = 2$ for the former. That is,

$$[X, Y]_\alpha \neq [Y, X]_\alpha, \quad 1 < \alpha < 2 \tag{13.96}$$

Definition 13.6 (Fractional lower-order covariance (FLOC)): By defining two arbitrary and possibly heavy-tailed random variables X and Y of zero location parameter and finite p th- order moment

$$E[|X|^p]\rangle\infty, \quad E[|Y|^p]\rangle\infty \tag{13.97}$$

For some fractional order p, $0 \leq p \leq 2$, the FLOC that is the FLOM-based correlation function is defined as

$$<X, Y>_{p_1, p_2} = E\left(X^{\langle p_1 \rangle} Y^{\langle p_2 \rangle}\right) \tag{13.98}$$

When the random variables X and Y are SaS, the parameters p_1 and p_2 should be chosen to satisfy $p_1 + p_2 \leq 2$ so that the expression in eq. (13.98) is finite. Choosing $p_1 = p_2 \leq \frac{\alpha}{2}$ is a good choice. Clearly, when $p_1 = p_2 = 1$, the aforementioned definition of FLOC is the usual correlation function between X and Y.

13.3.4 General Properties

Property 13.8 (The stability property): A random variable X is stable (or has an alpha-stable distribution) if and only if for any positive real numbers A and B, there exists a positive C and a real D such that the distribution of the random variable $AX_1 + BX_2$ is equal to that of $CX + D$, whenever X_1 and X_2 are independent random of X.

Property 13.9 (Generalized central limit theorem (GCLT)): X is the limit in the distribution of normalized sums of the form

$$S_n = \frac{X_1 + X_2 + \ldots + X_n}{a_n} - b_n$$

where X_1, X_2, $\ldots$, X_n are i.i.d. and $n \to \infty$, if and only if the distribution of X is stable, wherein a_n is a positive real number, and b_n is a real number.

For the special case where all random variables have finite variances, the sum possesses a Gaussian distribution, due to the well-known central limit theorem.

Property 13.10: A random variable X is stable if and only if for any integer $n \geq 2$ there exists a positive real C_n and a real D_n such as the probability distribution of the sum $X_1 + X_2 + \ldots + X_n$ is the same as that of $C_n X + D_n$. Whenever $X_1, X_2, \ldots, X_n$ are independent random copies of X.

It is classically known that the above three properties are equivalent.

Property 13.11: If $X \sim S_\alpha(\gamma, 0, \mu)$ and $\alpha \neq 2$, then

$$\lim_{t\to\infty} t^\alpha Pr(|X| > t) = \gamma C_\alpha \tag{13.99}$$

where C_α is a constant that depends only on α.

For this reason, the p th-order moments of the $S\alpha S$ pdfs are finite only for $0 < p < \alpha$ (except for the limiting case of $\alpha = 2$). As a result, for α strictly less than 2 (i.e., $0 < \alpha < 2$), alpha-stable random variables have infinite variance and more generally infinite moments for orders larger than α.

Property 13.12: Let X_1 and X_2 be two $S\alpha S$ variables of dispersions γ_1 and γ_2 and pdfs $f_1(.)$ and $f_2(.)$, respectively. Then, for $k \geq \alpha$, we have

$$\frac{E(|X_1|^k)}{E(|X_2|^k)} \overset{\Delta}{=} \lim_{T\to\infty} \frac{\int_{-T}^T |x|^k f_1(x)dx}{\int_{-T}^T |u|^k f_2(u)du} = \frac{\gamma_1}{\gamma_2} \tag{13.100}$$

Property 13.13: Let X be a $S\alpha S$ variable of dispersion γ and pdfs $f(.)$ and $f_2(.)$. Then, for $k > \alpha$, we have

$$\frac{(E|X|^k)^2}{E|X|^{2k}} \overset{\Delta}{=} \lim_{T\to\infty} \frac{(\int_{-T}^T |x|^k f(x)dx)^2}{\int_{-T}^T |x|^{2k} f(x)dx} = 0 \tag{13.101}$$

Property 13.14: Let X_1 and X_2 be two independent alpha-stable random variables, respectively, whose distributions are $S_\alpha(\gamma_1, \beta_1, a_1)$ and $S_\alpha(\gamma_2, \beta_2, a_2)$, and then the random variable $Y = X_1 + X_2$ is also an alpha-stable distribution as $S_\alpha(\gamma, \beta, a)$ with

$$\gamma^\alpha = \gamma_1^\alpha + \gamma_2^\alpha, \ \beta = \frac{\beta_1\gamma_1^\alpha + \beta_2\gamma_2^\alpha}{\gamma_1^\alpha + \gamma_2^\alpha}, \ a = a_1 + a_2 \tag{13.102}$$

Property 13.15 (Infinite variance): For any alpha-stable random variable with characteristic exponent α, all moments or order $p > \alpha$ are infinite, that is,

$$E[|X|^p] = \infty, \ p > \alpha \tag{13.103}$$

The FLOMs are the only moments ($p \leq \alpha$) that exist and are finite, that is,

$$E[|X|^p] < \infty, \ 0 \leq p < \alpha \tag{13.104}$$

Especially, when $\alpha = 2$,

$$E[|X|^p] < \infty, \ p \geq 0 \tag{13.105}$$

For the case, $0 < \alpha \leq 1$, the alpha-stable random variable does not have a definable mean. And for $0 < \alpha \leq 2$, the alpha-stable random variable's variance is infinite. As a consequence of an infinite variance, we cannot define a value to noise power if we are modeling impulsive noise as alpha-stable distribution. This means that the traditional signal-to-noise ratio (SNR) cannot be used. The generalized SNR (GSNR), defined as the ratio of the signal average power to the noise dispersion in the finite interval of interest, is

$$GSNR_{dB} = 10\log_{10}\left\{\frac{1}{\gamma M}\sum_{t=1}^{M}|s(t)|^2\right\} \tag{13.106}$$

As a result of Property 13.11, we can determine whether a given set of samples, $\{X_k\}$, is Gaussian or non-Gaussian alpha-stable distributed. To do this, the sample variance of this set should be computed.

Set $X_k(k=1, \ldots, N)$ to be samples of the normal distribution; the definition of sample variance, $\bar{\sigma}_s^2$, is

$$\bar{\sigma}_s^2 = \frac{1}{N}\sum_{k=1}^{N}(X_k - \bar{X}_N)^2 \tag{13.107}$$

where $\bar{X}_N = \frac{1}{N}\sum_{k=1}^{N} X_k$ is the mean.

If the sample variance converges to a value, then that value is the variance and the set of samples is Gaussian. If, however, the sample variance diverges, the sample set is non-Gaussian with infinite mean. Figure 13.7 shows the sample variance for $\alpha = 2$ and $\alpha = 1.5$. Note that for $\alpha = 2$, the sample variance converges to the expected variance, while it diverges for $\alpha = 1.5$.

In addition, some signal processing methods based on variance or second-order statistics (such as spectral analysis and least squares) would result in poor performance. And fractional lower-order statistics is a powerful tool to study it.

Property 13.16: If $X_-S_\alpha(\gamma, \beta, a)$ with $\alpha > 1$, then we have $E(X) = a$.

Property 13.17: Linearity exists for the first term: if X_1, X_2 and Y are jointly $S\alpha S$, then

$$[aX_1 + bX_2, Y]_\alpha = a[X_1, Y]_\alpha + b[X_2, Y]_\alpha \tag{13.108}$$

where a and b are any real numbers.

Property 13.18: When $\alpha = 2$, that is, to say, when X and Y are jointly distributed Gaussian random variants with zero mean, the covariation degenerates to the covariance:

$$[X, Y]_\alpha = E(XY) \tag{13.109}$$

Property 13.19: Pseudo-linearity exists for the second term: If Y_1 and Y_2 are independent and X, Y_1 and Y_2 are jointly $S\alpha S$ distributed, then

$$[X, aY_1 + bY_2]_\alpha = a^{<\alpha-1>}[X, Y_1]_\alpha + b^{<\alpha-1>}[X, Y_2]_\alpha \tag{13.110}$$

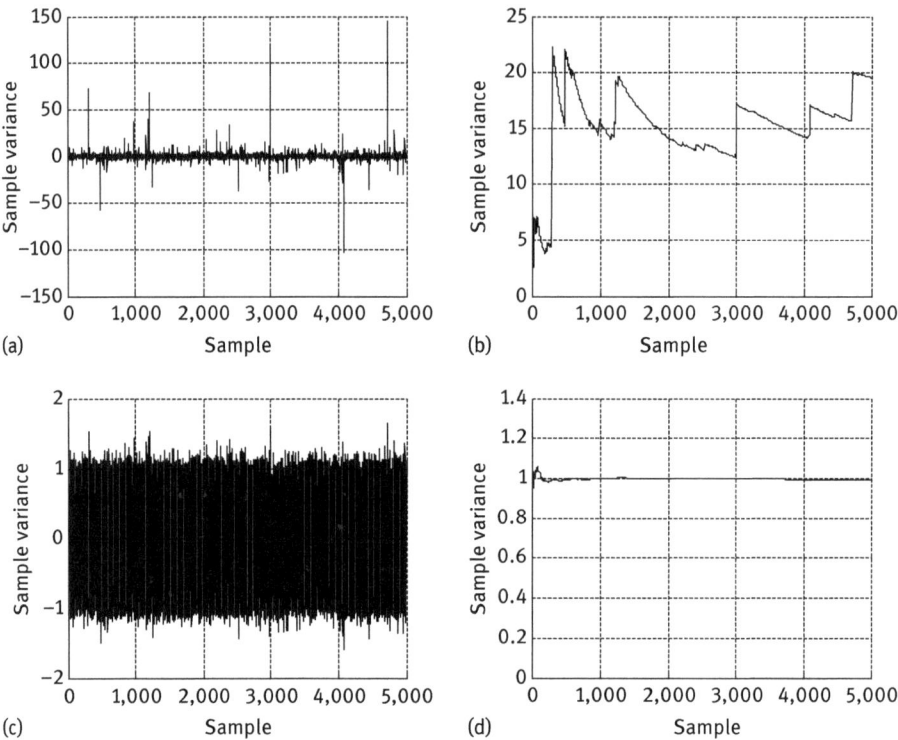

Figure 13.7: The sample variance: (a) non-Gaussian sample with $\alpha = 1.5$; (b) the sample variance for $\alpha = 1.5$; (c) non-Gaussian sample with $\alpha = 1.5$; and (d) the sample variance for $\alpha = 2$.

Property 13.20: If X and Y are independent, then

$$[X, Y]_\alpha = 0 \tag{13.111}$$

Property 13.21: If X and Y are jointly $S\alpha S$ distributed, then

$$|[X, Y]_\alpha| \leq \|X\|_\alpha \|Y\|_\alpha^{<\alpha - 1>} \tag{13.112}$$

If X and Y have unite dispersion coefficient, then

$$|[X, Y]_\alpha| \leq 1 \tag{13.113}$$

13.3.5 Characteristic Exponent Estimation of $S\alpha S$

The problem of estimation of parameters is one of the most important issues in mathematical statistics, motivated by the practical necessity to describe the probability law governing the experimental data. This problem has a simple solution in the case of the known analytical expression for the pdf. However, such situation does not take place for SaS. Only three alpha-stable distributions can be expressed explicitly in terms of elementary functions: the Levy distribution ($\alpha = 1/2, \beta = 1$), the Cauchy distribution ($\alpha = 1/2, \beta = 1$), and the normal distribution ($\alpha = 2, \beta = 1$). In other cases, the density distributions are expressed in terms of special functions. This leads to some difficulties in the construction of the estimators. Several estimators, compromising optimality for the sake of computational efficiency, have been proposed in the past. Among them, maximum likelihood methods are asymptotically efficient, but difficult to compute. The estimation of the stable parameters by fitting the Fourier transform of the data to the characteristic function is also a computationally intensive procedure. Hence, a number of suboptimal but simple methods have been devised. Let $X_1, \ldots, X_N$ be a sample of identically distributed random variables. The distribution of each random variable belongs to the domain of attraction of an SaS. The parameters are estimated from the sample $X_1, \ldots, X_N$. In this chapter, two suboptimal methods to estimate α and γ of SaS are summarized below.

1. Positive- and negative-order function estimator
The FLOM of an SaS variable with zero location parameter is shown in eq. (13.90), which gives rise to

$$E(|X|^p)E(|X|^{-p}) = \frac{2\tan(p\pi/2)}{\alpha\sin(p\pi/\alpha)} \tag{13.114}$$

where $0 < p < \min(\alpha, 1)$ and therefore

$$\sin c\left(\frac{p\pi}{\alpha}\right) = \frac{2\tan(p\pi/2)}{p\pi E(|X|^p)E(|X|^{-p})} \tag{13.115}$$

From eq. (13.115), α is given by

$$\alpha = p\pi\left\{\sin c^{-1}\left[\frac{2\tan(p\pi/2)}{p\pi E(|X|^p)E(|X|^{-p})}\right]\right\}^{-1} \tag{13.116}$$

Once the value of α is estimated from eq. (13.116), then γ can be estimated as

$$\gamma = \left(\frac{E(|X|^p)}{C_1(p, \alpha)}\right)^{\alpha/p} \tag{13.117}$$

2. Logarithm
Defining a variable Y as $Y = \log|X|$, it can be shown that the mean of Y is given by

$$E(Y) = C_e\left(\frac{1}{\alpha} - 1\right) + \frac{1}{\alpha}\log(\gamma) \tag{13.118}$$

where $C_e = 0.57721566\ldots$ is the Euler constant. The variance of Y is

$$\text{Var}(Y) = E\left\{[Y - E(Y)]^2\right\} = \frac{\pi^2}{6}\left(\frac{1}{\alpha^2} + \frac{1}{2}\right) \tag{13.119}$$

The estimation process involves solving eq. (13.119) for α and substituting it back into eq. (13.118) to find γ.

13.3.6 Multivariate Alpha-Stable Random Variables

Definition 13.7: A random vector $X = (X_1, \ldots, X_d)$ is said to be alpha-stable distributed in $\mathbb{R}^d$ if for any positive numbers A and B, there is a positive number C and a vector D in $\mathbb{R}^d$ such as $AX^{(1)} + BX^{(2)} \overset{d}{=} CX + D$, where $X^{(1)}$ and $X^{(2)}$ are independent copies of the vector X. It is strictly stable law.

Property 13.22: Let $X = (X_1, \ldots, X_d)$ be a stable random vector, then we have the following:
1. Any linear combination of the components of X of form $Y = \sum_{k=1}^{d} b_k X_k$ is an alpha-stable variable.
2. There is a unique $\alpha \in [0, 2]$ such that constants A, B, and C of definition 13.7 satisfy: $A^\alpha + B^\alpha = C^\alpha$.
3. A random vector $X = (X_1, \ldots, X_d)$ is alpha-stable in $\mathbb{R}^d$ with $0 < \alpha < 2$ if and only if for any $n \geq 2$ there is a vector D_n of $\mathbb{R}^d$ such as $X^{(1)} + \cdots + X^{(n)} \overset{d}{=} n^{\frac{1}{\alpha}}X + D_n$, where $X^{(1)} + \ldots + X^{(n)}$ are independent copies of X.

Proposition 13.23 Let X be a random vector in $\mathbb{R}^d$, then we have the following:
1. If any linear combination $Y = \sum_{k=1}^{d} b_k X_k$ is strictly stable, then the random vector X is also strictly stable.
2. If any linear combination Y is symmetric stable, then X is also symmetric stable.
3. If any linear combination Y is alpha-stable with a stability index greater or equal to 1, then the vector X is also alpha-stable.
4. If X is infinitely divisible and if any linear combination of Y is stable, then the vector X is also stable.

13.3.7 Generation of Random Samples

For the simulation of the various techniques presented, a way is needed to generate large numbers of alpha-stable distributed random numbers. The transformation formulas can be used to obtain alpha-stable random variables from uniform and exponential random variables.

1. Chambers–Mallows–Stuck method

1. Generate a random number U with uniform distribution on $(-\pi/2, \pi/2)$.
2. Generate another random variable W distributed with exponential distribution with mean 1.
3. If $\alpha \neq 1$, calculate $X = S_{\alpha,\beta} \times \dfrac{\sin(\alpha(U - B_{\alpha,\beta}))}{(\cos U)^{1/\alpha}} \times \left(\dfrac{\cos(U - \alpha(U - B_{\alpha,\beta}))}{W}\right)^{[1-\alpha]/\alpha}$, where

$B_{\alpha,\beta} = \dfrac{\arctan(\beta \tan\frac{\pi\alpha}{2})^{1/2\alpha}}{\alpha}$, $S_{\alpha,\beta} = (1 + \beta^2 \tan^2\frac{\pi\alpha}{2})^{1/2\alpha}$.

4. Else if $\alpha = 1$, compute $X = \dfrac{2}{\pi}\left[\left(\dfrac{\pi}{2} + \beta U\right)\right]U - \beta\log\left(\dfrac{W \cos U}{\pi/2 + \beta U}\right)$.
5. Random number X generated above is a standard random variable, that is, $X \sim S_\alpha(\beta, 1, 0)$. For nonstandard stable random variable $Y \sim S_\alpha(\beta, \gamma, \mu)$, modify X as follows:

$$Y = \begin{cases} \gamma X + \mu, & \alpha \neq 1 \\ \gamma X + \frac{2}{\pi}\beta\gamma\log\gamma + \mu, & \alpha = 1 \end{cases}$$

Here, a simple MATLAB code for the generation of stable random variables with arbitrary parameters is given.

```
Function Y=alpharnd(n,alpha,beta,gama,miu)
% … uniform random variable …
nuni=rand(1,n);
U=(nuni*pi)-pi/2;
% … exponential random variable …
zhi=rand(1,n);
W=-log(zhi);
if alpha~=1
    X1=S(alpha,beta);
    X2=sin(alpha*(U+B(alpha,beta)))./(cos(U)).^(1/alpha);
    X3=(cos(U-alpha*(U+B(alpha,beta)))./W).^(1/alpha-1);
    X=X1.*X2.*X3;
else
    X=(2/pi)*((pi/2+beta*U).*tan(U)-beta*log(0.5*pi*W.*cos(U)./(pi/2+
      beta*U)));
end
if alpha~=1
    Y=gama*X+miu;
else
    Y=gama*X+miu+(2/pi)*beta*gama*log(gama);
end
function y=B(alpha,beta)
y=atan(beta*tan(pi*alpha/2))/alpha;
function y=S(alpha,beta)
y=(1+beta^2*(tan(pi*alpha/2))^2)^(1/(2*alpha));
```

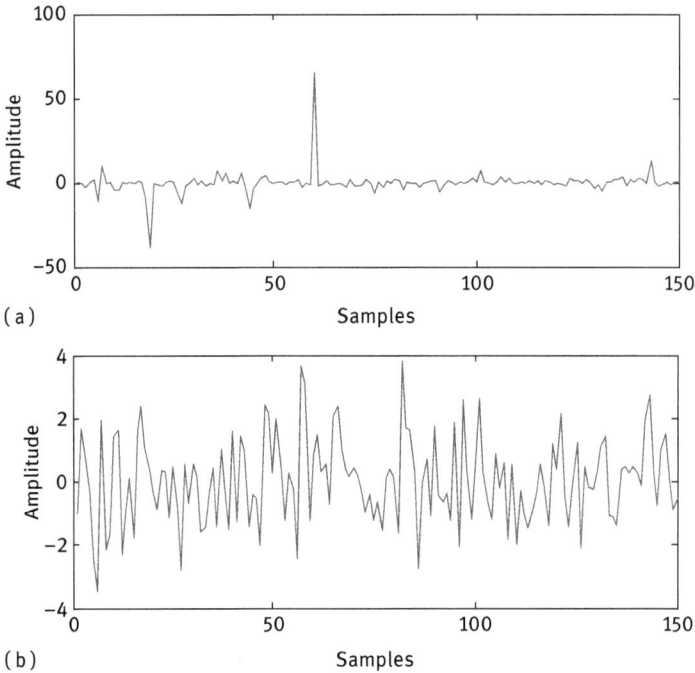

Figure 13.8: Generated samples with different characteristic exponents α :(a) $\alpha = 1.5$ and(b) $\alpha = 2$.

2. Deroye method

This method generates random samples according to SaS when $\alpha \neq 1$.

1. Generate a random variable U with uniform distribution on $(0, \pi)$.
2. Generate another random variable W with exponential distribution with mean 1.
3. Define another variable $G = \frac{\sin{(\alpha U)^{\frac{\alpha}{1-\alpha}}}}{\cos{(U)^{\frac{\alpha}{1-\alpha}}}} \cos[U(1-\alpha)]$.
4. The SaS stable distribution random sample is $X = \left(\frac{G}{W}\right)^{\frac{\alpha}{1-\alpha}}$.

Figure 13.8 shows signals with different values of characteristic exponent α.We can observe that the signal with a smaller α exhibits a heavy impulsive effect. This means that the degree of impulsiveness can be easily controlled by adjusting the parameter α. Many experiments show that most of the impulsive noise in the real applications can be described when $\alpha = 1.2\text{\_}1.6$.

13.4 Application of Fractional Lower-Order Statistics to Time Delay Estimation

Time delay refers to the time difference caused by different signal transmission distances between the different receivers in the receiver array. From the delay estimates, the time delay estimation (TDE) is useful in other important applications such as synchronization in wireless communications, echo cancellation, speech enhancement, and multimedia.

According to the difference between the target source and the detection system, the TDE problem can be divided into two types: active TDE and passive TDE. Radar and active sonar are examples of active TDE problem. A radar (or an active sonar) emits electromagnetic waves (or acoustic waves) to search targets, and when these signals meet the target, some of them return to radar (or sonar) receiving systems. According to the time delay between signal sending and receiving, the azimuth, distance, and velocity of the target can be determined. Unlike the active TDE, the passive TDE problem does not send out any signal initiatively, and it only uses the signal received from the target to search and locate the target. This method cannot control the magnitude of the received signal energy, but its concealment is strong, which is of great significance for military applications.

The basic model of TDE is the double bases model, as shown in Figure 13.9, where A and B are two receivers with a distance of L, and S is the target source. Let the received signals at A and B receivers are $x_1(t)$ and $x_2(t)$,then the signal model of TDE problem is

$$\begin{aligned} x_1(t) &= s(t) + v_1(t) \\ x_2(t) &= s(t-D) + v_2(t) \end{aligned} \tag{13.120}$$

wherein $s(t)$ is the received source signal, and D is the propagation time difference between the two receivers. $v_1(t)$ and $v_2(t)$ are superimposed noise. For the convenience of analysis, it is assumed that $s(t)$, $v_1(t)$, and $v_2(t)$ are normal stationary random processes, and they are uncorrelated. The discrete-time sensor outputs can be represented by

$$\begin{aligned} x_1(n) &= s(n) + v_1(n) \\ x_2(n) &= s(n-D) + v_2(n) \end{aligned} \tag{13.121}$$

In Figure 13.9, $\triangle AGB$ is a right triangle, so

$$BG = AB \cos \beta \tag{13.122}$$

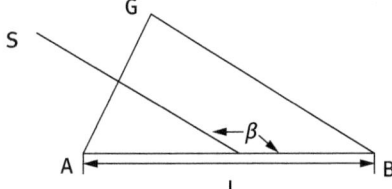

Figure 13.9: Double bases model.

Because $AB = L$, $BG = vD$, the azimuth angle of the target is

$$\beta = \arccos(-vD/L) \qquad (13.123)$$

From eq. (13.123), it is known that, as long as the accurate TDE is obtained, the location of the target source can be accurately determined.

Many TDE methods have been proposed. One commonly used approach to find the delay is to prefilter $x_1(n)$ and $x_2(n)$ and then perform cross correlation. An approximation of D is given by the time lag at which the cross-correlation function peaks. The purpose of prefiltering is to accentuate the signal passed to the correlator at those frequencies at which the SNR is highest. In some applications, the delay is time varying and adaptive techniques are required to keep track of it. When the time delay is approximated by multiples of sampling period, the simple adaptive methods have been derived and studied to track a time-varying delay. By modeling the delay as an FIR filter, the TDE problem can be transformed into one of estimating the coefficients of the FIR filter. Adaptive algorithms can then be applied directly to update the filter coefficients for obtaining a time shift. In particular, Widrow's least mean square algorithm is widely used and this method for delay estimation is usually referred to as least mean square TDE. Figure 13.10 is the diagram of adaptive TDE system.

A common deficiency of these two approaches is that they generally assume the disturbances in the sensor outputs to be Gaussian, although the noise components in practice often exhibit non-Gaussian properties. The traditional algorithms can work

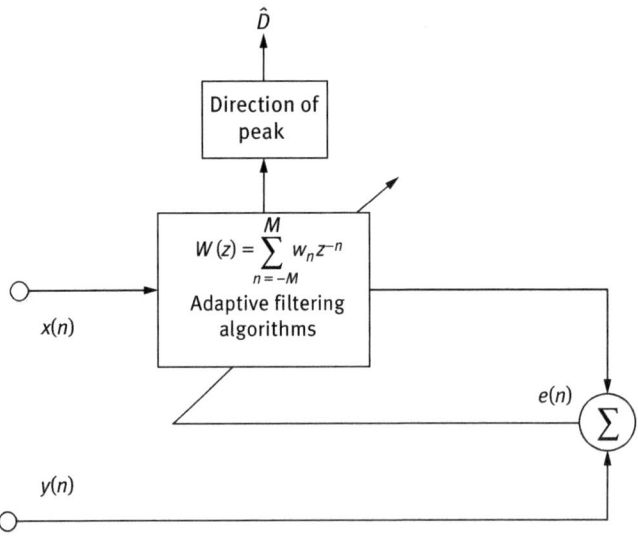

Figure 13.10: Adaptive TDE system.

normally in the condition of Gaussian noise. However, they will be a failure if the noise is alpha-stable distributed because all of them are based on the second-order statistics of signal and noise, while the secondary moment of alpha-stable distribution process does not exist.

To inhibit the peak impact oise more effectively, and guarantee the algorithm has good toughness in the alpha-stable distribution noise environment, according to the MD (minimum dispersion) criterion, LMP (Least Mean P-norm) algorithm is proposed.

Defining the cost function as

$$J_p(\mathbf{w}) \overset{\Delta}{=} E\{|x_2(n) - \sum_{m=-M}^{M} w(m)x_1(n+m)|^p\}, 1 < p < \alpha \tag{13.124}$$

Using the steepest descent method to get the minimum value of cost function, and then getting the adaptive iteration formula of LMP shown as

$$\mathbf{w}(n+1) = \mathbf{w}(n) - \mu p |e(n)| \mathbf{x}_1(n) \text{sgn}(e(n)) \tag{13.125}$$

So the estimation value of delay is

$$\hat{D} = \arg[\max_m(\mathbf{w})], \quad m = -M, -M+1, \ldots, 0, \ldots, M \tag{13.126}$$

In LMP algorithm, p is a key parameter. Its value must be $p \in [1, \alpha)$ to meet the demand of convex of the cost function and the limited moment of SaS process. Thus, LMP can only be used in the condition of $1 < \alpha \leq 2$.

Another adaptive TDE algorithm is LETDE (low-order explicit time delay estimation). LETDE algorithm uses α norm of $e(n)$, $J = \|e(n)\|_\alpha$, to represent the cost function of an adaptive system. This can avoid performance degradation caused by least mean square algorithm. According to the FLOM theory, as long as $0 < p < \alpha$, α norm of SaS process is in direct proportion to its p th-order moment. Thus, we can get

$$J = E[|e(n)|^p] \tag{13.127}$$

Using the instantaneous value in the gradient method to replace the statistical average value of the error signal, we get the gradient estimation as

$$\hat{\nabla}_p(n) = \frac{\partial |e(n)|^p}{\partial \hat{D}} = p(e(n))^{\langle p-1 \rangle} (\sum_{k=-M}^{M} f(k - \hat{D}(n))x(n-k)) \tag{13.128}$$

wherein $1 \leq p < \alpha \leq 2$. Using the identity $X^{\langle p-1 \rangle} = |X|^{p-1}\text{sgn}(X)$, where X is a real stable distribution variable, we can get the adaptive iterative formula of LETDE as

$$\hat{D}(n+1) = \hat{D}(n) - \mu p|e(n)|^{p-1}\mathrm{sgn}(e(n)) \sum_{k=-M}^{M} x(n-k)f(k-\hat{D}(n)) \qquad (13.129)$$

wherein μ is the step factor, eq. (13.129) can also be written as

$$\hat{D}(n+1) = \hat{D}(n) - 2\frac{\mu p|e(n)|^{p-2}}{2} e(n) \sum_{k=-M}^{M} x(n-k)f(k-\hat{D}(n)) \qquad (13.130)$$

LETDE algorithm can be looked as ETDE (explicit time delay estimation) algorithm with time-varying step size adjustment, that is,

$$\mu(n) = \frac{\mu p|e(n)|^{p-2}}{2} \qquad (13.131)$$

When $p=2$, the LETDE algorithm turns to ETDE algorithm. Thus, ETDE algorithm is a special case of LETDE algorithm. As ETDE, the selection of original value is very important.

To increase the stability and the convergence speed of the algorithm, normalization method must be used. And to avoid a big step when the denominator is very small, a parameter must be added:

$$\hat{D}(n+1) = \hat{D}(n) - \mu p|e(n)|^{p-1}\mathrm{sgn}(e(n)) \frac{\sum_{k=-M}^{M} x(n-k)f(k-\hat{D}(n))}{\left\| \sum_{k=-M}^{M} x(n-k)f(k-\hat{D}(n)) \right\|_p^p + \lambda} \qquad (13.132)$$

In the experiments, $s(k)$ is a zero mean Gaussian process with white spectrum. The delay signal is generated when $s(n)$ passes through an FIR filter of order 31 whose impulse response is $\sum_{k=-M}^{M} \mathrm{sinc}(k-D)z^{-k}$. Noise $v_1(n)$ and $v_2(n)$ are alpha stable distribution processes. Each experiment has 15,000 data points, and 500 Monte Carlo simulations are averaged. And at the same time, the error energy $\sigma_D^2 = E[\hat{D}(k) - D]^2$ is defined to evaluate the algorithm.

To fairly compare the performance of LETDE and ETDE, we set the true delay value as $D=1.7$ and original estimation value to be zero and make the two algorithms close to the critical state of divergence by adjusting μ to get the equal error energy of both algorithms. With this special μ (which is called equivalent convergence factor), we evaluate the performance of these algorithms.

Figure 13.11 shows the convergence curve of the two algorithms with $\alpha=1.5$. Obviously, LETDE converges more quickly and steadily.

Under the condition of impulsive noise, we detect the tracing ability of delay mutation of the two algorithms by using equivalent convergence factor (the true

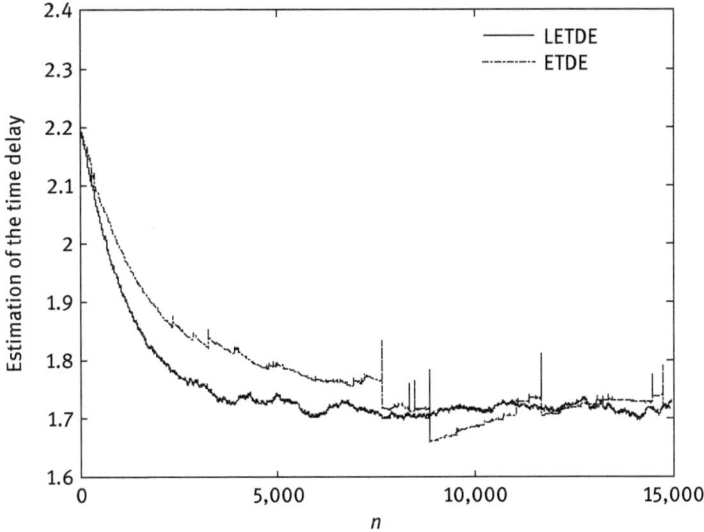

Figure 13.11: TDE convergent curve of two algorithms (true delay value 1.7, $\alpha = 1.5$,GSNR = 0 dB).

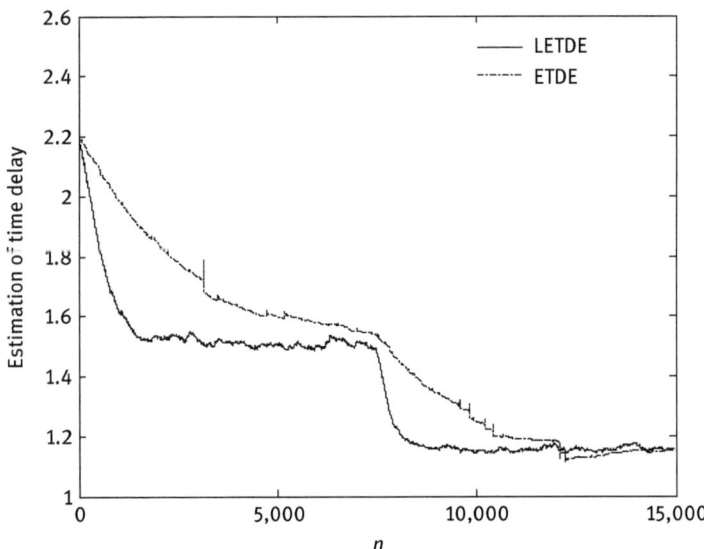

Figure 13.12: Detecting tracing ability of delay mutation by equivalent convergence factor (sudden change in true delay value from 1.5 to 1.2).

delay value sudden changes from 1.5 to 1.2, as shown in Figure 13.12. It shows that, although ETDE has some tracing ability, but its tracing ability and convergent speed are not as good as that of LETDE.

According to the performance comparison of ETDE and LETDE, we see that both of them can work properly when $\alpha = 2$ (Gaussian noise). But when the noise is more impulsive, LETDE algorithm performs better. Hence, LETDE not only retains the advantages of traditional ETDE algorithm but also improves the performance degradation under the non-Gaussian noise.

ℹ **Exercises**

13.1 Explain the application occasion of fractional lower-order statistics and HOS.

13.2 Explain the difference between HOS and the second-order statistics.

13.3 Explain the definition and characteristic of higher-order spectrum.

13.4 Explain the motivation for using HOS in signal processing.

13.5 There is a SISO system without noise, please describe bispectrum, trispectrum, and the relation of their power spectrum when the input of system is a stationary non-Gaussian random process.

13.6 Explain the effect of the selection of window function on higher-order spectrum estimation.

13.7 Please describe the process of the nonparametric method in higher-order spectrum estimation.

13.8 Explain the characteristic parameter estimation method of $S\alpha S$.

13.9 Describe the SNR setting method when the noise under the condition of alpha-stable distribution.

13.10 There is a sinusoidal signal $x(t) = a \sin(2\pi f t + \theta)$, where amplitude a and frequency f are constants, phase θ is a random variable with independent identically distributed in $[-\pi, \pi]$. Please figure out the autocorrelation function, the second-order cumulant, the third-order cumulant, the power spectrum, the second-order spectrum, and the third-order spectrum of $x(t)$.

13.11 There is an FIR system, its impulse response is $h(n) = \delta(n) - a\delta(n-1)$, its transfer function is $H(\omega) = 1 - ae^{-j\omega}$, when the input $x(n)$ is a non-Gaussian white noise with zero mean value, its variance is r_2^x, and the skewness of it is r_3^x. Please figure the out output, second-order spectrum, and third-order spectrum of this system.

13.12 There is an observing sequence, $y(n) = x(n) + w(n)$, $n = 0, 1, \ldots, N-1$. $x(n)$ is a sinusoidal wave with random phase, and $x(n) = A \cos(2\pi \times 0.21 \times n + \theta)$, where the normalized frequency of $x(n)$ is 0.21, A is the amplitude, θ is a random variant with uniform distribution in $[0, 2\pi]$, and $w(n)$ is an additive white Gaussian noise with zero mean. When $N = 4096$,

(1) Make $SNR = +\infty$, use numerical computation method to estimate second-, third-, and the fourth-order cumulant of $y(n)$.

(2) Make $SNR = 20, 0, -3dB$ separately, use the second-, third-, and fourth-order cumulant of $y(n)$, or use the second-, third-, and fourth-order spectrum of $y(n)$ to estimate the frequency of sine wave in Gaussian noise.

13.14 Gaussian distribution, Cauchy distribution, and Pearson distribution are stable distribution, and they have closed pdf. Try to deduce their pdf, mean value, and variance if they exist. (1) Gaussian distribution, $\alpha = 2, \beta = 0$, (2) Cauchy distribution $\alpha = 1, \beta = 0$, and (3) Pearson distribution $\alpha = 0.5, \beta = -1$.

13.15 The definition of coefficient of covariation is $\lambda_{XY} = \dfrac{E(XY^{\langle p-1 \rangle})}{E(|Y|^p)}, \quad \forall p, \quad 0 \le p < \alpha,$

as far as the variance of FLOM estimator is concerned. What is the influence of p on λ_{XY}.

14 Introduction of Modern Signal Processing

In this chapter, some basic theories, methods, and applications in the field of modern signal processing are given. The examples of time–frequency analysis, wavelet analysis, and Hilbert–Huang transform are introduced.

14.1 Time–Frequency Analysis

14.1.1 Fundamental Concept of Time–Frequency Analysis

The aim of signal analysis is to extract interesting information from the mixture of components that form signal variability. This extraction can be carried out by looking for a representation of the signal whose components are well separated. In theory, random signals can be divided into two categories: stationary and nonstationary. For a long time, due to the limitations of theoretical research and analysis tools, it is generally supposed that the signals are all stationary. For stationary signals, the mapping relation between frequency domain and time domain is established by Fourier transform and its inverse transformation. But the Fourier transform is a global transformation, which means that the characteristic of signals cannot be depicted in both time domain and frequency domain. Therefore, it cannot analyze the case when the characteristic of signal frequency varies with time. In fact, many signals of interest, such as speech, music, images, and medical signals, have changing frequency characteristic. The two signals $x_1(t)$ and $x_2(t)$ in Figure 14.1 illustrate the limitation of Fourier transform:

$$x_1(t) = \sin(6\pi t) + \sin(12\pi t) + \sin(18\pi t) \quad 0 \leq t \leq 4\,\text{s} \tag{14.1}$$

$$x_2(t) = \begin{cases} 2\sin(6\pi t) + \sin(12\pi t) & 0 \leq t < 2\,\text{s} \\ \sin(12\pi t) + 2\sin(18\pi t) & 2 \leq t \leq 4\,\text{s} \end{cases} \tag{14.2}$$

These two signals are both composed of three frequency components, but their waveforms are totally different. These three frequency components always exist in $x_1(t)$. However, in $x_2(t)$, only one frequency component always exists, and the other two only exist in the first half or the second half of the whole process. Figure 14.2 shows the spectra of signals $x_1(t)$ and $x_2(t)$.

Obviously, these two different signals have the similar spectra, which means that Fourier analysis cannot distinguish some different characteristics of the two signals. In order to overcome the limitation of the global transformation of the Fourier transform, it is necessary to introduce the time–frequency analysis of signals. The time–frequency analysis technique is especially effective in analyzing

https://doi.org/10.1515/9783110465082-014

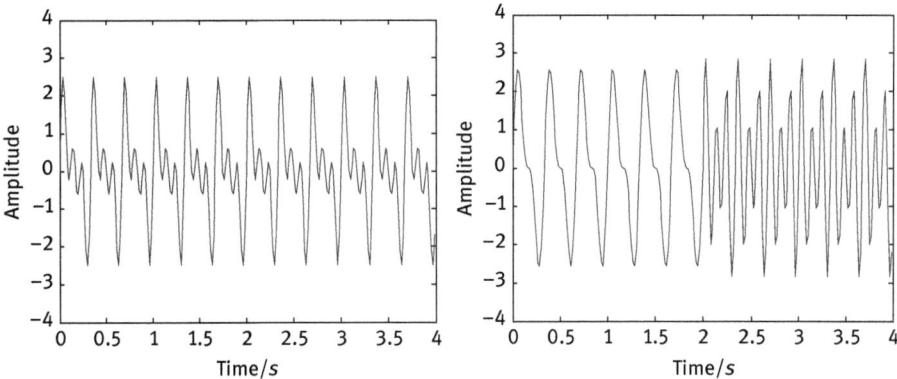

Figure 14.1: The waveform of two signals: (a) the waveform of $x_1(t)$ and (b) the waveform of $x_2(t)$.

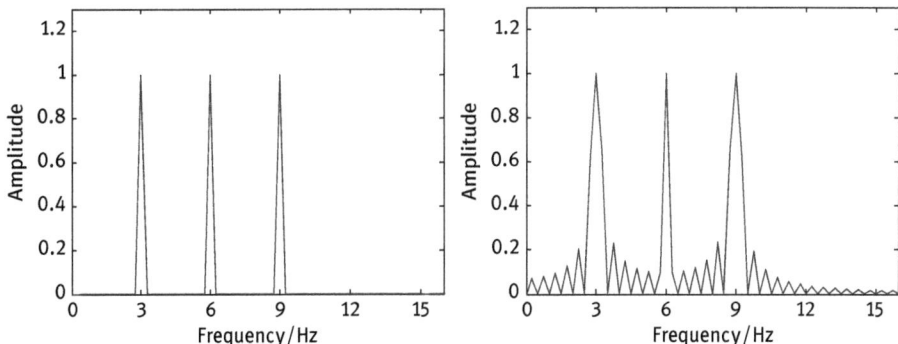

Figure 14.2: The spectra of the two signals: (a) the spectrum of $x_1(t)$ and (b) the spectrum of $x_2(t)$.

nonstationary signals, whose frequency distribution and magnitude vary with time. It maps one-dimensional time domain signal to a two-dimensional time–frequency plane, so as to provide joint distribution information in both time and frequency domains.

The methods of time–frequency analysis can be divided into linear time–frequency representation and bilinear time–frequency representation.

All linear time–frequency representations evolve from Fourier transform and satisfy the linear superposition principle. Assuming $x(t) = ax_1(t) + bx_2(t)$, the linear time–frequency representations of $x(t)$, $x_1(t)$, and $x_2(t)$ are $P(t,f)$, $P_1(t,f)$, and $P_2(t,f)$, respectively, and then we have

$$P(t,f) = aP_1(t,f) + bP_2(t,f) \tag{14.3}$$

Three typical linear time–frequency representations are the short-time Fourier transform (STFT), the Gabor transform, and the wavelet transform. The procedure for computing STFT is to divide a long signal into short segments with equal length and then compute the Fourier transform separately on each short segment. The Gabor transform is a special case of the STFT, which is used to determine the sinusoidal frequency and phase content of local sections of a signal when it changes over time. In Gabor transform, the signal to be transformed is first multiplied by a Gaussian function, which can be regarded as a window function, and the multiplied result is then transformed with the Fourier transform to derive the time–frequency analysis. In STFT and the Gabor expansion, the width of the window function is fixed, while the wavelet transform is a time–frequency representation with an adjustable window function.

Bilinear time–frequency representation is also called quadratic time–frequency representation and is defined as the two-dimensional Fourier transform of its kernelled function. Bilinear time–frequency representation does not satisfy the linear superposition property. Assuming that $x(t) = ax_1(t) + bx_2(t)$, the linear time–frequency representations of $x(t)$, $x_1(t)$, and $x_2(t)$ are $P(t,f)$, $P_1(t,f)$, and $P_2(t,f)$, respectively, and then we have

$$P(t,f) = |a|^2 P_1(t,f) + |b|^2 P_1(t,f) + ab^* P_{12}(t,f) + ba^* P_{21}(t,f) \qquad (14.4)$$

where $P_1(t,f)$ and $P_2(t,f)$ are called signal items, and $P_{12}(t,f)$ and $P_{21}(t,f)$ are called cross-items or interference items. Accordingly compared with STFT, the bilinear transformation suffers from an inherent cross-term contamination when analyzing multicomponent signals. By carefully choosing the window function, the interference can be significantly mitigated, but at the expense of resolution. The bilinear time–frequency representation mainly includes Cohen's class of time–frequency distribution and affine time–frequency distribution. The most famous bilinear time–frequency representation is Wigner–Ville distribution (WVD).

14.1.2 Short-Time Fourier Analysis

1. Continuous-time STFT

(1) Definition
Fourier transform cannot explicitly show the time location of frequency components. In order to overcome this shortcoming, Gabor who invented the holographic technology and won the Nobel Prize proposed the STFT in 1946.

The idea of STFT is to regard the nonstationary process as a series of short-time stationary signal superposition, and the short time (or short term) is achieved by suitably prewindowing the signal $x(t)$, so it is called windowed Fourier transform. In the continuous-time case, the signal to be transformed is multiplied by a window function which is nonzero for only a short period of time. With the window sliding along the time axis, the Fourier transform (a one-dimensional function) of the signal is performed, thus resulting in a two-dimensional representation of the signal.

The continuous-time STFT of the signal $x(t)$ is defined as

$$\text{STFT}_x(t,f) = X(t,f) = \int_{-\infty}^{\infty} x(\tau)g^*(\tau - t)e^{-j2\pi f\tau}d\tau \tag{14.5}$$

where the superscript *denotes complex conjugation; $g(t)$ is the window function, commonly a Hanning window or Gaussian window centered around zero; $x(t)$ is the signal to be transformed; $X(t,f)$ is essentially the Fourier transform of $x(\tau)g(\tau - t)$, a complex function representing the phase and magnitude of the signal over time and frequency.

That is, the STFT in the time domain is the Fourier transform of the signal multiplied by a shifted "analysis window" $g^*(\tau - t)$ centered around t. Because multiplication by the relatively short window $g^*(\tau - t)$ effectively suppresses the signal outside the neighborhood around the analysis time point $t' = t$, the STFT is simply a "local spectrum" of the signal $x(t)$ around the "analysis time" t.

It segments the signal into narrow time intervals (i.e., narrow enough to be considered stationary) and then takes the Fourier transform of each segment.

(2) Choosing window function
In the definition of STFT, the Fourier transform of $g_{t,f}(\tau) = g(\tau - t)e^{j2\pi ft}$ can be seen as

$$G_{t,f}(v) = \int_{-\infty}^{+\infty} g(u - t)e^{j2\pi vu}du = e^{-j2\pi(v-f)t}\int g(t') - j2\pi(v-f)t'dt' \tag{14.6}$$

$$= G(v - f)e^{-j2\pi(v-f)t}$$

where v is a frequency variable that is equivalent to Ω. Equation (14.5) can be rewritten as

$$\text{STFT}_x(t,f) = \int_{-\infty}^{\infty} x(\tau)g^*(\tau - t)e^{-j2\pi f\tau}d\tau$$
$$= \int_{-\infty}^{\infty} x(\tau)[g(\tau - t)e^{j2\pi f\tau}]^*d\tau \tag{14.7}$$

Equation (14.7) is the inner product of two signals $x(\tau)$ and $g(\tau - t)e^{j2\pi f\tau}$. Because the inner product in the time domain is equal to the inner product in the frequency domain, we have

$$\text{STFT}_x(t,f) = e^{-j2\pi ft}\frac{1}{2\pi}\int_{-\infty}^{\infty} X(v)G^*(v - f)e^{j2\pi vt}dv \tag{14.8}$$

It means that adding window to $x(\tau)$ in the time domain guides out windowed $X(v)$ in the frequency domain. Therefore, STFT is the analysis of the local signal spectrum. The characteristics of the local spectrum are determined by the shape (such as rectangular, Gaussian, and elliptic) and the width of the window function. As far as precision is concerned, the window should be narrow enough to make sure that the portion of the signal falling within the window is stationary. But a too narrow window does not offer good localization in the frequency domain.

For example, signal $x(t)$ is superimposed by two Gaussian amplitude modulation signals with different time periods. The time center of one signal is $t_1 = 40\,\text{s}$, and the time duration is 20 s. The time center of another signal is $t_2 = 80\,\text{s}$ and its time duration is also 20 s. Normalized frequency of modulation signal is 0.25.

Figures 14.3 and 14.4 show the STFT results of $x(t)$. Figure 14.4 demonstrates the influence of window width on time–frequency resolution.

In Figure 14.3, when the window function width is 65, the frequency resolution is better, but two components of the signal in the time domain cannot be distinguished. When window function's width is reduced to 25, the frequency resolution is reduced, but the time resolution is improved, that is, good frequency resolution results in poor time resolution, and good time resolution leads to poor frequency resolution. This is the uncertainty principle which states that both time and frequency cannot be measured exactly at the same time.

Figure 14.4 illustrates the influence of the shapes of the windows on the time–frequency resolution, even with the same window function width.

(3) Steps of calculating STFT
Step 1: Choose a window function of finite length.
Step 2: Place the window on top of the signal at $t = 0$.
Step 3: Truncate the signal using this window.
Step 4: Perform the Fourier transform to the truncated signal and then save the results.
Step 5: Incrementally slide the window to the right.
Step 6: Go to step 3, until the window reaches the end of the signal.

2. Discrete-time STFT

When the STFT of a signal is implemented on a computer, the signal must be discrete and finite. In the discrete-time case, the signal to be transformed could be broken up into frames, which usually overlap each other. Each chunk is taken as the Fourier transform, and the complex result is added to a matrix, which records magnitude and phase for each point in time and frequency. Assuming that the given signal is $x(n), n = 0, 1, ..., L-1$, we have

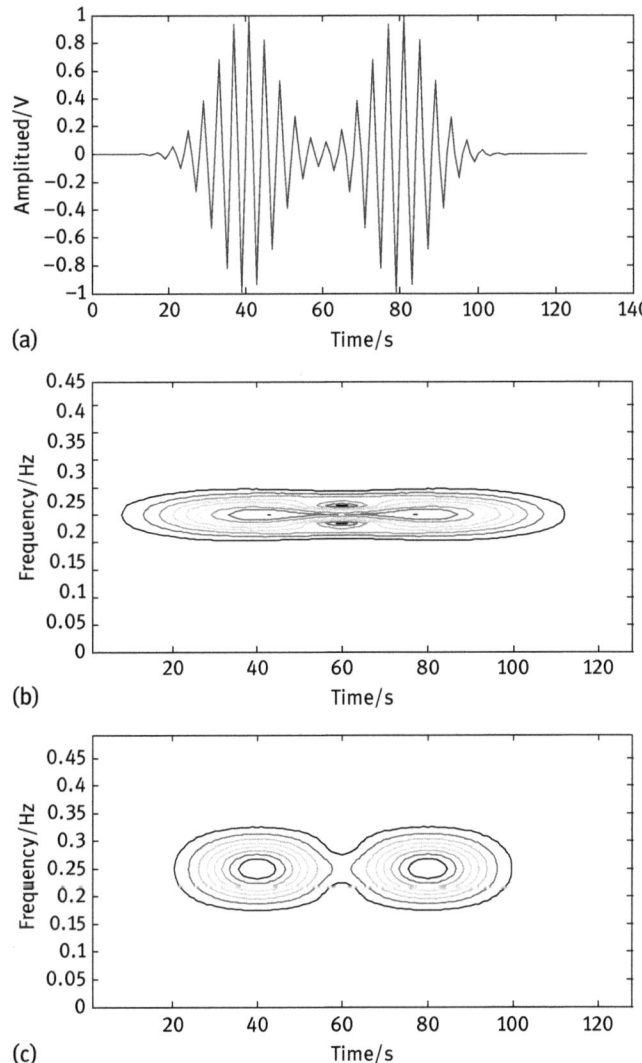

Figure 14.3: The influence of window function width on the time–frequency resolution: (a) nonstationary signal $x(t)$; (b) Hamming window function, the width is 65; and (c) Hamming window function, the width is 25.

$$\text{STFT}_x(m, e^{j\omega}) = \sum_n x(n)g^*(n-mN)e^{-j\omega n}$$

$$= \langle x(n), g(n-mN)e^{j\omega n} \rangle \tag{14.9}$$

where N is the moving step size of the window function on the time axis, and ω is the circular frequency, that is, $\omega = \Omega T$, in which T is the sampling interval of $x(n)$ obtained by $x(t)$. This equation corresponds to the discrete-time Fourier transform

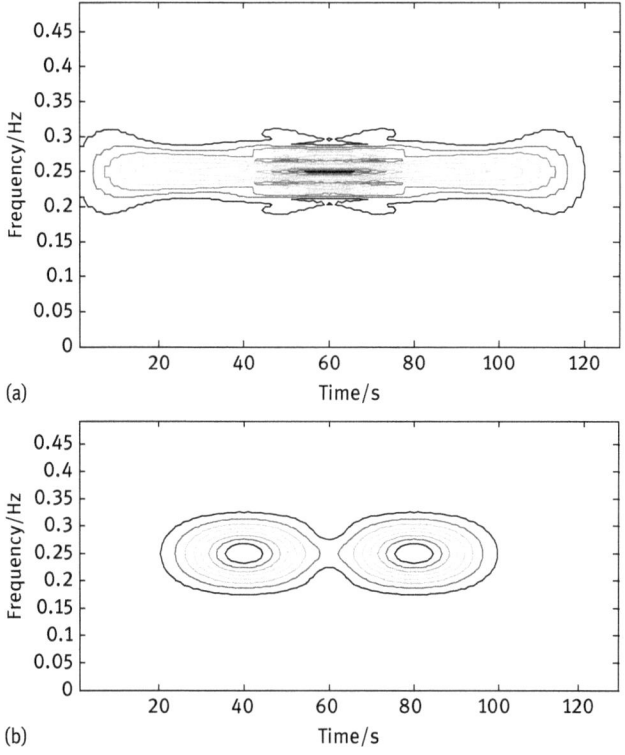

(a)

(b)

Figure 14.4: Influence of different shaped window functions on time–frequency resolution: (a) for rectangular window function, the width is 25; and (b) for Hamming window function, the width is 25.

(DTFT) in which the time is discrete and the frequency is continuous. For the implementation on a computer, the frequency ω should be discretized. Let $\omega_k = (2\pi/M)k$, then we have

$$\mathrm{STFT}_x(m, \omega_k) = \sum_n x(n)g^*(n - mN)e^{-\mathrm{j}\frac{2\pi}{M}nk} \tag{14.10}$$

where the period 2π in the frequency domain is divided into M parts. If the width of window function $g(n)$ is exact M points, then eq. (14.10) can be written as

$$\mathrm{STFT}_x(m, k) = \sum_{n=0}^{M-1} x(n)g^*(n - mN)e^{-\mathrm{j}\frac{2\pi}{M}nk}, \quad k = 0, 1, K, M - 1 \tag{14.11}$$

The value of N determines the distance that the window function moves along the axis of time. The smaller N results in more m values, so that the time–frequency curve is more intensive.

Equation (14.11) can be realized by the time–frequency analysis toolbox function tfrstft.m in MATLAB. The inverse transform of eq. (14.11) is

$$x(n) = \frac{1}{M} \sum_{m} \sum_{k=0}^{M-1} \text{STFT}_x(m, k) e^{j\frac{2\pi}{M}nk} \tag{14.12}$$

In eq. (14.12), the range of m depends on the length of the data L and the step size of the window function N.

3. Resolution issues
Is the window as small as possible in STFT? For this purpose, the concept of phase space is introduced.

The so-called phase space refers to the Euclidean space, which takes "time" as the abscissa and "frequency" as the coordinate. The limited area in phase space is called window. The effect of phase space is used to describe a certain physical state, so it has a strong engineering background.

Mathematically, if $g(t) \in L^2(\mathbb{R})$ (the space $L^2(\mathbb{R})$ of measurable functions $f(t)$ is defined on the real line $\mathbb{R}$, which satisfy $\int_{-\infty}^{\infty} |f(t)|^2 dt < \infty$) and $tg(t) \in L^2(\mathbb{R})$, then $g(t)$ is called window function. The point (t_0, Ω_0) in phase space is

$$\begin{cases} t_0 = \frac{1}{\|g(t)\|^2} \int_{-\infty}^{\infty} t|g(t)|^2 dt \\ \Omega_0 = \frac{1}{\|G(j\Omega)\|^2} \int_{-\infty}^{\infty} \Omega|G(j\Omega)|^2 d\Omega \end{cases} \tag{14.13}$$

where $G(j\Omega)$ is the Fourier transform of window function $g(t)$, t_0 is called the time center, and Ω_0 is called the frequency center. Let us define the time width of window function Δg and the frequency width of window function ΔG as follows:

$$\begin{cases} \Delta g = \left(\frac{1}{\|g(t)\|^2} \int_{-\infty}^{\infty} (t - t_0)^2 |g(t)|^2 dt \right)^{1/2} \\ \Delta G = \left(\frac{1}{\|G(j\Omega)\|^2} \int_{-\infty}^{\infty} (\Omega - \Omega_0)^2 |G(j\Omega)|^2 d\Omega \right)^{1/2} \end{cases} \tag{14.14}$$

Δg is the time resolution, which means how well two spikes in time can be separated from each other in the time domain. ΔG is frequency resolution, which means how well two spectral components can be separated from each other in the transform domain.

For STFT, in order to get better frequency resolution, the condition that needs to be met is

$$\Delta g \Delta G \geq 1/2 \tag{14.15}$$

This is the Heisenberg uncertainty principle. It is a restriction relationship between Δg and ΔG. It tells us that there is no such a window function with both arbitrary small time width and arbitrary small bandwidth. It means that all the time–frequency representation can only approximate the energy density of the signal (t, f) in different degrees. When $g(t)$ is a Gaussian function, the equal sign in eq. (14.15) holds good.

Obviously, once the window function $g(t)$ is selected, Δg and ΔG are also determined. Thus, for any given t_0 and Ω_0, the time–frequency resolution of STFT can be represented by $[(t_0 \pm \Delta g) \times (\Omega_0 \pm \Delta G)]$, that is, the information about the signal $x(t)$ given by STFT at any point (x_0, Ω_0) in the phase space is defined by the two indeterminate quantities of Δg and ΔG.

Let us use STFT to analyze signals $x_1(t)$ and $x_2(t)$ in eqs. (14.1) and (14.2), respectively. Figures 14.5 and 14.6 show the transformation results.

Because the size and the shape of the window are fixed, the resolution in the time domain and frequency domain is fixed. In Figure 14.5, it can be seen that the three different frequency components in the signal exhibit the same bandwidth. In order to illustrate the interdependence between Δg and ΔG, we use window functions with different widths to analyze $x_1(t)$ and $x_2(t)$. The results are shown in Figure 14.6.

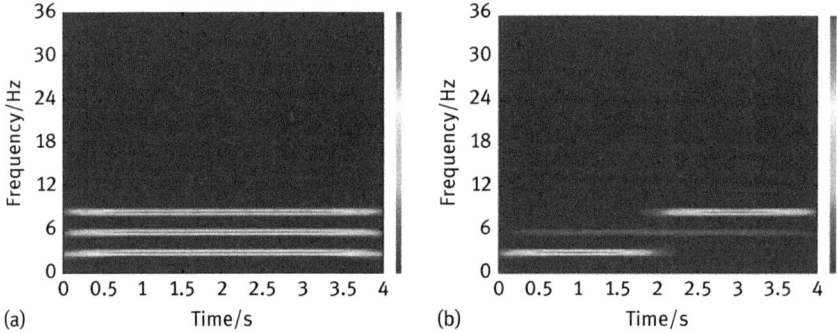

Figure 14.5: The STFT of signals (the window length is one-eighth of the signal length): (a) $x_1(t)$ and (b) $x_2(t)$.

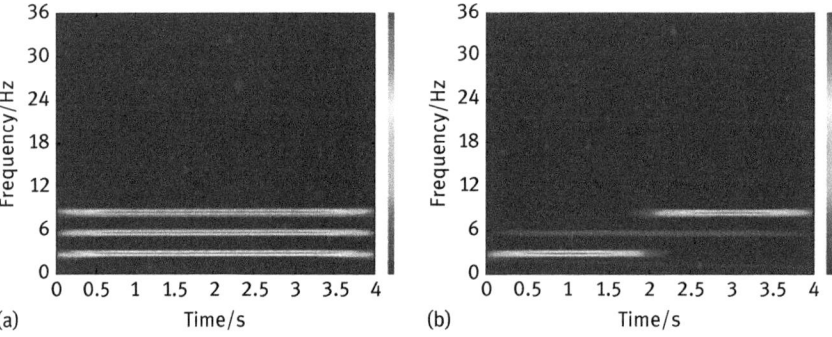

Figure 14.6: The STFT of signals (the window length is one-fourth of the signal length): (a) $x_1(t)$ and (b) $x_2(t)$.

Compared with Figure 14.5, the width of window in Figure 14.6 is changed from one-eighth of the signal length to one-fourth of the signal length. It is obvious that the frequency resolution in Figure 14.6 is higher than that in Figure 14.5, but its resolution in time axis decreases due to the two overlapped components of $x_2(t)$.

From the above analysis we can see that if we want to obtain good time domain resolution of the high-frequency components, we must select the wideband and short-time window. If we want to obtain good time domain resolution of the low-frequency components, we must select the narrowband and wide-time window. These two objectives cannot be achieved at the same time.

The disadvantage of STFT is that its discrete form does not have an orthogonal expansion. Therefore, it is difficult to implement efficient algorithms, which greatly limits its scope of application.

14.1.3 Gabor Expansion and Gabor Transform

One of the fundamental issues in the time–frequency analysis is obtaining the distribution of signal energy over joint time–frequency plane with sufficient time and frequency resolution. In 1946, D. Gabor proposed to use time- and frequency-shifted Gaussian windows as basis functions because of their optimal time–frequency concentration. Hence the Gabor expansion represents a signal as a combination of time- and frequency-translated basis functions. And the integral expression of Gabor expansion coefficients is called Gabor transform. Gabor transform is a special case of STFT, where he used translations and modulations of a single Gaussian function to represent one-dimensional signal.

1. Gabor expansion
The Gabor expansion represents a signal in terms of time- and frequency-shifted basis functions, and has been used in various applications to analyze the time-varying frequency content of a signal.

Suppose $f(t), g(t) \in L^2(\mathbb{R})$ and $\{g_{m,n}(t) : g_{m,n}(t) = g(t - ma)e^{j2\pi mbt}, a, b > 0, ab \leq 2\pi, m, n \in \mathbb{Z}\}$ constitute a frame, that is, there exist constants A and B with $A \geq B > 0$, such that

$$B|\Phi(t)|^2 \leq \sum_m \sum_n \left| \langle |\Phi(t), g_{mn}(t)| \rangle \right|^2 \leq A|\Phi(t)|^2 \qquad \forall \Phi(t) \in L^2(\mathbb{R}) \tag{14.16}$$

where $\langle \Phi(t), g_{mn}(t) \rangle = \int_R \Phi(t) g_{mn}^*(t) dt$ is the inner product of $\Phi(t)$ and $g_{mn}(t)$. $g_{mn}^*(t)$ is the complex conjugate of $g_{mn}(t)$, and $|\Phi(t)| = \sqrt{\langle \Phi(t), \Phi(t) \rangle}$ is the norm of $\Phi(t)$ in $L^2(\mathbb{R})$. Gabor expansion of $f(t)$ under the frame $\{g_{mn}(t)\}$ can be expressed as

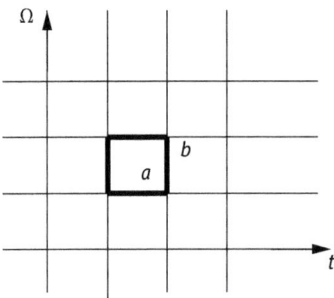

Figure 14.7: Gabor expansion of the sampling grid.

$$f(t) = \sum_{m=-\infty}^{+\infty} \sum_{n=-\infty}^{+\infty} C_{m,n} g_{m,n}(t) = \sum_{m=-\infty}^{+\infty} \sum_{n=-\infty}^{+\infty} C_{m,n} g(t-ma) e^{j2\pi nbt} \qquad (14.17)$$

where a represents the time interval of the signal, b represents the frequency sampling interval of the signal, $\{g_{mn}(t)\}$ is called an (m, n)-order Gabor basis function or a Gabor atom, $C_{mn} = \langle f(t), \gamma_{mn}(t) \rangle$ is referred to as Gabor coefficient, and $\{g_{mn}(t)\}$ is the dual frame of $\{\gamma_{mn}(t)\}$. Signal $\gamma(t)$ is called the analysis window function and $g(t)$ the synthesis window function. Equation (14.17) is the atomic decomposition of signal $f(t)$, which indicates that the superposition of locally vibrating waves at lattice (m, n) is equal to $f(t)$. This implies all information contained in signal $f(t)$ can be reconstructed from Gabor coefficients $\{C_{mn}\}$. Figure 14.7 shows the diagram of the sampling grid in (Ω, t) plane.

Obviously, Gabor basis function $g_{m,n}(t)$ in eq. (14.17) is constructed by the translation and modulation of the mother function $g(t)$. So, if $g(t)$ is a nonorthogonal function, Gabor basis function $g_{m,n}(t)$ is also nonorthogonal. Thus, Gabor expansion is a kind of nonorthogonal series expansion. In mathematics, the nonorthogonal expansion of a function is called atomic expansion. That is the reason why the Gabor basis function is also called Gabor atom. The significance of Gabor expansion is that it can construct Gabor basis function $g_{m,n}(t)$, which is easy to locate and highly aggregate according to time and frequency. And it can show time–frequency characteristics near the time–frequency point (ma, nb).

When $ab = 1$ is satisfied, it is called critical sampling. When $ab < 1$ is satisfied, it is called oversampling. When $ab > 1$ is satisfied, it is called undersampling. It has been shown that the undersampling Gabor expansion can cause numerical instability. In fact, $ab = 1$ is a bound on the density of the time–frequency sampling grid and is the minimum sampling frequency that can be permitted by stable Gabor expansion. In the critical or extreme case, if the band-limited signal is sampled at the minimum sampling rate, the number of Gabor expansion coefficients $a_{m,n}$ is exactly equal to the number of samples of the original signal. That means Gabor expansion coefficient $a_{m,n}$ matches the number of signal samples exactly, without any redundancy. In other words, the Gabor expansion coefficients are uniquely determined. At the same time, in this case, the Gabor basis functions are linearly independent.

In the case of critical sampling, when the basis function set of Gabor expansion $\{g_{m,n}(t) = g(t-ma)e^{j2\pi nbt}, m \in \mathbb{Z}, n \in \mathbb{Z}\}$ can constitute a framework, there must exist a dual function $\gamma(t)$. And the shift and modulation set of $\gamma(t)$: $\{\gamma_{m,n}(t) = \gamma(t-ma)e^{j2\pi nbt}, m \in \mathbb{Z}, n \in \mathbb{Z}\}$ constituted a dual framework. Then the Gabor expansion coefficient $a_{m,n}$ can be calculated using $\gamma_{m,n}(t)$, that is,

$$a_{m,n} = \int_{-\infty}^{+\infty} f(t)\gamma_{m,n}^*(t)dt = \int_{-\infty}^{+\infty} f(t)\gamma_{m,n}^*(t-ma)e^{-j2\pi nb}dt$$
$$= \langle f(t), \gamma_{m,n}(t)\rangle = \text{STFT}(ma, nb)$$
(14.18)

Equation (14.18) is called the continuous Gabor transform which is nothing but a simple localization of the Fourier transform via the introduction of a sliding window function. The existence of this window makes this transform a function of two parameters: a time parameter giving the location of the center of the window and a frequency parameter for the computation of the Fourier transform of the windowed signal. Obviously, it shows that Gabor expansion coefficient $a_{m,n}$ can be obtained instantly by Gabor transform when the signal and the auxiliary function $\gamma(t)$ are known.

Actually, the original function and its dual function can be exchanged. Therefore, as long as one of them is taken as an analytical function, another one is taken as an integrated function, that is,

$$f(t) = \sum_{m=-\infty}^{+\infty}\sum_{n=-\infty}^{+\infty} \langle f(t), \gamma_{m,n}(t)\rangle g_{m,n}(t) = \sum_{m=-\infty}^{+\infty}\sum_{n=-\infty}^{+\infty} \langle f(t), g_{m,n}(t)\rangle \gamma_{m,n}(t)$$
(14.19)

In practical applications, we need to select an original function and obtain its dual function. Substituting eq. (14.18) into eq. (14.19), and exchanging the order of the summation and integration,
we can get

$$f(t) = \int_{-\infty}^{+\infty} f(t') \sum_{m=-\infty}^{+\infty}\sum_{n=-\infty}^{+\infty} \gamma_{m,n}^*(t')g_{m,n}(t)dt'$$
(14.20)

This is the reconstruction equation of the signal. If the above equation is true for all $t \in R$, we call this is a perfect construction of the signal. In order to make the signal reconstruction perfectly, $g_{m,n}(t)$ has to meet the following condition:

$$\sum_{m=-\infty}^{+\infty}\sum_{n=-\infty}^{+\infty} \gamma_{m,n}^*(t')g_{m,n}(t) = \delta(t-t')$$
(14.21)

If $g_{m,n}(t)$ satisfies the above condition, it is known as complete. An important and practical equation describing the relationship between $g(t)$ and $\gamma(t)$ is

$$\int_{-\infty}^{\infty} \gamma(t) g^*(t - ma) e^{-j2\pi nb} dt = \delta(m)\delta(n) \tag{14.22}$$

This is referred to as the biorthogonal relationship. As long as any one value in m or n is not zero, $\gamma(t)$ is orthogonal to $g(t)$. Therefore, we call $\gamma(t)$ a biorthogonal function of the window function $g(t)$.

In summary, after selecting the appropriate Gabor basis function $g_{m,n}(t)$, the analytical method for determining the Gabor expansion coefficients can be divided into two steps. The first step is to solve the biorthogonal equation (14.22) and obtain the auxiliary function $\gamma(t)$. The second step is to calculate the Gabor integral equation (14.18) and obtain the Gabor expansion coefficient $a_{m,n}$.

When $ab < 1$, it is called oversampling. There are more dense sampling grids in this case, which means that more data are collected. Thus, there is a certain redundancy in the Gabor expansion of the signal. Just like the Gabor transform in critical sampling case, the calculation process in oversampling case helps to determine the dual function according to the basis function. When the window function is selected, the kernel function $\gamma(t)$ of the Gabor transform in the oversampled case is not uniquely defined by the complete reconstruction condition. But the optimal biorthonormal function $\gamma(t)$ in the sense of minimum norm can be obtained, and it is uniquely determined.

2. Discrete Gabor transform

Sampling of time variables leads to periodicity in the frequency domain, and sampling of the frequency variable leads to periodicity in the time domain. In the Gabor expansion, both time and frequency have been discretized, so that both signal sequence and coefficient sequence are periodic. So, this part focuses on the discrete-time periodic signals. If the period is L, then the discrete-time periodic signal $\tilde{x}(k)$ can be expressed as $\tilde{x}(k) = \tilde{x}(k + L)$. Discrete sequences are defined only when k is an integer. It can be regarded as the result of sampling at equal intervals for continuous signals. For writing convenience, the sampling interval is l.

(1) Discrete Gabor transform under critical sampling

For the discrete periodic sequence, if it is periodic with L, then the frequency spectrum of it is also periodic with 2π. Since L and 2π should be integer multiples of the time interval grid a and frequency interval grid b, respectively, they can be set

$$a = L/M, \quad b = 1/N \tag{14.23}$$

In the case of critical sampling, $ab = 1$, $L = MN$. Substituting $L = MN$ into eq. (14.23), we can get $a = N$, that is, the time grid divides the signal of length L into M segments, and each segment has N discrete points. Usually, the frequency period 2π is divided into L discrete points, and the L discrete points are divided into N segments by the frequency grids, and each segment has M discrete points. Therefore, the coefficients $a_{m,n}$ of eq. (14.18) should be two-dimensional periodic, that is, $a_{m,n} = a_{m+M,\, n+N}$.

We define the periodic basis function $\tilde{g}(t)$ as the periodic sum of the basis function $g(t)$ (its period is L), that is,

$$\tilde{g}(t) = \sum_{i=-\infty}^{\infty} g(t - iL) \tag{14.24}$$

Combining the Poisson summation formula

$$\sum_{k=-\infty}^{\infty} e^{j2\pi kt} = \sum_{k=-\infty}^{\infty} \delta(t - k)$$

we get

$$x(t) = \sum_{m=0}^{M-1}\sum_{n=0}^{N-1} a_{m,n}\tilde{g}_{m,n}(t) \sum_{k=-\infty}^{\infty} \delta(t - k) \tag{14.25}$$

The discrete Gabor expansion equation is represented in the discrete form:

$$\tilde{x}(k) = \sum_{m=0}^{M-1}\sum_{n=0}^{N-1} a_{m,n}\tilde{g}_{m,n}(k) \tag{14.26}$$

where $\tilde{g}_{m,n}(k) = g(k - mM)W_N^{nk}$, $W_N = e^{j\frac{2\pi}{N}}$.

$\tilde{\gamma}(k)$ is defined as the periodic (its period is L) sum of $\gamma(k)$, that is, $\tilde{\gamma}(t) = \sum_{i=-\infty}^{\infty} \gamma(t - iL)$. Therefore, Gabor transform can be written as

$$a_{m,n} = \sum_{k=0}^{L-1} \tilde{x}(k)\tilde{\gamma}_{m,n}(k) \tag{14.27}$$

where $\tilde{\gamma}_{m,n}(k) = \tilde{\gamma}(k - mM)W_N^{nk}$.

Substituting eq. (14.27) into eq. (14.26), we obtain the complete reconstruction condition of discrete periodic sequence:

$$\sum_{m=0}^{M-1}\sum_{n=0}^{N-1} \tilde{g}_{m,n}(k)\tilde{\gamma}_{m,n}^{*}(k') = \delta(k - k') \tag{14.28}$$

Using the derivation method which is similar to that of the continuous Gabor transform, we can get the biorthogonal condition:

$$\sum_{k=0}^{L-1} \tilde{g}_{m,n}(k)\tilde{\gamma}^*(k) = \delta(m)\delta(n) \tag{14.29}$$

It can also be written as

$$\sum_{k=0}^{L-1} [\tilde{g}^*(k+mN)W_N^{nk}]\tilde{\gamma}(k) = \delta(m)\delta(n) \tag{14.30}$$

Since there are many choices for M and N to satisfy $L = MN$, the discrete Gabor expansion and Gabor transform are not uniquely defined in the case of critical sampling. However, the continuous Gabor expansion and Gabor transform under critical sampling are uniquely determined. The biorthogonal condition (14.30) can also be written in matrix form:

$$\tilde{\mathbf{G}}\tilde{\gamma} = \mathbf{e}_l \tag{14.31}$$

where $\tilde{\gamma} = [\tilde{\gamma}(0), \tilde{\gamma}(1), \cdots, \tilde{\gamma}(L-1)]^T$ and $\mathbf{e}_l = [1, 0, \cdots, 0]^T$ are the column vectors of $L \times 1$, $\tilde{\mathbf{G}}$ is the Hankel matrix which can be written as

$$\tilde{G} = \begin{bmatrix} \tilde{G}(0) & \tilde{G}(1) & \cdots & \tilde{G}(M-1) \\ \tilde{G}(1) & \tilde{G}(2) & \cdots & \tilde{G}(0) \\ \vdots & \vdots & \ddots & \vdots \\ \tilde{G}(M-1) & \tilde{G}(0) & \cdots & \tilde{G}(M-2) \end{bmatrix} \tag{14.32}$$

$\tilde{G}(i)$ is an $N \times N$ matrix which is written as

$$\tilde{G}(i) = \begin{bmatrix} W_N^0 & W_N^0 & \cdots & W_N^0 \\ W_N^0 & W_N^1 & \cdots & W_N^{N-1} \\ \vdots & \vdots & \ddots & \vdots \\ W_N^0 & W_N^{N-1} & \cdots & W_N^1 \end{bmatrix} \bullet \text{diag}[\tilde{g}^*(iN), \tilde{g}^*(iN+1), \dots, \tilde{g}^*(iN+N+1)] \tag{14.33}$$

Since $\tilde{G}$ is a block Hankel matrix, it is an invertible matrix. Thus, we obtain the least-squares solution of eq. (14.31) as

$$\tilde{\gamma} = \tilde{G}^{-1}\mathbf{e}_l \tag{14.34}$$

Once $\tilde{\gamma}$ is obtained, the Gabor expansion coefficient $a_{m,n}$ can be determined.

(2) Discrete Gabor transform under oversampling

In the case of oversampling $MN > L$, we decompose the period L of the periodic function $\tilde{x}(t)$ into $L = \bar{N}M = N\bar{M}$. $\bar{N}, N, M, \bar{M}$ are integers and satisfy $\bar{N} < N, \bar{M} < M$. Under this decomposition condition, the Gabor expansion of the periodic signals is

$$\tilde{x}(k) = \sum_{m=0}^{M-1} \sum_{n=0}^{N-1} a_{m,n}\, \tilde{g}_{m,n}(k) \tag{14.35}$$

The Gabor expansion coefficients are obtained by the discrete Gabor. They are

$$a_{m,n} = \sum_{k=0}^{L-1} \tilde{x}(k) \bar{\gamma}_{m,n}(k) \tag{14.36}$$

where $\tilde{g}_{mn}(k) = g(k - m\bar{M})e^{j2\pi nk/\bar{N}}$, $\tilde{\gamma}_{mn}(k) = \tilde{\gamma}(k - m\bar{M})e^{-j2\pi nk/\bar{N}}$.

The discrete sequences $g(k)$ and $\gamma(k)$ obey orthogonal condition:

$$\sum_{k=0}^{L-1} [\tilde{g}^*(k + m\bar{M}) W_{\bar{N}}^{nk}] \tilde{\gamma}(k) = \frac{L}{M\bar{N}}\delta(m)\delta(n) \tag{14.37}$$

where $0 \le m \le \bar{M} - 1,\ 0 \le n \le \bar{N} - 1$. They can be written in the matrix form as

$$\tilde{G}\tilde{\gamma} = \mathbf{b} \tag{14.38}$$

where $\tilde{\gamma} = [\tilde{\gamma}(0), \tilde{\gamma}(1), \ldots, \tilde{\gamma}(L-1)]^T$, $\mathbf{b} = [L/M\bar{N}, 0, \ldots, 0]^T$. $\tilde{G}$ is an $\bar{M}\bar{N} \times L$ matrix, and it can be written as the block matrix form

$$\tilde{G} = \begin{bmatrix} \tilde{G}(0) & \tilde{G}(1) & \cdots & \tilde{G}(\bar{M}-1) \\ \tilde{G}(1) & \tilde{G}(2) & \cdots & \tilde{G}(0) \\ \vdots & \vdots & \ddots & \vdots \\ \tilde{G}(\bar{M}-1) & \tilde{G}(0) & \cdots & \tilde{G}(\bar{M}-2) \end{bmatrix} \tag{14.39}$$

$\tilde{G}(i)$ is an $\bar{N} \times \bar{N}$ matrix:

$$\tilde{G}(i) = \begin{bmatrix} W_{\bar{N}}^0 & W_{\bar{N}}^0 & \cdots & W_{\bar{N}}^0 \\ W_{\bar{N}}^0 & W_{\bar{N}}^1 & \cdots & W_{\bar{N}}^{\bar{N}-1} \\ \vdots & \vdots & \ddots & \vdots \\ W^0 & W_{\bar{N}}^{\bar{N}-1} & \cdots & W_{\bar{N}}^1 \end{bmatrix} \bullet \operatorname{diag}[\tilde{g}^*(i\bar{N}), \tilde{g}^*(i\bar{N}+1), \ldots, \tilde{g}^*(i\bar{N}+\bar{N}+1)] \tag{14.40}$$

Therefore, when $L < MN$, we can know $L > \bar{M}\bar{N}$ from $L^2 = \bar{M}\bar{N}MN$. $\tilde{\mathbf{G}}$ is a flat matrix (the number of rows is less than the number of columns), so the matrix equation (14.38) is underdetermined with infinite solutions. The minimal norm solution is

$$\tilde{\gamma} = \tilde{\mathbf{G}}^{\mathbf{H}}(\tilde{\mathbf{G}}\tilde{\mathbf{G}}^{\mathbf{H}})^{-1}\mathbf{b} \tag{14.41}$$

This is uniquely determined so that the Gabor expansion coefficients can be calculated.

3. Selection of Gabor basis functions

The two basic problems in Gabor expansion are how to choose the Gabor basis function and how to compute the Gabor expansion coefficients. In the previous section, we have introduced how to calculate Gabor expansion coefficients. In fact, from the definition of Gabor expansion, we can see that when the window function has been selected, the selection of time sampling interval and frequency sampling interval will affect the completeness, uniqueness, and data integrity of Gabor expansion.

For the selection of Gabor basis functions, the basis function can be any form as long as the time–frequency sampling grids are enough. There are many good window functions which can be used to construct Gabor basis functions. Among them, the most common window functions are the rectangular function and the Gaussian function.

The rectangular window function is

$$g(t) = \left(\frac{1}{T}\right)^{1/2} p\left(\frac{2t}{T}\right) \tag{14.42}$$

where $p(x) = \begin{cases} 1, & -1 \le x \le 1 \\ 0, & \text{others} \end{cases}$. Its corresponding dual function is

$$\gamma(t) = \left(\frac{1}{T}\right)^{1/2} p\left(\frac{2t}{T}\right) \tag{14.43}$$

The Gaussian window function is

$$g(t) = \left(\frac{\sqrt{2}}{T}\right)^{1/2} e^{-\pi\left(\frac{1}{T}\right)^2} \tag{14.44}$$

Its corresponding dual function is

$$\gamma(t) = \left(\frac{\sqrt{2}}{T}\right)^{1/2} e^{-\pi\left(\frac{1}{T}\right)^2} \tag{14.45}$$

If $ab = 1$, then $g_{mn}(t)$ is linearly independent, and its dual function $\gamma_{mn}(t)$ is unique, and the sum $g_{mn}(t)$ is biorthogonal.

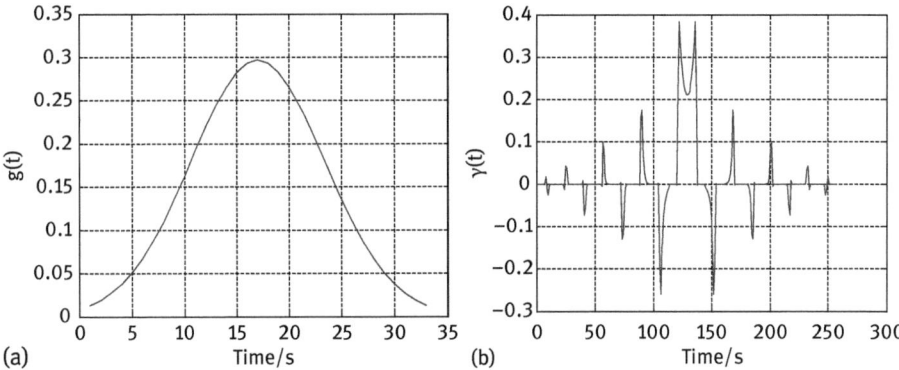

Figure 14.8: Dual functions of Gaussian when $ab = 1$: (a) $g(t)$ and (b) $\gamma(t)$.

Figure 14.8 shows a Gaussian window $g(t)$ and its dual window $\gamma(t)$. It can be seen that when the oversampling factor is 1, the energy concentration is completely lost.

In the case of oversampling, for the Gaussian function $g(t)$, $\gamma(t)$ becomes more and more closer to $g(t)$ with the increase in oversampling rate. And when the sampling rate is greater than or equal to 3, $g(t)$ and $\gamma(t)$ are almost the same. In this condition, the Gabor coefficients have good time–frequency analysis performance. The above conclusions can be well described in Figure 14.9.

In brief, the difference and relationship between the Gabor transform and STFT are as follows: (1) the window function of STFT must be a narrow window, while the window function of Gabor transform has no such limit. We can take the Gabor transform as a windowed Fourier transform, which is more applicable than the STFT. (2) STFT(t, f) is the two-dimensional time–frequency representation of the signal, while Gabor transformation coefficients are the two-dimensional representation of time shift and frequency modulation of the signal.

14.1.4 Cohen's Class Time–Frequency Distribution

Both STFT and Gabor expansion are linear time–frequency representations. The linearity of time–frequency representation is an important property that we expected. But when using time–frequency representation to describe the time–frequency energy distribution (i.e., the instantaneous power spectral density), the quadratic time–frequency representation is more intuitive and reasonable, because the energy itself is a nonlinear representation. The quadratic time–frequency representation is also known as a bilinear form. That is to say, the studied signal appears twice in a multiplicative form in the mathematical expression of time–frequency distribution. Compared with other time–frequency analysis techniques, such as STFT, the bilinear transformation (or quadratic time–frequency distributions) may not have higher

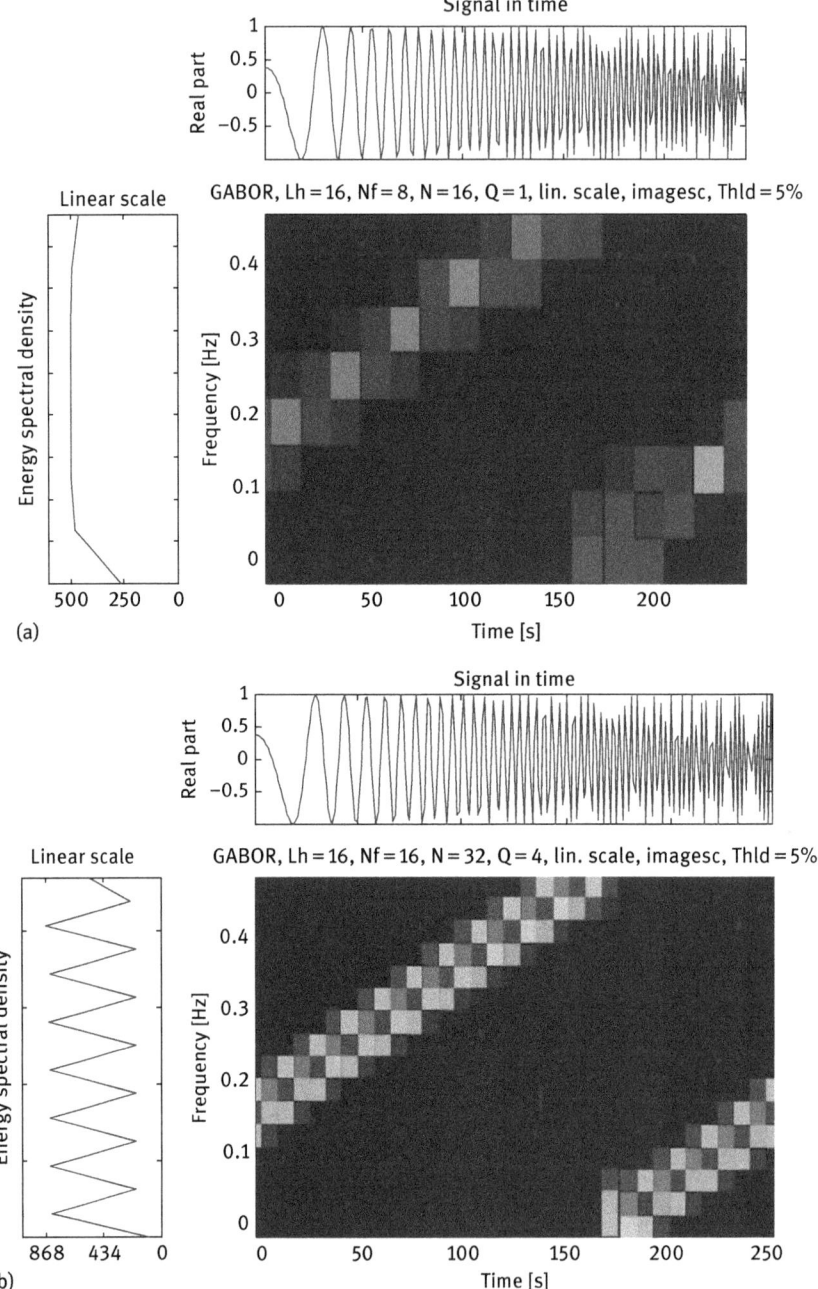

Figure 14.9: The Gabor transform of a linear frequency modulation signal: (a) oversampling factor $q = 1$ and (b) oversampling factor $q = 4$.

clarity for most practical signals, but it provides an alternative framework to investigate new definitions and new methods. While it does suffer from an inherent cross-term contamination when analyzing multicomponent signals, by using a carefully chosen window function, the interference can be significantly mitigated at the expense of resolution.

1. Mathematical definition

For signal $z(t)$, the definition of Cohen's class of bilinear (or quadratic) time–frequency distributions is as follows:

$$P(t,f) = \int_{-\infty}^{\infty} \int_{-\infty}^{\infty} \int_{-\infty}^{\infty} z\left(u + \frac{1}{2}\tau\right) z^*\left(u - \frac{1}{2}\tau\right) \phi(\tau,v) e^{-j2\pi(vt + f\tau - vu)} du\, dv\, d\tau \qquad (14.46)$$

where "*" represents the plural conjugate, $\phi(\tau,v)$ is Cohen's kernel function, which is often a low-pass function, τ is the time delay, and v is the frequency delay. The performance of the bilinear time–frequency representation is determined by the kernel function. The requirements for the kernel function contain both to minimize the cross-term and have a good feature.

Equation (14.46) can also be abbreviated as

$$P(t,f) = \int_{-\infty}^{\infty} \int_{-\infty}^{\infty} A_z(\tau,v) \phi(\tau,v) e^{-j2\pi(vt + f\tau)} dv\, d\tau \qquad (14.47)$$

$A_z(\tau,v) = \int_{-\infty}^{\infty} z(t + \frac{1}{2}\tau) z^*(t - \frac{1}{2}\tau) e^{j2\pi vt} dt$ is called an ambiguity function. It is a two-dimensional time–frequency distribution function obtained by inverse Fourier transform of dual time signal $r_z(t,\tau) = z(t + \frac{1}{2}\tau) z^*(t - \frac{1}{2}\tau)$ with respect to time t.

There are three main advantages of using kernel function to characterize time–frequency distribution of signals. First, the characteristics of the time–frequency distribution can be easily determined by examining the kernel function. Second, for a given kernel function, the time–frequency distribution of the signal can be easily obtained. Third, different time–frequency representations and their properties can be connected together.

If the kernel function $\phi(\tau,v) = 1$ in eq. (14.46), we get another important time–frequency distribution, namely, Wigner–Ville distribution. When the kernel function $\phi(\tau,v)$ is not equal to 1, it can be considered as a filter of ambiguity function.

2. WVD

WVD is a particular case of Cohen's class distributions which yields a time–frequency energy density computed by correlating the signal with a time and frequency translation of itself.

(1)   Definition

The WVD of an analytic signal $z(t)$ is mathematically defined as

$$W_z(t, \Omega) = \frac{1}{2\pi} \int_{-\infty}^{\infty} z\left(t + \frac{1}{2}\tau\right) z^*\left(t - \frac{1}{2}\tau\right) e^{-j\Omega\tau} d\tau \qquad (14.48)$$

which can be considered as a Fourier transform of the instantaneous autocorrelation $r_z(t, \tau) = z(t + \frac{1}{2}\tau) z^*(t - \frac{1}{2}\tau)$. $z(t)$ appears twice in eq. (14.48). It avoids the problem of constraining frequency and time resolution in the linear time–frequency distribution, because it does not contain any window function. This procedure avoids any loss of time–frequency resolution, for instance when performing the windowing in STFT. It has Hermitian symmetry in τ, which remains always real.

In the frequency domain, WVD can also be represented by the spectrum $Z(\Omega)$ of the analytic signal $z(t)$:

$$W_Z(t, \Omega) = \int_{-\infty}^{\infty} Z^*\left(\Omega + \frac{1}{2}v\right) Z\left(\Omega - \frac{1}{2}v\right) e^{-j\Omega v} dv \qquad (14.49)$$

WVD has the desirable property of fulfilling the marginal conditions, thus total signal energy can be calculated in time or in frequency using the Plancherel formula:

$$\|z\|^2 = \int_{-\infty}^{+\infty} |z(t)|^2 dt = \frac{1}{2\pi} \int_{-\infty}^{+\infty} |Z(\Omega)|^2 d\Omega \qquad (14.50)$$

the values $\|z\|^2$ and $|Z(\Omega)|^2$ can be interpreted as energy densities in time and frequency, respectively. This enables the computation of energy present at a given time–frequency box directly from the WVD output.

Another important characteristic of the WVD is its first conditional moment for a given time t_c, which mathematically equals the instantaneous frequency:

$$\frac{d\phi}{dt} = \langle \Omega \rangle_{t_c} = \frac{1}{|s(t_c)|^2} \int_{-\infty}^{+\infty} \Omega W_Z(t_c, \Omega) d\Omega \qquad (14.51)$$

Therefore, it can be computed as the average of all frequencies Ω presented in the time–frequency plane at time t_c.

(2)   Cross-terms

It can be seen from the definition that the WVD is bilinear in nature. Therefore, it deteriorates from the presence of cross-terms, if the signal under analysis is either multicomponent signal or nonlinear frequency-modulated monocomponent signal. For example, a signal $x(t)$ consists of $x_1(t)$ and $x_2(t)$, that is,

$$x(t) = x_1(t) + x_2(t) = A_1 e^{j\phi_1(t)} + A_2 e^{j\phi_2(t)} \tag{14.52}$$

where $\phi(t)$ represents the phase of the signal. The WVD for signal $x(t)$ can be given as

$$\begin{aligned} W_z(t,f) &= \int [z_1(t+\tau/2) + z_2(t+\tau/2)] \left[z_1^*(t-\tau/2) + z_2^*(t-\tau/2) \right] e^{-j2\pi f\tau} d\tau \\ &= \underbrace{W_{z_1}(t,f) + W_{z_2}(t,f)}_{\text{Part 1}} + \underbrace{2\mathrm{Re}\left[W_{z_1,z_2}(t,f) \right]}_{\text{Part 2}} \end{aligned} \tag{14.53}$$

where $f = \Omega/2\pi$; $z_1(t)$ and $z_2(t)$ are the analytical signals of $x_1(t)$ and $x_2(t)$, respectively; and $\mathrm{Re}(g)$ represents the real part. In this equation, Part 1 represents WVDs of the monocomponent signals, and Part 2 represents the WVDs due to cross-component signals.

3. Pseudo-Wigner–Ville distribution (PWVD)

The PWVD is a prominent member of the class of quadratic time–frequency representations. It satisfies a large number of desirable mathematical properties, such as high resolution, energy centrality, and the time–frequency margin property.

WVD is defined on the full time axis ($-\infty < t < \infty$). Although it can be used as an energy distribution to represent the instantaneous characteristics of the signal, it is not easy for real-time signal processing. In order to overcome the shortcoming, a window function can be used to move over the original signal, that is, $x'(t) = x(t)w(t-\tau)$, where $w(t-\tau)$ is a symmetric function centered on τ. WVD of the windowed signal $x'(t)$ is

$$PW_x(t, \Omega) = \int_{-\infty}^{\infty} x\left(t + \frac{1}{2}\tau\right) x^*\left(t - \frac{1}{2}\tau\right) w\left(t - \tau + \frac{1}{2}\tau\right) w^*\left(t - \tau - \frac{1}{2}\tau\right) e^{-j\Omega\tau} d\tau$$

$$\tag{14.54}$$

Equation (14.54) can also be written as

$$PW_x(t, \Omega) = \frac{1}{2\pi} \int_{-\infty}^{\infty} W_x(t, \eta) W_w(t - \tau, \Omega - \eta) d\eta \tag{14.55}$$

$W_w(t, \Omega)$ is the WVD of the window function $w(t)$, that is

$$W_w(t, \Omega) = \int_{-\infty}^{\infty} w\left(t + \frac{\tau}{2}\right) w^*\left(t - \frac{\tau}{2}\right) e^{-j\Omega\tau} d\tau$$

But we only need to know $W_{x'}(t, \Omega)$ when $t = \tau$ (i.e., the center of window function). So we can write

$$PW_x(t, \Omega) = \int_{-\infty}^{\infty} x\left(t + \frac{1}{2}\tau\right) x^*\left(t - \frac{1}{2}\tau\right) w\left(\frac{1}{2}\tau\right) w^*\left(-\frac{1}{2}\tau\right) e^{-j\Omega\tau} d\tau \qquad (14.56)$$

$$PW_x(t, \Omega) = \frac{1}{2\pi} \int_{-\infty}^{\infty} W_x(t, \eta) W_w(0, \Omega - \eta) d\eta \qquad (14.57)$$

Equations (14.56) and (14.57) are called PWVD. It can be seen that $PW_x(t, \Omega)$ is the convolution of $W_x(t, \Omega)$ in the frequency domain. So the result of sliding window is to shorten the data in the time domain and to smooth $W_x(t, \Omega)$ in the frequency domain. Smoothing in the frequency domain degrades the resolution of $W_x(t, \Omega)$ on the frequency axis. By controlling the width of the window function, the degree of reduction of the cross-items can be adjusted.

For instance, the PWVD of the double sinusoidal signal is

$$PW_x(t, \Omega) = \frac{1}{\sqrt{2\pi a}} \left[A_1^2 e^{-(\Omega - \Omega_1)/(2a)} + A_2^2 e^{-(\Omega - \Omega_2)/(2a)} \right]$$
$$+ \frac{2A_1 A_2}{\sqrt{2\pi a}} \cos((\Omega_2 - \Omega_1)t) e^{-(\Omega - (\Omega_1 - \Omega_2)/2)^2/(2a)} \qquad (14.58)$$

It can be seen that if a becomes smaller, then the cross-term becomes smaller. But the true component of the signal also becomes smaller. So the cross-term reduction effect of PWVD distribution on the frequency axis is not very obvious.

PWVD can suppress the cross-terms to some extent and improve the readability of the results. But it all loses some of the fine mathematical properties of WVD, such as the edge characteristics.

4. The realization of WVD

Just as many other signal processing algorithms, the final goal of WVD is to apply it to scientific research or engineering practice. The problems encountered at this time are the discretization of signals and the finite length of data.

(1) The implementation of discrete WVD
The time domain discrete equation obtained from eq. (14.54) is

$$W_x(nT_s, \Omega) = \int_{-\infty}^{\infty} x\left(nT_s + \frac{1}{2}\tau\right) x^*\left(nT_s - \frac{1}{2}\tau\right) e^{-j\Omega\tau} d\tau \qquad (14.59)$$

where $\Omega = 2\pi f$. Let the sampling interval for signal $x(t)$ be T_s, that is, $t = nT_s$ and $\tau/2 = kT_s$, and then $\tau = 2kT_s$. Thus, the integral of τ in eq. (14.73) becomes the sum of k, that is,

$$W_x(nT_s) = 2 \sum_{k=-\infty}^{+\infty} x(n+k)x^*(n-k)e^{-j\frac{2\pi}{M}nk} \tag{14.60}$$

The discrete WVD in the time domain is a function of frequency w, whose period is π. If the sampling period T_s is normalized to 1, then discretization of the frequency means taking M frequency points in the frequency period of the signal. Thus the discrete WVD in the frequency domain is represented as

$$W_x(t, w) = 2 \sum_{k=-\infty}^{\infty} x(n+k)x^*(n-k)e^{-j\frac{2\pi}{M}mk} \tag{14.61}$$

In eq. (14.61), variables of the WVD in both time and frequency domains are all discretized and implemented by using discrete Fourier transform (DFT).

It should be noted that when $x(t)$ becomes $x(n)$, the spectrum $X(j\Omega)$ of $x(t)$ becomes the periodic spectrum $X(e^{jw})$. The period is 2π, and the corresponding sampling frequency of 2π is f_s. In the same way, $W_x(n, w)$ becomes a periodic function with the period π. In order to avoid aliasing, the sampling frequency has to satisfy

$$f_s \geq 4f_{max} \tag{14.62}$$

(2) Discretization of PWVD

The discrete form of eq. (14.56) in the time domain is

$$PW_x(n, w) = 2 \sum_{m=-L}^{L} x(n+m)x^*(n-m)w(k)w^*(-m)e^{-j2mw} \tag{14.63}$$

The width of the window function is $2L - 1$. In order to calculate WVD by the base 2 fast Fourier transform (FFT), it is necessary to perform discretization in the frequency domain. From eq. (14.63) we know that the period of $PW_x(n, w)$ in the frequency domain is π. If the number of sampling points in one period is M, then the sampling interval is $\Delta\Omega = \pi/M$. To calculate easily, zeroes should be padded to make $M = 2L$. So, let

$$G(n, -L) = 0 \tag{14.64}$$

where $G(n, m) = w(m)x(n+m)$, $m = -L+1, \ldots, L-1$. Thus, we have

$$PW_x(n, k) = 2 \sum_{m=-L+1}^{L-1} G(n, m)G^*(n, -m)e^{-j\frac{2\pi}{M}mk} \qquad (14.65)$$

Since FFT is usually calculated in the range of $0 : L - 1$ which is different from that in eq. (14.65), we have to reorder the data sequence. Let

$$f(n, m) = \begin{cases} G(n, m)G^*(n, -m), & m = 0, 1, \ldots, L-1 \\ G(n, m-2L)G^*(n, -m+2L), & m = L, \ldots, 2L-1 \end{cases} \qquad (14.66)$$

So eq. (14.65) can be written as

$$PW_x(n, k) = 2 \sum_{m=0}^{L-1} f(n, m)e^{-j\frac{2\pi}{M}mk} \qquad (14.67)$$

which can be calculated by FFT. In MATLAB, WVD is implemented by the function tftwv, and PWVD is implemented by the function tfrpwv of time–frequency toolbox.

Let's look at the signal $x(t)$ composed of several atomic signals. The so-called atomic signals are usually referred to those signals whose information is concentrated at the atom point in both time and frequency domains. For example, $h(t - t_0)e^{j\Omega_0 t}$ is a typical atomic signal whose atom is (t_0, Ω_0), where $h(t)$ is a finite window function in the time domain. Let

$$x(t) = x_1(t) + x_2(t) + x_3(t) + x_4(t) \qquad (14.68)$$

As shown in Figure 14.10(a), the four atoms of the signals are $(t_1, \Omega_1) = (28, 0.1)$, $(t_2, \Omega_2) = (28, 0.4)$, $(t_3, \Omega_3) = (100, 0.1)$, $(t_4, \Omega_4) = (100, 0.4)$, respectively, and WVD of these signals is shown in Figure 14.10(b).

It can be seen that four terms of the WVD of $x(t)$ are caused by signals themselves. Therefore, their concentration locations in the time–frequency domain are the time positions and the modulation frequencies of $x_1(t)\tilde{x}_4(t)$, respectively. In Figure 14.10, there are also five cross-terms with the center of $(28, 0.25)$, $(64, 0.1)$, $(64, 0.25)$, $(64, 0.4)$, and $(100, 0.25)$, respectively. It can be said roughly that the concentration position of the cross-term is approximately $\left((t_i + t_j)/2, (\Omega_i + \Omega_j)/2\right)$, where t_i and t_j are the concentration positions of cross-component signals in the time domain, and Ω_i and Ω_j are the concentration positions of cross-component signals in the frequency domain.

We can also use PWVD to analyze the signal, which means adding a window on $r_x(t, \Omega) = x(t + \tau/2)x^*(t + \tau/2)$. The result is shown in Figure 14.10(c). Obviously, the cross-items are effectively suppressed, that is, the number of cross-terms is reduced from six to two. At the same time, it can be found that, in Figure 14.10(b), the amplitude of each time–frequency curve is triangular. The reason is that the numbers of data points used to calculate each curve are totally different. In fact,

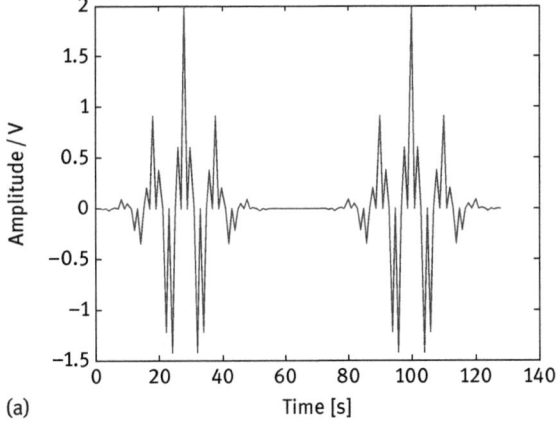

(a)

WV, lin. scale, mesh, Threshold = 5%

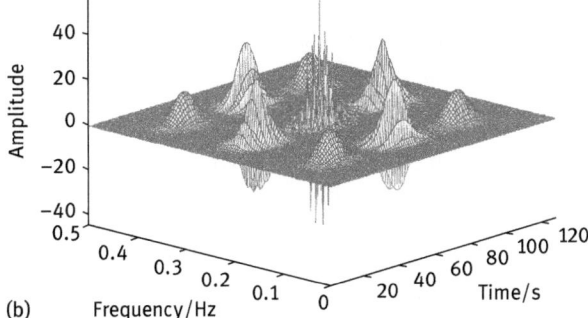

(b)

PWV, Lh = 16, Nf = 128, lin. scale, mesh, Threshold = 5%

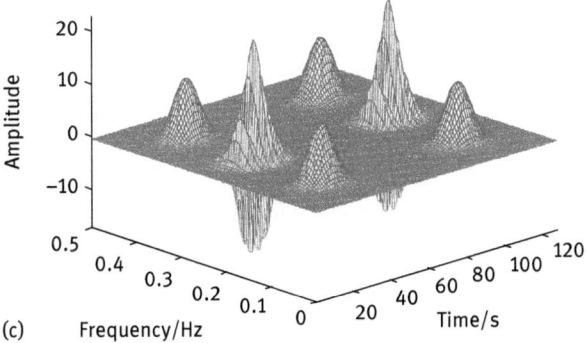

(c)

Figure 14.10: WVD of four atomic signal superposition: (a) real part of the signal; (b) WVD without window; and (c) PWVD.

in order to get WVD of a signal, the signal is first formed into its analytic signal and then a window with length $2L - 1$ is preceded on the signal, and finally, the WVD of the analytic signal is obtained.

14.1.5 The Application of Time–Frequency Distribution in Electrocardiogram Signal Processing

1. Introduction of electrocardiogram (ECG) signal

The heart is a significant organ in the human body, and heart disease is one of the main causes for the death of humans. ECG signal is generated by the action electric potential of the heart. The analysis and detection of ECG signals play an important role in preventing heart diseases.

Normal ECG waveforms are composed of a series of regular wave groups. Through the analysis of ECG waveforms, some basic characteristics of the heart can be obtained. In general, the typical ECG waveform is shown in Figure 14.11. We can see that there are P wave, QRS interval, ST segment, T wave, and U wave in a typical ECG signal from left to right.

(1) P wave is caused by the activation of the atrium. The left part of P wave is caused by the activation of the right atrium, and the corresponding right part of P wave is caused by the activation of the left atrium.

(2) PR interval (or PQ interval) represents the time interval between the start of P wave and the start of QRS interval. It represents the period from the beginning of atrial depolarization to the beginning of ventricular depolarization. The length of PR interval is related to age and heart rate.

(3) QRS interval is caused by the activation of left and right ventricles. And QRS interval generally contains Q wave, R wave, and S wave. The Q wave is the first downward trough after the P wave, the R wave is the crest with a general value below 5 mV after P wave, and the S wave is the first trough after R wave. Sometimes, QRS interval can also have no Q wave or S wave.

(4) ST segment is a relatively smooth ECG signal after QRS interval. In ST segment, by the amplitude measurement, the morphological identification, and classification, we can clinically diagnose myocardial ischemia, myocardial infarction, and coronary heart disease.

(5) T wave is the waveform produced during ventricular depolarization. In most cases, T wave is an upright wave. Because T wave is a relatively high crest after QRS interval, it is always not less than 0.1 times of R wave. And in ECG detecting

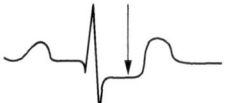

Figure 14.11: The typical ECG waveform.

algorithms, this feature can help to eliminate the crests that are not T wave after QRS interval effectively.

2. The source of data

The data of ECG signal used in this chapter are all from the Massachusetts Institute of Technology standard ECG database, MIT-BIH (Massachusetts Institute of Technology-Beth Israel Hospital). The database collects tens of thousands of dynamic ECGs from Beth Hospital. It includes the data of different status such as at rest and during exercise of normal people and various patients (e.g., sudden cardiac death, heart failure, arrhythmia, epilepsy, and sleep apnea syndrome). Most of the samples are annotated in detail. These data are divided into three categories, namely, Class 1, 2, and 3. Class 1 is the data that the experts have made a note. Class 2 is the original data. Class 3 is the data still in research progress. Each data record in the database includes at least three types of files: a head file (.hea), a data file (.dat), and a comment file (. atr, . al, . aiM, etc.). The head file is a text file describing the data attributes. It contains some important information, such as the record name, the number of signals, the storage format, the number and type of signals, the sampling frequency, the digitization feature, the duration of record, and the start time.

3. The analysis of ECG signal with WVD

In the experiments, two normal sinus rhythm signals are used, and they are obtained from no. 16420 data of MIT-BIH Normal Sinus Rhythm RR Interval Database with a sampling frequency of 128 Hz, as shown in Figure 14.12. Calculating their mutual WVD, we can get the results as shown in Figure 14.13, while the arrhythmia signal is

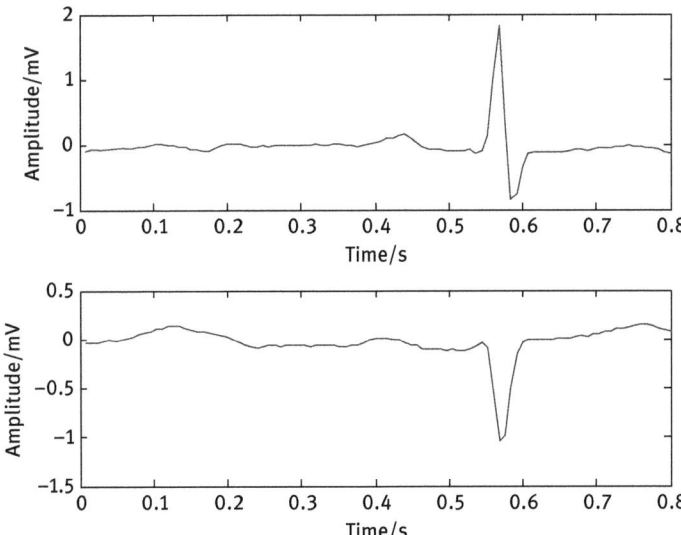

Figure 14.12: Two normal ECG signal waveforms in the time domain.

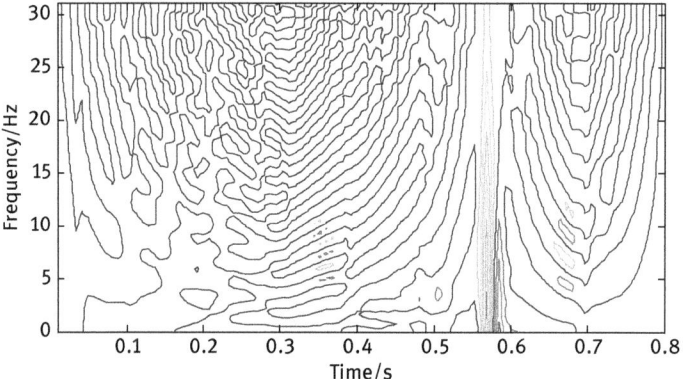

Figure 14.13: The Wigner–Ville distribution of normal heart rate signal.

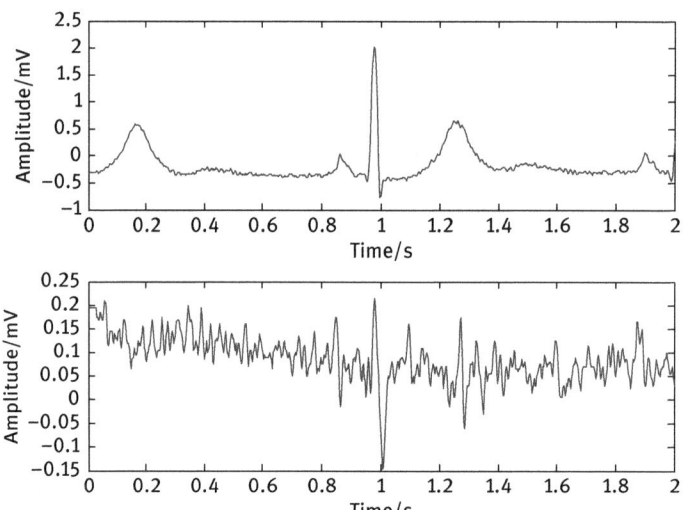

Figure 14.14: Two arrhythmia signals in the time domain.

from no. 106 data of MIT-BIH Arrhythmia Database with a sampling frequency of 360 Hz. The waveform in the time domain and two cross WVDs are shown in Figures 14.14 and 14.15, respectively.

These results shown in the figures demonstrate that the difference between the normal signal and the abnormal signal is obvious; not only the peak value but also the frequency range has changed.

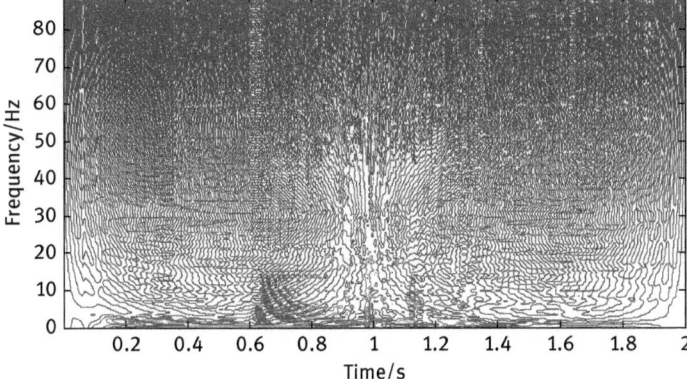

Figure 14.15: The Wigner–Ville distribution of arrhythmia signal.

14.2 The Basic of Wavelets Analysis

14.2.1 Overview

Wavelets analysis theory is proposed by French geophysicist J. Morlet in 1984, which can represent the partial feature of the signal in both time domain and frequency domain. Wavelet analysis has higher frequency resolution and lower time resolution in the low-frequency part, while higher time resolution and lower frequency resolution in the high-frequency part, so it is called mathematical microscope. Some view wavelets as a new basis for representing functions or signals, some consider it as a kind of technique for time–frequency analysis, and others take it as a new mathematical subject. Absolutely, all of them are right, since "wavelets" is a versatile tool with very rich mathematical content and great potential for applications. It is extensively applied in many fields such as time–frequency analysis of signals, signal compression, signal denoising, singularity analysis, and features extraction.

In terms of the transformation, wavelets analysis is similar to Fourier analysis whose nature is to represent a signal by basis functions/signals. In mathematics, a basis function is an element of a particular basis for a function space. Every continuous function in the function space can be represented as a linear combination of basis functions, just as every vector in a vector space can be represented as a linear combination of basis vectors.

Let us analyze the Fourier analysis from the "basis" point of view. Let $L^2(0, 2\pi)$ denote the collection of all measurable functions $f(x)$ defined on the interval $(0, 2\pi)$ with $\int_0^{2\pi} |f(x)|^2 dx < \infty$. It is always assumed that functions in $L^2(0, 2\pi)$ are extended periodically to the real line $\mathbb{R} := (-\infty, \infty)$, that is, $f(x) = f(x - 2\pi)$ for all x. The collection $L^2(0, 2\pi)$ is called the space of 2π-periodic square-integrable function.

Any function $f(x)$ in $L^2(0, 2\pi)$ has a Fourier series representation: $f(x) = a_0 + a_1 \cos x + a_2 \sin x + a_3 \cos 2x + a_4 \sin 2x + \ldots$, where a_n, $n = 0, 1, 2, \ldots$, is called the Fourier coefficients of $f(x)$. This means that $f(x)$ is decomposed into a linear sum of infinitely many mutually orthogonal basis (sine or cosine), where orthogonality means that $\langle f, g \rangle = f^T g = \int f(x)g(x)\mathrm{d}x = 0$. This orthogonality is a distinct feature of the Fourier series representation, and it is useful for determining the coefficients a_0, a_1, a_2, $\ldots$. For example, if the signal $f(x)$ and its Fourier series are known, we can compute inner product between all parts of $f(x)$ and $\cos x$ to obtain a_1:

$$\langle f(x), \cos x \rangle = \langle a_0, \cos x \rangle + \langle a_1 \cos x, \cos x \rangle + \langle a_2 \sin x, \cos x \rangle + \ldots$$

Due to the orthogonal properties, all the non-a_1 items on the right side of the above equation disappear, and only a_1 is left.

We emphasize that the basis is generated by "dilation" of a single function: $w(t) = e^{jx} = \cos x + j \sin x$, which is called integral dilation. This remarkable fact can be summarized by saying that every 2π-periodic square-integral function is generated by a "superposition" of integral dilations of the basic function $w(t) = e^{jx}$ which is a "sinusoidal wave."

We next consider the space $L^2(\mathbb{R})$ of measurable functions $f(x)$, defined on the real line $\mathbb{R}$, which satisfy $\int_{-\infty}^{\infty} |f(x)|^2 \mathrm{d}x < \infty$. Clearly, the two function spaces $L^2(\mathbb{R})$ and $L^2(0, 2\pi)$ are quite different. Since every function (wave) in $L^2(\mathbb{R})$ must "decay" to zero at $L^2(\mathbb{R})$, the sinusoidal (wave) functions do not belong to $L^2(R)$. For all practical purposes, the decay should be very fast. That is, we should look for small waves that are referred to as "wavelets," to generate $L^2(\mathbb{R})$. In other words, basis functions of the wavelet transform are small waves located in different times, which are obtained using scaling function and wavelet function. Therefore, the wavelet transform is located in both time and frequency. It is very useful for analyzing instantaneous time-varying signals, while the Fourier analysis needs information about the whole time domain so that it provides a signal representation localized only in the frequency domain. It does not give any information of signal in the time domain. Therefore, it is less suitable for instantaneous processing. For example, when reconstructing a rectangular wave in Figure 14.16, the Fourier analysis uses a large number of sine waves to approximate the wave, which leads to the Gibbs phenomenon. Because wavelets tell us about variations in local averages, wavelet analysis is superior to Fourier analysis in dealing with this kind of abrupt changing signal. A wave (e.g., sine and cosine) is usually smooth and regular in shape, and can be everlasting, while in contrast, a wavelet may be irregular in shape, and normally lasts only for a limited period of time. A wave is typically used as a deterministic template in Fourier transform to represent a time-invariant or stationary

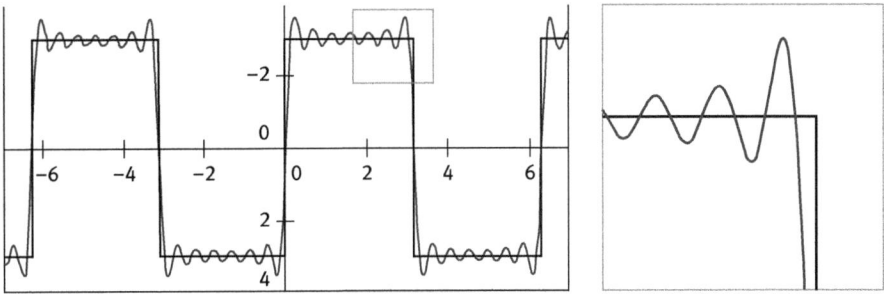

Figure 14.16: The Gibbs phenomenon brought by Fourier analysis.

signal. In comparison, a wavelet can serve as both deterministic and nondeterministic templates to represent a time-varying or nonstationary signal.

14.2.2 Continuous Wavelet Transform

As in the situation of $L^2(0, 2\pi)$, where one single function $w(t) = e^{jx}$ generates the entire space, we also prefer to have a single function, named mother wavelet $\psi(t)$ to generate all of $L^2(\mathbb{R})$. But if $\psi(t)$ decay very fast, how can it cover the whole real line? The obvious way is to shift $\psi(t)$ along $\mathbb{R}$. Therefore, the recognition of continuous wavelet transform (CWT) should start from the mother wavelet first.

1. Mother wavelet
The mother wavelet is also called the basic wavelet or wavelet mother function. In brief, a mother wavelet $\psi(t) \in L^2(\mathbb{R})$ is an oscillation that decays quickly. The equivalent mathematical conditions are admissibility conditions:

$$\int_{-\infty}^{\infty} \frac{\Psi(\Omega)}{\Omega} d\Omega < \infty; \quad \int_{-\infty}^{\infty} \Psi(t)dt = 0; \quad \int_{-\infty}^{\infty} |\Psi(t)|^2 dt < \infty \qquad (14.69)$$

where $\Psi(\Omega)$ is the Fourier transform of $\psi(t)$. If $\psi(t)$ satisfies this condition, it is called the mother wavelet (also called prototype wavelet, basic wavelet, and wavelet prototype function). When the wavelet transform is used to analyze the specific signals in engineering applications, it is often preferred to use the ready-made classical wavelets with good characteristics (such as Morlet wavelet, Mayer wavelet, and spline wavelet) as the mother wavelets. It is also possible to generate a mother wavelet or a corresponding filter through the construction of specific algorithm (such as the construction algorithm of compactly supported orthogonal wavelets). In this case, the admissible condition must be satisfied, so the admissible condition is important to the theoretical analysis. Generally, the mother wavelet has the following characteristics:

(1) Band-pass property

When $\Omega \to 0$, $\Psi(\Omega)/\Omega$ must be meaningful, so $\lim_{\Omega \to \infty}\Psi(\Omega) \to 0$. Therefore, the mother wavelet is not a low-pass filter, and it is only a band-pass filter or high-pass filter. This conclusion is very important in multiresolution analysis (MRA).

(2) Zero mean

$$\Psi(\Omega=0) = \Psi(\Omega)|_{\Omega=0} = 0 \Rightarrow \int_R \psi(t)e^{-j\Omega t}dt\bigg|_{\Omega=0} = 0 \Rightarrow \int_R \psi(t)dt = 0 \qquad (14.70)$$

This indicates the mother wavelet $\psi(t)$ balances itself above/below 0.

(3) Mother wavelets are "small waves"

When selecting mother wavelet, besides the requirement of eq. (14.69), its energy must be concentrated in both time domain and frequency domain. Compared with the sine wave and cosine wave used in Fourier transform, mother wavelets are so small, which is the origin of the name of "wavelet."

In general, the mother wavelet used in wavelet transform is not unique, while the Fourier transform only uses the basis function $\sin(\Omega t)$, $\cos(\Omega t)$, or $\exp(j\Omega t)$. The same problem analyzed by different mother wavelet may lead to different results, so the mother wavelet selection is a very important part of the wavelet analysis. It has a direct effect on the signal processing quality and the calculation complexity.

2. Wavelet basis function

Basis functions of wavelet transform consist of the dilated and translated versions of the mother wavelet. For a given mother wavelet $\psi(t)$, it can be dilated by replacing t with t/a, and translated by replacing t with $t - \tau$ or $t + \tau$. Thus, we get a new function, which is similar in shape but different in "fatness-thinness" and "position," compared with mother wavelet. This function can be written as

$$\psi_{a,\tau}(t) = \frac{1}{\sqrt{a}}\psi\left(\frac{t-\tau}{a}\right), \quad a > 0 \quad \tau \in \mathbb{R} \qquad (14.71)$$

Here, the function $\psi_{a,\tau}(t)$ is called wavelet basis function (also called wavelet in brief or son wavelet) whose a, τ are real numbers. And a presents the scale or dilation parameter and is analogous to frequency. τ is the translation parameter localizing the wavelet basis function in the time domain. As a and τ are continuous, this wavelet is called continuous wavelet. When the scale parameter a increases, the mother wavelet is expanded and changes slowly so that $\psi_{a,\tau}(t)$ can be understood as a larger time window. In this case, $\psi_{a,\tau}(t)$ is a low-frequency signal. Hence, the larger the scale factor a is, the lower the temporal resolution will be. On the contrary, when the scale parameter a reduces, the

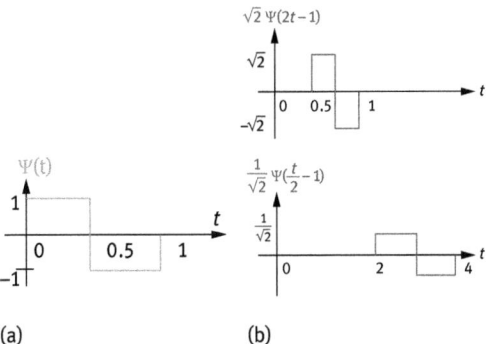

Figure 14.17: (a) Mother wavelet (Haar function); (b) son wavelets that are generated by dilatation and translation from the mother wavelet.

(a)

(b)

mother wavelet is contracted and the waveform changes quickly, which means that $\psi_{a,\tau}(t)$ is narrow in the time domain and corresponds roughly to higher frequencies. So in wavelet analysis, high-frequency components are analyzed with the sharper time resolution. Figure 14.17(a) shows the original mother wavelet; Figure 14.17(b) shows the scale parameter transformation and the translation parameter transformation.

It is worth noting that the factor $1/\sqrt{a}$ before $\psi_{a,\tau}(t)$ is used to ensure energy preservation. If $\varepsilon = \int |\psi(t)|^2 dt$ is the energy of the mother wavelet, then the energy of $\psi_{a,\tau}(t)$ is

$$\varepsilon' = \int \left| \frac{1}{\sqrt{a}} \psi\left(\frac{t}{a}\right) \right|^2 dt = \frac{1}{a} \int \left| \psi\left(\frac{t}{a}\right) \right|^2 dt = \varepsilon \tag{14.72}$$

3. CWT

For $x(t) \in L^2(\mathbb{R})$, taking an inner product (cross-correlation) between the input function $x(t)$ and one of the basis functions $\psi_{a,\tau}(t)$, the CWT is defined as

$$WT_x(a,\tau) = \langle \psi_{a,\tau}(t), x(t) \rangle = \int_{-\infty}^{\infty} \psi_{a,\tau}(t)x(t)dt = \frac{1}{\sqrt{a}} \int_{-\infty}^{\infty} x(t)\psi\left(\frac{t-\tau}{a}\right)dt \tag{14.73}$$

The result $WT_x(a,\tau)$ is called the wavelet transform coefficient. The scale a corresponds to frequency in the Fourier transform.

There are two ways to view eq. (14.73):

(1) It computes the inner product (or cross-correlation) of $x(t)$ with $\psi_{a,\tau}(t)$, at shift τ/a. Therefore, it means, for every (a,τ), the scaled wavelet is similar to the function $x(t)$ at location $t = \tau/a$.

(2) It is the output of a band-pass filter of impulse response $\psi_{a,\tau}(t)$ of input $x(t)$, at the instant $t = \tau/a$.

It can be proved that, under permissive conditions, $x(t)$ can be reconstructed by a double integral as follows:

$$x(t) = \frac{1}{C_\psi} \int_0^\infty \int_{-\infty}^\infty \frac{1}{a^2} \mathrm{WT}_x(a, \tau)\psi_{a,\tau}(t)\mathrm{d}\tau\mathrm{d}a, \quad C_\Psi = 2\int_0^\infty \frac{|\Psi(\Omega)|^2}{\Omega}\mathrm{d}\Omega \tag{14.74}$$

It is called inverse CWT. From eq. (14.74), we can think that $x(t)$ is decomposed into basis function $\psi_{a,\tau}(t)$ and the coefficient are obtained by wavelet transformation of $x(t)$. However, parameters a and τ change continuously, so basis functions $\psi_{a,\tau}(t)$ with different a and τ are not linearly independent. That is to say, the redundancy exists among basis functions, which leads to the correlation among $\mathrm{WT}_x(a, \tau)$

In fact, the CWT is the sum over all time of the signal, multiplied by scaled and shifted versions of the mother wavelet. Its calculation process can be summarized as follows:

Step 1: Select a mother wavelet and place the mother wavelet at the start of the signal $t = 0$. Using eq. (14.73) to calculate the wavelet transform coefficient with $\tau = 0$.

Step 2: Shift the mother wavelet to the right with one time unit, which means to change the translation parameter τ, and repeat step 1.

Step 3: Calculate the wavelet transform coefficients $\mathrm{WT}_x(a, \tau)$ for each different translation parameter τ in the whole signal to be analyzed.

Step 4: Change the scale parameter to expand the mother wavelet by a unit, and repeat steps 1–3 until the whole signal is covered.

Step 5: Calculate all the mother wavelets with different scale parameter, and get the continuous wavelet coefficients of different translation parameters in the whole signal to be analyzed.

Originally, according to the definition of CWT, a and τ must be changed continuously. But the above calculation process has to be accomplished using computer, so a and τ must be increased with a small step size. This is equivalent to sampling the time-scale plane.

Here is a simple example of CWT. Given $x(t) = 3\sin(100\pi t) + 2\sin(68\pi t) + 5\cos(72\pi t)$ which is contaminated by a white Gaussian noise, 3 db is chosen as the wavelet, with different scale parameter ($a = 1, 1.2, 1.4, 1.6, \ldots, 3$), and the CWT obtained by MATLAB is:

```
t = 0:0.01:1;
x = 3*sin(100*pi*t) + 2*sin(68*pi*t) + 5*cos(72*pi*t) + randn(1 , length
    (t));
coefs = cwt(x , [ 1:0.2:3 ] , 'db3' , 'plot');
title(' wavelet transform coefficient with different scale factor');
ylabel('scale');
xlabel('time');
```

Similar to the definition of inner product type (also known as correlation type) given by eq. (14.73), there is also a convolution type definition:

$$\text{WT}_x(a, \tau) = x(\tau) * \psi_a(\tau) = \frac{1}{a} \int\limits_{-\infty}^{\infty} x(t)\psi\left(\frac{\tau - t}{a}\right) dt \tag{14.75}$$

where $\psi_a(t) = (1/a)\psi(t/a)$. Convolution type and inner product type definitions are essentially equivalent, if we set $h(t) = (1/\sqrt{a})\psi*(-t)$, and the inner product type is easy to be defined by the convolution type definition.

14.2.3 Discrete Wavelet Transform

1. The concept of discrete wavelet transform (DWT)

Although the discretized CWT enables the computation of CWT by computers, it is not a true discrete transform. It is just simply a sampled version of the CWT, and the information it provides is highly redundant as far as the reconstruction of the signal is concerned. This redundancy, on the other hand, requires a significant amount of computation time and resources. DWT, on the other hand, provides sufficient information for both analysis and synthesis of the original signal, with a significant reduction in the computation time.

DWT is an implementation of the wavelet transform using a discrete set of the wavelet scales and translations obeying some defined rules. Theoretically, we can discrete the scale factor a and the displacement factor τ in any way. Among them, a more typical and generally accepted discrete way is shown as follows:

(1) Dyadic wavelet transform

The scale factor a is discretized to $a = a_0^j$ ($j = 0, \pm 1, \pm 2, \ldots$, $a_0 > 0$ is the sampling interval and $a_0 \in \mathbb{R}$). Usually $a = 2^j$, and the corresponding wavelet function is $\psi_{a,\tau}(t) = a_0^{-j/2}\psi[a_0^{-j}(t - \tau)]$. In this case, the wavelet transform is referred to as dyadic wavelet transform, unsampled DWT, static wavelet transform, or stationary wavelet transform.

(2) DWT

At the same scale, the translation factor is discretized to $\tau = ka_0^j\tau_0$ ($k = 0, \pm 1, \pm 2, \ldots$, $\tau_0 > 0$ is the sampling interval, and $\tau_0 \in \mathbb{Z}$). The discrete wavelet function is $\psi_{a_0^j, k\tau_0} = a_0^{-(j/2)}\psi[a_0^{-j}(t - ka^j\tau_0)] = a_0^{-(j/2)}\psi(a_0^{-j}t - k\tau_0), j, k \in \mathbb{Z}$.

Adjusting the time axis to set $k\tau_0$ as an integer k on the axis (normalized), we can define the DWT basis as

$$\psi_{j,k}(t) = a_0^{-\frac{j}{2}}\psi(a_0^{-j}t - k), j, k \in \mathbb{Z} \tag{14.76}$$

where $\psi_{0,0}(t)$ is the mother wavelet. With real parameters $a_0 > 0$ and $\tau_0 > 0$, the DWT of $x(t)$ is defined as

$$d(j,k) = \int_R x(t)\psi_{j,k}^*(t)dt \tag{14.77}$$

where $d(j,k)$ is also called DWT coefficient.

The DWT, based on subsamples of the CWT, makes the analysis much more efficient, which is easy to implement and fast to compute. At the same time, by using DWT, the original signal can be recovered fully without loss of data. We notice that a continuous-time signal can be represented in a discrete form as long as the sampling frequency is chosen properly. This is done by using the sampling theorem, termed as the Nyquist theorem: the sampling frequency used to turn the continuous signal into a discrete signal must be twice as large as the highest frequency component present in the signal.

If the set $\left\{\psi_{j,k}(t)\right\}$ forms a frame, any $x(t) \in L^2(\mathbb{R})$ can be recovered from $d(j,k)$ by

$$x(t) = \sum_{j=-\infty}^{+\infty} \sum_{k=-\infty}^{+\infty} d_{j,k}\hat{\psi}_{j,k}(t) \tag{14.78}$$

where $\left\{\hat{\psi}_{j,k}(t)\right\}$ denotes a dual frame of $\left\{\psi_{j,k}(t)\right\}$.

A central issue of the wavelet transform is how to build $\left\{\psi_{j,k}(t)\right\}$ and its dual frames $\left\{\hat{\psi}_{j,k}(t)\right\}$ with desired properties. One can design a tight and exact frame leading to an orthonormal basis in $L^2(\mathbb{R})$:

$$\int \psi_{j,k}(t)\psi_{r,s}(t)dt = \begin{cases} 1, & j=r, k=s \\ 0, & \text{otherwise} \end{cases} \tag{14.79}$$

Under this orthonormal frame, eq. (14.78) becomes

$$x(t) = \sum_{j=-\infty}^{+\infty} \sum_{k=-\infty}^{+\infty} d_{j,k}\psi_{j,k}(t) \tag{14.80}$$

Equation (14.80) is the inverse of DWT, the wavelet series. The equivalent is Fourier series, where only the frequency is the discrete parameter. For computational efficiency, $a_0 = 2$ and $\tau_0 = 1$ are commonly used, resulting in a binary dilation of 2^{-j} and a dyadic translation of 2^{-jk}:

$$\psi_{j,k}(t) = a_0^{-\frac{j}{2}}\psi(a_0^{-j}t - k), j,k \in \mathbb{Z} \tag{14.81}$$

These wavelets for all integers j and k produce an orthonormal basis.

2 . Wavelet frame

As mentioned earlier, wavelet transform is, from an analytical point of view, an inner product between signal $x(t)$ and wavelets. The inner product is the measure of similarity between the signal and the function $\psi_{j,k}(t)$, and is proportional to the degree of closeness $x(t)$ and $\psi_{j,k}(t)$. The more the signal is similar to the function $\psi_{j,k}(t)$, the bigger the inner product. If $x_1(t)$ is close to $x_2(t)$, then the corresponding inner products $\langle x_1(t), \psi_{j,k}(t) \rangle = \int\limits_{-\infty}^{\infty} x_1(t)\psi_{j,k}^*(t)dt$ and $\langle x_2(t), \psi_{j,k}(t) \rangle = \int\limits_{-\infty}^{\infty} x_2(t)\psi_{j,k}^*(t)dt$ should be close to each other too, for all j, k. Mathematically, this property can be written as

$$\left| \sum_{j,k} \left\langle x_1(t), \psi_{j,k}(t) \right\rangle - \left\langle x_2(t), \psi_{j,k}(t) \right\rangle \right|^2 B\|x_1(t) - x_2(t)\|^2, \quad 0 < B < \infty \tag{14.82}$$

where B is a nonnegative real number. Let $x(t) = x_1(t) - x_2(t)$, eq. (14.82) reduces to

$$\sum_{j,k} \left| \left\langle x(t), \psi_{j,k}(t) \right\rangle \right|^2 \le B\|x\|^2, \quad 0 \le B < \infty \tag{14.83}$$

Conversely, if $\langle x_1(t), \psi_{j,k}(t) \rangle$ is close to $\langle x_2(t), \psi_{j,k}(t) \rangle$, then $x_1(t)$ and $x_2(t)$ should be close too, for all j, k. Analogous to eq. (14.83), this property can be described by

$$A\|x\|^2 \le \sum_{j,k} \left| \left\langle x, \psi_{j,k}(t) \right\rangle \right|^2, \quad 0 < A < \infty \tag{14.84}$$

where B is a nonnegative real number.

Combining eqs. (14.82) and (14.83) yields

$$A\|x\|^2 \le \sum_{j,k} \left| \left\langle x(t), \psi_{j,k}(t) \right\rangle \right|^2 \le B\|x\|^2, \quad 0 < A \le B < \infty \tag{14.85}$$

In this case, $\psi_{j,k}(t)$ is called frame, a family of vectors that characterizes any signal from inner product $\langle x(t), \psi_{j,k}(t) \rangle$. For $A = B$, $\psi_{j,k}(t)$ forms a tight frame. If there is a set of dual function $\hat{\psi}_{j,k}(t)$ for the frame $\psi_{j,k}(t)$, then we can further recover the original signal from the set of inner product, that is

$$x(t) = \sum_{j,k} \langle x(t), \psi_{j,k}(t) \rangle \cdot \hat{\psi}_{j,k}(t) \tag{14.86}$$

The dual functions form a dual frame, that is

$$\frac{1}{A}\|x\|^2 \le \sum_{j,k}\left|\left\langle x(t), \hat{\psi}_{j,k}(t)\right\rangle\right|^2 \le \frac{1}{B}\|x\|^2, \quad 0 < A \le B < \infty \tag{14.87}$$

In the N-dimensional space, frame $\mathbf{T}$ is a matrix. Dual frames come from columns $\mathbf{v}_n$ of $\mathbf{T}^{-1}$ and rows $\mathbf{r}_n$ of $\mathbf{T}:\mathbf{T}^{-1}\mathbf{T}=\mathbf{I}$.

As mentioned earlier, we have introduced the fundamental of the DWT. According to frame theory, nonorthogonal bases do not reproduce the signal energy exactly, and reconstructing a signal from the coefficients may amplify any error. Therefore, the remaining question is how to build a set of orthogonal wavelets $\{\psi_{j,k}(t)\}$ to perform wavelet analysis. To construct the DWT effectively, particularly the orthogonal wavelets, an approach called the Mallat algorithm or Mallat's MRA can be used.

14.2.4 Multiscale (Resolution) Analysis

The reason that wavelet transform is widely used in applications is the wavelet transform that has the characteristics of multiresolution. When we look at an object from far away, only the general view of things can be observed and the local details cannot be seen; if we are near to the object, we can observe the local details of things but cannot overview the whole scene. Therefore, if we want to know not only the overall outline of the object but also its local details, we must choose a different distance to observe the object. The analysis of things from coarse to fine or from fine to coarse at different scales (resolution) is called multiscale analysis, also known as MRA. Multiscale analysis is first used in the field of computer vision. When dividing the edges of the image, researchers found that the boundaries between edges and textures depend on the scale of observation and analysis, which inspired them to detect the peak point of the image at different scales. In 1987, Mallat introduced this idea to the wavelet analysis theory in order to study the structure of wavelet function and the decomposition/reconstruction of the signal.

The idea of MRA is to decompose the signal into subsignals corresponding to different frequency contents. This decomposition sometimes is called as subband coding depicted in Figure 14.18.

As illustrated in it, in the time domain, we convert the time resolution to half of the original resolution every time. This means that a larger sampling interval is adopted. For a digital signal, the operation is downsampled by two. While in the frequency domain, the signal bandwidth is divided into two parts, such as high-pass (actually band-pass) and low-pass every time, which means the frequency resolution has been changed to two times of the original. This splitting does not continue until a certain frequency is achieved. All remaining low-frequency components are represented by a low-pass filter. And this low-pass filter is scaling function $\phi(t)$, which is also named father wavelet. It is obvious that the scale function is nothing more than a

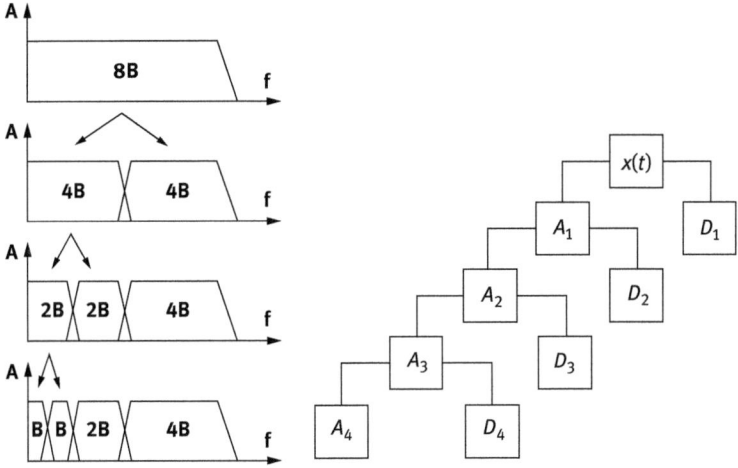

Figure 14.18: Schematic diagram of subband coding.

low-pass filter at a certain scale. Similarly, the complementary basis of DWT called scaling functions of MRA also satisfy the following orthonormality condition within the same scale:

$$\int \phi_{j,k}(t)\phi_{j,s}(t)dt = \delta(k-s) \tag{14.88}$$

Moreover, we can obtain the scaling coefficients of $x(t)$ through the projection

$$c_{j,k} = \langle x(t), \phi_{j,k}(t)\rangle \tag{14.89}$$

The orthonormal wavelet and scaling bases jointly satisfy their complementary basis property as follows:

$$\int \psi_{j,k}(t)\phi_{r,s}(t)dt = 0 \quad \text{for } \forall j, k, r, s \tag{14.90}$$

From Figure 14.19, we can also find that the high-pass part remained each time because the details of the signal are concentrated in it. While most energy of the signal is in the low-frequency part which still has more detail to be partitioned, the splitting should be continued in the low-pass part. Therefore, it is concluded that the result of the DWT is a multilevel decomposition, in which the signal is decomposed in "approximation" (low-pass filtering) and "detail" (high-pass filtering) coefficients at each level.

It is noted that one must use infinite scales in the case of representing $x(t)$ in terms of discrete wavelet functions as shown in eq. (14.80). In contrast, we can

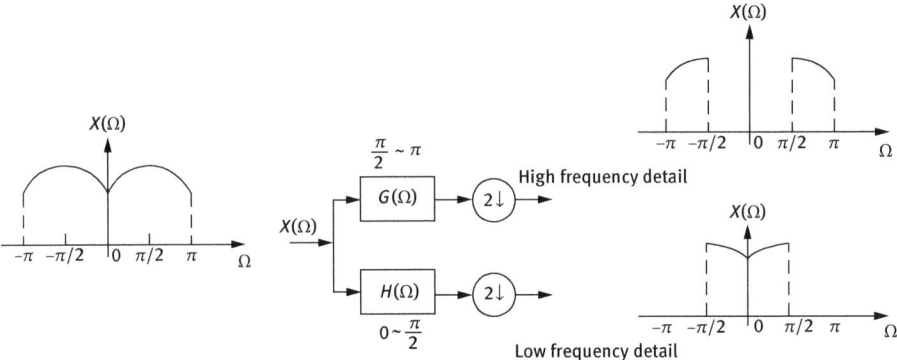

Figure 14.19: Decomposition of the signal in the frequency domain.

express $x(t)$ in a finite number of resolutions if we jointly utilize the scaling and wavelet basis functions as follows:

$$x(t) = \sum_{k=-\infty}^{\infty} c_{j,k} \phi(t-k) + \sum_{k=-\infty}^{\infty} \sum_{j=1}^{L} d_{j,k} c_k \psi(2^j t - k) \tag{14.91}$$

$\phi(t)$ is called the father wavelet, $\psi(t)$ is the mother wavelet, j means the scale (scaling), k means the translation, $c_{j,k}$ is the scale coefficient, representing the discrete approximation of the signal, and $d_{j,k}$ is the wavelet coefficient, representing the discrete details of the signal. This expression is indeed a combination of a low-pass approximation to $x(t)$ utilizing the scaling function at scale L as expressed in the first term and the wavelet representation of the detail signal or approximation error given in the latter.

The scale function (father wavelet) is introduced to facilitate multiscale analysis. The scaling function and the wavelet function are the basis functions of the approximation space and the detail space, respectively. The Fourier transform of the scaling function can generally be treated as a low-pass filter, whereas the Fourier transform of the wavelet mother function is generally used as a band-pass or high-pass filter. This representation is clearly more efficient than the wavelet representation and also highlights the significant role of the scaling basis in the multiresolution signal decomposition framework based on wavelet transforms.

Next, let us interpret the multiscale and scaling functions from the filter and function space in detail.

1. Understanding the MRA from the ideal filter banks

When the sampling frequency of the signal meets the Nyquist requirements, the normalized bandwidth is $-\pi \sim \pi$. As shown in Figure 14.19, it can be decomposed into (positive frequency part) the low-frequency part of $0\sim(\pi/2)$, which depicted approximations, and high-frequency part of $(\pi/2)\sim\pi$, which depicted details. Because the bands do not overlap, it can be said that the two outputs are orthogonal. Since the bandwidth of both outputs is halved, the sampling rate can be halved without causing loss of information. In Figure 14.19, $2\downarrow$ means "decimation by 2."

The two-channel decomposition in Figure 14.19 can be regarded as a primary processing module, and multistage process can be cascaded. The low-frequency part after each decomposition is inputted to the next stage, and it will be redecomposed by the filters $H(\Omega)$ and $G(\Omega)$ so that the output is the low-frequency part $H(\Omega)$ and the high-frequency part $G(\Omega)$ of the next stage. And the sampling rate of each output can be further reduced by half, as shown in Figure 14.20(a). If band $0\sim\pi$ of the original signal $f(t)$ is defined as space V_0, after the first decomposition, V_0 is divided into the low-frequency V_1 (frequency band $0\sim(\pi/2)$) and high frequency W_1 (frequency band $(\pi/2)\sim\pi$). After the second decomposition, V_1 is then divided into V_2 (frequency band $0\sim(\pi/4)$) and W_2 (frequency band $(\pi/4)\sim(\pi/2)$), which is shown in Figure 14.20(b).

In the decomposition process of the signal, we can see that the decomposition divides the original signal $x(n)$ into some "subband" signals with different band-width. Intuitively, if we perform DFT to the subband signals with the same length, then the frequency resolution of each subband signal should not be the same. This analysis process is a process of "coarse" to "fine." Therefore, this method is called

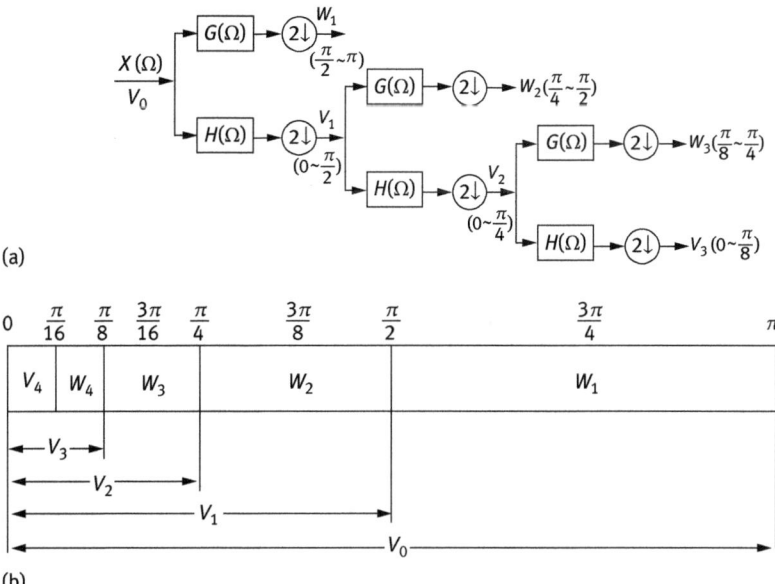

(a)

(b)

Figure 14.20: Cascade schematic of signal decomposition: (a) the level-wise decomposition of signal and (b) partition of frequency space.

"multiresolution analysis (or decomposition)." Each processing stage corresponds to a scale, and multistage cascading generates multiscale or multiresolution for the original signal. The scaling function represents the low-frequency characteristics of the signal, while wavelet function approximates the high-frequency part. Note that only the low-frequency part of upper stage is inputted to the next stage for redecomposition, and the high-frequency part remained at each decomposition stage. If decomposition is preceded in both high-frequency part and low-frequency part at the same stage, it becomes wavelet packet decomposition.

2. From filter bank to space theory

The above-mentioned subspace decomposition can also be described by the mathematical space theory. In the mathematical definition, there is an important Lebesgue space for signal processing. It can be represented by $L^p(\mathbb{R})$ whose meaning is a function space composed of p-integrable functions. The signals studied in wavelet transform belong to $L^2(\mathbb{R})$ space. This space is a set of all square-integrable, measurable functions over R. Essentially, this means that the signal energy must be finite, otherwise it is not integral. This is a requirement of the wavelet transform definition, but this is not the case in practical applications.

Define a space V_0 in $L^2(\mathbb{R})$, which is spanned by $V_0 = \overline{\mathrm{span}\{\phi(t-k)\}}$, where "—" $k \in Z$
indicates that the space is a closed space (i.e., for any $x(t) \in V_0$, there is $x(t) = \sum_k a_k \phi(t-k)$). $\phi(t)$ is called scale function. For the dilation of scaling function, we have

$$\phi_{j,k}(t) = 2^{-\frac{j}{2}}\phi(2^{-j}t-k) = \phi_k(2^{-j}t) \tag{14.92}$$

For each j, $\phi_k(2^{-j}t)$ can span the space V_j, that is, $V_j = \overline{\mathrm{span}_{k \in Z}\phi_k(2^{-j}t)}$.

The process in Figure 14.21 can be described by space $L^2(\mathbb{R})$, which is shown in Figure 14.21.

Obviously, $V_0 \supset V_1 \supset V_2 \ldots, V_0 = V_1 \oplus W, V_1 = V_2 \oplus W_2, \ldots, V_{j-1} = V_j \oplus W_j$.

W_j is the space of detail (high-pass) sequences, so it is called successive detail space. While V_j is the space of coarse (low-pass) sequences V_{j-1} and it is called the successive approximate space. As $j \to \infty$, we get finer and finer approximation. V_{j-1} can be decomposed into two subspaces V_j and W_j as follows:

$$V_{j-1} = V_j \oplus W_j \tag{14.93}$$

W_j and V_j are complementary spaces to each other. The symbol $\oplus$ "direct sum" means that those smaller spaces meet only in the zero function. Figure 14.22 shows the meaning of eq. (14.93).

An MRA is an increasing sequence of subspaces $\{V_n\} \subset L^2(\mathbb{R})$, $n \in Z$ with the fundamental characteristics:

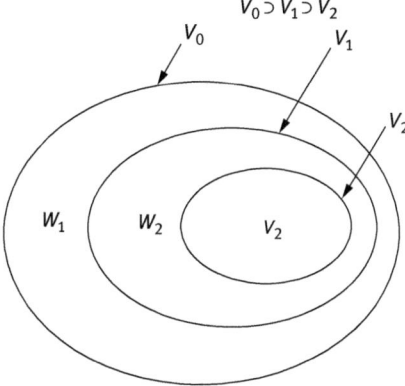

$$V_0 \supset V_1 \supset V_2$$

Figure 14.21: Multiresolution space.

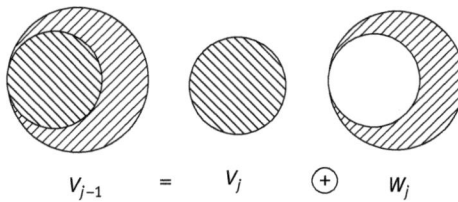

$$V_{j-1} \quad = \quad V_j \quad \oplus \quad W_j$$

Figure 14.22: Interpretation of $V_{j-1} = V_j \oplus W_j$.

Embedding: $V_{-\infty} \supset \cdots \supset V_{-1} \supset V_0 \supset V_1 \supset V_2 \supset \cdots \supset V_\infty$

Upward and downward completeness: $\overset{\infty}{\underset{j=-\infty}{\cup}} V_j = L^2(\mathbb{R})$ and $\overset{\infty}{\underset{j=-\infty}{\cap}} V_j = \{0\}$.

Shift invariance: $x(t) \in V_j \Leftrightarrow x(t - 2^j k) \in V_j$, and scale invariance: $x(t) \in V_j \Leftrightarrow x(2^m t) \in V_{j-m}$.

3. Multiscale analysis from the functional space

Next, the scale space, wavelet space,, and corresponding basis functions are discussed in detail. According to the above discussion, if we want to analyze $L^2(\mathbb{R})$, $V_j|_{j\to\infty}$ is needed. But in practice, $j \to \infty$ is impossible. So we can discuss from the subspace V_0.

(1)   Scale space and scale function

① Subspace V_0

Let's recall the definition of the subspace $V_0 = \overline{\text{span}_{k\in Z}\{\phi(t-k)\}}$, that is, the space V_0 formed by $\phi_k(t) = \phi(t-k)$ (which is called scale function). This means that $\phi(t-k)$ forms a set of orthogonal bases, and any function $x(t)$ in V_0 can be expressed as a linear combination of $\phi(t-k)$, that is

$$x(t) = \sum_{k\in Z} a_k \phi(t-k) \tag{14.94}$$

According to the orthogonal property, it is easy to find the coefficient a_k as

$$a_k = \langle x(t), \phi(t-k) \rangle \tag{14.95}$$

So,

$$x(t) = \sum_{k \in Z} a_k \phi(t-k) = \sum_{k \in Z} \langle x(t), \phi(t-k) \rangle \phi(t-k) \tag{14.96}$$

② Subspace V_1

According to the definition, if $\phi(t) \in V_0$, then according to the property of shift invariance, there must be $\phi(t/2) \in V_1$. After translating it many times, we get $\{\phi(t/2-k)\}_{k \in Z}$. It is easy to prove that $\{\phi(t/2-k)\}_{k \in Z}$ must be a set of orthogonal base in V_1 and meets

$$\langle \phi(t/2-k), \phi(t/2-m) \rangle = \delta(k-m) \tag{14.97}$$

$\forall g(t) \in V_1$; it can be expressed as a linear combination of $\{\phi(t/2-k)\}_{k \in Z}$, that is,

$$g(t) = \sum_{k \in Z} b_k \phi(t/2-k) \tag{14.98}$$

where $b_k = \langle g(t), \phi(t/2-k) \rangle$.

③ Subspace V_j

After stepwise decomposition, the whole space is divided into a series of subspaces, and the orthogonal basis corresponding to each subspace can be obtained. The embedding property of the nested spatial sequence indicates that $\{V_j\}_{j \in Z}$ is a scaled version of V_0. In other words, for any upper spaces that contain V_0, their bases can be obtained by enlarging or narrowing the scale function $\phi(t)$ in the frequency domain, and then make integer transformations in the time domain, that is:

$$\phi_{j,k}(t) = 2^{-\frac{j}{2}} \phi(2^{-j}t - k) \tag{14.99}$$

j is called the scale factor and k is called the translation factor, which is the same as those in the discrete wavelet basis function. V_j is defined as $V_j = \overline{\mathrm{span}_k\{\phi_{j,k}(t)\}}$ for $j \in Z$ which represents the scale space under the scale j. For $j > 0$, V_j is a stretched version of V_0, and for $j < 0$, V_j is a compressed version of V_0 (both by 2^j). This also means that for any function $x(t)$ belongs to space V_j, it can be expressed as $x(t) = \sum_{k \in Z} a_k \phi_{j,k}(t)$. It can be seen that when j is larger, the magnitude of the integer translation is smaller. Thus, the finer detail of $x(t)$ in this j subspace is shown, and vice versa. That is to say, scaling and translation of scale functions can bring different subspaces, and different subspaces will bring different resolutions, so we can use different resolutions to observe the target signal, and this is the meaning of multi-scale analysis. According to the properties of MRA, $f(t) \in V_0 \Leftrightarrow f(2^{-j}t) \in V_j$. And we can say there exists the following relationship between the subspaces in $L^2(\mathbb{R})$:

$$V_{-\infty} \supset \cdots \supset V_{-1} \supset V_0 \supset V_1 \supset \cdots \supset V_\infty$$

$$\llcorner \ = \{0\}$$

$$V_1 = \left\{ f(x) : f(2x) \in V_0 \right\}$$

$$V_0 = \left\{ f(x) \right\}$$

$$V_{-1} = \left\{ f(x) : f(2^{-1}x) \in V_0 \right\}$$

$$\text{dense in } L^2(R)$$

(2)   Wavelet space and wavelet function

In this part, the decomposition and inclusion relationship of scale space V_j is discussed emphatically, and the scaling function is derived. The high-frequency detail space W_j, which is the difference between the adjacent scale spaces, is obtained in each decomposition, that is,

$$W_j = V_{j-1} - V_j \tag{14.100}$$

On the other hand, we can say that W_j is the orthogonal complement of V_j in V_{j-1}:

$$V_{j-1} = V_j \oplus W_j \tag{14.101}$$

Using eq. (14.101) and iterating it n times, we get $V_j = W_{j+1} \oplus W_{j+2} \oplus \cdots \oplus W_{j+n} \oplus V_{j+n}$. As $n \to \infty$ and due to the property of embedding, we obtain

$$V_j = \overset{\infty}{\underset{i=j+1}{\oplus}} W_i \tag{14.102}$$

Let $j \to \infty$, then we have

$$L^2(\mathbb{R}) = \underset{j \in Z}{\oplus} W_j \tag{14.103}$$

Thus we obtain a split of $L^2(\mathbb{R})$ into a collection $\{W_j\}_{j \in Z}$. This is a good complementary to the concept of MRA. Usually, we call $\{W_j\}_{j \in Z}$ as wavelet space.

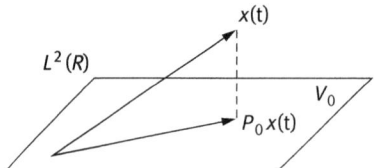

Figure 14.23: Schematic diagram of signal projection.

(3)   Function approximation

In the above part, we have discussed MRA from the perspective of the whole space. Actually, a function in $L^2(\mathbb{R})$ can be regarded as a signal vector and projected into scale space. For example, for $x(t) \in L^2(\mathbb{R})$, it can be projected onto scale space V_0, which is denoted as $P_0 x(t)$. From the point of view of the signal vector, $P_0 x(t)$ is also a kind of decomposition. The decomposed information of signal $x(t)$ is entirely located in the space V_0, which is shown in Figure 14.23.

Obviously, the left part of the signal can be represented as

$$P_0^\perp x(t) = x(t) - P_0 x(t) \tag{14.104}$$

that is,

$$x(t) = P_0 x(t) + P_0^\perp x(t) \tag{14.105}$$

$P_0 x(t) \in V_0$ can be expressed as a linear combination of $\{\phi(t-k)\}_{k \in Z}$:

$$P_0 x(t) = \sum_{k \in Z} a_{0,k} \phi(t-k) \tag{14.106}$$

The coefficient $a_{0,k}$ is

$$a_{0,k} = \langle P_0 x(t), \phi(t-k) \rangle \tag{14.107}$$

Since $P_0^\perp x(t)$ is orthogonal to V_0, it is also orthogonal to the orthogonal base $\{\phi(t-k)\}_{k \in Z}$ in V_0, that is

$$\langle P_0^\perp x(t), \phi(t-k) \rangle = 0 \tag{14.108}$$

Therefore, eq. (14.107) can be rewritten as

$$a_{0,k} = \langle P_0 x(t), \phi(t-k) \rangle = \langle P_0 x(t) + P_0^\perp x(t), \phi(t-k) \rangle = \langle x(t), \phi(t-k) \rangle \tag{14.109}$$

In mathematics, $P_0 x(t)$ is called the smooth approximation of $x(t)$ in V_0, which means the approximation of $x(t)$ at scale $j = 0$, and a_k is called the approximation coefficient (also known as discrete approximation).

Similarly, it can be deduced that if $P_j x(t)$ is the projection of $x(t)$ in V_j, the smooth approximation in V_j can be represented as

$$P_j x(t) = \sum_{k \in Z} a_{j,k} \phi_{j,k}(t) = 2^{-j/2} \sum_{k \in Z} a_{j,k} \phi(2^{-j}t - k) \tag{14.110}$$

The coefficient $a_{j,k}$ is

$$a_{j,k} = \langle P_0 x(t), \phi_{j,k}(t) \rangle = \langle x(t), \phi_{j,k}(t) \rangle \tag{14.111}$$

Combining eqs. (14.110) and (14.111), we get

$$P_j x(t) = \sum_{k \in Z} \langle x(t), \phi_{j,k}(t) \rangle \cdot \phi_{j,k}(t) \tag{14.112}$$

(4)   Two-scale equation

The two-scale equation is of great importance to MRA and wavelet analysis. It represents the relations of these orthogonal basis functions in the neighbor scaling spaces V_{j-1} and V_j, and the wavelet function $\psi(t)$ can be represented by the scale function $\phi(t)$.

As $V_0 \subset V_{-1}$, $W_0 \subset V_{-1}$, $\phi(t)$, and $\psi(t)$ must belong to V_{-1} space, then $\phi(t)$ (the orthonormal basis function of scale space V_0) and $\psi(t)$ (the orthonormal basis function of wavelet space W_0) can be expanded in orthonormal basis of space V_{-1}:

$$\phi(t) = \sum_n h(n)\phi_{-1,n}(t) = \sqrt{2} \sum_n h(n)\phi(2t - n) \tag{14.113}$$

$$\psi(t) = \sum_n g(n)\phi_{-1,n}(t) = \sqrt{2} \sum_n g(n)\phi(2t - n) \tag{14.114}$$

where the coefficients $h(n)$ and $g(n)$ depend on $\phi(n)$ and $\psi(n)$, and not on the scale j, that is,

$$h(n) = \langle \phi(t), \phi_{-1,n}(t) \rangle \tag{14.115}$$

$$g(n) = \langle \psi(t), \phi_{-1,n}(t) \rangle \tag{14.116}$$

Equations (14.115) and (14.116) describe the relationship between the basis functions of the two neighbor scaling spaces; therefore, they are called two-scale equations.

The two-scale equation relationship exists in any adjacent scaling space:

$$\phi_{j,0}(t) = \sum_n h(n)\phi_{j-1,n}(t) \tag{14.117}$$

$$\psi_{j,0}(t) = \sum_n g(n)\phi_{j-1,n}(t) \tag{14.118}$$

It can be seen that the right side of eqs. (14.113) and (14.114) are the convolution at n. From the viewpoint of digital filter processing, this is a generalized digital filter. Low-frequency smoothing approximation $\phi_{0,n}(t)$ can be obtained by inputting $\phi_{-1,n}(t)$ into $h(n)$, so $h(n)$ is called low-pass filter. Similarly, high-frequency smoothing approximation $\psi_{0,n}(t)$ can be obtained by inputting $\phi_{-1,n}(t)$ into $g(n)$, so $g(n)$ is called high-pass filter.

14.2.5 Implementation Technology of Wavelet Transform

1. Mallat algorithm

As described in the previous sections, a signal is decomposed using the wavelet transform technique into two sets of coefficients called approximations $c_{j,k}$ and details $d_{j,k}$, that is,

$$x(t) = \sum_{k=-\infty}^{\infty} c_{j,k}\phi(t-k) + \sum_{k=-\infty}^{\infty}\sum_{j=0}^{\infty} d_{j,k}c_k\psi(2^j t - k) \tag{14.119}$$

The scale factors $c_{j,k} = \langle x(t), \phi_{j,k}(t)\rangle$ represent the low frequency. The wavelet coefficients $d_{j,k} = \langle x(t), \psi_{j,k}(t)\rangle$ represent the high-frequency signal components. Calculating the detail and approximation coefficients through such inner production is time consuming, especially when the decomposition algorithm is applied repeatedly to intermediate sets of coefficients (known as multilevel decomposition). A more efficient algorithm results from calculating $c_{j,k}$ and $d_{j,k}$ through convolution of the input signal with the scaling filter and wavelet filter, respectively. This recursive decomposition algorithm is sometimes referred to as the cascade algorithm or the pyramid algorithm. It is similar to the role of FFT in Fourier analysis, so it is the key to the fast wavelet transform algorithm. But it should be pointed out that the algorithm is only suitable for the orthogonal wavelet; if the wavelet is not orthogonal (such as B spline wavelet), then the algorithm is invalid.

Using $c_{j,k} = \langle x(t), \phi_{j,k}(t)\rangle$ and the dilation equation in two-scale equations, an efficient decomposition algorithm for computing the $c_{j,k}$ coefficients is obtained:

$$c_{j,k} = \sum_{m\in Z} h(m-2k)c_{j-1,m} \tag{14.120}$$

where j is the level (or scale), k is the translation index, and $h(m-2k)$ is the scaling function filter coefficients. This equation says that lower-level approximation coefficients $c_{j,k}$ are computed recursively by the approximation coefficients at a higher level c_{j-1}. Similarly, $d_{j,k}$ is obtained:

$$d_{j,k} = \sum_{m \in Z} g(m - 2k)c_{j-1,m} \tag{14.121}$$

where $g(m - 2k)$ is the wavelet filter coefficient. Recursive application of these decomposition equations provides a means for calculating lower-level detail and approximation coefficients once the higher-level approximation coefficients are obtained.

Equations (14.120) and (14.121) are the fast algorithm of Mallat. Conversely, $c_{j-1,k}$ can also be reconstructed from $c_{j,k}$ and $d_{j,k}$; this is the reconstruction algorithm of Mallat, and the reconstruction formula is

$$c_{j-1,k} = \sum_{m} c_{j,m} h(k - 2m) + \sum_{m} d_{j,m} g(k - 2m) \tag{14.122}$$

The schematic diagram of the algorithm is shown in Figure 14.24.

In the Mallat coefficient decomposition algorithm, the starting point is the coefficient c_0. But when it is applied, the sequence $x(n)$ which is obtained by the continuous signal $x(n)$ can be directly used as c_0, that is, $c_0 = x(n)$.

The Mallat fast algorithm (14.120) and (14.121) can also be described by the digital filter. In this case, the inner product of $x(t)$ with $\psi_{j,k}(t)$ and $\phi_{j,k}(t)$ is converted into the filtering of $x(t)$ with $h(n)$ and $g(n)$. Equations (14.120) and (14.121) are rewritten as follows:

$$c_j(k) = \sum_{m \in Z} h(m - 2k)c_{j-1}(m) \tag{14.123}$$

$$d_j(k) = \sum_{m \in Z} g(m - 2k)c_{j-1}(m) \tag{14.124}$$

In eqs. (14.123) and (14.124), changing the translation index, k, by 1 results in the indices of $h(n)$ and $g(n)$ sequences being offset by 2. Thus, there are half as many coefficients at level j as those at level $j-1$. The result is a downsampling of the coefficient vectors by a factor of 2 in the decomposition algorithm. So eqs. (14.123) and (14.124) are equivalent to a digital filter that $c_{j-1}(m)$ goes through the impulse responses of $h(-n)$ and $g(-n)$, and then they are carried on the "two extraction,"

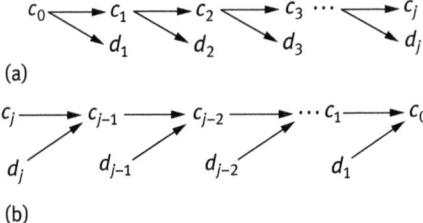

(a)

(b)

Figure 14.24: Schematic diagram of the Mallat algorithm: (a) decomposition algorithm and (b) reconstruction algorithm.

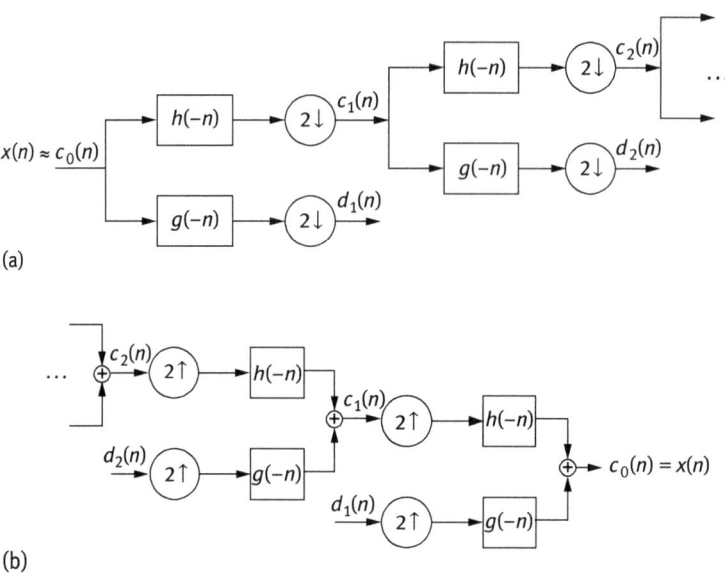

(a)

(b)

Figure 14.25: Equivalent schematic Mallat's algorithm from filtering perspective. (a) Decomposition process and (b) reconstruction process.

respectively. The block diagram is shown in Figure 14.25. In the diagram, $2\downarrow$ means "two extraction" and $2\uparrow$ means "two interpolation."

2. À-trous algorithm

À-trous algorithm is a fast algorithm that uses the Mallat algorithm structure to calculate the wavelet transform. À-trous is a French word and "trous" means holes in English. It refers the insertion of zeros in the low-pass filter $h(n)$ and the high-pass filter $g(n)$. The à-trous algorithm is a nonorthogonal, shift-invariant, dyadic, symmetric, undecimated, redundant DWT algorithm. The à-trous algorithm is more suitable for image data fusion and image enhancement.

By means of the equivalent translocation of decimation and filtering, the à-trous algorithm can be derived from the Mallat algorithm. Assuming that $X(z)$ is the z-transform of the sequence $x(n)$, $H(z)$ and $G(z)$ are the z-transforms of the filter $h(n)$ and the filter $g(n)$, and the process of the decomposition and reconstruction of the à-trous algorithm is shown in Figure 14.26. $H(z^{2^j})$ represents the z-transform of the filter obtained by inserting $2^j - 1$ zeros between any two points of the filter $h(n)$, so the à-trous algorithm is to insert $2^j - 1$ zero points between any adjacent points of the filters $h(n)$ and $g(n)$, and then convolve with the low-frequency signal.

Obviously, this algorithm does not carry out the two decimation, but only the two interpolation for the filters, so that all the wavelet transform values at each sampling point can be calculated. And it makes the decomposed signal to have the same data length as the original input signal, and the algorithm has translation invariance.

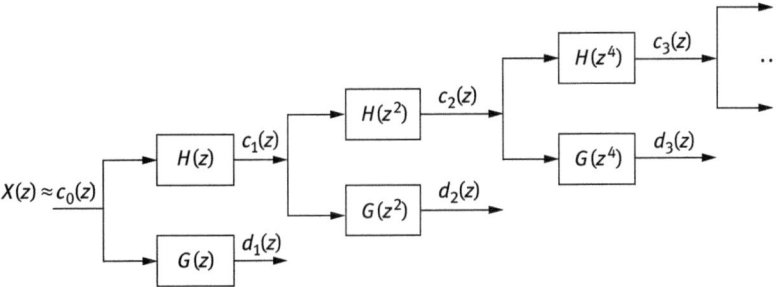

Figure 14.26: Diagram of the signal decomposition of à trous algorithm.

Compared with other wavelet transform algorithms, it has the following characteristics:

1. The computational space and time requirements are reasonable and easy to implement.
2. This transformation process can be achieved by filtering.
3. The calculation does not need sampling and interpolation, which is beneficial to obtain the details of the signal.

14.2.6 Common Wavelet Basis Functions

Compared with the standard Fourier transform, the choice of wavelet basis function determines the effect and efficiency of wavelet transform. Analyzing the same problem with different wavelet basis functions can produce different results. In general, the choice of wavelet basis function can be considered from the following aspects:

1. Complex-valued wavelet and real-valued wavelet
Using the complex-valued wavelet to analyze the signal can simultaneously obtain the amplitude information and the phase information of the signal, which are mainly suitable for analysis and calculation of the normal characteristics of the signal. The real-valued wavelet is better for peak or discontinuity detection.

2. Support length
The longer the support length is, the more time will be spent, which makes it more likely to generate wavelet coefficients with higher amplitude. Most applications choose wavelets whose support length is between 5 and 9. If the support length is too long, boundary problems will arise. If the length of the support is too short, the vanishing moment is too low for the concentration of the signal energy.

3. Symmetry
The corresponding filters to this kind of wavelets have linear phases, which can avoid phase distortion in the image processing.

4. Regularity

Regularity is generally used to describe the smoothness of the function; the higher the regularity, the better the smoothness of the function. The regularity of the wavelet basis mainly affects the stability of reconstruction. Under normal circumstances, the better the regularity, the longer the support length, and thus longer the calculation time; therefore there are a lot of compromises.

5. Similarity

The similarities are not the absolute equality or extreme approaching, but just represent a trend. Selecting a wavelet that is similar to the signal waveform is valuable for compression and denoising.

However, there is not a good method or even a well-known formula or an explicit standard for the selection of the wavelet basis function. Therefore, choosing the optimal wavelet basis function in the practical engineering applications usually comes from experience.

Some commonly used wavelet transform basis functions utilized in the literature include Daubechies, Coiflets, Symlets, and so on.

14.2.7 Application of Wavelet Analysis in Signal Denoising

The application of wavelet analysis is closely related to the theory of wavelet analysis. When the wavelet transform is applied to the signal processing, the general steps are as follows:

First, the signals are decomposed by wavelet transformation to get wavelet coefficients; then, according to the requirements of the applications, the wavelet coefficients are processed; finally, the signals are reconstructed by the inverse wavelet transformation.

In Figure 14.27, the decomposition and reconstruction processes are the same for different applications, but the middle "processing" is unequal for different applications. The application of wavelet transform in signal denoising is described as follows.

If the noise is stationary, an empirically recorded signal that is corrupted by additive noise can be represented as

$$x(t) = s(t) + \sigma n(t) \tag{14.125}$$

where $x(t)$ is noisy signal, $s(t)$ is noise-free actual signal, $n(t)$ is independently normal random signal, and σ represents the intensity of the noise in $x(t)$. The noise is usually modeled as stationary-independent zero-mean white Gaussian variables.

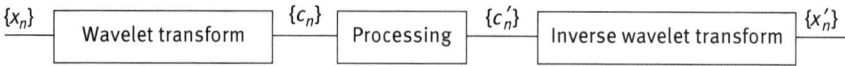

Figure 14.27: The process of signal analysis using wavelet transform.

The objective of noise removal is to reconstruct the original signal $s(t)$ from a finite set of $x(t)$ values without assuming a particular structure for the signal. The usual approach to noise removal models noise as high-frequency signal added to an original signal. The wavelet-based noise removal techniques can be divided into three categories:

(1) Denoising by modulus maximum principle of wavelet transformation

The signal and noise have different propagation characteristics at each scale of wavelet transform. So according to the variation of modulus maxima with scales, the modulus maxima generated by noise are eliminated. In this method, the selection of propagation neighborhood is very crucial.

(2) Denoising based on correlation of wavelet coefficients

After wavelet transformation of noisy signals, the correlations of wavelet coefficients between adjacent scales are calculated, the types of wavelet coefficients are distinguished according to correlation, and then the signal is reconstructed to realize denoising.

(3) Denoising by thresholding

Compared with the above-mentioned methods, this is a rather simple way to remove noise. So, we focus on it here. Indeed, the main idea of the wavelet denoising to obtain the ideal components of the signal from the noisy signal requires the estimation of the noise level. The estimated noise level is used in order to threshold the small coefficient assumed as noise. The procedure of the signal denoising based on DWT consists of three steps; decomposition of the signal, thresholding, and reconstruction of the signal. The procedure is shown in Figure 14.28.

In MATLAB, the signal denoising of the one-dimensional wavelet functions of the wavelet analysis toolbox are wden.m and wdencmp.m.

The core of wavelet denoising is the selection of threshold value because it directly affects the quality of noise reduction. Many theoretical and empirical models of threshold selection have been developed in literatures. But there is no universal model to solve this problem since each has its own limitation. Normally, the threshold is selected by σ which represents the noise intensity in eq. (14.125). After the noise intensity is obtained, the threshold can be determined by different mathematical models. In MATLAB, the functions to determine the threshold are ddencmp, wdcbm and wdcbm2, and thselect.m . Next, we take thselect.m as an example to describe the selection rules.

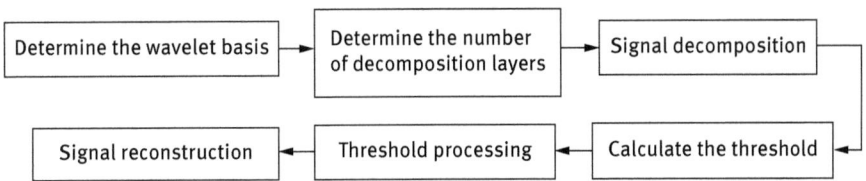

Figure 14.28: Procedure of denoising by thresholding.

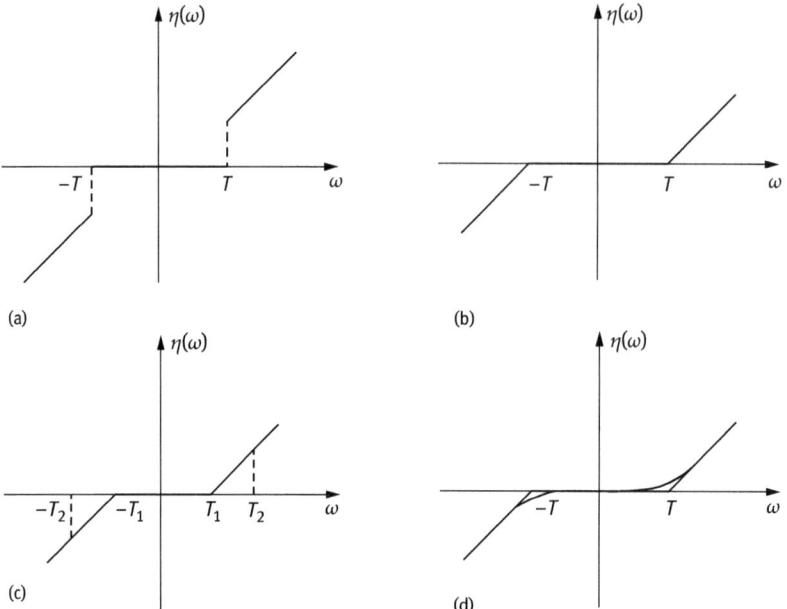

Figure 14.29: Threshold types: (a) hard threshold; (b) soft threshold; (c) semisoft threshold; and (d) improved soft threshold.

The thresholding can be applied by implementing four methods (as shown in Figure 14.29). Let ω be the original wavelet coefficient, $\eta(\omega)$ be the wavelet coefficients after threshold, and T be the threshold:

$$I(x) = \begin{cases} 1, & x \text{ is true} \\ 0, & x \text{ is false} \end{cases} \tag{14.126}$$

(1) Hard threshold: In this method, the wavelet coefficients below a given value are set to zero. And it does not affect the detail coefficients that are greater than the threshold level and they may be unstable and are sensitive even to small changes in the signal:

$$\eta(\omega) = \omega I(|\omega| > T) \tag{14.127}$$

(2) Soft threshold: In this method, the wavelet coefficients are reduced to a quantity or the threshold value. It can create unnecessary bias when the true coefficients are large:

$$\eta(\omega) = (\omega - \text{sgn}(\omega)T)I(|\omega| > T) \tag{14.128}$$

(3) Semisoft threshold: It takes into account the advantages of the soft threshold and hard threshold together:

$$\eta(\omega) = \text{sgn}(\omega)\frac{T_2(|\omega| - T_1)}{T_2 - T_1}I(T_1 < |\omega|\langle T_2) + \omega I(|\omega|\rangle T_2) \qquad (14.129)$$

(4) The improved soft threshold: It has a smooth transition between the wavelet coefficients representing the noise and the wavelet coefficients representing the signal:

$$\eta(\omega) = \begin{cases} \omega + T - \frac{T}{2k+1}, & \omega < -T \\ \frac{1}{(2k+1)T^{2k}}\omega^{2k+1}, & |\omega| \le T \\ \omega - T + \frac{T}{2k+1}, & \omega > T \end{cases} \qquad (14.130)$$

The following is a concrete example to illustrate the effect of wavelet denoising. MIT-BIH arrhythmia database directory of ECG signals from PhysioNet has been utilized. This database includes 48 files of 30 minutes recording separated into two parts: the first one contains 23 files (numbered from 100 to 124 with some missing numbers), and the second one contains 25 files (numbered from 200 to 234, again with few missing numbers).

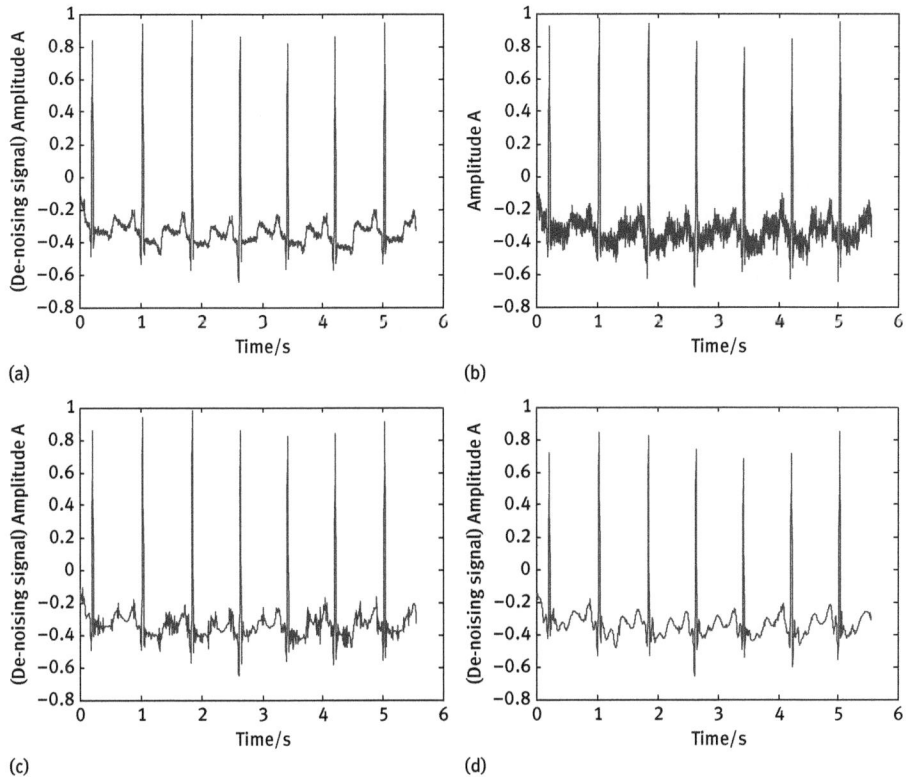

Figure 14.30: Effects of various threshold methods on MIT-BIH115 with white noise: (a) ECG raw signal; (b) ECG plus Gaussian white noise; (c) ECG hard threshold wavelet filtering results; and (d) ECG soft threshold wavelet filtering results.

The noise is not very large, so the noise is artificially added for denoising experiments. Figure 14.30(a)–(d) shows the original signal of MIT-BIH115 signal, noisy signal, hard threshold, and soft threshold denoising results.

It can be seen that the effect of soft threshold denoising method for noisy ECG is better than that of hard threshold denoising method. The amplitudes of P wave and T wave of the signal no. 115 are lower and their time durations are large. Due to the discontinuity of the hard threshold method in the preservation of the wavelet coefficients, the smooth low-frequency part of the signal is prone to distortion, causing a slight concussion, as shown in Figure 14.30(c).

14.3 Hilbert–Huang Transformation

14.3.1 Introduction

In order to analyze and deal with nonstationary and nonlinear signal, a series of time–frequency analysis theory are proposed and developed to describe the law of frequency changing with time. As mentioned in the previous sections, the time–frequency analysis methods are mainly divided into linear time–frequency analysis and nonlinear time–frequency analysis methods. The linear time–frequency analysis is converted from Fourier transform spectrum analysis, including STFT, Gabor transform, and wavelet transform. Nonlinear time–frequency analysis mainly refers to the quadratic time–frequency analysis, which is represented by the Cohen class distribution, and the common ones are WVD, and so on.

The basic idea of linear time–frequency analysis is realized by window function which cannot be changed after determined. The chosen wavelet basis function in wavelet transform can only be used to analyze the whole signal, thus making the adaptability worse. The nonlinear time–frequency analysis is always subject to interference of cross-terms. At the same time, the time and frequency resolution of time–frequency analysis is not ideal because of the uncertainty principle. Therefore, there is an urgent need for an adaptive analysis method to accurately reflect the variation of frequency with time. At the same time, because of the limitation of the principle of uncertainty, the resolution of time and frequency is not ideal. Therefore, there is an urgent need for an adaptive method to accurately reflect the variation of frequency with time.

Against such a background, Norden E. Huang et al. put forward the HHT (Hilbert–Huang transformation) theory to analyze the nonlinear and nonstationary data. HHT is a designated name for the combination of the empirical mode decomposition (EMD) and the Hilbert spectral analysis. Based on EMD, it decomposes the signal into the sum of a series of IMF (intrinsic mode function), and then performs Hilbert transform on each IMF to obtain its instantaneous frequency. The final presentation of the results is an energy–frequency–time distribution, designated as the Hilbert spectrum. Compared with the FFT method, each amplitude and frequency of the Hilbert–Huang transform is changed over time, which eliminates the redundancy and no physical harmonics introduced to reflect nonlinear and nonstationary processes. Compared with the wavelet analysis method, HHT can eliminate the ambiguity of the wavelet analysis, and have a more accurate spectral structure, which means that the analysis results of HHT can reflect the original physical characteristics of the system more accurately. The application field of HHT is very wide, including the analysis signals of medicine, acoustics, vibration and noise, and financial management.

14.3.2 The Basic Concepts and Theory of Hilbert–Huang Transform

The method of using HHT to analyze signals consists of three steps:
1. According to the characteristics of the signal, the signal is adaptively decomposed into a series of IMF from high to low frequencies.
2. For each IMF, the Hilbert transform is used to obtain the time–frequency representation of the signal, that is, the Hilbert spectrum.
3. The Hilbert spectrum of all IMF is aggregated, and the time–frequency energy distribution of the original signal is obtained.

From this process, we can see that the HHT is more like an algorithm, rather than a theoretical tool. The difference between Hilbert spectrum and other time–frequency analysis spectra, such as STFT and wavelet transform, lies in the following: HHT can analyze the instantaneous frequency and instantaneous energy of the signal. The next section elaborates the definition of instantaneous frequency.

1. The definition of instantaneous frequency
The instantaneous frequency is an intuitive concept in physical phenomena. The voice with changing tone and the phenomenon of periodic variation reflect its existence. It is an important parameter to describe nonstationary signals. But it is meaningless to obtain instantaneous frequency from the original signal directly. The necessary condition for a meaningful instantaneous frequency is that the function is symmetric and zero mean in the local range and has the same extreme point and zero point. When the Hilbert transform is introduced, the problem is well solved. The important meaning of HHT is that IMF components are obtained by EMD and the instantaneous information is obtained for each intrinsic mode function .

2. The Hilbert transform and its properties

(1) Definition
Hilbert transform of a signal $x(t)$ is defined as the transform in which phase angle of all components of the signal are shifted by $\pm 90°$:

$$\hat{x}(t) = \frac{1}{\pi} \int\limits_{-\infty}^{\infty} \frac{x(\tau)}{t-\tau} d\tau \qquad (14.131)$$

It can be seen that $\hat{x}(t)$ is the convolution of $x(t)$ and $1/\pi t$. This means that the Hilbert transform $\hat{x}(t)$ can be seen as the output of a linear time-invariant system whose input is $x(t)$.

The impulse response of the system is $h(t) = 1/\pi t$, whose frequency response is $H(j\Omega) = -j\text{sign}(\Omega)$. The Hilbert transform can be regarded as an all-pass filter with amplitude–frequency characteristic of 1, because

$$\text{sign}(\Omega) = \begin{cases} 1, & \Omega \geq 0 \\ -1, & \Omega < 0 \end{cases} \qquad (14.132)$$

According to the convolution theorem, due to the $\hat{x}(t) = x(t) \star 1/\pi t$, the spectrum of $\hat{x}(t)$ is $\hat{X}(j\Omega) = X(j\Omega)H(j\Omega)$. Thus $Y(j\Omega) = \begin{cases} -jX(j\Omega), & \Omega \geq 0 \\ jX(j\Omega), & \Omega < 0 \end{cases}$. That is to say, the Hilbert transform has the effect of shifting the phase of the negative frequency components of $x(t)$ by +90° ($\pi/2$ radians) and the phase of the positive frequency components by −90°. Therefore, the Hilbert transform is also called phase shift filter. And it can also be seen that the Hilbert transform, like the Fourier transform is a global integration of signals. The difference is that the Hilbert transform does not transform from time domain to frequency domain. The Hilbert transformation results are still in the time domain.

(2)   The Hilbert transformations of a few common signals
The Hilbert transformations of a few common signals are shown in Table 14.1.

When a real signal $x(t)$ and its Hilbert transform $\hat{x}(t)$ are used to form a new complex signal $z(t) = x(t) + j\hat{x}(t) z(t) = x(t) + j\hat{x}(t)$, the signal $z(t)$ is named as the (complex) analytic signal corresponding to the real signal $x(t)$:

Table 14.1: Selected Hilbert transforms.

$x(t)$	The Hilbert transform of $\hat{x}(t)$
$\cos(2\pi ft)$	$\sin(2\pi ft)$
$\sin(2\pi ft)$	$-\cos(2\pi ft)$
$\delta(t)$	$\frac{1}{\pi t}$
$\frac{1}{\pi t}$	$-\delta(t)$
$x_B(t)\cos(2\pi ft)$, $x_B(t)$ is a band-limited signal	$x_B(t)\sin(2\pi ft)$, $x_B(t)$ is a band-limited signal
$\cos(2\pi ft) + a$	$\sin(2\pi ft)$

$$z(t) = x(t) + j\hat{x}(t) = a(t)e^{j\theta(t)} \tag{14.133}$$

where $a(t) = [x^2(t) + y^2(t)]^{1/2}$, $\theta(t) = \arctan(\hat{x}(t)/x(t))$. $a(t)$ is the instantaneous amplitude of complex signal $z(t)$, $\theta(t)$ is called instantaneous phase.

The spectrum of $z(t)$ is

$$Z(j\Omega) = \int_{-\infty}^{+\infty} z(t)e^{-j\Omega t}\,dt \tag{14.134}$$

It is easy to prove that the spectrum of the analytic signal exists only in the positive frequency part. The average frequency can be calculated by the following formula:

$$\langle \Omega \rangle = \int \Omega |Z(j\Omega)|^2 d\Omega = \int \theta'(t)a^2(t)\,dt \tag{14.135}$$

Instantaneous frequency is defined as

$$\Omega(t) = \frac{d\theta(t)}{t} \tag{14.136}$$

This means that the instantaneous frequency is generated by the derivation of the phase function. The instantaneous frequency thus obtained is local and has high accuracy, while the obtained frequency of Fourier transform is global, and the obtained frequency of wavelet transform is regional.

(3) The restrictions of signal caused by the physical significance of the instantaneous frequency

A large number of experiments show that Hilbert transform is not suitable for all kinds of analytical signals. If the signal is not constrained, the instantaneous frequency obtained by this method has several contradictory problems in understanding. The signal must satisfy some restrictions so that the meaningful instantaneous frequency can be obtained.

It can be seen from eq. (14.136) that, for a given moment, only a unique instantaneous frequency value corresponds to the signal. So the Hilbert transform can only represent a single component signal. The meaning of a "single component signal" is that at any point in time the signal has only one frequency value, which represents a component. But so far, there is no definite definition on the "single component signal," and it can only be intuitively judged whether some signals with explicit analytic expressions are single component. For most of the signals in nature, such as seismic acceleration records, it is impossible to obtain their analytic formula, and it is impossible to judge whether it is a "single component function." In fact, such kind of complicated signal cannot be a "single component."

Because of the lack of a clear definition of this concept, bandwidth is often used as a constraint. In order to make the instantaneous frequency based on Hilbert transform has a physical meaning, the signal must be narrowband. In addition, the symmetry of time domain waveform of the signal about the zero axis plays an important role in the Hilbert transform.

① Bandwidth limitations

Although it is not mathematically possible to prove that a single component signal is a local narrowband signal, the local narrowband signal has obvious physical meaning. Moreover, the linear combination of local narrowband signals can represent most of the nonstationary signals, so from the practical point of view for signal analysis, the concept of single component can be limited to local narrowband signals.

② Symmetry conditions

The restriction condition is that the waveform of the signal should be axially symmetric about zero axis in order to make the instantaneous frequency based on the Hilbert transform meaningful. A simple example is used to illustrate this restriction.

Consider a simple signal with an offset and a fixed frequency, namely

$$x(t) = \alpha + \sin(t) \tag{14.137}$$

Its Hilbert transform is

$$y(t) = -\cos(t) \tag{14.138}$$

The resulting equation is

$$(x(t) - \alpha)^2 + y^2(t) = 1 \tag{14.139}$$

This equation shows that the signal and its Hilbert transform correspond to a unit circle in the Cartesian coordinate plane. The physical interpretation of the instantaneous frequencies corresponding to different α is shown in Figure 14.31. When $\alpha = 0$, the instantaneous frequency is a horizontal line whose value remains constant. This is consistent with our intuitive perception. Because even without the Hilbert transform, we can intuitively judge the frequency of the signal from eq. (14.137). When $0 < \alpha < 1$, the number of zero intersections of the signal is equal to the extreme number of points, that is, it satisfies the first kind of "narrowband" condition. But because the symmetry axis of the waveform has offset to the zero axis (time axis), the instantaneous phase and the instantaneous frequency of the Hilbert transform have obvious fluctuation; when $\alpha > 1$, the results of the Hilbert transform present a more serious problem: not only the instantaneous frequency has appeared the

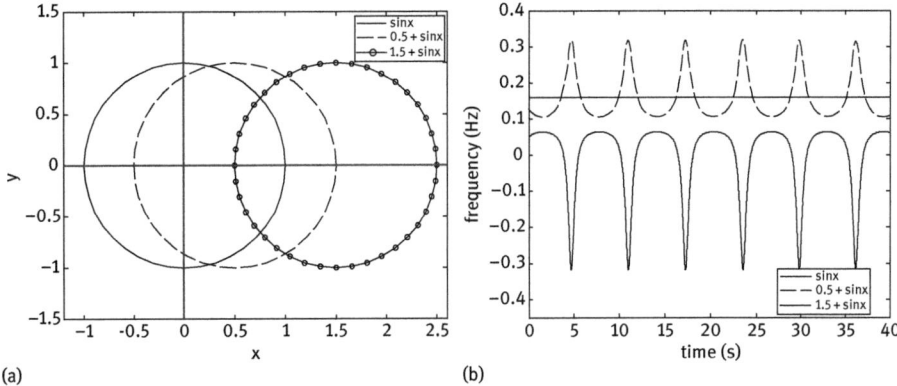

Figure 14.31: Instantaneous frequency of physical interpretation: (a) of the signal and Hilbert transform in rectangular coordinate plane and (b) the corresponding instantaneous frequency.

fluctuation but also there are some negative values in some time. While as a physical concept, the negative value of instantaneous frequency is meaningless.

Through this example, we can see that even for a simple cosine signal, if we translate the waveform so that it is no longer symmetrical about the zero axis, then the instantaneous frequency obtained by the Hilbert transform will also be meaningless.

A waveform will not change its frequency due to a simple translation. In the case of $0 < \alpha < 1, \alpha = 0, \alpha > 1$, the instantaneous frequency of the signal given by eq. (14.137) should be kept as a constant. Therefore, in order to make the instantaneous frequency obtained by the Hilbert transform meaningful, the waveform of the signal must be symmetrical about zero axis. In other words, the instantaneous frequency described above has a physical meaning only when the signal is locally symmetrical for zero mean.

Therefore, before the Hilbert transform, the signal must be properly processed to produce a new class of function definitions. The definition is based on the local characteristics of the signal called the IMF, and defines the function as a single component function. For an intrinsic mode function, the instantaneous frequency can be defined at any point.

14.3.3 Intrinsic Mode Function

The key part of the HHT is the EMD method, which can decompose any complicated data into a finite number of components called intrinsic mode functions). In Hilbert–Huang transform, the signal $x(t)$ can be decomposed into the sum of several IMF components $c_j(t)$ and residual signals $r_n(t)$, namely, $x(t) = \sum_{j=1}^{n} c_j(t) + r_n(t)$. The purpose of this is to perform time–frequency analysis. It is explained in Section 14.3.2 that Hilbert time–frequency analysis is suitable for the local narrowband signal. As a single component signal, the local narrowband signal is also a nonstationary signal. Its linear combination can represent most of the nonstationary signals. Based on this understanding, to some extent, the essence of IMF can be considered as local narrowband signals

1. Definition

The requirements for defining IMF are as follows:

1. In the whole data set, the number of extrema and the number of zero crossings must either be equal or differ at most by one.
2. At any point, the mean value of the envelope is zero.

These requirements are necessary for IMF because they represent the oscillation mode embedded in the data. The number of extremes here represents the sum of the number of peaks (maximum value) and trough (minimum value) in the whole data set, and the zero crossing indicates the time point at which the signal crosses the zero coordinate line.

The first condition in the IMF definition is equivalent to traditional narrowband requirements for a stationary Gaussian process. The second condition is a new concept that replaces the global requirement with local restrictions, which means that the decomposition is based on the local characteristic timescale of the data, namely, the local mean of the data is replaced by the local mean of the envelope. These envelopes are defined by local maxima and local minima. The upper envelope can be obtained by connecting all the maxima by curve fitting. The lower envelope can be obtained by connecting all the minima. The mean value of the envelope is the mean of the upper envelope and the lower envelope.

By definition, an IMF is any function with the same number of extrema and zero crossings, whose envelopes are symmetric with respect to zero. Therefore, an IMF represents a simple oscillatory mode as a counterpart to the simple harmonic function, but it is much more general. Instead of constant amplitude and frequency in a simple harmonic component, an IMF can have variable amplitude and frequency along the time axis. This decomposition method operating in the time domain is adaptive, and, therefore, highly efficient.

2. Decomposition

The EMD decomposes data in terms of IMFs. The decomposition is based on the assumptions:

1. The signal has at least two extrema: one maximum and one minimum.
2. The characteristic timescale is defined by the time difference between the extrema.

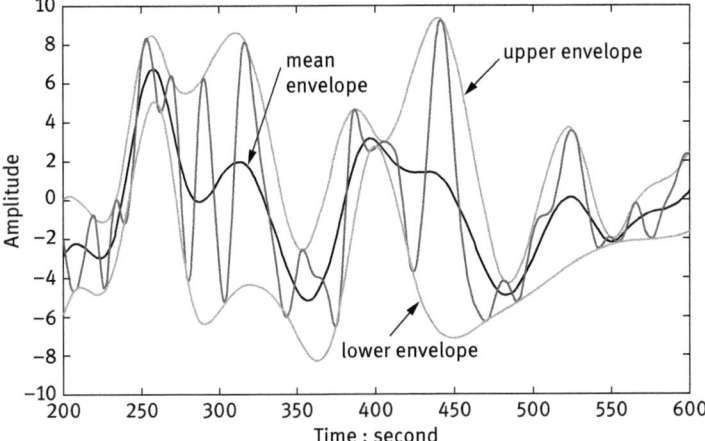

Figure 14.32: Test data and its envelope diagram.

3. If the data set only contains inflection points without any extrema, then it can be differentiated once or more times to reveal the extrema.

The decomposing process is as follows:

(1) Identify all the local extrema, connect all the local maxima by a cubic spline line as the upper envelope, and repeat the procedure for the local minima to produce the lower envelope.

Figure 14.32 shows the test data and its envelope. In fact, the curve can be fitted with other methods. The basic requirement is to ensure that the fitted curve can completely envelop the signal, and has certain smoothness, and can be adaptively adjusted according to the characteristics of the signal.

(2) The mean envelope is obtained by averaging the upper and lower envelopes, which is designated as m_1. The difference between the data and m_1 is the first component h_1, that is,

$$x(t) - m_1 = h_1 \tag{14.140}$$

Ideally, h_1 should satisfy the definition of an IMF. After the first round of sifting, the hump may become a local maximum. New extrema generated in this way actually reveal the proper modes lost in the initial examination. In the subsequent sifting process, h_1 can only be treated as a proto-IMF.

(3) h_1 is treated as the data, then

$$h_1 - m_{11} = h_{11} \tag{14.141}$$

If it still does not satisfy the requirements for the IMF, then the sifting process is performed again until the h_{1k} is the IMF, that is, $h_{1k-1} - m_{1k} = h_{1k}$.

(4) The first IMF component from the data is designated as

$$h_{1k} = c_1 \tag{14.142}$$

Historically, two different criteria have been used. The first stoppage criterion is determined by a Cauchy type of convergence test, that is

$$SD = \frac{1}{T} \int_0^T \frac{|h_{i,k}(t) - h_{i,k-1}(t)|^2}{h_{i,k-1}^2(t)} dt \tag{14.143}$$

which stops the sifting process when it becomes smaller than a predetermined value. Another criterion is based on the agreement of the numbers of zero crossings and extrema. Specifically, the sifting process will stop only if the numbers of zero crossings and extrema are equal or at most different by one. Once a stoppage criterion is selected, the first IMF, c_1, can be obtained. Overall, c_1 should contain the finest scale or the shortest period component of the signal.

(5) Separate c_1 from the rest of the data by

$$x(t) - c_1 = r_1 \tag{14.144}$$

Since the residue, r_1, still contains longer period variations in the data, it is treated as the new data and subjected to the same sifting process as described above.

(6) This procedure can be repeated to all the subsequent r_j, and the result is

$$\begin{aligned} r_1 - c_2 &= r_2 \\ r_2 - c_3 &= r_3 \\ &\vdots \\ r_{n-1} - c_n &= r_n \end{aligned} \tag{14.145}$$

r_n is called trend term.

(7) The sifting process stops finally when the residue r_n becomes a monotonic function from which no more IMF can be extracted. By summing up eqs. (14.144) and (14.145), it follows that

$$x(t) = \sum_{i=1}^{n} c_i + r_n \tag{14.146}$$

The residue r_n is the mean trend of $x(t)$. Thus, a decomposition of the data into n empirical modes is achieved. The IMFs $c_1, c_2, \ldots, c_n$ include different frequency bands ranging from high to low. The frequency components contained in each frequency band are different and they change with the variation of signal $x(t)$. It is shown that the EMD is equivalent to a dyadic filter bank.

14.3.4 Hilbert Spectrum and Marginal Hilbert Spectrum

1. Hilbert spectrum

After obtaining each IMF with EMD, in order to get the relationship between frequency and time, it is necessary to submit the IMFs to the Hilbert transformation process. The Hilbert spectrum is a spectrogram that represents the joint distribution of time, frequency, and energy.

After the EMD, original signal can be expressed as

$$x(t) = \sum_{i=1}^{n} c_i + r_n \tag{14.147}$$

where c_i is IMF components, r_n represents the central tendency of signal $x(t)$.

This is still a signal representation in the time domain. In order to convert the signal from time domain to frequency domain, each IMF component needs to perform the Hilbert transform. For one IMF $c_i(t)$ in eq. (14.147), we can always have its Hilbert transform as

$$\tilde{c}_i(t) = \frac{1}{\pi} \int_{-\infty}^{\infty} \frac{c_i(\tau)}{t-\tau} d\tau \tag{14.148}$$

With this definition, we can have an analytic signal as

$$z_i(t) = c_i(t) + j\tilde{c}_i(t) = a_i(t)e^{j\phi_i(t)} \tag{14.149}$$

in which

$$a_i(t) = [c_i^2(t) + j\tilde{c}_i^2(t)]^{1/2} \tag{14.150}$$

$$\phi_i(t) = \arctan \frac{c_i(t)}{\tilde{c}_i(t)} \tag{14.151}$$

From eq. (14.151), we can have the instantaneous frequency as

$$\Omega_i(t) = \frac{\mathrm{d}\phi_i(t)}{\mathrm{d}t} \tag{14.152}$$

$\Omega_i(t)$ is the instantaneous frequency function of the i th component $c_i(t)$. After performing the Hilbert transform to each IMF component, the original signal can be expressed as the real part in the following form:

$$x(t) = \mathrm{Re}\left(\sum_{i=1}^{n} a_i(t)\mathrm{e}^{\mathrm{j}\int \Omega_i(t)\mathrm{d}t}\right) \tag{14.153}$$

Here the residue r_n is left out, for it is either a monotonic function or a constant. Eq. (14.153) gives both amplitude and frequency of each component as functions of time. This frequency–time distribution of the amplitude is designated as the Hilbert spectrum $H(\Omega, t)$:

$$H(\Omega, t) = \sum_{i=1}^{n} a_i(t)\mathrm{e}^{\mathrm{j}\int \Omega_i(t)\mathrm{d}t} \tag{14.154}$$

If the square of the amplitude of the Hilbert spectrum is integrated, then the Hilbert energy spectrum can be defined as $\mathrm{Er}(\Omega)$:

$$\mathrm{Er}(\Omega) = \frac{1}{T}\int_0^T (H(\Omega, t))^2 \mathrm{d}t \tag{14.155}$$

Figure 14.33 shows the flowchart of the process of HHT.

2. Marginal spectrum
With the defined Hilbert spectrum, we can also define the marginal spectrum, $h(\Omega)$ as

$$h(\Omega) = \frac{1}{T}\int_0^T H(\Omega, t)\mathrm{d}t \tag{14.156}$$

where T is the total data length. The Hilbert spectrum offers a measure of amplitude contribution from each frequency and time, while the marginal spectrum offers a measure of the total amplitude contribution from each frequency. Its physical meaning is the average frequency in the whole data time.

After performing the Hilbert transform on some of the IMF components that we are interested in, we can also define the local Hilbert spectrum by

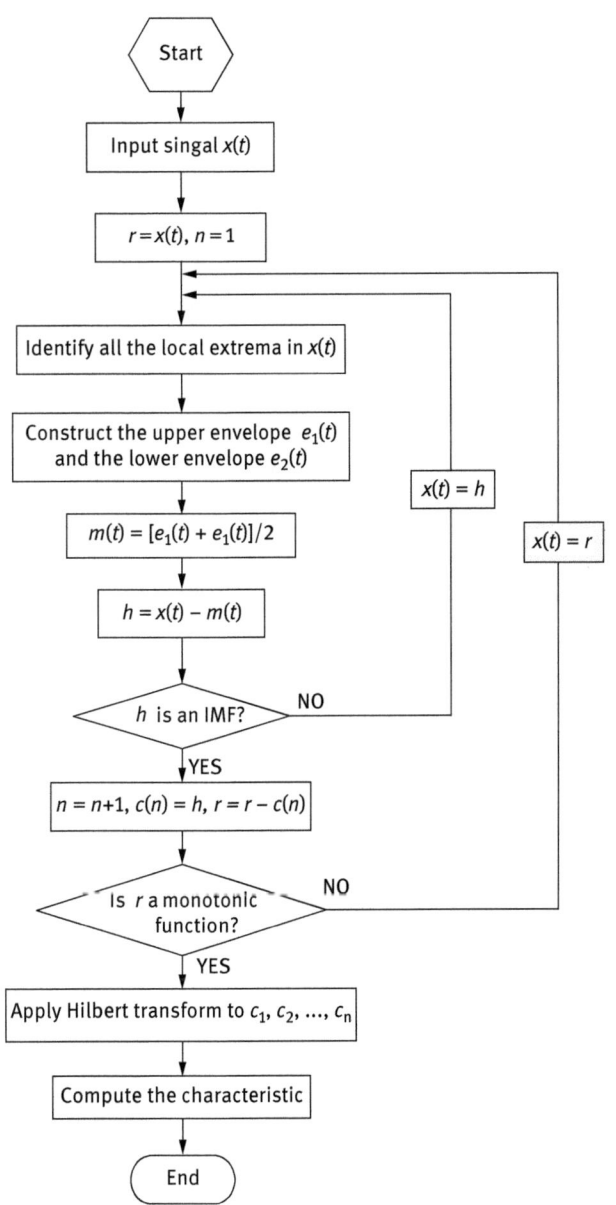

Figure 14.33: Flowchart of the Hilbert–Huang transform algorithm.

$$H'(\Omega, t) = \mathrm{Re}(\ldots + a_i(t)e^{j\int \Omega_i(t)dt} + \ldots + a_k(t)e^{j\int \Omega_k(t)dt} + \ldots) \qquad (14.157)$$

and the local Hilbert marginal spectrum by

$$h'(\Omega) = \int_0^T H'(\Omega, t)\mathrm{d}t \tag{14.158}$$

The $H'(\Omega, t)$ offers a measure of the amplitude contribution at each time and frequency. The local marginal $h'(\Omega)$ spectrum offers a measure of the total amplitude contribution from some frequencies that we are interested in.

14.3.5 Applications

The Hilbert spectrum has many practical applications. One application example is the use of the Hilbert spectrum for the analysis of the abrupt change information, so as to achieve the analysis, judgment, and control of the running fault. The profile of multifrequency mutation signal is given by

$$x(t) = \begin{cases} \sin(2\pi \times 10t), & 1 \le t \le 350 \text{ s} \\ \sin(2\pi \times 40t), & 351 \le t \le 700 \text{ s} \\ \sin(2\pi \times 80t), & 701 \le t \le 1,000 \text{ s} \end{cases}$$

The sampling frequency is $f_s = 1,000\,\mathrm{Hz}$. The signal waveform and its Fourier spectrum are shown in Figures 14.34 and 14.35, respectively.

As shown in Figure 14.34, the signal shows an obviously discontinuous frequency at 350 and 750 points. In Figure 14.35, the frequency spectrum of the signal shows the three frequency components of the signal: 10, 40, and 80 Hz, but it is

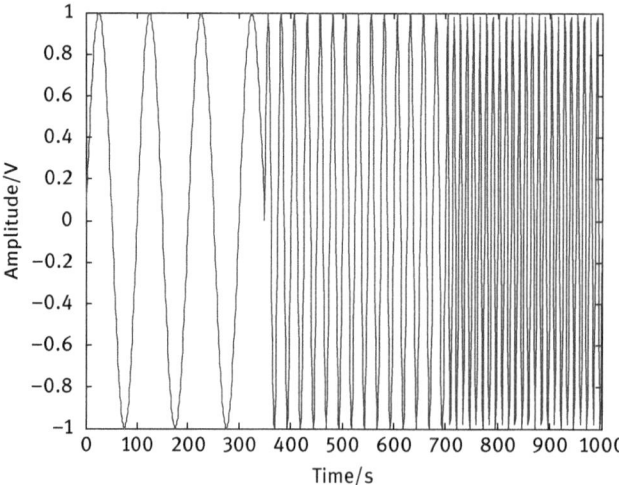

Figure 14.34: Time domain waveform of the signal.

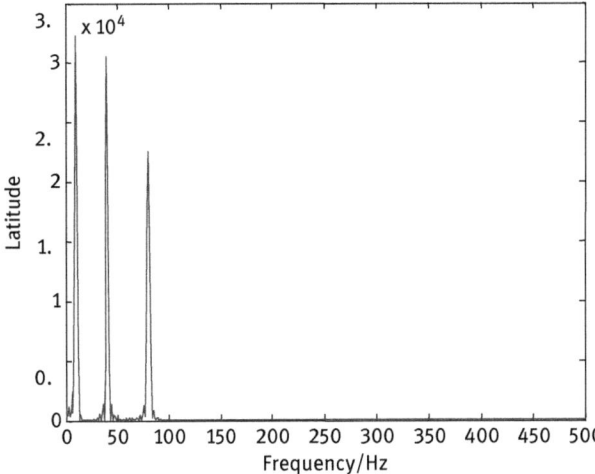

Figure 14.35: Fourier power spectrum.

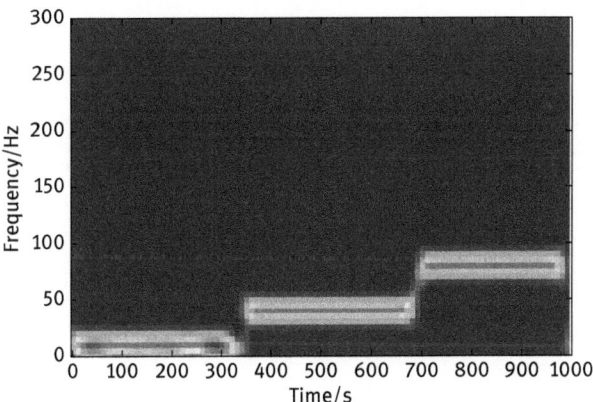

Figure 14.36: Gabor transform of the signal.

unable to identify the transient characteristics of the frequency. This proves that the Fourier transform spectrum is a global transformation, and it cannot reflect the change of frequency, so it cannot detect the abrupt change point of the signal.

After performing the Gabor transform, wavelet transform, and Hilbert–Huang transform on the signal, we can get the results shown in Figures 14.36–14.38, respectively.

The frequency variations of the signal are clearly shown in Figures 14.36, 14.37, and 14.38. But the time–frequency resolution of the Gabor transform in

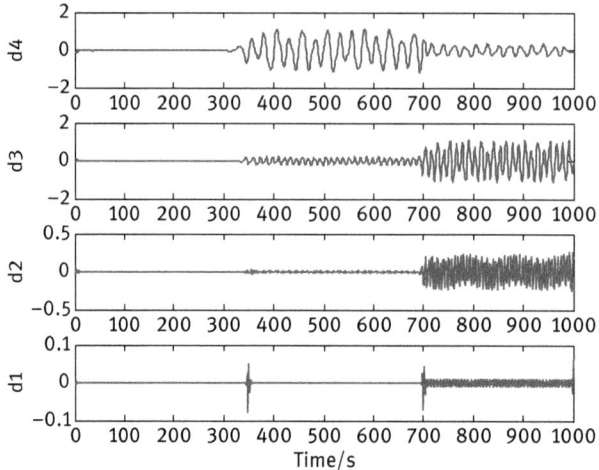

Figure 14.37: Wavelet transform of the signal.

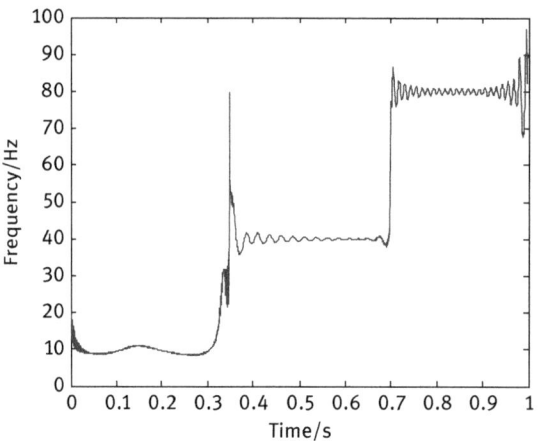

Figure 14.38: Signal Hilbert–Huang transform.

Figure 14.36 shows low-frequency resolution and cannot accurately detect the time of frequency discontinuity. It can be seen that the resolution of windowed Fourier transform and STFT is fixed in time–frequency domain and it is not suitable for simultaneous analysis of transient signals with varying speed and range. Figures 14.37 and 14.38 clearly show the time of the frequency change. But in wavelet transform, how to select wavelet basis and decomposition scales has a great influence on signal detection results. While in Hilbert–Huang transform, the result obtained by IMF1 can determine the moments of sudden changes. It is simple and reliable, and overcomes the difficulties of choosing wavelet basis and decomposing scales.

It is worth mentioning that the MATLAB package named plot_hht developed by Alan Tan is used in this section.

ℹ️ Exercises

14.1 Explain the impact of the window function on the frequency resolution in STFTs.

14.2 Illustrate the relationship between Gabor transform and STFT.

14.3 Illustrate the difference between wavelet analysis and Fourier analysis.

14.4 Describe the definition of the wavelet basis function and how to choose the wavelet basis function.

14.5 Explain the concept of multiscale analysis from the perspective of spatial theory.

14.6 Explain the concept and significance of the two-scale equations.

14.7 Describe the process of the Mallat algorithm.

14.8 Explain the process of wavelet denoising using wavelet analysis.

14.9 Explain the definition and nature of the Hilbert transformation.

14.10 Explain the definition of the IMF.

14.11 Explain EMD process.

14.12 Explain the concept of Hilbert spectrum and marginal spectrum.

14.13 Give the signal $x(\tau) = (\beta/\pi)^{1/4} e^{-\beta\tau^2/2}$, and the window function is $g(\tau) = (a/\pi)^{1/4} e^{-a\tau^2/2}$. Prove that the STFT of the signal is

$$STFT(t, \omega) = \sqrt{\frac{2\sqrt{a\beta}}{a+\beta}} \exp\left[-\frac{a\beta}{2(a+\beta)} t^2 - \frac{1}{2(a+\beta)} \omega^2 + j\frac{a}{a+\beta} \omega t\right].$$

14.14 Calculate the WVD of $x(t) = (a/\pi)^{1/4} \exp\left[-a(t-t_1)^2/2 + j\omega_1 t\right] + (a/\pi)^{1/4} \exp\left[-a(t-t_2)^2/2 + j\omega_2 t\right]$.

14.15 Set the signal $x(t) = A\delta(t-t_1) + B\delta(t-t_2)$, and calculate the CWT and $WT(a, b)$ of $x(t)$.

14.16 Let the CWT of $x(t)$ be represented as $WT(a, b)$, and describe the CWT of $x(2t-1)$.

14.17 A scale function in the MRA is defined as

$$\varphi(t) = \begin{cases} t+1, & -1 \leq t \leq 0 \\ t-1, & 0 < t \leq 1 \\ 0, & |t| > 1 \end{cases}$$

where $\{\varphi(t-k), k \in Z\}$ is a nonorthogonal basis of subspace V_0.

1. Describe the two scale equations
2. Try to prove that the Fourier transform of the scale function is

$$\hat{\varphi}(\omega) = 2\sqrt{\frac{2}{\pi}} \frac{\sin^2 \omega/2}{\omega^2}.$$

14.18 Let the signal be $x(t) = 1 + \sin(2\pi t)$, and use Haar wavelet to get the scale function and perform the process by MATLAB.

Bibliography

G. S. Hu, Digital Signal Processing: Theories, Algorithms, and Implementations (Third Edition), Beijing: Tsinghua University Press, 2012 (in Chinese)

P. Q. Cheng, A Course in Digital Signal Processing (Third Edition), Beijing: Tsinghua University Press, 2011 (in Chinese)

H. Y. Wang, Random Digital Signal Processing, Beijing: Science Press, 1980 (in Chinese)

H. Y. Wang, Modern Spectral Estimation, Nanjing: Southeast University Press, 1990 (in Chinese)

J. L. Zheng, Q. H. Ying, W. L. Yang, Signals and Systems (Third Edition), Beijing: Higher Education Press, 2011 (in Chinese)

Z. Z. Guan, G. K. Xia, Signals and Linear Systems (Fifth Edition), Beijing: Higher Education Press, 2011 (in Chinese)

A. V. Oppenheim, A. S. Willsky, S. H. Nawab, Signals and Systems (Second Edition), Englewood Cliffs, NJ: Prentice-Hall Inc, 1997

A. V. Oppenheim, R. Shafer, Discrete-Time Signal Processing, Englewood Cliffs, NJ: Prentice-HallInc, 1989

J. S. Lim, Two-Dimensional Digital Signal Processing, Englewood Cliffs, NJ: Prentice-Hall, 1989

T. S. Qiu, H. Tang, H.L. Liu, Statistical Signal Processing: Medical Signal Analysis and Processing, Beijing: Science Press, 2012 (in Chinese)

S. R. Qin, Z. Ji, A. J. Yin, Engineering Signal Processing, Beijing: Higher Education Press, 2008 (in Chinese)

H. Y.Wang, T. S. Qiu, Z. Chen, Nonstationary Random Signal Analysis and Processing (Second Edition), Beijing: National Defense Industry Press, 2008 (in Chinese)

S. J. Zhang, S. D. Zhang, Statistical Signal Processing, Beijing: Machinery Industry Press, 2003 (in Chinese)

M. W. Zhu, Y. X. Li, X. Z. Bo, Test Signal Processing and Analysis, Beijing: Beihang University Press, 2006 (in Chinese)

L. Li, Mechanical Signal Processing and Its Applications, Wuhan: Huazhong University of Science and Technology Press, 2007 (in Chinese)

Y. T, Fei, Error Theory and Data Processing (Sixth Edition), Beijing: Machinery Industry Press, 2010 (in Chinese)

Z. Qian, G. X. Jia, Error Theory and Data Processing, Beijing: Science Press, 2013 (in Chinese)

D. J. Dong, L. Qiao, L. Dong, Error Analysis and Data Processing, Beijing: Tsinghua University Press, 2013 (in Chinese)

D. C. Tong, B. H. Yao, Engineering Signal Processing and Equipment Diagnosis, Beijing: Science Press, 2008. (in Chinese)

T. S. Qiu, D. X. Wei, H. Tang, A. Q. Zhang, Adaptive Signal Processing in Communications, Beijing: Publishing House of Electronics Industry, 2005

X. D. Zhang, Modern Signal Processing, Beijing: Tsinghua University Press, 1995 (in Chinese)

T. S. Qiu, X. X. Zhang, X. B. Li, Y. M. Sun, Statistical Signal Processing Non-Gaussian Signal Processing and Its Applications, Beijing: Publishing House of Electronics Industry, 2004 (in Chinese)

D. F. Zhang, MATLAB Digital Signal Processing and Applications, Beijing: Tsinghua University Press, 2010 (in Chinese)

K. Zou, J. Q. Yuan, X. Y. Gong, MATLAB6.X Signal Processing, Beijing: Tsinghua University Press, 2002 (in Chinese)

https://doi.org/10.1515/9783110465082-015

X. H. Tang, H. L, Yue, X. F. Zheng, MATLAB and Its Application in the Course of Electronic Information, Beijing: Publishing House of Electronics Industry, 2006 (in Chinese)

H. Y. Wang, Statistical Signal Processing Theory and Solution of Problems, Beijing: National Defense Industry Press, 1996 (in Chinese)

H. L. Liu, Biomedical Signal Processing, Beijing: Chemical Industry Press, 2006 (in Chinese)

G. Li, X. Zhang, Biomedical Electronics, Beijing: Publishing House of Electronics Industry, 2008 (in Chinese)

W. J. Tompkins, Biomedical Digital Signal Processing, Englewood Cliffs, NJ: PTR Prentice-Hall, 1993

T. G. Zhuang, Computer Applications in Biomedicine, Beijing: Science Press, 2000 (in Chinese)

J. X. Cai, W. Z. Zhang, Biomedical Electronics, Beijing: Peking University Press, 1997 (in Chinese)

M. Owen, Practical Signal Processing, Cambridge University Press, 2007

J. W. Cooley, J. W. Tukey, An algorithm for the machine computation of complex Fourier series, Mathematics of Computation, 19(4): 297–301, 1965

X. D. Zhang, Z. Bao, Communication Signal Processing, Beijing: National Defense Industry Press, 2000 (in Chinese)

D. G. Lu, Random Process and Its Applications, Beijing: Tsinghua University Press, 1986 (in Chinese)

F. S. Yang, Y. S. Lv, Biomedical Signal Processing and Identification, Tianjin: Tianjin Science and Technology Translation Publishing Company, 1997 (in Chinese)

N. Nie, D. Z. Yao, Z. X. Xie, Digital Processing Technology and Applications of Biomedical Signals, Beijing: Science Press, 2005

F. L. Shen, H. Y. Chen, Biomedical Random Signal Processing, Hefei: China University of Science and Technology Press, 1999 (in Chinese)

B. H. Wang, Biomedical Measurement and Instruments, Shanghai: Fudan University Press, 2003 (in Chinese)

L. C. Ludeman, Random Processes Filtering, Estimation and Detection, John Wiley and Sons, Inc., 2003

S. V. Vaseghi, Advanced Digital Signal Processing and Noise Reduction (Third Edition), West Sussex: John Wiley and Sons, Inc., 2006

S. J. Zhang, S. D. Zhang, Statistical Signal Processing, Beijing: Machinery Industry Press, 2003 (in Chinese)

F. S. Yang, Random Signal Analysis, Beijing: Tsinghua University Press, 1990 (in Chinese)

B. Widrow, Adaptive sampled-data systems – a statistical theory of adaptation, IRE WESCON Convention Record, 4:74–85, 1959

V. Albrecht, P. Lfinsk, M. Indra, T. Radii-Weiss, Wiener filtration versus averaging of evoked responses, Biological Cybernetics 27:14–154, 1977

X. D. Zhang, M. Q. Lu, Discrete Random Signal Processing, Beijing: Tsinghua University Press, 2005 (in Chinese)

W. X. Liu, Cardiac Auscultation, Beijing: People's Military Medical Publishing House, 2006 (in Chinese)

X. H. Bi, Underdetermined Blind Source Separation and Its Application in Signal Extraction, Dalian: Dalian University of Technology, 2008 (in Chinese)

J. Zhang, M. F. Shen, J, Song, Application of three time-frequency analysis methods in heart sound signal analysis, Journal of Shantou University, 18(2):62–69, 2003 (in Chinese)

G. S. Hu, J. F. Wang, Studies on Behavior of EEG before epileptic explosion by higher order statistics, Chinese Journal of Biomedical Engineering, 17(2):111–118, 1998 (in Chinese)

H. Tang, Study on Theory of Generalized Normal-distribution Signal Processing and Its Application in Communications, Dalian: Dalian University of Technology, 2006 (in Chinese)

H. Y. Wang, Synthesis and Unification of Signal Processing Theory. Beijing: National Defense Industry Press, 2005 (in Chinese)

H. Tang, T. Li, Y. Park, T.S. Qiu, Separation of heart sound signal from noise in joint cycle frequency-time-frequency domains based on fuzzy detection, IEEE Transactions on Biomedical Engineering, 57(10):2438–2447, 2010

F. Wang, Z. Z. Luo, Based on the power-spectrum to classify the pattern of the surface electromyography, Journal of Hangzhou Dianzi University, 25(2):37–40, 2006

J. F. Wu, Research on Human Lower-limb Motion Information Acquisition Technology based on EMG, Hangzhou: Zhejiang University, 2008 (in Chinese)

J. M. Mendel, Tutorial on higher-order statistics (spectra) in signal processing and system theory: theoretical results and some applications, Proceedings of the IEEE, 79(3):278–305, 1991

L. J. Hadjileontiadis, Lung sounds: an advanced signal processing perspective. Synthesis Lectures on Biomedical Engineering, 3(1): 1–100, 2008

Z. S. Chen, Statistical Information Processing based on Matlab7.0, Changsha: Hunan Science and Technology Press, 2005 (in Chinese)

X. D. Zhang, Modern Signal Processing, Beijing: Tsinghua University Press, 2005 (in Chinese)

T. R. Yao, H. Sun, Modern Digital Signal Processing, Wuhan: Huazhong University of Science and Technology, 1999 (in Chinese)

H. Y. Wang, T. S. Qiu, Adaptive Noise Cancellation and Time Delay Estimation, Dalian: Dalian University of Technology Press, 1999 (in Chinese)

Y. Guo, The Study on Novel Time Delay Estimation Methods based on Stable Distribution, Dalian: Dalian University of Technology, 2009 (in Chinese)

Y. M. Sun, Study on the Theory and Application of Parameter Estimation and Spectral Analysis of α-stable Distribution, Dalian: Dalian University of Technology, 2006 (in Chinese)

N. E. Huang, The Hilbert-Huang Transform in Engineering. Boca Raton: Taylor & Francis Ltd, 2006

N. E. Huang, S. S. P. Shen, Hilbert-Huang transform and its applications, Singapore: World Scientfic Publishing, pp. 1–24, 2005

R. Cristi (translated by S. Xu), Modern Signal Processing, Beijing: Machinery Industry Press, 2005 (in Chinese)

D. K. Wang, J. Y. Peng, Wavelet Analysis and Its Application in Signal Processing, Beijing: Publishing House of Electronics Industry, 2005 (in Chinese)

D. F. Mix, K. J. Olejniczak (translated by Z. H. Yang, L. H. Yang), Wavelet Basis and Application Tutorial, Beijing: Machinery Industry Press, 2005 (in Chinese)

T. Liu, X. L. Zeng, J. Zeng, Introduction to Practical Wavelet Analysis, Beijing: National Defense Industry Press, 2006 (in Chinese)

H. Q. Wang, Wavelet Analysis and Applications, Beijing: Beijing University of Posts and Telecommunications Press, 2011 (in Chinese)

Z. X. Ge, Z. S. Chen, Matlab Time Frequency Analysis Technology and Its Applications, Beijing: Posts and Telecom Press, 2006 (in Chinese)

B. Han, J. Wu, H. Xu, Improved SNR estimation algorithm of higher-order statistics, Journal of Data Acquisition and Processing, 27(5):576–580, 2012 (in Chinese)

N. Madhavan, A.S. Madhukumar, A.P. Krishna, Spectrum sensing and modulation classification for cognitive radios using cumulants based on fractional lower order statistics, International Journal of Electronics and Communications, 67(6):479–490, 2013

S. Li, L. M. Song, T. Qiu, Steady-state and tracking analysis of fractional lower-order constant modulus algorithm, Circuits, Systems, and Signal Processing, 30(6):1275–1288, 2011

X.H. Zhong, A. B. Premkumar, A. S. Madhukumar, Particle filtering for acoustic source tracking in impulsive noise with Alpha-stable process, IEEE Sensors Journal, 13(2): 589–600. 2013

C. Conlin, B. Balas, Invariant texture recognition depends on high-order statistics. Journal of Vision, 14(10):1424–1424, 2014

S.H. Zhu, J.Hu,B.Y.Wang, et al., Image annotation using high order statistics in non-Euclidean spaces, Journal of Visual Communication and Image Representation, 24(8):1342–1348, 2013

C. Chang, W. Xiong, C.H. Wen. A theory of high-order statistics-based virtual dimensionality for hyperspectral imagery, IEEE Transactions on Geoscience and Remote Sensing, 52(1): 188–208, 2014

Y. S. Zhang, F. X. Zhao, Validation of non-stationary ground motion simulation method based on Hilbert transform, Acta Seismologica Sinica, 36(4):686–697, 2014 (in Chinese)

W. Wei, H. S. Zhang, The Hilbert-Huang transform technique and its applications to the study of the turbulence boundary layer, Acta Meteorologica Sinica, 71(6):1183–1193, 2013 (in Chinese)

X. F. Yang, W. Zhang, Y. X. Yang, Denoising technology of radar life signal based on lifting wavelet transform, Acta Optica Sinica, 34(3):1–6, 2013 (in Chinese)

L. Ma, J. S. Kang, Y. Meng, Research on feature extraction of rolling bearing incipient fault based on Morlet wavelet transform, Chinese Journal of Scientific Instrument, 34(4):920–926, 2013 (in Chinese)

N. Wang, Y. Wu, Y. Zhang, The Application of time-frequency analysis methods in deformation data, China Earthquake Engineering Journal, 36(2):413–420, 2014 (in Chinese)

Z. C. Sha, Z. T. Huang, Y. Y. Zhou, Time-frequency analysis of frequency-hopping signals based on sparse recovery, Journal on Communications, 34(5):107–112, 2014 (in Chinese)

D. Zhang, J. Z. He, Y. P. Jiang, et al., Analysis and classification of heart sounds with mechanical prosthetic heart valves based on Hilbert-Huang transform, International Journal of Cardiology, 151(1):126–127, 2011

J. Gilles, Empirical wavelet transform. IEEE Transactions on Signal Processing, 61(16):3999–4010, 2013

A. Pande, J. Zambreno, The secure wavelet transform, Journal of Real-Time Image Processing, 7(2):131–142, 2012

P. Kaikkonen, E. Hynynen, H. Rusko, et al., Heart rate variability is related to training load variables in interval running exercises, European Journal of Applied Physiology, 112(3):829–838, 2012

H. Giv. Directional short-time Fourier transform, Journal of Mathematical Analysis and Applications, 399(1):100–107, 2013

Z. Qian, G. X. Jia, Error Theory and Data Processing, Beijing: Science Press, 2013 (in Chinese)

D. J. Dong, Error Analysis and Data Processing, Beijing: Tsinghua University Press, 2013 (in Chinese)

T. Nedunthally, Alpha-Stable, Normal Inverse Gaussian and Multi-Factor Models for Spot and Futures modeling in Natural Gas, University of Calgary, 2009

A. E. Ackie, Performance Analysis of Spectrum Sensing Schemes Based on Fractional Lower Order Moments for Cognitive Radios in Alpha-stable Noise Environments, Florida Atlantic University, 2016

M. Sahmoudi, K. Abed-Meraim, M. Benidir, Blind separation of heavy-tailed signals using normalized statistics, International Conference on Independent Component Analysis and Signal Separation, ICA 2004, LNCS 3195, pp. 113–120, 2004

E. Alameda-Hernández, F. G. Montoya, M. J. Mercado-Vargas, et al., Higher-order statistics for power systems: Effects of the sampling frequency on ergodicity, Applied Mathematical Modelling, 40:6924–6933, 2016

J. J. de la Rosa, A. A. Perez, J. C. Palomares-salas, et al., Higher-order statistics: Discussion and interpretation, Measurement 46:2816–2827, 2013

H. S. Lui, N. V. Shuley, On the Analysis of electromagnetic transients from radar targets using smooth pseudo Wigner-Ville distribution (SPWVD), Proceedings of the IEEE Antenna and Propagation Society International Symposium, 5701–5704, 2007

Z. Sun, S. Kong, W. Wang, Sparse representation of higher-order statistics, Pattern Recognition, 2017:1–11 (DOI: 10.1016/j.patcog.2017.07.022.)

S. Haykin, Adaptive Filter Theory (Fourth Edition), Englewood Cliffs, NJ: Prentice Hall, 2002

Index

https://doi.org/10.1515/9783110465082-016